State-of-the-Art Polymer Science and Technology in Mexico

State-of-the-Art Polymer Science and Technology in Mexico

Guest Editors

Emilio Bucio
Lorenzo Antonio Picos Corrales
Elizabeth Carvajal-Millán
Rubén González-Núñez
Norma-Aurea Rangel-Vázquez

MDPI

Basel • Beijing • Wuhan • Barcelona • Belgrade • Novi Sad • Cluj • Manchester

Guest Editors

Emilio Bucio
Instituto de Ciencias Nucleares
Universidad Nacional Autónoma de México
México
Mexico

Lorenzo Antonio Picos Corrales
Facultad de Ingeniería Culiacán
Universidad Autónoma de Sinaloa
Culiacán
Mexico

Elizabeth Carvajal-Millán
Biopolymers Laboratory
Research Center for Food and Development
Hermosillo
Mexico

Rubén González-Núñez
Departamento de Ingeniería Química
Universidad de Guadalajara
Guadalajara
Mexico

Norma-Aurea Rangel-Vázquez
Division of Graduate Studies and Research
Technological Institute of Aguascalientes
Aguascalientes
Mexico

Editorial Office

MDPI AG
Grosspeteranlage 5
4052 Basel, Switzerland

This is a reprint of the Special Issue, published open access by the journal *Polymers* (ISSN 2073-4360), freely accessible at: https://www.mdpi.com/journal/polymers/special_issues/964O0F30NP.

For citation purposes, cite each article independently as indicated on the article page online and as indicated below:

Lastname, A.A.; Lastname, B.B. Article Title. *Journal Name* **Year**, *Volume Number*, Page Range.

ISBN 978-3-7258-7306-7 (Hbk)
ISBN 978-3-7258-7307-4 (PDF)
https://doi.org/10.3390/books978-3-7258-7307-4

Contents

About the Editors

Emilio Bucio

Emilio Bucio is a Research Scientist at the National University of Mexico. His fields of expertise include radiation chemistry, fluorinated polymers, and the synthesis of smart polymers and drug-delivery systems. Recently, his research team has been working on the application of sensitivity copolymers for drug delivery. As a Doctor of Sciences from the National University of Mexico (UNAM), he graduated in Chemical Physics from the Chemistry Faculty of UNAM in 1999. Since then, he has worked at the Nuclear Science Institute at UNAM. He was awarded the "Gustavo Baz Prada" for his BS in 1993 from UNAM and the "Alfonso Caso" medal for best student during his PhD in 1999. In addition to this, he received the LAS/ANS award for best publication of the year in 2010 from the American Nuclear Society, Latin American Section. He is a member of the National System of Researchers in Mexico, the American Chemical Society, the Mexican Polymer Society, the Mexican Chemical Society, and the Mexican Membranes Society. His publications and presentations include 193 journal publications; 69 chapters in books; 2 books; 340 oral papers, invited lectures, poster presentations, and short courses; and 79 thesis directions (BS, MSc, and PhD). He collaborates as an Editor-in-Chief for the *Open Journal of Polymer Chemistry*, is Associate Editor for the journal *Radiation Physics and Chemistry*, is Executive Editor-in-Chief for the *Drug Design Development and Delivery Journal*, and is an Editorial Board Member of more than twenty other international journals.

Lorenzo Antonio Picos Corrales

Lorenzo Antonio Picos Corrales is a Research Professor at the Autonomous University of Sinaloa (UAS, since 2012), Mexico, with scientific interests focused on the synthesis and characterization of polymers for drug delivery systems and water treatment. He studied Chemical Engineering at the Faculty of Chemical-Biological Sciences, UAS. He received his Master's in Chemistry from the Technological Institute of Tijuana (IT-Tijuana), Mexico, receiving the 2008 Best Master's Thesis Award from the National Technological System. In 2012, he received his PhD in Chemistry (Polymer Science) at IT-Tijuana, Mexico, obtaining a scholarship in 2010 from the German Academic Exchange Service for the "Sandwich PhD Model" at Technische Universität Dresden, Germany. He has also received research training at Leibniz-Institut für Polymerforschung (IPF) in Dresden, Germany, and at the Autonomous University of Baja California in Tijuana, Mexico. At present, Dr. Lorenzo A. Picos-Corrales is a full-time Professor at the Faculty of Engineering Culiacan, UAS, and a member of the Basic Academic Nucleus of the Postgraduate Program in Biological Sciences at the Faculty of Biology-UAS. He is also a member of the National System of Researchers in Mexico. Additionally, he collaborates as a Special Guest Editor for the journal *Polymers*, Assistant Editor for the journal *Hybrid Advances*, and Associate Editor for the journal Frontiers in *Water*, and is an Editorial Board Member of the journal *npj Emerging Contaminants*.

Elizabeth Carvajal-Millán

Elizabeth Carvajal-Millán has held the position of Research Scientist at the Research Center for Food and Development (CIAD) in Hermosillo, Mexico, since 2005. Her research interests are centered on the chemistry, physics, and biochemistry of polysaccharides, as well as the evaluation of polysaccharide functionality in biomedicine, food, and health. She obtained her PhD in France at École Nationale Supérieure Agronomique à Montpellier, her M.Sc. degree at CIAD, and her undergraduate degree at the University of Sonora in Mexico. Her research encompasses the extraction and characterization of gelling and bioactive polysaccharides that have potential applications in the food, biomedical, and pharmaceutical sectors. She also focuses on the development and characterization of biomaterials based on ferulated and sulfated polysaccharides (gels, micro- and nanospheres, films, and nanofibers), with an emphasis on systems for the controlled release of biomolecules and cells. Dr. Carvajal-Millan is a member of the National System of Researchers in Mexico (level 3, the maximum granted), a member of the Mexican Academy of Sciences, and part of the Scientific Committee of La Maison Universitaire Franco-Mexicaine (MUFRAMEX). She has published 145 indexed articles, 49 book chapters, and holds 6 granted patents (2 internationally). Dr. Carvajal-Millan has an h-index of 35, and her publications have been cited 3419 times (without self-citations). She has supervised 11 PhD. theses, 11 MSc theses, and 8 bachelor's degree theses, some of which received national or international awards.

Rubén González-Núñez

Rubén González-Núñez has been a professor at Universidad de Guadalajara since 1987. His research interests are related to polymer processing, multicomponent materials, biopolymers, and recycling. He obtained his B.Sc. degree (1985) from Universidad Autónoma de Nayarit (Mexico), M.Sc. degree (1988) from Universidad de Guadalajara (Mexico), and PhD (1994) from Sherbrooke University (Canada), all in Chemical Engineering. He has supervised more than 100 theses and has produced over 100 scientific publications. He is a member of the National System of Researchers in Mexico and of the Polymer Society of Mexico, for which he served as National President.

Norma-Aurea Rangel-Vázquez

Norma-Aurea Rangel-Vázquez (PhD) is a Research Professor at Instituto Tecnológico de Aguascalientes, with scientific interests focused on different areas such as drug delivery systems, hydrogels, biopolymers, simulations, materials science, and mechanical engineering. She is level 1 in the National System of Researchers in Mexico, has developed 25 projects, and has published 3 books in molecular simulation, 47 papers (JCR), 96 conferences (Internationals and Nationals), and 12 book chapters. She completed a research stay at the University of Minnesota, USA. She has served as a reviewer for several journals, including the *Asian Journal of Research in Medical and Pharmaceutical Sciences, Carbohydrate Polymers, International Journal of Nanomedicine, Journal of Advances in Biology & Biotechnology, Journal of Applied Polymer Science, Journal of Applied Physical Science International, Journal of Applied Life Sciences International, Journal of Complementary and Alternative Medical Research, Journal of Molecular Structure, Journal Polymer Engineering & Science, Surface and Coatings Technology, Materials Letters, Polymer Bulletin, Spectrochimica Acta Part A: Molecular and Biomolecular Spectroscopy,* and *Thermochimica Acta*. She has worked on the organizing boards of 15 international congresses. She is on the Editorial Board of several journals, including the *Journal of Pharmaceutics & Drug Development, Material Science and Engineering with Advanced Research, Journal of Nanomedicine Research, Nanomedicine Research Journal, BAOJ Pharmaceutical Sciences, Independent Journal of Management & Production,* and *Global Journal of Nanomedicine*. Additionally, she has collaborated as a Special Guest Editor of the journals *Polymers* and *Applied Science*.

Preface

Mexico is one of the primary economies of the American continent. The research on polymers within the country is of an outstanding quality, following world trends, developing materials and methods that respond to the needs of today's society and environmental conservation. Mexican researchers with a focus on polymer science have contributed significantly to this interdisciplinary area at national and international levels, promoting collaboration among scientists, academics, students, and industry professionals. Currently, research on polymers in Mexico is diverse and covers different aspects of polymer science, sub-disciplines, and key areas. The studies conducted involve explorations of the synthesis, extraction from natural sources, processing, characterization, theoretical calculations, and the evaluation of outstanding materials for applications in areas such as medicine, water treatment, food industry, packaging, textiles, agriculture, energy storage, bioelectronics, cosmetics, and sensors, among others. Materials made from modified natural polymers have been widely tested due to their biocompatibility, biodegradability, or eco-friendly nature. Hence, this Reprint aims to provide an update on the efforts of Mexican colleagues who are developing polymer science and technology. The articles in this collection seek to serve the researchers and students working on polymeric materials, as well as industrialists facing polymer-related challenges.

Emilio Bucio, Lorenzo Antonio Picos Corrales, Elizabeth Carvajal-Millán, Rubén González-Núñez, and Norma-Aurea Rangel-Vázquez
Guest Editors

polymers

MDPI

Article

Silver Nanoparticles Loaded on Polyethylene Terephthalate Films Grafted with Chitosan

Guadalupe Gabriel Flores-Rojas [1], Felipe López-Saucedo [2], Ricardo Vera-Graziano [3], Héctor Magaña [4], Eduardo Mendizábal [1,*] and Emilio Bucio [2,*]

1 Departamento de Química e Ingeniería Química, Centro Universitario de Ciencias Exactas e Ingenierías, Universidad de Guadalajara, Blvd. M. García Barragán # 1451, Guadalajara 44430, Jalisco, Mexico
2 Departamento de Química de Radiaciones y Radioquímica, Instituto de Ciencias Nucleares, Universidad Nacional Autónoma de México, Circuito Exterior, Ciudad Universitaria, Mexico City 04510, Mexico
3 Instituto de Investigaciones en Materiales, Universidad Nacional Autónoma de México, Circuito Exterior, Ciudad Universitaria, Mexico City 04510, Mexico
4 Facultad de Ciencias Químicas e Ingeniería, Universidad Autónoma de Baja California, Calzada Universidad # 14418, Parque Industrial Internacional Tijuana, Tijuana 22390, Mexico
* Correspondence: lalomendizabal@hotmail.com (E.M.); ebucio@nucleares.unam.mx (E.B.)

Abstract: Currently, polyethylene terephthalate (PET) is one of the most widely used polymeric materials in different sectors such as medicine, engineering, and food, among others, due to its benefits, including biocompatibility, mechanical resistance, and tolerance to chemicals and/or abrasion. However, despite all these excellent characteristics, it is not capable of preventing the proliferation of microorganisms on its surface. Therefore, providing this property to PET remains a difficult challenge. Fortunately, different strategies can be applied to remove microorganisms from the PET surface. In this work, the surface of the PET film was functionalized with amino groups and later with a dicarboxylic acid, allowing a grafting reaction with chitosan chains. Finally, the chitosan coating was loaded with silver nanoparticles with an average size of 130 ± 37 nm, presenting these materials with an average cell viability of 80%. The characterization of these new PET-based materials showed considerable changes in surface morphology as well as increased surface hydrophilicity without significantly affecting their mechanical properties. In general, the implemented method can open an alternative pathway to design new PET-based materials due to its good cell viability with possible bacteriostatic activity due to the biocidal properties of silver nanoparticles and chitosan.

Keywords: chitosan; silver; nanoparticles; polyethylene terephthalate; graft; coating

1. Introduction

The presence and proliferation of microorganisms as biofilm [1] on polymeric materials for medical use and food containers can cause adverse effects on healthcare as well as for the stored food [2,3]. Therefore, polymeric materials capable of inhibiting the proliferation of microorganisms have gained great relevance. Therefore, it is necessary to develop more durable and efficient materials capable of inhibiting the proliferation not only of bacteria but also of fungi and even viruses [4] that can cause infections in open wounds, as well as avoiding the modification of the organoleptic properties of food and the production of secondary compounds that can be harmful to health [2,5,6].

One of the most widely used materials in food and medical device packaging is polyethylene terephthalate (PET) due to its excellent mechanical properties and chemical stability under several conditions [7], availability, and low production cost. Therefore, PET is an ideal candidate for its use as antibacterial surfaces [8]. Approaches to providing antimicrobial activity can be classified into two types: the first can be achieved by incorporating antimicrobial agents into the polymeric matrix, and the second by modifying the surface and coating it with antimicrobial agents [9,10]. The latter approach is one of the most widely used since it significantly maintains the polymeric matrix's original properties. This approach can be carried out through different methodologies, such as using covalently bonded coatings, which make them more stable in different physical–chemical environments. Antimicrobial coatings can be polymers from a natural or synthetic origin, providing new properties to the original material such as stimulus-response [11], bacteriostatic effects [12], biocompatibility [13], or allowing the loading and stabilization of nanomaterials [14]. In this context, chitosan, which is a polymer of natural origin, has shown antimicrobial activity against different pathogenic microorganisms with a high mortality rate and low toxicity to humans in various studies, making this polymer suitable for different applications in food [15], drug-delivery [16,17], biomedical devices [18,19], and other chemistry fields [20]. Chitosan is also used to stabilize and load nanoparticles synthesized using different methods [21,22] and to add new functional properties to coated nanoparticles by using chitosan as a carrier, such as in magnetic drug delivery systems [23,24]. Moreover, some nanoparticles are capable of providing antimicrobial activity through the release of metallic ions that affect the integrity of microorganisms, causing their death, as is the case with silver nanoparticles [25].

Chitosan has been widely modified to meet various biological and medical needs due to its active functional groups. Chemical modification is the commonly used method because its amino groups can participate in chemical reactions such as alkylation, quaternization, and condensation reactions with aldehydes and ketones. The hydroxyl group also gives rise to hydrogen bonding and some reactions, such as o-acetylation, cross-linking, and grafting [26,27]. Currently, the condensation reaction between the amino group and the carboxyl group is mainly used to modify the chemical properties of chitosan, improving its effect on metal nanoparticles, and reducing its possible toxic effects [28–30].

Chitosan is used as a coating on different materials, such as polyethylene terephthalate (PET), which is a slightly polar material and lacks active functional groups, making the adhesion, coating, or covalent grafting of chitosan difficult. In this research work, the surface of PET films was modified to perform the covalent grafting of chitosan, providing a suitable surface for loading silver nanoparticles. Additionally, the grafted chitosan chains possibly provide a bacteriostatic effect and a sustainable release of silver ions capable of inhibiting the proliferation of microorganisms and increasing the cytocompatibility of the final material.

2. Materials and Methods

2.1. Materials

Silver nitrate (99.9%), chitosan (75–85% deacetylated; 50–190 kDa), itaconic acid (99%), acetic acid, sodium carbonate, hydrochloric acid, ethylenediamine, and polyethylene terephthalate (PET) films (200 μm thickness; average molecular weight 18 kDa; 34.3% crystallinity) were purchased from Aldrich Chemical Co. (St. Louis, MO, USA). Ethanol and distilled water were acquired from Baker (Mexico City, Mexico).

2.2. Methods

2.2.1. Aminolysis Reaction (PETN)

Three previously weighed PET films were placed in 10 mL of ethylenediamine at a temperature of 40 °C and reacted for different reaction times (0.5–4 h). Subsequently, the films were sonicated for 1 h and thoroughly washed with distilled water, removing the abraded surface of the film and residues from the reaction medium. Experiments were carried out in triplicate. The loss percentage was calculated according to the following equation (Equation (1)):

$$\text{Weight loss (\%)} = 100[(W_{PET} - W_{PETN})/W_{PET}] \quad (1)$$

where W_{PET} is the weight of the unmodified PET film and W_{PETN} is the weight of the film once it has undergone the aminolysis reaction.

The quantification of the amino groups present on the surface of the film was carried out by acid-base titration. Three aminolysis reaction (PETN) films were placed in 10 mL of a hydrochloric acid solution [0.02 M], keeping them constantly stirred for 24 h. Subsequently, the excess of hydrochloric acid was titrated with a sodium carbonate solution [0.001 M]. The number of amino groups was calculated according to the following equation (Equation (2)):

$$\text{Ethylenediamine modification (\%)} = 100[6.1(V_{HCl}C_{HCl} - 2V_TC_{NaC})/W_{PETN}] \quad (2)$$

where V_{HCl} and C_{HCl} are the volume and concentration of the hydrochloric acid solution, V_T is the volume of the titration and C_{NaC} is the concentration of the sodium carbonate solution and W_{PETN} is weight of the PETN film.

2.2.2. Michael Addition Reaction (PETI)

Three previously weighed PETN films were made to react at reflux for 48 h in 10 mL of ethanol and an excess of itaconic acid (200 mg). Subsequently, the obtained films were exhaustively washed with distilled water. The incorporated itaconic acid was calculated according to the next equation (Equation (3))

$$\text{Itaconic acid modification (\%)} = 100[(W_{PETN} - W_{PETI})/W_{PETN}] \quad (3)$$

where W_{PETN} is the weight of the PETN film and W_{PETI} is the weight of the film after the Michael addition reaction.

2.2.3. Chitosan Grafting (PETC)

Three previously weighed Michael addition reaction (PETI) films were placed in 10 mL of an aqueous solution of chitosan 1% (*w*/*v*) and 0.5% (*v*/*v*) acetic acid for 48 h at 25 °C. After this time, the films were removed from the solution and placed under vacuum at a temperature of 90 °C for 6 h. Finally, the obtained films (PETC) were exhaustively washed with distilled water and dried.

The percentage of chitosan graft on the surface of the film was calculated according to the following equation (Equation (4)):

$$\text{Chitosan graft (\%)} = 100[(W_{PETI} - W_{PETC})/W_{PETI}] \quad (4)$$

where W_{PETI} and W_{PETC} are the weight of the PETI and PETC films.

2.2.4. Load of Silver Nanoparticles on PETC Film (PETCAg)

Three previously weighed PETC films were placed in 10 mL of an aqueous solution of silver nitrate with different concentrations (500, 1000, 3000, and 5000 ppm) for 72 h at 25 °C in the presence of sunlight. After the reaction time, the films were removed from the solution and thoroughly washed with distilled water and dried. The number of nanoparticles loaded on the PETCAg film was determined by film calcination using the weight difference of the residues yield at 800 °C (Equation (5)).

$$\text{Silver load (\%)} = C_{PETAg} - C_{PETC} \quad (5)$$

where C_{PETCAg} and C_{PETC} are the residue percentage of the PETCAg and PETC films provided by the TGA instrument.

2.2.5. Cell Viability Study

Cell viability tests were performed using a BALB/3T3 fibroblast line (mouse). These experiments were performed to assess the toxicity of the films in an in vitro model for potential biomedical applications. For the experiments, 25 mg of each type of film were cut and placed in 96-well plates containing 3000 cells (fibroblasts), with Dulbecco's modified Eagle's medium (DMEM), penicillin–streptomycin, gentamicin, and Fetal Bovine Serum (FBS); the films were in contact with the cell medium for 24 h in an incubator at 37 °C (5%

CO_2). Subsequently, the films were removed from the cell medium, and MTT kit reagent (Roche, Switzerland) was added and incubated again for 4 h; later, they were solubilized and incubated for 12 h. The 96-well plates were plated in a Multiskan FC spectrophotometer, Thermo Fisher Scientific (Waltham, MA, USA); the absorbances from each film were read at a wavelength of 620 nm. Finally, cell activity (viability) was determined, making a comparison with control cells, using the following equation (Equation (6)), and the results obtained were statically analyzed by analysis of variance (ANOVA) using Fisher's comparison (Figures S1–S7, Tables S1–S4):

$$\text{Cytocompatibility } (\%) = 100(\text{Abs}_{\text{Sample}}/\text{Abs}_{\text{Control}}) \tag{6}$$

2.3. Instrumental

Kruss DSA 100 drop shape analyzer (Matthews, NC, USA) was used to measure the contact angle on the surface of films at time 1 min.

Fourier Transform Infrared Attenuated Total Reflection (FTIR ATR) of dry samples was analyzed using a Perkin–Elmer Spectrum 100 spectrometer (Norwalk, CT, USA) of 16 scans.

Scanning electron microscope (SEM) images were acquired by the Zeiss Evo LS15 instrument (Jena, Germany), and small pieces of 0.5 cm in length were cut and coated with gold and analyzed under a high vacuum.

Thermogravimetric analysis (TGA) data of the weight loss and decomposition of films were heated at a rate of 10 °C min^{-1} and run from 20 to 800 °C under nitrogen flow at 100 cm^3/min in a TGA instrument Q50 TA Instruments (New Castle, DE, USA).

Differential scanning calorimetry (DSC) runs were recorded under a nitrogen flow at 100 cm^3/min using a DSC 2010 calorimeter (TA Instruments, New Castle, DE, USA) from 25 to 250 °C at a heating rate of 10 °C min^{-1}.

Mechanical properties of the films were studied by applying a uniaxial tension test, as described in ASTM D1708. All tests were carried out on an INSTRON 1125 (Instron Inc., Norwood, MA, USA) universal tensile testing machine at a crosshead speed of 10 mm/min, and all experiments were carried out in triplicate.

3. Results and Discussions

3.1. Chitosan Grafting on PET Film

PET films were functionalized by aminolysis using ethylenediamine as a reagent and solvent. This reaction was studied at different times at a temperature of 40 °C. The results showed a weight loss of the PET film in the functionalization process with amino groups (Figure 1a). The weight loss increased with the reaction time. In the IR-ATR spectra (Figure 1b), the intensity bands of amines barely changed, where PETN-0.5 corresponds to the sample with the shorter time of reaction (30 min), while PETN-4 corresponds to the longer time of reaction (4 h). This result was confirmed in the acid-base titration (Table 1), since the increment in the number of amino groups on the film surface was small. When reaction times longer than 4 h were used, film disintegration occurred when the films were washed.

According to the results obtained at different reaction times and infrared studies, the modified films with a reaction time of 1 h were selected to continue with the next chemical activation step of the PET film, which presented an average amount of primary amino groups on the surface of 0.0078 mmol quantified by acid-base titration, with a percentage of the modification of the film with ethylenediamine by a weight of 1.25 ± 0.5%. The next activation step was carried out through the Michael addition reaction, between the amino groups present on the surface film and itaconic acid; this reaction was carried out in ethanol at reflux for 48 h. The resulting films were exhaustively washed with methanol and water, providing a weight-average surface modification of the 0.8 ± 0.2%. Subsequently, the obtained films were placed in 10 mL water at 5% acetic acid with 1% chitosan for 48 h to 25 °C. Finally, the films were removed from the chitosan solution and incubated at 90 °C for

6 h under vacuum, showing grafting of 1.38 ± 0.2% according to the weight of the chitosan. Once the films were obtained, they were placed in a solution of silver nitrate (3000 ppm) in the presence of sunlight to promote the reduction in silver ions and nucleation of silver to nanoparticles (Figure 2), resulting in an average load of 6.3 ± 1.3% by weight of silver nanoparticles concerning the matrix.

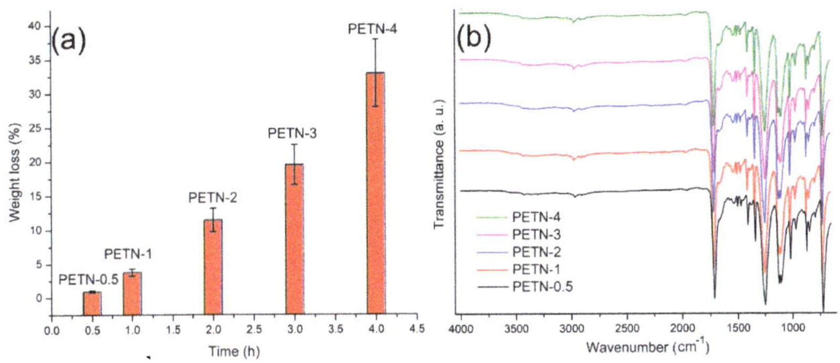

Figure 1. (**a**) Weight loss of PET films in the aminolysis reaction at different times, (**b**) IR-ATR spectra of PET films after aminolysis reaction at different times.

Table 1. Modification percentage of the PET film by the aminolysis reaction at different times, quantified by acid-base titration.

Time (h)	0.5	1	2	3	4
Modification PET with ethylenediamine (% w.)	1.18 ± 0.75	1.25 ± 0.5	1.31 ± 0.25	1.23 ± 0.37	1.39 ± 0.12

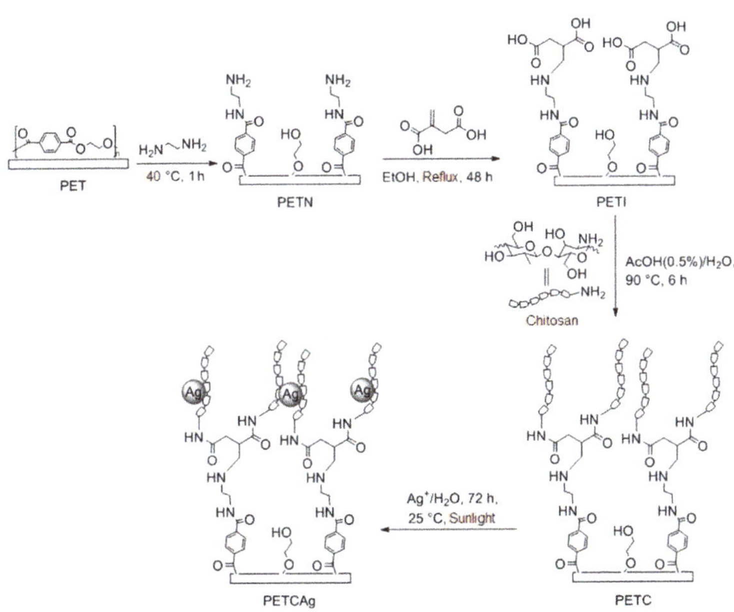

Figure 2. Scheme of chitosan graft on the functionalized PET film.

3.2. Contact Angle Study

Contact angle studies showed changes in the hydrophilicity of the films. The PET films had a contact angle of 90° which decreased to 78° when modified with ethylenediamine; moreover, upon reaction with itaconic acid, the angle slightly increased to 88°. These results are in accordance with the new hydrophilic groups present on the film surface. When grafting the chitosan chains, the contact angle increased to 110°. Finally, the loading of the silver nanoparticles caused a slight increase in the hydrophilicity of the film (Figure 3).

Figure 3. Contact angle of superficially modified films.

3.3. FTIR-ATR Analysis

TIR-ATR spectra (Figure 4a) were recorded to obtain information about the chemical groups incorporated into the PET films. The bands that evidenced the incorporation of the amino groups on the PET film surface were found at 3356 cm^{-1} (Figure 4a, PETN), which were assigned to the stretching vibration of the $-NH_2$ and –OH groups, as well as the corresponding bands of the ester carbonyl groups at 1713 cm^{-1} and the amide carbonyl at 1676 cm^{-1}. The spectra of the PETI film (Figure 4b, PETI) did not show the presence of double bonds, indicating the successful Michael addiction reaction and showing an increase in the band at 1724 cm^{-1}, as well as a greater amplitude of the band located at 3356 cm^{-1} corresponding to the carboxylic groups from the addition of itaconic acid and the secondary amino groups formed after the reaction.

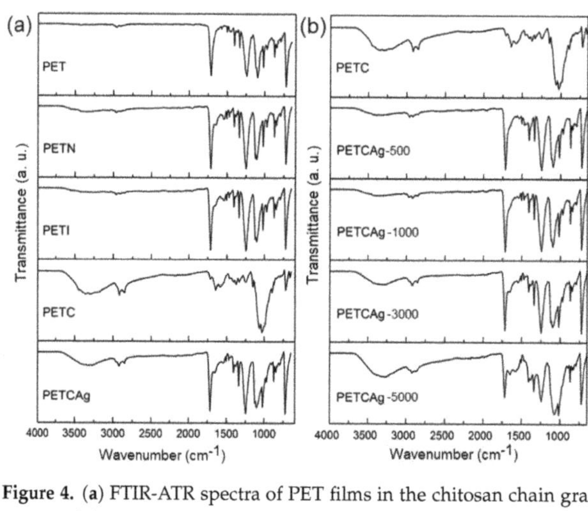

Figure 4. (**a**) FTIR-ATR spectra of PET films in the chitosan chain grafting process; (**b**) FTIR-ATR spectra of the PETC films loaded with silver nanoparticles in different silver nitrate concentrations.

Spectra of the PETC film indicated an adequate grafting of the chitosan polymer chains on the functionalized PET film, showing a decrease in the bands corresponding to the carbonyl groups coming from the matrix due to the coverage of grafted chitosan and showing an increase in the bands corresponding to the –OH and $-NH_2$ groups from the chitosan chains grafted at 3368 and 3299 cm^{-1} and amide carbonyl at 1645 cm^{-1}, as well as the vibrations of the C–H bonds at 2850–2920 cm^{-1} and C–O–C at 1098–1121 cm^{-1}. Finally, the load of silver nanoparticles caused a broad band at 3300 cm^{-1}, probably due to the interactions of the –OH and $-NH_2$ groups with the silver nanoparticles.

This study indicated that silver nanoparticles were obtained through a reduction in silver nitrate using chitosan. The FTIR of PETCAg shows the presence of new bands at 1715 and 1250 cm^{-1} due to the carboxylic and carbonyl groups confirming the oxidation of the hydroxyl groups of chitosan by the silver reduction. Likewise, the infrared spectrum shows that the carboxylic and carbonyl groups' bands increase as the silver nitrate concentration increases (Figure 4b) [31,32].

3.4. Thermal Analysis

The thermogravimetric analysis (Figure 5a and Table 2) shows that the films have a decomposition temperature range of 436 to 448 °C. However, after 460 °C, PETC and PETCAg films show a second decomposition range, which is attributed to the decomposition of the chitosan grafted onto the film. The amount of the residue yield obtained is shown in Table 2. DSC studies show that the melting temperature of the PET film (246 °C) was not affected by the grafting process. The PET film's glass transition (Tg) shows an increase in the grafting process. When itaconic acid was incorporated, the Tg of PETI increased from 78.6 to 95.6 °C; the addition of chitosan raised the Tg to 100.9 °C. Loading of silver nanoparticles elevated the Tg of the PETCAg to 110.7 °C, probably due to the strong interactions of the polymer chains with the silver nanoparticles (Figure 5b) [33].

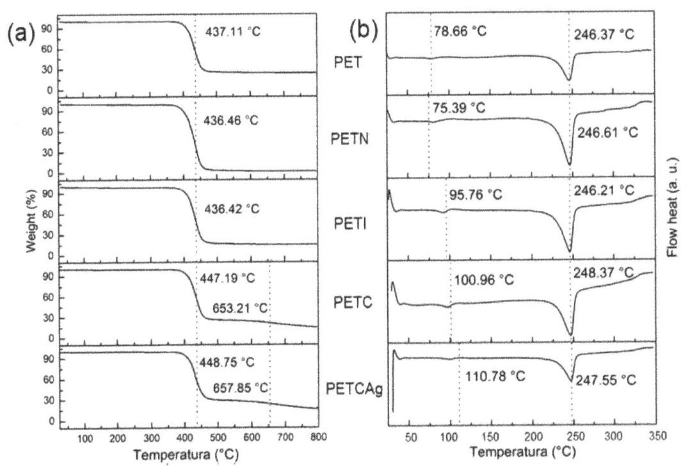

Figure 5. (**a**) TGA and (**b**) DSC studies under nitrogen flow on PET films obtained in the chitosan chain grafting process.

Table 2. Decomposition temperatures of PET films obtained in the chitosan chain grafting process.

Sample	Glass Transition Temperature (°C)	Decomposition Temperature (°C)	Residue Yield (800 °C, w.%)
PET	78.6	437.11	24.33
PETN	75.3	436.46	2.03
PETI	95.7	436.42	15.41
PETC	100.9	447.1, 653.21	7.33
PETCAg	110.7	448.75, 657.85	13.13

Silver quantification in PETC films was carried out by residual weight at 800 °C. PETC films not exposed to a silver nitrate solution presented carbonization with an average residue of 7.33%. Therefore, the residual weight difference of the total mass of the films incubated in the silver solutions (500, 1000, 3000, and 5000 ppm) can be attributed to the silver nanoparticles loaded on films, resulting in a residual weight difference of 1.2 ± 0.4, 2.1 ± 0.6, 6.3 ± 1.3, and 7.1 ± 1.4%, respectively (Figure 6). As a result, a clear correlation was observed between the amount of silver nanoparticles and the concentration of silver nitrate used in the reaction medium [34].

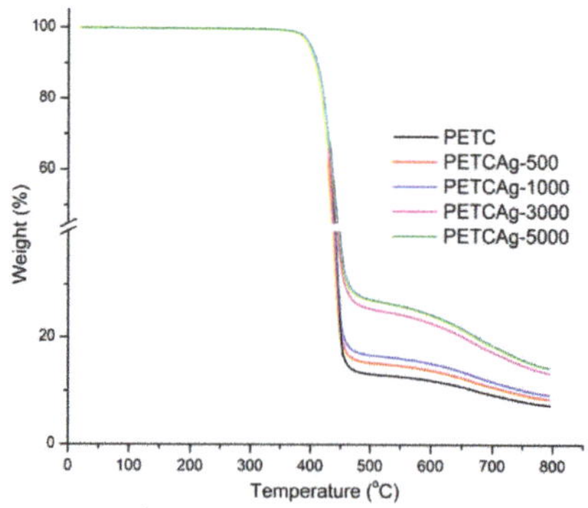

Figure 6. Silver quantification on PETCAg films by TGA under nitrogen flow.

3.5. Study of Mechanical Properties

The mechanical properties of the films were not significantly affected when grafting the chitosan polymer chains (Table 3), indicating that these were only grafted on the surface of the material, findings which were confirmed by the SEM studies, and the differences are possibly attributed to the wear of the material in the surface modification process, increasing the deformation for the PETN and PETI films, decreasing again, and becoming very similar to the original matrix for the films of PETC, with this finding being attributed to the grafting of the chitosan polymeric chains (Figure 7).

Table 3. Mechanical properties of PET films in the grafting process.

Sample	Elastic Modulus (MPa)	Stress Rupture (MPa)
PET	1144.762 ± 216.627	122.163 ± 20.3536
PETN	990.817 ± 104.047	136.780 ± 7.45101
PETI	1247.815 ± 188.311	140.172 ± 13.7654
PETC	1188.304 ± 122.115	103.506 ± 17.6539

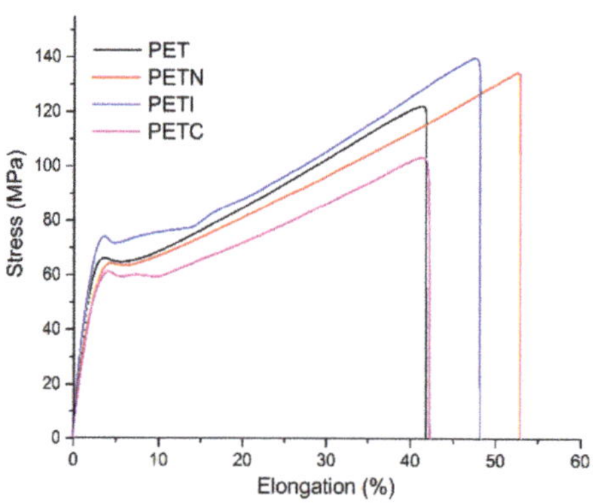

Figure 7. Tensile stress–strain graph of PET films in the process of grafting chitosan polymer chains.

3.6. SEM Study

SEM studies of the modified films showed relevant changes with respect to the PET films (Figure 8). The magnification showed that the film modification process caused the formation of elongated structures in the form of layers across the surface. The formation of the layers and their thickness increased with the grafting process of the chitosan chains. Finally, the SEM images also showed the presence of silver nanoparticles on the formed chitosan layer with a mean size of 130 ± 37 nm.

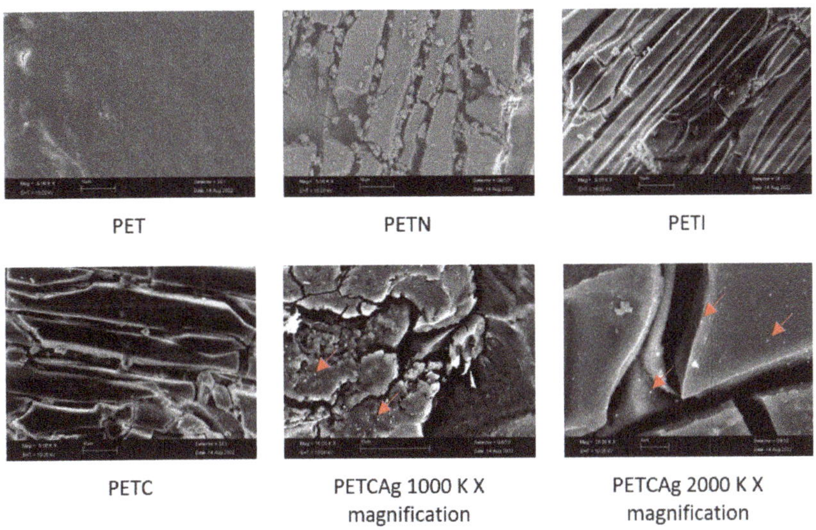

Figure 8. SEM images of PET films at different stages of the grafting process of chitosan polymer chains and with silver nanoparticles (marked with red arrows).

3.7. Cell Viability Study

Cell viability was evaluated against 3T3 fibroblasts (Figure 9). The cell viability of the films decreased by 15–25% compared to the control, showing no differences among these samples, as indicated by statistical analysis (Figures S1–S7, Tables S1–S4). The exception was the PETCAg-1000 film, for which viability was 59 ± 9.2%. This finding indicates that the films, except for the PETC-1000 film, have good compatibility regardless of the modification and percentage of the silver loaded.

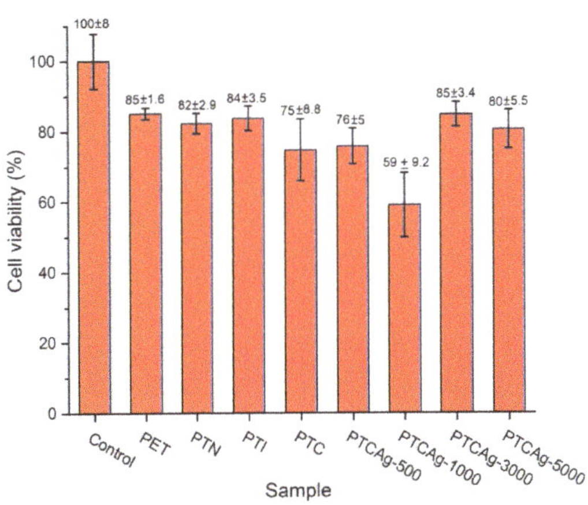

Figure 9. Cell viability along the grafting process (PET, PETN, PETI, and PETC) and silver-loaded PETC with different silver nanoparticles percentages (PETAg-500 (1.2 ± 0.4%), PETAg-1000 (2.1 ± 0.6%), PETAg-3000 (6.3 ± 1.3%), and PETAg-5000 (7.1 ± 1.4%)).

4. Conclusions

The grafting of chitosan polymer chains was successfully carried out by employing adequate surface functionalization of the PET films. The grafted chitosan allowed the nucleation and loading of silver nanoparticles with different percentages (1.2–7.1% with respect to the matrix) without using other reducing agents. Silver nanoparticles with an average size of 130 $\pm$ 37 nm were obtained. The films with a silver-coated surface obtained in this work are a good candidate for use as part of antimicrobial biomedical devices and disposable medical devices

Supplementary Materials: The following supporting information can be downloaded at: https://www.mdpi.com/article/10.3390/polym15010125/s1, Table S1: Analysis of Variance; Table S2: Model Summary; Table S3: Means; Table S4: Grouping Information Using the Fisher LSD Method and 95% Confidence; Figure S1: Fisher individual 95%; Figure S2: Interval plot of control PET; Figure S3: Individual value plot of control PET; Figure S4: Boxplot of control PET; Figure S5: Normal Probability Plot; Figure S6: Versus fits; Figure S7: Histogram of residual values.

Author Contributions: Conceptualization, G.G.F.-R. and F.L.-S.; methodology, G.G.F.-R. and E.M.; software, G.G.F.-R.; validation, E.M., R.V.-G. and E.B.; formal analysis, G.G.F.-R.; investigation, G.G.F.-R., H.M. and E.M.; resources, R.V.-G., E.M. and E.B.; data curation, H.M.; writing—original draft preparation, G.G.F.-R., F.L.-S. and E.M.; writing—review and editing, E.M.; visualization, H.M.; supervision, E.M., R.V.-G., E.B.; project administration, E.M., R.V.-G. and E.B.; funding acquisition, R.V.-G., E.M. and E.B. All authors have read and agreed to the published version of the manuscript.

Funding: This work was supported by the Dirección General de Asuntos del Personal Académico (DGAPA), Universidad Nacional Autónoma de México under Grant IN202320. University of Guadalajara under PROSNI 2021. Support Program for Technological Research and Innovation Projects under PAPIIT IG100220 and National Council of Science and Technology under CONACyT CF-19 No 140617. Call for basic scientific research CONACYT 2017–2018 del "Fondo Sectorial de Investigación para la educación CB2017-2018" (Grant A1-S-29789).

Institutional Review Board Statement: Not applicable.

Informed Consent Statement: Not applicable.

Data Availability Statement: The data presented in this study are available on request from the corresponding author.

Acknowledgments: Thanks to CONACyT for the postdoctoral fellowship awarded to G.G. Flores-Rojas (CVU 407270). In addition, the authors thank M.C.E. Hernández-Mecinas from the Materials Research Institute of UNAM for his technical assistance.

Conflicts of Interest: The authors declare no conflict of interest.

References

1. Berlanga, M.; Guerrero, R. Living together in biofilms: The microbial cell factory and its biotechnological implications. *Microb. Cell Factories* **2016**, *15*, 165. [CrossRef] [PubMed]
2. López-Saucedo, F.; Flores-Rojas, G.; Varca, J.; Varca, G.; Bucio, E. Antimicrobial materials and devices for biomedical applications. In *Frontiers in Clinical Drug Research: Anti-Infectives*; Rahman, A., Ed.; Bentham Science: Singapore, 2020; pp. 78–126. [CrossRef]
3. Zemljič, L.F.; Tkavc, T.; Vesel, A.; Šauperl, O. Chitosan coatings onto polyethylene terephthalate for the development of potential active packaging material. *Appl. Surf. Sci.* **2013**, *265*, 697–703. [CrossRef]
4. Mallakpour, S.; Azadi, E.; Hussain, C.M. Recent breakthroughs of antibacterial and antiviral protective polymeric materials during COVID-19 pandemic and after pandemic: Coating, packaging, and textile applications. *Curr. Opin. Colloid Interface Sci.* **2021**, *55*, 101480. [CrossRef] [PubMed]
5. Ana, P.; Bortoleto, J.; Cruz, N.; Rangel, E.; Durrant, S. Surface Properties of PET Polymer Treated by Plasma Immersion Techniques for Food Packaging. *Int. J. Nano Res.* **2018**, *1*, 33–41.
6. Sanchez-Garcia, M.D.; Gimenez, E.; Lagaron, J.M. Novel PET Nanocomposites of Interest in Food Packaging Applications and Comparative Barrier Performance with Biopolyester Nanocomposites. *J. Plast. Film Sheeting* **2007**, *23*, 133–148. [CrossRef]
7. Sulyman, M.; Haponiuk, J.; Formela, K. Utilization of Recycled Polyethylene Terephthalate (PET) in Engineering Materials: A Review. *Int. J. Environ. Sci. Dev.* **2016**, *7*, 100–108. [CrossRef]
8. Çaykara, T.; Sande, M.G.; Azoia, N.; Rodrigues, L.R.; Silva, C.J. Exploring the potential of polyethylene terephthalate in the design of antibacterial surfaces. *Med. Microbiol. Immunol.* **2020**, *209*, 363–372. [CrossRef]

9. Sun, W.; Liu, W.; Wu, Z.; Chen, H. Chemical Surface Modification of Polymeric Biomaterials for Biomedical Applications. *Macromol. Rapid Commun.* **2020**, *41*, e1900430. [CrossRef]
10. Ratner, B.; Castne, D. *Surface Modification of Polymeric Biomaterials*, 1st ed.; Springer: Boston, MA, USA, 1997. [CrossRef]
11. Zander, Z.K.; Becker, M.L. Antimicrobial and Antifouling Strategies for Polymeric Medical Devices. *ACS Macro Lett.* **2018**, *7*, 16–25. [CrossRef]
12. Bian, N.; Yang, X.; Zhang, X.; Zhang, F.; Hou, Q.; Pei, J. A complex of oxidised chitosan and silver ions grafted to cotton fibres with bacteriostatic properties. *Carbohydr. Polym.* **2021**, *262*, 117714. [CrossRef]
13. Aravamudhan, A.; Ramos, D.; Nada, A.; Kumbar, S. *Natural Polymers: Polysaccharides and Their Derivatives for Biomedical Applications*; Elsevier Inc.: Amsterdam, The Netherlands, 2014. [CrossRef]
14. Novakovic, D.; Peltonen, L.; Isomäki, A.; Fraser-Miller, S.J.; Nielsen, L.H.; Laaksonen, T.; Strachan, C.J. Surface Stabilization and Dissolution Rate Improvement of Amorphous Compacts with Thin Polymer Coatings: Can We Have It All? *Mol. Pharm.* **2020**, *17*, 1248–1260. [CrossRef] [PubMed]
15. Stoleru, E.; Munteanu, S.B.; Dumitriu, R.P.; Coroaba, A.; Drobotă, M.; Zemljic, L.F.; Pricope, G.M.; Vasile, C. Polyethylene materials with multifunctional surface properties by electrospraying chitosan/vitamin E formulation destined to biomedical and food packaging applications. *Iran. Polym. J.* **2016**, *25*, 295–307. [CrossRef]
16. Peers, S.; Montembault, A.; Ladavière, C. Chitosan hydrogels for sustained drug delivery. *J. Control. Release* **2020**, *326*, 150–163. [CrossRef] [PubMed]
17. Wang, F.; Li, J.; Tang, X.; Huang, K.; Chen, L. Polyelectrolyte three layer nanoparticles of chitosan/dextran sulfate/chitosan for dual drug delivery. *Colloids Surfaces B Biointerfaces* **2020**, *190*, 110925. [CrossRef] [PubMed]
18. Zhao, D.; Yu, S.; Sun, B.; Gao, S.; Guo, S.; Zhao, K. Biomedical Applications of Chitosan and Its Derivative Nanoparticles. *Polymers* **2018**, *10*, 462. [CrossRef] [PubMed]
19. Satitsri, S.; Muanprasat, C. Chitin and Chitosan Derivatives as Biomaterial Resources for Biological and Biomedical Applications. *Molecules* **2020**, *25*, 5961. [CrossRef] [PubMed]
20. Ji, J.; Wang, L.; Yu, H.; Chen, Y.; Zhao, Y.; Zhang, H.; Amer, W.; Sun, Y.; Huang, L.; Saleem, M. Chemical Modifications of Chitosan and Its Applications. *Polym. Technol. Eng.* **2014**, *53*, 1494–1505. [CrossRef]
21. Franconetti, A.; Carnerero, J.M.; Prado-Gotor, R.; Cabrera-Escribano, F.; Jaime, C. Chitosan as a capping agent: Insights on the stabilization of gold nanoparticles. *Carbohydr. Polym.* **2019**, *207*, 806–814. [CrossRef]
22. Phan, T.T.V.; Hoang, G.; Nguyen, V.T.; Nguyen, T.P.; Kim, H.H.; Mondal, S.; Manivasagan, P.; Moorthy, M.S.; Lee, K.D.; Junghwan, O. Chitosan as a stabilizer and size-control agent for synthesis of porous flower-shaped palladium nanoparticles and their applications on photo-based therapies. *Carbohydr. Polym.* **2019**, *205*, 340–352. [CrossRef]
23. Flores-Rojas, G.G.; López-Saucedo, F.; Vera-Graziano, R.; Mendizabal, E.; Bucio, E. Magnetic Nanoparticles for Medical Applications: Updated Review. *Macromol* **2022**, *2*, 374–390. [CrossRef]
24. Yeamsuksawat, T.; Zhao, H.; Liang, J. Characterization and antimicrobial performance of magnetic Fe3O4@Chitosan@Ag nanoparticles synthesized via suspension technique. *Mater. Today Commun.* **2021**, *28*, 102481. [CrossRef]
25. Yin, I.X.; Zhang, J.; Zhao, I.S.; Mei, M.L.; Li, Q.; Chu, C.H. The Antibacterial Mechanism of Silver Nanoparticles and Its Application in Dentistry. *Int. J. Nanomed.* **2020**, *15*, 2555–2562. [CrossRef] [PubMed]
26. El Knidri, H.; Belaabed, R.; Addaou, A.; Laajeb, A.; Lahsini, A. Extraction, chemical modification and characterization of chitin and chitosan. *Int. J. Biol. Macromol.* **2018**, *120*, 1181–1189. [CrossRef] [PubMed]
27. Sahariah, P.; Másson, M. Antimicrobial Chitosan and Chitosan Derivatives: A Review of the Structure–Activity Relationship. *Biomacromolecules* **2017**, *18*, 3846–3868. [CrossRef] [PubMed]
28. Qian, L.; Zheng, J.; Wang, K.; Tang, Y.; Zhang, X.; Zhang, H.; Huang, F.; Pei, Y.; Jiang, Y. Cationic core–shell nanoparticles with carmustine contained within O6-benzylguanine shell for glioma therapy. *Biomaterials* **2013**, *34*, 8968–8978. [CrossRef] [PubMed]
29. Stephen, Z.R.; Kievit, F.M.; Veiseh, O.; Chiarelli, P.A.; Fang, C.; Wang, K.; Hatzinger, S.J.; Ellenbogen, R.G.; Silber, J.R.; Zhang, M. Redox-Responsive Magnetic Nanoparticle for Targeted Convection-Enhanced Delivery of O^6-Benzylguanine to Brain Tumors. *ACS Nano* **2014**, *8*, 10383–10395. [CrossRef] [PubMed]
30. Veiseh, O.; Sun, C.; Gunn, J.; Kohler, N.; Gabikian, P.; Lee, D.; Bhattarai, N.; Ellenbogen, R.; Sze, R.; Hallahan, A.; et al. Optical and MRI Multifunctional Nanoprobe for Targeting Gliomas. *Nano Lett.* **2005**, *5*, 1003–1008. [CrossRef]
31. Tran, H.; Chiang, K.; Scott, J.; Amal, R. Understanding selective enhancement by silver during photocatalytic oxidation. *Photochem. Photobiol. Sci.* **2005**, *4*, 565–567. [CrossRef]
32. Wang, X.; Cheng, F.; Gao, J.; Wang, L. Antibacterial wound dressing from chitosan/polyethylene oxide nanofibers mats embedded with silver nanoparticles. *J. Biomater. Appl.* **2015**, *29*, 1086–1095. [CrossRef]
33. Vázquez, E.; Duarte, L.; López-Saucedo, F.; Flores-Rojas, G.; Bucio, E. Cellulose-based antimicrobial materials. In *Advanced Antimicrobial Materials and Applications*; Inamuddin, Ahamed, M.I., Prasad, R., Eds.; Springer: Singapore, 2021; pp. 61–85. [CrossRef]
34. López-Saucedo, F.; Flores-Rojas, G.G.; López-Saucedo, J.; Magariños, B.; Alvarez-Lorenzo, C.; Concheiro, A.; Bucio, E. Antimicrobial silver-loaded polypropylene sutures modified by radiation-grafting. *Eur. Polym. J.* **2018**, *100*, 290–297. [CrossRef]

polymers

MDPI

Article

Binary Graft of Poly(acrylic acid) and Poly(vinyl pyrrolidone) onto PDMS Films for Load and Release of Ciprofloxacin

Belén Santillán-González [1], Lorena Duarte-Peña [2,*] and Emilio Bucio [2,*]

[1] División de Ciencias Biológicas y de la Salud, Unidad Xochimilco, Universidad Autónoma Metropolitana, Calzada del Hueso 1100, Col. Villa Quietud, Delegación Coyoacán, Ciudad de México C.P. 04960 CDMX, Mexico

[2] Departamento de Química de Radiaciones y Radioquímica, Instituto de Ciencias Nucleares, Universidad Nacional Autónoma de México, Circuito Exterior, Ciudad Universitaria, Ciudad de México C.P. 04510 CDMX, Mexico

* Correspondence: lorena.duarte@correo.nucleares.unam.mx (L.D.-P.); ebucio@nucleares.unam.mx (E.B.)

Abstract: Polymers are versatile compounds which physical and chemical properties can be taken advantage of in wide applications. Particularly, in the biomedical field, polydimethylsiloxane (PDMS) is one of the most used for its high biocompatibility, easy manipulation, thermal, and chemical stability. Nonetheless, its hydrophobic nature makes it susceptible to bacterial pollution, which represents a disadvantage in this field. A potential solution to this is through the graft of stimuli-sensitive polymers that, besides providing hydrophilicity, allow the creation of a drug delivery system. In this research, PDMS was grafted with acrylic acid (AAc) and vinyl pyrrolidone (VP) in two steps using gamma radiation. The resulting material was analyzed by several characterization techniques such as infrared spectroscopy (FTIR), swelling, contact angle, critical pH, and thermogravimetric analysis (TGA), demonstrating the presence of both polymers onto PDMS films and showing hydrophilic and pH-response properties. Among the performed methods to graft, the loading and release of ciprofloxacin were successful in those samples obtained by direct irradiation method. Furthermore, the antimicrobial assays showed zones of inhibition for microorganisms such as *Staphylococcus aureus* and *Escherichia coli*.

Keywords: polydimethylsiloxane; acrylic acid; vinyl pyrrolidone; ciprofloxacin; gamma radiation; drug-load and release

1. Introduction

The development of new drug-modified delivery systems is booming in the pharmaceutical industry due to them offering greater advantages than conventional systems. For example, they can improve the solubility and absorption of active pharmaceutical ingredients (API) in the organism, and decrease the dosage, diminishing the side effects. Additionally, they can modify the API's rate and site of action, and thus, the main objective is achieved. Thereby, it is important to create drug transporters, which help to achieve the objectives mentioned before. In this sense, polymers are proving to be one of the most desired materials for carrier design, especially in the field of nanotechnology as nanocarriers, because they can be used for the delivery of small therapeutic molecules to genes, and proteins, also in cancer therapy, diabetes treatment, anti-infections, etcetera, helping to improve their therapeutic efficiency. Generally, we can find a variety of carriers. Among the main polymeric nanocarriers, there are polymeric micelles, polymersomes, polymeric nanogels, polymeric nanocapsules, and dendrimers. In addition, there are lipid-based nanocarriers as nanoemulsions, phospholipid micelles, liposomes, solid lipid nanocarriers, and inorganic nanocarriers that include quantum dot, carbon nanotubules, gold nanoparticles, magnetic nanoparticles, and silicon nanoparticles. They all have different sizes, structures, and morphology and, therefore, different biomedical applications [1–5].

Polymers are macromolecules with a structure consisting of repeating units and their properties depend on the chemical structure (the nature of end groups, possible branches, and tacticity), chain length, molar weight distribution pattern, cohesive forces, density, crystallinity, molecular mobility, morphology, thermochemical and calorimetric properties, degradation, etcetera [6].

Polydimethylsiloxane (PDMS), also known as silicone, is a polymer, whose physical and chemical properties allow it to have widespread use in the biomedical industry. These properties include biocompatibility, thermal and chemical stability, chemically inert, gas permeability, good mechanical properties, simplicity to handle and manipulate, and thermal and electrical insulator. For all of the above, it's a material that has been widely used in the development of micropumps, microvalves, catheter surfaces, dressings and bandages, optical systems, implants, microfluidic devices, devices for drug delivery, DNA sequencing, and clinical diagnostic, etcetera [7–12]. However, PDMS in biological environments can result in bacterial fouling or incompatibility due to its hydrophobic characteristics that help the adhesion of microorganisms [13,14]. The surface modification of this material is an alternative to improve their properties and create more complex systems with prophylactics characteristics that can take part in an active way in medical treatments. A promising alternative is the surface modification with materials that allow the controlled load and release of active agents through supramolecular interactions.

Poly acrylic acid (PAAc) is an anionic polymer, highly hydrophilic, sensitive to pH, and biocompatible [15], which has been used for the carrier systems fabrication of active agents as metallic nanoparticles or drugs [16]. On the other hand, polyvinyl pyrrolidone (PVP) is a highly biocompatible polymer used widely in the pharmaceutical industry for the development of various formulations, from tablets to nanofibers and contact lenses [17]. Additionally, due to its amphiphilic structure, it can interact with both hydrophilic and hydrophobic molecules; nonetheless, its application is limited for the low mechanical properties that show in its homopolymer form, so it tends to copolymerize or be used with other polymers, thus obtaining materials with better properties. The combination of PAAc and PVP has shown favorable results in the fabrication of hydrogels and the stabilization of drugs [18–21].

This work aims to propose the synthesis of a binary graft in two steps of acrylic acid (AAc) and 1-vinyl-2-pyrrolidone (VP) onto silicone films (PDMS) using gamma radiation as a physical initiator of the polymerization, to provide the surface hydrophilicity, biocompatibility and pH sensitivity, for its application in the load and release of ciprofloxacin. The ionizing radiation allows the formation of active dots along the polymer matrix, favoring the graft, and is an initiator that does not generate pollution in the final product, achieving high-purity materials [22].

2. Materials and Methods

2.1. Materials

PDMS films with a thickness of 1 mm were from Good-fellow (Huntingdon, UK). AAc, VP and ethanol were purchased from Aldrich Chemical, Saint Louis, MO, USA. The monomers AAc and VP were purified by vacuum distillation. Phosphate-buffered saline was prepared with salts from Aldrich Chemical, Saint Louis, MO, USA. For all assays, it was used distilled water. The gamma-ray source was a ^{60}Co Gamma beam 651-PT (Nordion Ottawa, Ontario Canada Inc., Toronto, ON, Canada) proportioned by the Nuclear Science Institute at the National Autonomous University of México (ICN-UNAM).

2.2. Synthesis of PDMS-g-AAc

AAc grafted onto PDMS films was made by the direct irradiation method (DM), using gamma (γ) radiation. This method consists of the simultaneous irradiation of the polymeric matrix (PDMS) and the monomer to graft (AAc) under the atmosphere without oxygen.

PDMS films (6 $\times$ 1 cm) were weighed and placed in glass ampoules. AAc solutions at different concentrations were prepared: 20, 30, and 35% v/v, using ethanol: water (1:1) as solvent. To each ampoule 9 mL of AAc solution was added with the concentration chosen for each one and then they were bubbled with Ar for 20 min. Subsequently, the ampoules were sealed and subjected to γ radiation at doses of 20 and 30 kGy. After, the ampoules were opened, and the samples were removed to wash them with distilled water in order to remove the non-grafted homopolymer. Finally, the samples were dried in a vacuum oven, weighed, and the graft percentage was calculated according to Equation (1), where W_2 is the grafted sample weight (g), and W_1 is the initial weight of the sample (g).

$$\text{Graft } (\%) = \frac{(W_2 - W_1) \times 100}{W_1} \tag{1}$$

The graft procedure was taken from work carried out by Velazco-Medel, et.al., changing the parameters of monomer concentration and the applied dose [23].

2.3. *Synthesis of (PDMS-g-AAc)-g-VP*

VP grafted onto PDMS-*g*-AAc films was also made by employing the direct irradiation method with the methodology described in the previous Section 2.2. In this case, the factors to be modified were VP concentration [15, 20, 25, 30, 35, 40, 50% v/v] and the radiation dose applied (2.5, 5, 10, 15, 20, 25, 35 kGy). Distilled water was used as solvent.

Additionally, it was observed that PDMS-*g*-AAc showed interaction with VP, only using the factors of temperature and reaction time under and oxygen-free atmosphere. For which reason, tests without radiation (WR) were carried out, modifying the parameters: VP concentration [10, 20, 30% v/v], temperature (50, 60, 70 °C), and reaction time at 8, 16, and 24 h.

2.4. *Infrared Spectroscopy (FTIR-ATR)*

This technique was used to characterize the grafted samples and thus be able to determine the presence of the different functional groups. Analysis was performed using a PerkinElmer Spectrum 100 spectrometer (PerkinElmer Cetus Instruments, Norwalk, CT, USA) with 16 scans in the ATR module and in the 4000 to 650 cm^{-1} range.

2.5. *Swelling*

The dry samples were weighed and each one was placed in a glass with distilled water. A water bath at 25 °C was prepared, where the glasses were placed. Afterward, each certain time (5, 10, 15, 30 min., 1, 2, 4, 6, 8, 24, and 32 h), the samples were removed from the glasses and the excess of water was removed with blotting paper to be able to weigh them and return them to their corresponding glass. The procedure was repeated until it did not observe significant changes in the weight of the films. All the samples were measured in triplicate. Swelling percentage was determinate by Equation (2), where W_2 is the swelled sample weight (g), and W_1 is the initial weight of the sample (g).

$$\text{Swelling } (\%) = \frac{(W_2 - W_1) \times 100}{W_1} \tag{2}$$

2.6. *Contact Angle*

To confirm the hydrophilic or hydrophobic properties of grafts, their degree of wetting was measured by this technique. It was employed a goniometer DSA 100 Krüss GmbH, Hamburg, Germany. The angles of the drop onto the surface of the films were measured from time 0, 1, 5,10, 15, and 20 min until the moment which did not observe significant changes. All measurements were performed three times.

2.7. Critical pH

Phosphate buffer solutions were prepared at several pH (2.3, 3.3, 4.1, 5.5, 6.5, 8, and 10). The dry samples were weighed and each one was placed in a glass with a different pH solution for each day, starting with pH 2.3 and ending with pH 10. The glasses were placed in a water bath prepared at 25 °C. As in the swelling procedure, the samples were removed from the glasses and the excess of water was removed with blotting paper to be able to weigh them and return them to their corresponding glass. However, in this case, the procedure was repeated every 24 h (time limit of swelling found). All measurements were performed three times.

2.8. Thermogravimetric Analysis (TGA)

The analysis of thermal behavior of the PDMS and the grafted samples was monitored using a TGA Q50 (TA Instruments, New Castle, DE, USA), which works under a nitrogen atmosphere, the temperature range used was from 30 to 800 °C with a heating rate of 10 °C/min.

2.9. Loading and Release of Ciprofloxacin for Spectrophotometer UV-Vis

2.9.1. Ciprofloxacin Loading

The load of the drug was made for PDMS, [PDMS-g-AAc], [PDMS-g-AAc]-g-VP for DM, and [PDMS-g-AAc]-g-VP without radiation. The dry samples were weighed and placed in 3 mL of a solution of 12 μg/mL of ciprofloxacin in a water bath prepared at 25 °C. Monitoring of the loading was made with a spectrophotometer UV-Vis SPECORD 200 PLUS (Analytik Jena, Jena, Germany) at 266 nm, first at time 0 with the reference solution (distilled water) and then with the samples at time 2, 4, 6, 24, 30, 48, and 96 h. The quantification was performed with a calibration curve (Figure S1), and each measurement was performed in triplicate.

2.9.2. Ciprofloxacin Release

The loaded samples were washed with distilled water and placed in 3 mL of a phosphate buffer dissolution (pH 7.4), in a water bath prepared at 37 °C and with oscillation at 130 rpm. The release monitoring was performed with the spectrophotometer UV-Vis, at 266 nm measuring at times 0, 0.5, 1, 2, 4, 6, 24, and 30 h. Carrying out a calibration curve to the quantification of the released drug (Figure S2). Furthermore, the statistic program Excel DDSolver software was used for the analysis of the release model.

2.10. Antimicrobials Assay: Agar Diffusion or Kirby-Bauer Method

Samples of around 30 mg were inserted into a petri dish that contained Müeller-Hinton medium, sown with *Staphylococcus aureus* or *Escherichia coli*, at a concentration of 1.5×10^8 CFU/mL and 24 h of grown. The samples were incubated for 24 h at 37 °C, and the zone of inhibition was measured from the center of the sample until the periphery where the bacterial grew was observed.

Figure 1 shows a scheme of the experimental methodology developed during the work.

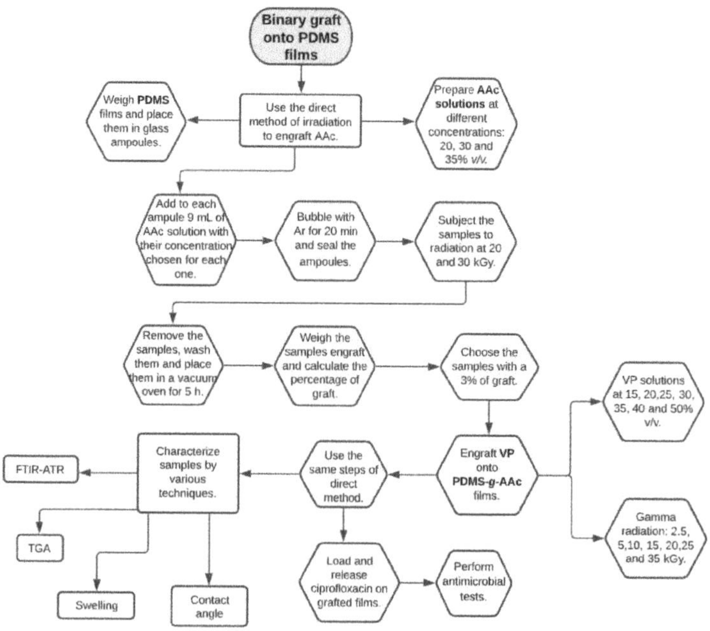

Figure 1. Experimental methodology.

3. Results

3.1. Synthesis of PDMS-g-AAc

In this synthesis, the factors that changed were the concentration of monomer to graft (AAc) at 20, 30, and 35% v/v and the dose of γ radiation at 20 and 30 kGy. Figure 2a shows that the higher concentrations increased the AAc graft. Nonetheless, the films modified with AAc concentrations of 35% presented a break when tried to manipulate. Concerning the dose, in Figure 2b, the graph exhibits that a significant difference regarding the percentage of graft for both doses (3%) did not exist, and this percentage allowed the manipulation of the films. With this in mind, it was decided to work with 20 kGy. The chosen final conditions were 20% v/v AAc concentration, Ar bubbling for 20 min, and a dose of 20 kGy to obtain materials with approximately 3% of the graft [PDMS-g-AAc (3%)].

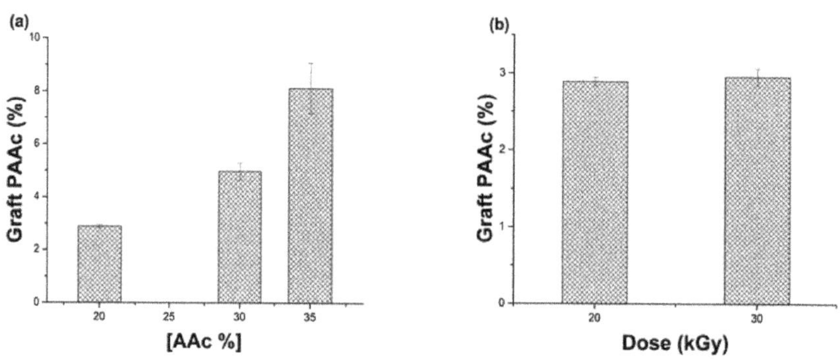

Figure 2. Percentages of PAAc graft onto PDMS films. (**a**) Effect of the monomer concentration and (**b**) Effect of the γ radiation doses. Reported: the mean ± standard error of the mean, n = 3.

3.2. Synthesis of [PDMS-g-AAc]-g-VP, Second Graft for the Direct Irradiation Method

In Figure 3a, it is noted that, as the concentration of VP increased, the higher the graft was. Concentrations of 30, 35, and 40 % showed graft values close to each other, 9.3 ± 0.3, 9.8 ± 0.4, and 10.5 ± 0.6%, respectively. On the other hand, the effect of the doses is shown in Figure 3b, the doses between 5 and 20 kGy did not present an apparent change as the graft remained at around 6 to 7%, but with a dose of 25 kGy there existed an increase until 11 ± 1.1% of graft; the opposite effect occurred when a dose of 35 kGy, where it was possible that there existed a matrix degradation, as the percentage decayed until 6%. The films tended to deform when the percentage of graft increases. Hence, it was decided to leave the conditions of VP to 30% and 10 kGy of dose, thereby this gives us results modifications around 10%.

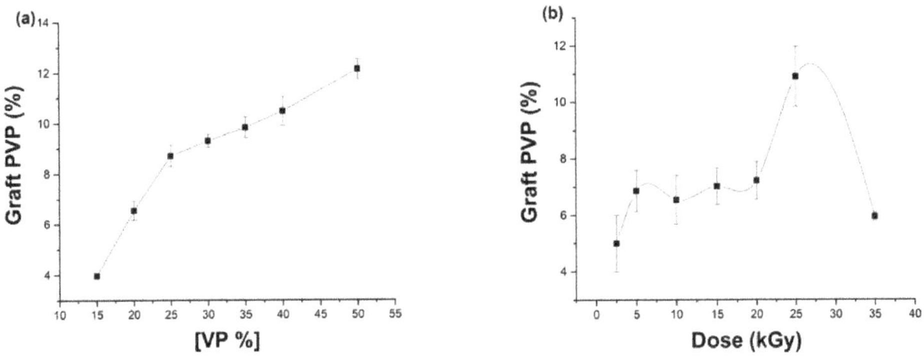

Figure 3. Synthesis of [PDMS-g-AAc]-g-VP by direct irradiation method (**a**) Effect of VP monomer concentration and (**b**) Effect of γ radiation doses. Reported: the mean ± standard error of the mean, $n = 3$.

3.3. Synthesis of [PDMS-g-AAc]-g-V, Second Graft for without Irradiation Method

Figure 4a shows that lower concentrations of VP (10 and 20% v/v) produced higher graft percentages between 8 and 9%, while with 30% v/v VP concentration, the graft was minimum (5.3% ± 1.2). However, maintaining the same concentration (30% v/v) but diminishing the temperature at 50 °C, the graft increased to 12.3 ± 0.04% (Figure 4b). Finally, Figure 4c shows that there was no apparent difference in the percentage of VP grafted by varying the reaction time between 8, 16, and 24 h, so the key factor was the temperature. The conditions to obtain materials [PDMS-g-AAc (3%)]-g-VP (10%) were 30% v/v VP concentration, 50 °C, and 16 h.

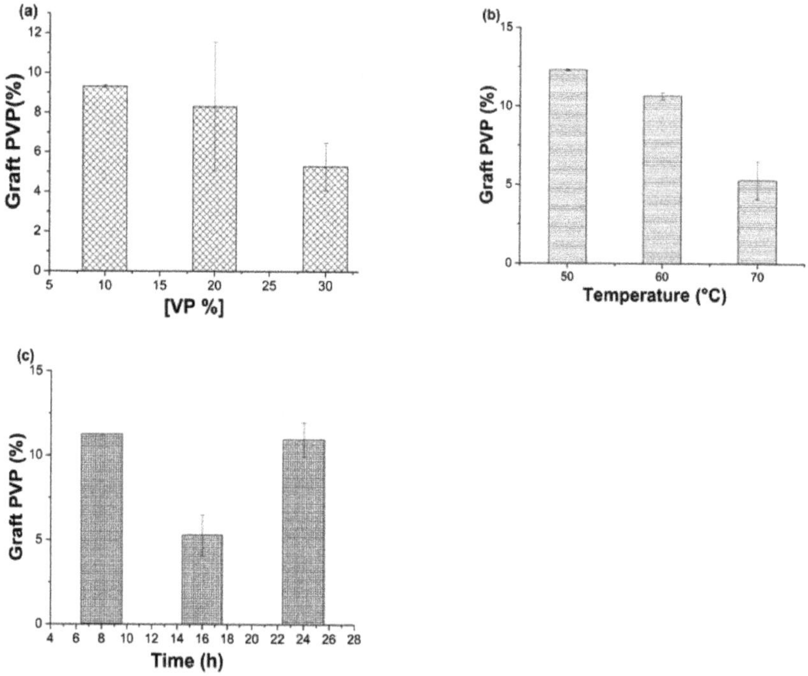

Figure 4. Percentages of graft of PVP onto films [PDMS-*g*-AAc], (**a**) Effect of VP concentration, (**b**) Effect of temperature, and (**c**) Effect of reaction time. Reported: the mean ± standard error of the mean, $n = 3$.

3.4. Infrared Spectroscopy (FTIR-ATR)

Figure 5 presents the infrared specters of different materials showing the functional groups expected for each sample. First, the polymeric matrix of PDMS showed bands between ~1000 y ~750 cm^{-1}, corresponding to the Si–O–Si stretching and Si–CH_3 vibrations. There was a short band attributed near 2900 cm^{-1}, which is the typical band for C–H stretching [23]. In the [PDMS-*g*-AAc] spectrum, the band in 1712 cm^{-1} corresponded to the carbonyl group of AAc, confirming its presence on the film, although the bands of PDMS still appeared [24]. The next specters belonged to binary graft [PDMS-*g*-AAc]-*g*-VP, although each one was obtained for a different method, the band to ~1650 cm^{-1} that belonged to the carbonyl group of pyrrolidone appeared in both specters [25].

The IR technique proved the presence of both monomers onto PDMS, maintaining the functional groups of these in all of the specters and exhibited the characteristic functional group of AAc and VP.

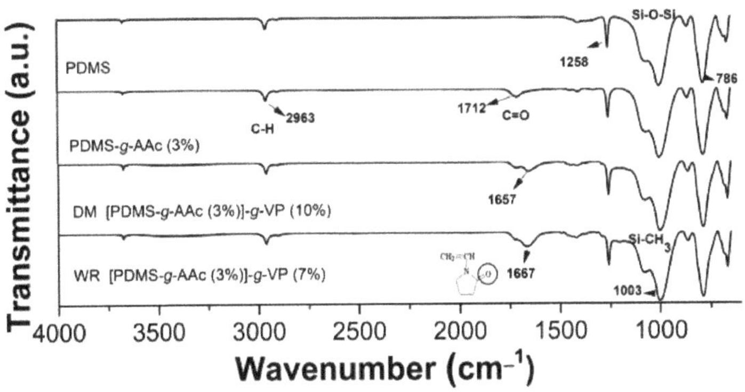

Figure 5. IR specters of samples PDMS without modification (up), first graft for DM (second from up to down), double graft for DM (third from up to down), and WR (down).

3.5. Swelling

Figure 6 indicates that the PDMS did not present swelling, which coincides with its hydrophobic character; PDMS is an organic polymer, which is impermeable to water but soluble in nonpolar organic solvents [26]. On the other hand, the [PDMS-g-AAc] graft showed swelling (approximately 3%) due to the hydrophilic character given by the AAc, a characteristic given by its molecular structure that has a carboxyl group. The [PDMS-*g*-AAc]-*g*-VP films by both methods (DM and WR) had higher swelling due to the properties of the VP monomer, which is a polymer with hydrophilic and hydrophobic groups in its molecular structure that makes it soluble in various solvents [27]. The limit of swelling was found at 24 h, after which there were no significant changes in the weight of the films.

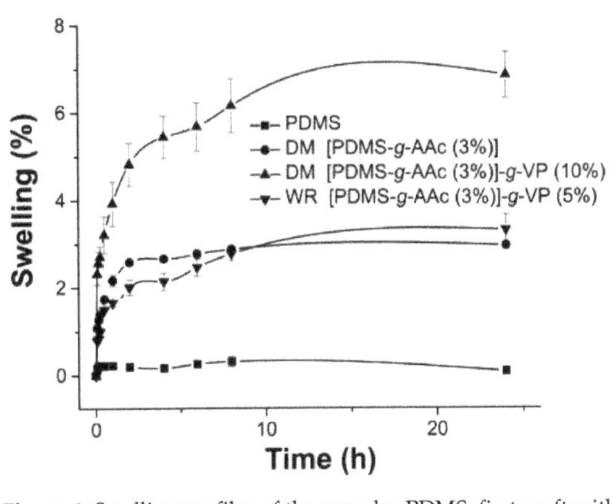

Figure 6. Swelling profiles of the samples PDMS, first graft with AAc and second graft obtained by DM and WR. Reported: the mean ± standard error of the mean, $n = 3$.

The fact that there is a double graft of AAc and VP on PDMS, increased its hydrophilic properties. Although the double grafts were obtained by both methods (DM and WR), it is observed that the swelling was more notorious in those samples obtained by DM. This fact shows that the employ of radiation allows a chemical copolymerization process to occur, whereas if not radiation is used, the result is a physical interaction.

3.6. Contact Angle

Figure 7 shows how to vary the water contact angles on the different materials. The PDMS changed from 103 ± 1° to 98 ± 2° in 10 min, so it did not change noticeably, confirming the hydrophobic property of this polymer [28]. In the sample with the first graft of AAc it was observed that the angle went from 93 ± 1° to 81 ± 3° in 15 min, which indicated that hydrophilicity was conferred to the PDMS film; however, the change was not high when compared to double grafts. The binary grafting by direct method showed a higher hydrophilic property than the other samples, the contact angle changed from 83 ± 1° to 67 ± 2°, due to the properties of both compounds on the base polymer giving it higher hydrophilicity. Something similar occurred for the double graft obtained without irradiation, but the hydrophilic property present in the film was slightly less, similar behavior to the swellings test, indicating a possible physical interaction.

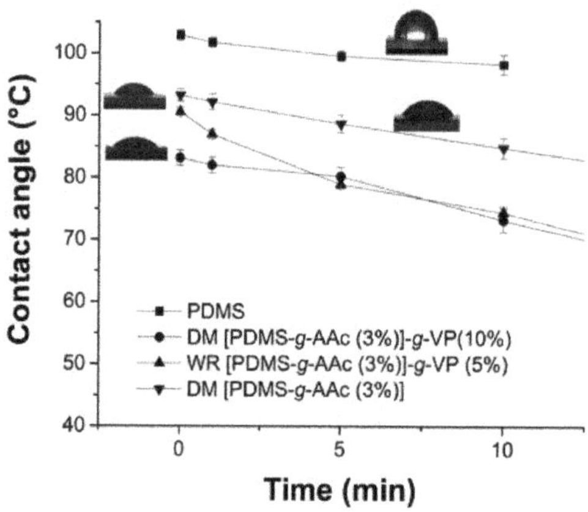

Figure 7. Contact angles of water for unmodified PDMS, PDMS-*g*-AAc (3%), and [PDMS-*g*-AAc (3%)]-*g*-VP (10%) with and without irradiation. Reported: the mean ± standard error of the mean, $n = 3$.

3.7. pH Sensitivity

The pH sensitivity is a characteristic of ionic polymers, which, when there is a shift in pH in the environment, responds to this and causes change in the charge on the polymer chains, leading to swelling and drug release [29]. In this case, the PAAc is an anionic polymer with a pKa close to 4.3 [30]. Figure 8 shows that PDMS films did not have this behavior due to their hydrophobic properties. The PDMS grafted with AAc showed pH sensitivity with a critical pH of 5.3, meaning, the pH to which it responds, being the same as the material with binary graft [PDMS-g-AAc]-*g*-VP modified by direct irradiation method. However, the material modified without radiation did not show this characteristic, but a weight loss when changing the pH, which confirmed the physical nature of the interaction by this synthesis method. When changing the pH, the electrostatic balance of the material was altered, weakening the physical interaction and leading to weight loss.

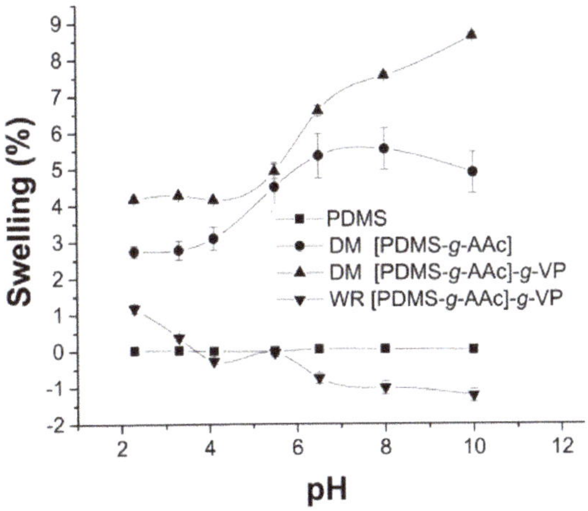

Figure 8. pH sensitivity tests after 24 h of swelling. Reported: the mean ± standard error of the mean, $n = 3$.

3.8. Thermogravimetric Analysis

PDMS is a polymer with high thermal resistance whose temperature at 10% weight loss was approximately 503 °C, the material modified with AAc showed a decrease in this temperature to 486 °C, and the modification with the binary graft of AAc and VP caused a decrease in the stability with, at a temperature of 10%, weight loss of 441 °C. Finally, the (PDMS-*g*-AAc)-*g*-VP material modified without radiation did not show a significant difference at this temperature concerning the PDMS-*g*-AAc material, indicating that the interaction may have been physical. The decomposition temperature of PDMS was 574 °C; the material modified with AAc showed two decomposition temperatures at 550 °C for the decomposition of AAc and 640 °C for silicone. Finally, the modification with VP led to the apparition of an additional decomposition temperature at 442 °C (Figure 9) [31]. The derivative thermogravimetry curves are presented in Figure S3.

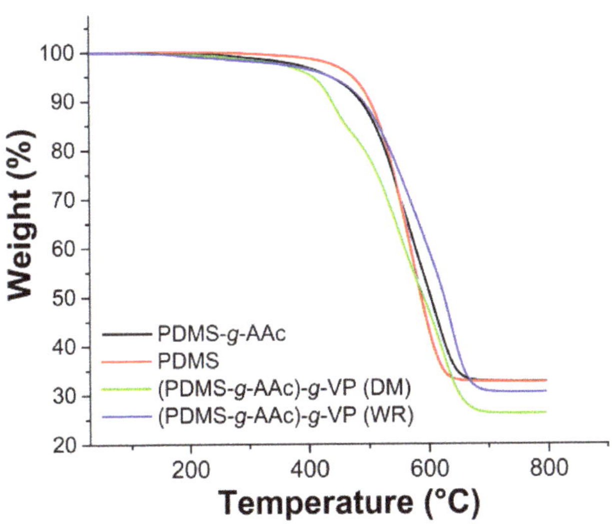

Figure 9. Thermograms of PDMS, PDMS-g-AAc (3%), and [PDMS-*g*-AAc (3%)]-*g*-VP (10%) with and without irradiation.

3.9. Loading and Release of Ciprofloxacin

Figure 10a shows the loading profiles for PDMS, PDMS-*g*-AAc (3%), and [PDMS-*g*-AAc (3%)]-*g*-VP (10%) with and without irradiation; it was observed that the PDMS during the 96 h loaded 38 ± 5.6 μg/g. When adding the first graft with AAc, the load increased to 348.14 ± 72.5 μg/g, and, finally, the double grafts had a higher drug load. However, the binary graft by the direct irradiation method reached a load of 378 ± 18 μg/g, and the one obtained by the method without irradiation had the highest load of 450 ± 18.5 μg/g. The behavior of the load profile indicated that, after 48 h, the load began to stabilize.

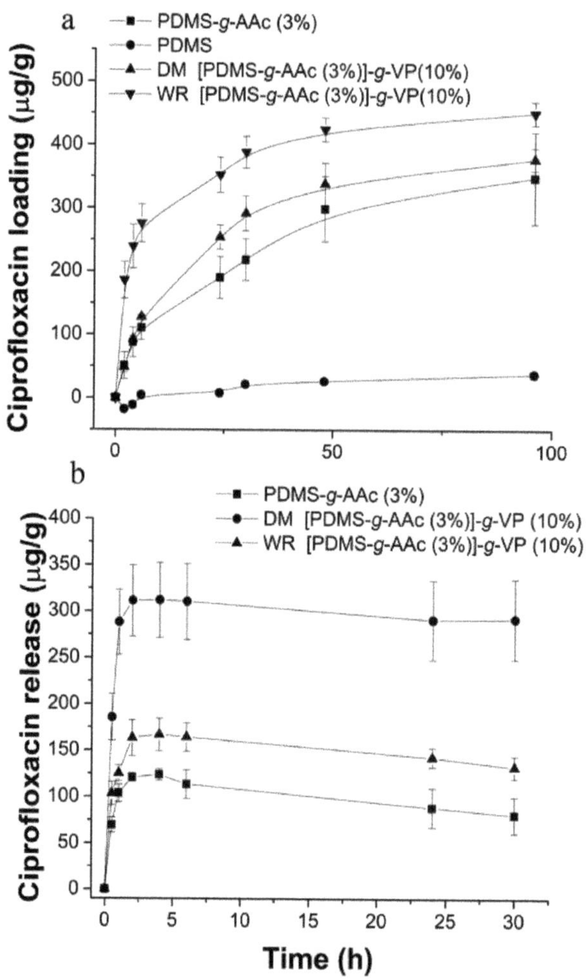

Figure 10. Drug delivery profiles, (**a**) the profiles of loading of ciprofloxacin, and (**b**) the profiles of the release of ciprofloxacin. By the UV-Vis spectrophotometry technique of the samples PDMS, PDMS-*g*-AAc, [PDMS-*g*-AAc]-*g*-VP by DM and WR. Reported: the mean ± standard error of the mean, $n = 3$.

The release profiles of ciprofloxacin are shown in Figure 10b; the sample [PDMS-*g*-AAc (3%)]-*g*-VP (10%) synthesized by the direct irradiation method had a release percentage close to 100%, while the sample synthesized without irradiation released only 37%, a value near to the release obtained from PDMS-*g*-AAc, which was ~33%. All the materials presented maximum release at around 6 h. The PDMS without modification had no release.

From the data obtained for each sample, the release model that best fits each profile was searched using the DDSolver software. The three materials analyzed indicated, by their values of correlation index R^2, Akaike information criterion (AIC), and the model selection criterion (MSC), that the best adjustment was to the Peppas-Sahlin model with T_{lag}, Equation (3). Where k_1 is related to the mechanism of diffusion, k_2 to the chain relaxation mechanisms, m is the Fick index, and T_{lag} refers to the delay time [32,33].

$$\frac{M_t}{M_\infty}=k_1(t-T_{\text{lag}})^m+k_2(t-T_{lag})^{2m} \tag{3}$$

Table 1 shows the parameters obtained for each sample. All the materials present values of k_1 greater than k_2, indicating that the drug diffusion mechanism in the films was of the Fick type and not due to the relaxation of the polymer chains.

Table 1. Parameters of Peppas-Sahlin release model.

Sample	k_1	k_2	M	T_{lag}
PDMS-*g*-AAc (3%)	44.7 ± 3.9	−14.4 ± 2.3	0.28 ± 0.02	0.29 ± 0.09
[PDMS-*g*-AAc (3%)]-*g*-VP (10%) DM	133.8 ± 14.3	−54.1 ± 4.6	0.14 ± 0.01	0.49 ± 0.01
[PDMS-*g*-AAc (3%)]-*g*-VP (10%) WR	39.5 ± 2.6	−10.8 ± 2.4	0.30 ± 0.07	xx

xx This material fits better for the traditional Peppas-Salhin model.

3.10. Agar Diffusion or Kirby-Bauer Method

The PDMS samples, [PDMS-*g*-AAc (3%)] and [PDMS-*g*-AAc (3%)]-*g*-VP (10%), both by the direct method and without radiation, did not show a zone of inhibition. However, the binary grated materials loaded with ciprofloxacin showed inhibition zones for both *S. aureus* and *E. coli*. For gram-positive bacteria (*S. aureus*), the sample without irradiation presented the greatest inhibition zone of approximately 17 mm, while for gram-negative bacteria (*E. coli*), the sample with the binary graft by direct irradiation presented the largest zone of inhibition of about 25 mm (Figure 11). Comparing these results with the zones of inhibition of ciprofloxacin antibiograms loaded with 5 μg of drug, it was observed that the zones of inhibition coincided with bacteria not resistant to this drug that presented zones of inhibition >21 mm for *E. coli*, and for intermediate resistance in the case of *S. aureus* with a zone of inhibition between 16–20 mm [34,35]. This demonstrated that the material loaded with ciprofloxacin had antimicrobial activity. However, the amount of ciprofloxacin loaded and released was approximately 11 μg. Future studies should try to find an optimal load-release ratio for the material.

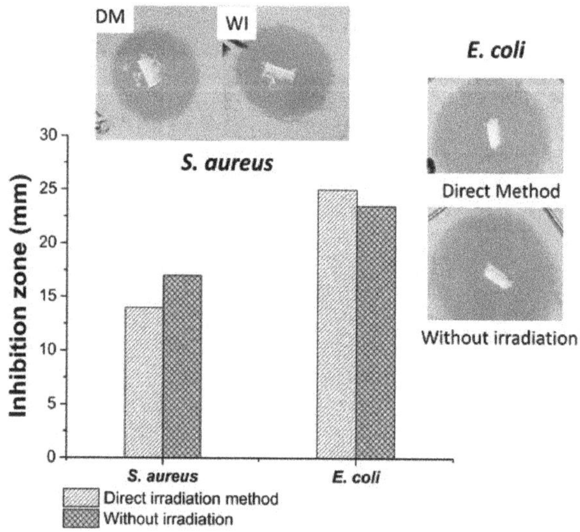

Figure 11. The zones of inhibition for [PDMS-*g*-AAc (3%)]-*g*-VP (10%) loaded with ciprofloxacin.

4. Discussion

Copolymerization is a process that improves the characteristics of a polymer. In this work, graft copolymerization with AAc and VP allowed us to provide PDMS with hydrophilic properties, giving it the ability to load and release ciprofloxacin, a broad-spectrum antimicrobial commonly used for the treatment of various infections, which gave it activity against *E. coli* and *S. aureus*.

The direct radiation method proved effective for performing binary grafting in two steps, with graft percentages that allowed for manipulation and provided the ability to load drugs without deforming the films. Grafting of 3% AAc and 10% VP allowed the loading of ciprofloxacin and the subsequent release in a period of around six hours. When PDMS was exposed to gamma radiation, break of bonds occurred to forming radicals, which later allowed the grafting polymerization. The C–H bonds were more likely to break since they have the lowest binding energy, Figure 12 shows a schematic of the possible structure of this graft (step 1) [36]. For the second graft, not only the PDMS was irradiated but also the PAAc, which was why radicals were also created on the PAAc structure where the VP could also be grafted (Figure 12, step 2) [37].

On the other hand, the binary graft by the method without radiation was not very viable despite providing similar VP grafted percentages to the direct method. The pH sensibility test showed that the material did not have a completed chemical copolymerization because the material lost mass into the buffer solution. Additionally, in the UV-vis specter of the drug release test, a signal appeared before the label for the ciprofloxacin at 266 nm, possibly due to the loss of VP monomer or oligomer. Which demonstrated that the second graft with VP on PDMS-*g*-AAc films with only heating was not feasible since it may have been a physical and not a chemical interaction; VP will interact with acrylic acid forming hydrogen bonds that allow a highly stable physical interaction, which is affected by changing the electrostatic environment of the medium, for example, by changing the pH [38]. The swelling, contact angle, and pH-sensitive tests confirmed the hydrophilic property of the modified films, an important aspect for the purposes of this research, the loading and release of a drug that avoids bacterial contamination.

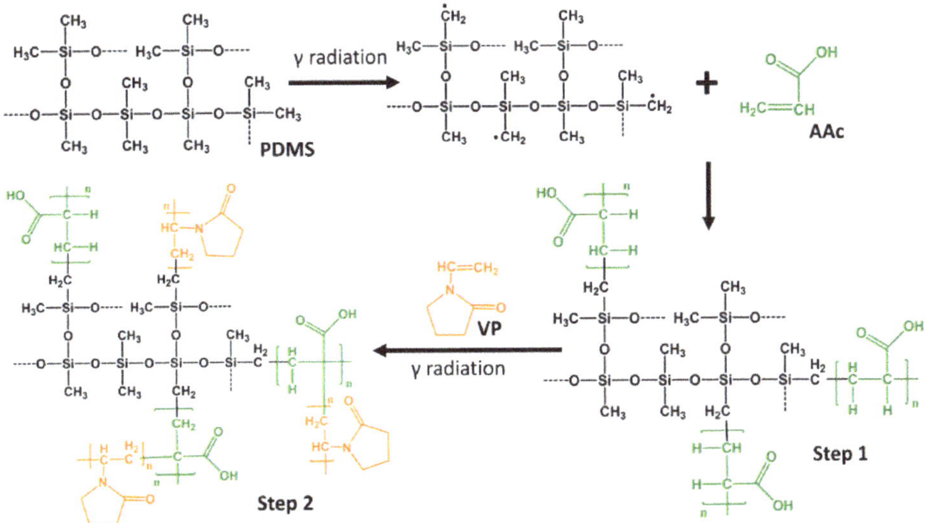

Figure 12. Schematic representation of the binary grafting formation. The black structure represents the PDMS, on the other hand, the green color refers to the AAc molecule and the orange color represents the VP molecule.

5. Conclusions

The binary grafting of AAc and VP on PDMS films was successfully performed using the direct irradiation method, obtaining percentages of 3% for the first graft and 10% for the second graft. The grafted material showed capability as a system for loading and releasing ciprofloxacin, which allowed it to have activity against *S. aureus* and *E. coli* in the agar diffusion assay with inhibition zones diameters of 14 and 25 mm for each bacterial strain, respectively. This system shows a potential application for the development of medical devices with antimicrobial characteristics; in addition, it gives the possibility of being a localized administration system since the drug is directly in the medical device.

Supplementary Materials: The following supporting information can be downloaded at: https://www.mdpi.com/article/10.3390/polym15020302/s1, Figure S1: Calibration curves to quantify load of ciprofloxacin. Figure S2: Calibration curves to quantify release of ciprofloxacin. Figure S3: DTG curve.

Author Contributions: Conceptualization, L.D.-P. and E.B.; methodology, B.S.-G.; formal analysis, L.D.-P. and B.S.-G.; investigation, B.S.-G. and L.D.-P.; resources, E.B.; writing—original draft preparation, B.S.-G. and L.D.-P.; writing—review and editing, E.B.; visualization, L.D.-P.; supervision, E.B.; project administration, E.B.; and funding acquisition, E.B. All authors have read and agreed to the published version of the manuscript.

Funding: This work was supported by the Dirección General de Asuntos del Personal Académico (DGAPA), Universidad Nacional Autónoma de México under Grant IN204223.

Institutional Review Board Statement: Not applicable.

Informed Consent Statement: Not applicable.

Data Availability Statement: Not applicable.

Acknowledgments: B. Santillán González acknowledges to ICN-UNAM for all the resources provided. L. Duarte Peña (887494) acknowledges CONACyT for the doctoral scholarship. B. Leal and M Cruz from ICN-UNAM and H. Magaña from UABC are acknowledged for his technical assistance.

Conflicts of Interest: The authors declare no conflict of interest.

References

1. De, R.; Mahata, M.K.; Kim, K.-T. Structure-Based Varieties of Polymeric Nanocarriers and Influences of Their Physicochemical Properties on Drug Delivery Profiles. *Adv. Sci.* **2022**, *9*, 2105373. [CrossRef] [PubMed]
2. Zheng, D.; Yang, K.; Nie, Z. Engineering heterogeneity of precision nanoparticles for biomedical delivery and therapy. *VIEW* **2021**, *2*, 20200067. [CrossRef]
3. Iyisan, B.; Landfester, K. *Polymeric Nanocarriers BT—Biological Responses to Nanoscale Particles: Molecular and Cellular Aspects and Methodological Approaches*; Gehr, P., Zellner, R., Eds.; Springer International Publishing: Cham, Switzerland, 2019; pp. 53–84, ISBN 978-3-030-12461-8.
4. Haider, M.; Zaki, K.Z.; El Hamshary, M.R.; Hussain, Z.; Orive, G.; Ibrahim, H.O. Polymeric nanocarriers: A promising tool for early diagnosis and efficient treatment of colorectal cancer. *J. Adv. Res.* **2022**, *39*, 237–255. [CrossRef] [PubMed]
5. De Lima, C.S.A.; Balogh, T.S.; Varca, J.P.R.O.; Varca, G.H.C.; Lugão, A.B.; Camacho-Cruz, L.A.; Bucio, E.; Kadlubowski, S.S. An updated review of macro, micro, and nanostructured hydrogels for biomedical and pharmaceutical applications. *Pharmaceutics* **2020**, *12*, 970. [CrossRef]
6. Cheremisinoff, N.P. Properties of Polymers. In *Polymer Characterization*; William Andrew Publishing: Westwood, NJ, USA, 1996; ISBN 978-0-8155-1403-9.
7. Miranda, I.; Souza, A.; Sousa, P.; Ribeiro, J.; Castanheira, E.M.S.; Lima, R.; Minas, G. Properties and applications of PDMS for biomedical engineering: A Review. *J. Funct. Biomater.* **2021**, *13*, 2. [CrossRef]
8. Dardouri, M.; Bettencourt, A.; Martin, V.; Carvalho, F.A.; Santos, C.; Monge, N.; Santos, N.C.; Fernandes, M.H.; Gomes, P.S.; Ribeiro, I.A.C. Using plasma-mediated covalent functionalization of rhamnolipids on polydimethylsiloxane towards the antimicrobial improvement of catheter surfaces. *Biomater. Adv.* **2022**, *134*, 112563. [CrossRef]
9. Yu, H.; Zhou, G.; Sinha, S.K.; Chau, F.S.; Wang, S. Lens integrated with self-aligned variable aperture using pneumatic actuation method. *Sens. Actuators A Phys.* **2010**, *159*, 105–110. [CrossRef]
10. Bozukova, D.; Pagnoulle, C.; Jérôme, R.; Jérôme, C. Polymers in modern ophthalmic implants—Historical background and recent advances. *Mater. Sci. Eng. R Reports* **2010**, *69*, 63–83. [CrossRef]
11. Kumar, R.; Kumar Sahani, A. Role of superhydrophobic coatings in biomedical applications. *Mater. Today Proc.* **2021**, *45*, 5655–5659. [CrossRef]
12. Shakeri, A.; Khan, S.; Didar, T.F. Conventional and emerging strategies for the fabrication and functionalization of PDMS-based microfluidic devices. *Lab Chip* **2021**, *21*, 3053–3075. [CrossRef]
13. Tan, S.H.; Nguyen, N.T.; Chua, Y.C.; Kang, T.G. Oxygen plasma treatment for reducing hydrophobicity of a sealed polydimethylsiloxane microchannel. *Biomicrofluidics* **2010**, *4*, 032204. [CrossRef]
14. Wu, M.H.; Urban, J.P.G.; Cui, Z.; Cui, Z.F. Development of PDMS microbioreactor with well-defined and homogenous culture environment for chondrocyte 3-D culture. *Biomed. Microdevices* **2006**, *8*, 331–340. [CrossRef] [PubMed]
15. Ramírez-Fuentes, Y.S.; Bucio, E.; Burillo, G. Thermo and pH sensitive copolymer based on acrylic acid and N-isopropylacrylamide grafted onto polypropylene. *Polym. Bull.* **2008**, *60*, 79–87. [CrossRef]
16. Cabana, S.; Lecona-Vargas, C.S.; Meléndez-Ortiz, H.I.; Contreras-García, A.; Barbosa, S.; Taboada, P.; Magariños, B.; Bucio, E.; Concheiro, A.; Alvarez-Lorenzo, C. Silicone rubber films functionalized with poly(acrylic acid) nanobrushes for immobilization of gold nanoparticles and photothermal therapy. *J. Drug Deliv. Sci. Technol.* **2017**, *42*, 245–254. [CrossRef]
17. Koczkur, K.M.; Mourdikoudis, S.; Polavarapu, L.; Skrabalak, S.E. Polyvinylpyrrolidone (PVP) in nanoparticle synthesis. *Dalt. Trans.* **2015**, *44*, 17883–17905. [CrossRef]
18. Miyazaki, T.; Yoshioka, S.; Aso, Y.; Kojima, S. Ability of polyvinylpyrrolidone and polyacrylic acid to inhibit the crystallization of amorphous acetaminophen. *J. Pharm. Sci.* **2004**, *93*, 2710–2717. [CrossRef]
19. Kadłubowski, S.; Henke, A.; Ulański, P.; Rosiak, J.M.; Bromberg, L.; Hatton, T.A. Hydrogels of polyvinylpyrrolidone (PVP) and poly(acrylic acid) (PAA) synthesized by photoinduced crosslinking of homopolymers. *Polymer* **2007**, *48*, 4974–4981. [CrossRef]
20. Rizwan, M.; Rubina Gilani, S.; Iqbal Durani, A.; Naseem, S. Materials diversity of hydrogel: Synthesis, polymerization process and soil conditioning properties in agricultural field. *J. Adv. Res.* **2021**, *33*, 15–40. [CrossRef]
21. Ito, T.; Yamaguchi, S.; Soga, D.; Yoshimoto, T.; Koyama, Y. Preparation of a bioadhesive Poly(Acrylic Acid)/Polyvinylpyrrolidone complex gel and its clinical effect on dental hemostasis. *Gels* **2022**, *8*, 462. [CrossRef] [PubMed]
22. Bucio, E.; Burillo, G. Radiation-induced grafting of sensitive polymers. *J. Radioanal. Nucl. Chem.* **2009**, *280*, 239–243. [CrossRef]
23. Velazco-Medel, M.A.; Camacho-Cruz, L.A.; Magaña, H.; Palomino, K.; Bucio, E. Simultaneous grafting polymerization of acrylic acid and silver sggregates formation by direct reduction using γ radiation onto silicone surface and their antimicrobial activity and biocompatibility. *Molecules* **2021**, *26*, 2859. [CrossRef]
24. Feairheller, W.R.; Katon, J.E. The vibrational spectra of acrylic acid and sodium acrylate. *Spectrochim. Acta Part A Mol. Spectrosc.* **1967**, *23*, 2225–2232. [CrossRef]
25. Casimiro, M.; Gomes, S.; Rodrigues, G.; Leal, J.; Ferreira, L. Chitosan/Poly(vinylpyrrolidone) Matrices Obtained by Gamma-Irradiation for Skin Scaffolds: Characterization and Preliminary Cell Response Studies. *Materials* **2018**, *11*, 2535. [CrossRef] [PubMed]
26. De Faria, P.; Martin, A.; Alves, N. Caracterização no Infravermelho (IV) e Eletrônica de superfície (MEV) de membranas assimétricas à base de Poli (acrilonitrila-co-acetato de vinila). *Matéria* **2017**, *22*. [CrossRef]

27. Haaf, F.; Sanner, A.; Straud, F. Polymers of N-Vinylpyrrolidone: Synthesis, characterization and uses. *Polym. J.* **1985**, *17*, 143–152. [CrossRef]
28. Gokaltun, A.; Yarmush, M.L.; Asatekin, A.; Usta, O.B. Recent advances in nonbiofouling PDMS surface modification strategies applicable to microfluidic technology. *Technology* **2017**, *5*, 1–12. [CrossRef] [PubMed]
29. Rizwan, M.; Yahya, R.; Hassan, A.; Yar, M.; Azzahari, A.D.; Selvanathan, V.; Sonsudin, F.; Abouloula, C.N. pH sensitive hydrogels in drug delivery: Brief history, properties, swelling, and release mechanism, material selection and applications. *Polymers* **2017**, *9*, 137. [CrossRef]
30. Pandey, S.P.; Shukla, T.; Dhote, V.K.; Mishra, D.K.; Maheshwari, R.; Tekade, R.K. *Use of Polymers in Controlled Release of Active Agents*; Elsevier Inc.: Amsterdam, The Netherlands, 2018; ISBN 9780128179093.
31. Chun, M.-K.; Cho, C.-S.; Choi, H.-K. Characteristics of poly(vinyl pyrrolidone)/poly(acrylic acid) interpolymer complex prepared by template polymerization of acrylic acid: Effect of reaction solvent and molecular weight of template. *J. Appl. Polym. Sci.* **2004**, *94*, 2390–2394. [CrossRef]
32. Zhang, Y.; Huo, M.; Zhou, J.; Zou, A.; Li, W.; Yao, C.; Xie, S. DDSolver: An add-in program for modeling and comparison of drug dissolution profiles. *AAPS J.* **2010**, *12*, 263–271. [CrossRef]
33. Siepmann, J.; Siepmann, F. Mathematical modeling of drug delivery. *Int. J. Pharm.* **2008**, *364*, 328–343. [CrossRef]
34. *CLSI Document M100-S21*; Performance Standards for Antimicrobial Susceptibility Testing; Twenty-First Informational Supplement. Clinical and Laboratory Standards Institute: Wayne, PA, USA, 2011; ISBN 1-56238-742-1.
35. Tascini, C.; Sozio, E.; Viaggi, B.; Meini, S. Reading and understanding an antibiogram. *Ital. J. Med.* **2016**, *10*, 289–300. [CrossRef]
36. Palsule, A.; Clarson, S.; Widenhouse, C. Gamma Irradiation of Silicones. *J. Inorg. Organomet. Polym. Mater.* **2008**, *18*, 207–221. [CrossRef]
37. Jeong, J.-O.; Park, J.-S.; Kim, E.J.; Jeong, S.-I.; Lee, J.Y.; Lim, Y.-M. Preparation of Radiation Cross-Linked Poly(Acrylic Acid) Hydrogel Containing Metronidazole with Enhanced Antibacterial Activity. *Int. J. Mol. Sci.* **2019**, *21*, 187. [CrossRef] [PubMed]
38. Lau, C.; Mi, Y. A study of blending and complexation of poly(acrylic acid)/poly(vinyl pyrrolidone). *Polymer* **2002**, *43*, 823–829. [CrossRef]

polymers

MDPI

Article

Sorption of Total Petroleum Hydrocarbons in Microplastics

Arely Areanely Cruz-Salas [1], Maribel Velasco-Pérez [1,*], Nayely Mendoza-Muñoz [1], Alethia Vázquez-Morillas [1], Margarita Beltrán-Villavicencio [1], Juan Carlos Alvarez-Zeferino [1] and Sara Ojeda-Benítez [2]

1 Departamento de Energía, Universidad Autónoma Metropolitana, Unidad Azcapotzalco, Ciudad de México 02200, Mexico
2 Laboratorio de Residuos Sólidos, Instituto de Ingeniería, Universidad Autónoma de Baja California, Campus Mexicali, Mexicali 21100, Mexico
* Correspondence: mvp@azc.uam.mx; Tel.: +52-55-53-18-90-57

Abstract: As is the case for many others in the world, Mexican seas face complex pollution challenges; two of the contaminants that require special attention for their prevalence, possible chemical interactions, and relation to the country's economy are leaked petroleum and microplastics (MP). This research assessed the sorption of total petroleum hydrocarbons (TPH) as fuel oil on microplastics in laboratory and field scenarios. Preliminary tests allowed the development and validation of a methodology to measure the sorbed fuel oil by Soxhlet extraction, with a 99.65% recovery rate. The amount of TPH sorbed in the lab followed the order LDPE > PS > PP > PVC > PET > HDPE, with the highest concentration found on LDPE. The sorption of fuel oil on microplastics is correlated to the surface area of the plastic particles and could also be related to the crystallinity of plastics. Sorption, for all plastics, was consistent with a second-order kinetic model. The analysis of field samples collected on beaches of the Gulf of Mexico varied from 1660 to 35,258 mg/kg MP. It must be noticed that, unlike others, this research quantified a family of contaminants, which could explain the high concentrations observed on microplastics.

Keywords: beach; Gulf of Mexico; kinetics; marine environments; organic pollutants

1. Introduction

Plastics are valuable materials for industrial, commercial, medical, and technological applications. In 2020, the world production reached 367 million tons [1]. The increasing production of plastics causes a corresponding growth in plastic waste. Europe reported a post-consumer plastic waste production of 29.5 million tons, of which 42% was used for energy recovery, 34.6% was recycled, and 23.4% was landfilled [1].

In 2019 in Mexico, the plastic industry contributed 0.4% to the gross domestic product, placing this country as the 12th producer and the 11th consumer of plastics globally [2]. In 2020, 43.8 million tons of municipal solid waste (MSW) were generated in Mexico, with 12.9% (18.6 million tons) of plastic waste [3]. This country does not produce official statistics for plastic waste treatment, including recycling and final disposal. However, considering that 71.9% of MSW was taken to final disposal sites (FDS) and that 47.8% of the FDS do not have the basic infrastructure to prevent pollution [3], contamination potential from waste management in Mexico is relevant.

Plastic waste can reach the environment due to mismanagement, transport by waterways, and natural phenomena. Plastics can enter the ocean through wind, inland waterways (rivers, lakes, or channels), or tides [4]. An environmental risk associated with plastics' presence in natural ecosystems is the formation of microplastics (MP), particles < 5 mm that can be produced in this size or formed by the fragmentation of discarded plastic waste. MPs vary in size, shape, color, density, and chemical composition [5,6]. Once in the ocean, MPs can be transported by marine currents [4], float, stay in the water column, or settle in beaches and marine sediments due to their density, which can be modified by biofilm formation [7]. Marine organisms can ingest MP because of their size. MPs sorb and accumulate persistent organic pollutants (POP) and other hydrophobic compounds in the surrounding environment [8,9]. Due to this, sometimes, MPs can be considered as a bioremediation strategy to eliminate contaminants present in the aqueous medium, in the same way composite materials are used to remove organic pollutants from aqueous media. For instance, cotton fiber covered with chitosan [10], the functional organic ligand of 4-tert-Octyl-4-((phenyl)diazenyl)phenol (TPDP) with mesoporous silica [11] and the chemical ligand of N,N–bis(salicylidene)1,2–bis(2–aminophenylthio)ethane (BSBAE) also with mesoporous silica [12] have shown high removal of Remazol Brilliant Red F3B (RR), copper (II) and cadmium (II), respectively. However, if the biota ingests MP with pollutants, they can introduce toxic compounds into the trophic chain, leading to chemical, ecological, and biological alterations in species and ecosystems [5,13], leading to bioaccumulation in organisms [14]. Examples of the adverse effects of MP include mechanical injuries and alteration in enzymatic biomarkers in *Scrobicularia plana* exposed to benzo(a)pyrene (BaP) and perfluorooctanesulfonic acid (PFOS) sorbed in low-density polyethylene (LDPE) [15]; decreases in embryo survival and growth, as well as behavior disorders for the fish *Oryzias melastigma* after exposition to PFOS, BaP, and benzophenone sorbed in polyethylene [16]. Although it has been proposed that this phenomenon occurs mainly in coastal waters [9], pollutants sorbed on MP can be transported to pristine ecosystems, where they can desorb. MP could increase the residence time of pollutants in organisms, contributing to the accumulation and chronic exposure of macro and microorganisms [17]. Due to the high surface/volume ratio of MP, concentrations of hydrophobic pollutants, such as polychlorinated biphenyls (PCB), dichlorodiphenyltrichloroethane (DDT), and polycyclic aromatic hydrocarbons (PAH), can be higher than those found on larger plastics in the surrounding water [18].

The presence of PAHs on MPs found in natural environments is well documented. Concentrations of 640 ng PAH/g MP and 1200–6200 ng PAH/g MP have been reported in albatrosses on Guadalupe Island in Mexico and San Gabriel River in California, United States [19]. In the Canary Islands, Spain, pellets presented 52.1–17,023.6 ng PAH/g MP, and plastic fragments presented 35.1–8725.8 ng PAH/g MP [9]. Analysis of MP collected on the Portuguese beaches of Cresmina and Thela showed 91.3–285.1 ng BaP/g MP [8]. On the MP found in surface waters in the Bohai and Huanghai seas, the concentration of PAH was 3400–119,000 ng PAH/g MP [20]; the primary source of pollutants was attributed to oil and gasoline. These data show that PAHs are found on MP in different types of ecosystems with a wide range of concentrations.

On the other hand, pollution by leaks, spills, and mismanagement of hydrocarbons during extraction and production is a relevant environmental problem in Mexico. There have been 3350 hydrocarbon pollution events from 2016 to 2020, with 931 in 2020 alone. In 2020, 77.4% of hydrocarbon pollution occurred during exploration and production [21].

Mixtures of hydrocarbons such as crude oil and some of its fractions are usually described and quantified as total petrol hydrocarbons (TPH). This group of compounds enters organisms through ingestion, inhalation, or dermal absorption. Due to their low water solubility and high lipid solubility, TPHs accumulate in organisms and organic matter in particles and sediments, becoming more susceptible to bioaccumulation [22]. TPHs include compounds such as PAH and diesel [20]. PAHs are formed during the incomplete combustion of organic compounds such as fossil fuels; these chemicals have carcinogenic and mutagenic effects in humans [23]. The individual effect of 18 aromatic hydrocarbons was studied on *Vibrio fischeri*, a bioluminescence bacteria found in symbiosis with marine animals such as mollusks [24]. It was found that the baseline toxicity model well explains luminescence inhibition by petroleum hydrocarbons. Luminescence inhibition of 50% was observed for individual petroleum hydrocarbons with chemical activities between 0.01 and 0.1. However, since baseline toxicity is additive in concentration, the toxicity cutoff observed for individual compounds may not be valid for mixtures.

This research aimed to evaluate the sorption of fuel oil on MP in a laboratory-scale simulated marine environment and to investigate sorption kinetics and the concentration of organic compounds on MP from beaches in the Gulf of Mexico (a region with a vast petroleum industry).

2. Materials and Methods

The methodology consisted of the following six steps: (1) the preparation of virgin microplastics (VMP), (2) determination of the concentration of organic compounds, (3) measurement of the surface area of VMP, (4) sorption of fuel oil on VMP in laboratory conditions, (5) determination of the sorption kinetics of fuel oil on VMP, and (6) determination of the concentration of organic compounds on MP collected on Mexican beaches (real microplastics, RMP). For the determination of the TPH concentration, a technique frequently used in many laboratories worldwide was used. This technique has the following advantages: it is easy to perform, does not require expensive equipment or chemical substances, and only requires equipment found in most laboratories (such as flasks, heating and stirring racks, stirrers, and universal supports, to name a few), and the analysis time is relatively short.

2.1. Preparation of Virgin Microplastics

Most studies on the MP sorption of pollutants from aqueous medium use reagent-grade microplastics [25]. However, this study uses MP from virgin post-consumer plastic. VMPs were cut and then sieved in a size mesh of around 4.75 mm. High-density polyethylene (HDPE) was obtained from yogurt containers, polypropylene (PP) from carrier bags from Walmart®, polyethylene terephthalate (PET) from disposable water bottles, low-density polyethylene (LDPE) from clear carrier bags, polyvinyl chloride (PVC) from crystal transparent film, and polystyrene (PS) from Styrofoam balls (Figure 1).

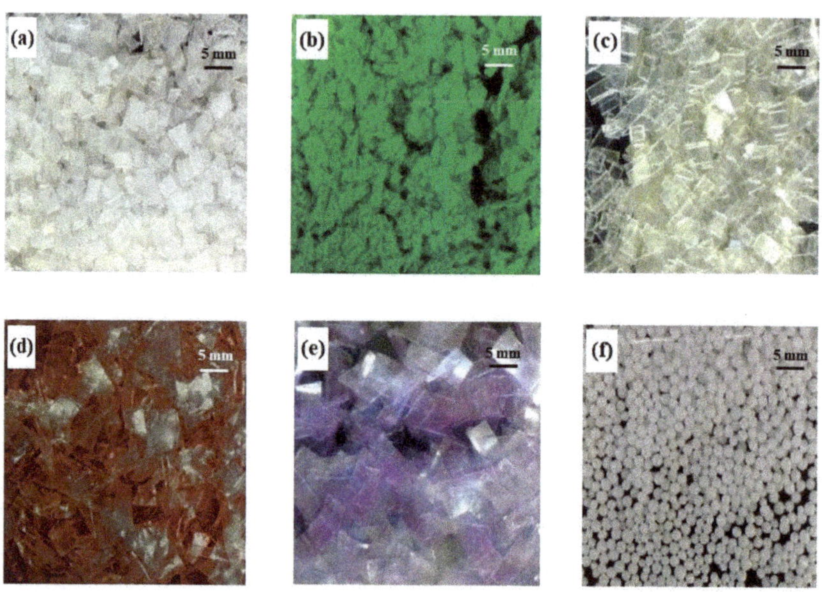

Figure 1. Types of virgin microplastics used in this research. (**a**) HDPE, (**b**) PP, (**c**) PET, (**d**) LDPE, (**e**) PVC, and (**f**) PS.

2.2. Method for the Determination of the Concentration of Organic Compounds

2.2.1. Determination of the Concentration of Fuel Oil in the Liquid Phase

The concentration of fuel oil, measured as the total organic carbon (TOC), in the liquid phase was determined by a TOC Analyzer with a Tekmar Dohrmann-Apollo 9000 model DRB-200 [26]. The operation conditions of the TOC analyzer were as follows: temperature combustion of 368.15 K and analysis time of 12 min. The volatility of the compounds was accounted for by determining the concentration of TOC in aerated and unaerated reactors.

2.2.2. Determination of the Concentration of Fuel Oil on Microplastics

The concentration of organic compounds (OC) on microplastics was determined with Soxhlet extraction [27]. In Soxhlet extraction, the solid matrix (MP in this case) is placed in a thimble, a solvent is heated under reflux, solvent vapors rise through the thimble, and then the solvent cools and condensates fall in the thimble.

This method has been used to measure PAH and TPH in soil samples [28], but also to quantify concentrations of various OC (PAH, PCB, pesticides, aliphatic hydrocarbons, dichlorodiphenyldichloroethylene, among others) on MP recovered from beaches [19,29]. Since a study found no difference in the concentration of hydrocarbons and petroleum fractions using extraction times of 4, 8, 12, and 16 h [30], the time selected for the Soxhlet extraction was 4 h. Extraction was performed with 1 g of MP, at 318.15 K and with 120 mL of solvent (dichloromethane).

The concentration of fuel oil on VPM and RMP was calculated with Equation (1),

$$q_e = \frac{V\,(\,Co - Ce\,)}{M} \tag{1}$$

where q_e is the sorption capacity at equilibrium (mg/kg), V is the volume of the liquid phase (L), C_0 is the initial liquid phase concentration of fuel oil at the initial time, C_e is the liquid phase concentration of fuel oil at equilibrium and M is the mass of MP (kg). C_e and C_0 were determined as described in Section 2.2.1.

It was verified with Equation (2) that the initial mass of fuel oil ($m_{initial}$) was equal to the sum of the mass of fuel oil sorbed on MP (m_{MP}) and the mass of fuel oil remaining in the liquid phase (m_{lp}).

$$\%\,Recovery = \frac{m_{lp} + m_{MP}}{m_{initial}} \times 100 \tag{2}$$

2.3. Determination of Surface Area, Average Pore Diameter, and Pore Volume of Virgin Microplastics

Surface area, average pore diameter, and pore volume of VMP were determined with physisorption analysis (N_2 and 77 K). The surface area, diameter, and volume of pores were calculated by observing the adsorption of a gas (N_2, for instance) on the particles of the solid (the sorbent) [31]. A Belsorp Max analyzer by MicrotracBEL Corp (Osaka, Japan) was used for physisorption on VMP; the surface area was calculated with the Brunauer–Emmett–Teller (BET) method, and pore size distribution with the Barrett–Joyner–Halenda (BJH) method. Preparation of samples consisted of drying VMP at 378.15 K for 24 h.

2.4. Evaluation of Sorption of Fuel Oil on Virgin Microplastics in Laboratory Conditions

The following procedure was undertaken for a fuel oil and synthetic seawater mixture without supernatant. Fuel oil (between 10 and 30 mL) was added to 500 mL of deionized water; then, this mixture was shaken for 2 h; after this time, the supernatant of the mixture was removed with a separation funnel. The remaining mixture was used to prepare emulsions with different concentrations of THP. The concentration of TPH (measured as TOC) of fuel oil and synthetic seawater emulsions was determined as described in Section 2.2.1.

Synthetic seawater was prepared with a method by Cifuentes et al. [32] and had the following chemical composition: 24 g/L sodium chloride (NaCl), 5 g/L magnesium chloride ($MgCl_2$), 4 g/L sodium sulfate (Na_2SO_4), 1.1 g/L calcium chloride ($CaCl_2$), 0.7 g/L potassium chloride (KCl), 0.2 g/L sodium bicarbonate ($NaHCO_3$), 0.096 g/L sodium bromide (NaBr), 0.026 g/L boric acid (H_3BO_3), 0.024 g/L strontium chloride ($SrCl_2$), and 0.003 g/L sodium fluoride (NaF) [32].

VMPs were placed in 500 mL glass reactors, and 150 mL of the fuel oil and synthetic seawater emulsion was added. Different reactors were used for each type of VMPs (HDPE, PP, PET, LDPE, PVC, and PS) and each initial concentration of fuel oil. VMPs were added at a 2% weight concentration. Each reactor was aerated for 8 h per day for seven days to simulate oxygenation conditions in real seawater; the air was added with the glass tubing. This experiment was performed in triplicate. No separation of the mixture of fuel oil and synthetic seawater was observed during the sorption experiments.

2.5. Sorption Kinetics of Fuel Oil on Microplastics

Sorption kinetics experiments were carried out with three types of plastic and one fuel oil concentration. The types of plastic selected for the kinetic study were the three with the highest sorption capacity in the sorption experiment conducted in Section 2.4. The fuel oil concentration used was the concentration that presented the highest sorption in the experiments conducted in Section 2.4. Each type of VMP was placed in a glass reactor with a mixture of fuel oil and synthetic seawater for 96 h. Samples of 25 mL of the emulsion were taken at 3, 6, 12, 24, 48, 72, and 96 h. This experiment was performed in triplicate. Experimental data were fitted to the pseudo first-, pseudo second-order, and intraparticle diffusion (ID) kinetic models (Table 1).

Table 1. Sorption kinetics models used in this research [33,34].

Model	Equation	Linear Equation
Pseudo first-order model	$\frac{dq}{dt} = k_1(q_e - q_t)$	$\log(q_e - q_t) = \log(q_e) - \frac{k_1}{2.3}t$
Pseudo second-order model	$\frac{dq}{dt} = k_2(q_e - q_t)^2$	$\frac{t}{q_t} = \frac{1}{k_2 {q_e}^2} + \frac{1}{q_e}t$
Intraparticle diffusion (ID)	-	$q_t = k_3 t^{0.5} + C$

k_1 = rate constant of the pseudo first-order model (1/h), k_2 = rate constant of the pseudo second-order model (kg/mg h), k_3 = rate constant of ID model (mg/kg $h^{0.5}$), t = time (h), C = constant of the ID model (mg/kg), and q_e = sorption of fuel oil at equilibrium in the solid phase (mg/kg).

2.6. Determination of the Concentration of Organic Compounds in Real Microplastics

2.6.1. Sampling of Microplastics

RMPs were sampled from the following beaches located in the Mexican Gulf: Tamiahua (21°17′23.4″ N, 97°25′12.06″ O), Tuxpan (20°58′28.5″ N, 97°18′24.84″ O), La Barra de Sontecomapan (18°33′45″ N, 95°0′28.74″ O), and Coatzacoalcos (18°9′7.3″ N, 94°27′7.75″ O) (Figure 2). Samples were collected in April 2018 at 6 a.m.

Sampling of RMPs was carried out with the following procedure. First, the beach's high tide line was identified in an area polluted with OC. Then, a rope with a longitude of 100 m was placed parallel to the high tide line. Next, RMPs were collected with metallic tweezers along the 100 m transect. The collected RMPs were stored in glass containers with cork stoppers. This process was repeated in a non-polluted region of the beach to collect RMPs free of OC (to be used as blanks).

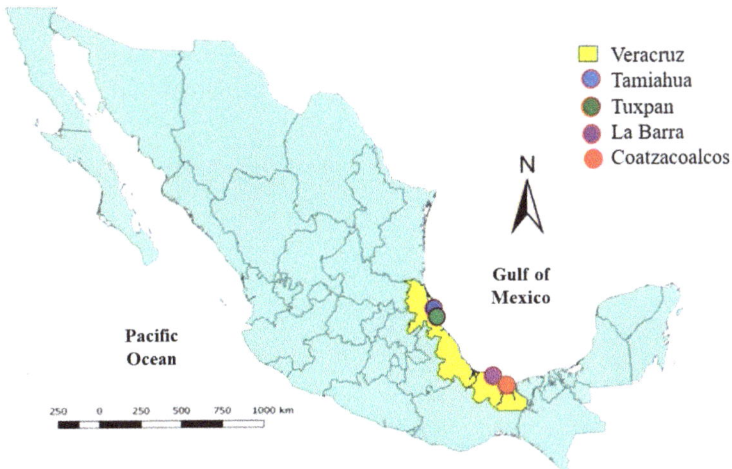

Figure 2. Location of the beaches where microplastics were sampled.

2.6.2. Extraction of Fuel Oil from Microplastics

The sand was removed from the RMPs with deionized water. Then, RMPs were dried in an oven at 333.15 K for 2 h. Next, Soxhlet extraction was carried out [27]. Dichloromethane was used as a solvent to extract OC from the RMP (120 mL of solvent per 1 g of MP), and extraction was carried out for 4 h after the first cycle.

Determination of the concentration of OC in the RMP was estimated with Equation (1). The concentration of OC in blanks (RMP collected in the non-polluted area of beaches) was deducted from the concentration of OC from the polluted areas of the beaches. This part was carried out to account for biofilm and other organic matter on the MP.

3. Results and Discussion

3.1. Validation of the Method Used for the Determination of the Concentration of Organic Compounds in Emulsions

The fuel oil saturation concentration in water was determined by emulsifying different amounts of fuel oil in 500 mL of deionized water. It was concluded that the maximum concentration of fuel oil that could be emulsified in water was 150 ± 5.98 mg/L. Hence, the highest concentration studied in this research was around that value. In addition, the percentage of recovery for fuel oil was determined in different types of plastics to validate the analytical methods used in this research. The experiment was conducted with six types of plastics (PS, HDPE, PET, PP, LDPE, PVC) and four concentrations of fuel oil (126.35 ± 10.12 mg/L, 14.38 ± 2.20 mg/L, 1.93 ± 0.30 mg/L, 0.23 ± 0.10 mg/L) in triplicate. Since the recovery rate ranged from 87.03 to 99.88%, it was concluded that the analytical methods were robust.

3.2. Sorption of Fuel Oil on Virgin Microplastics in Laboratory Conditions

Table 2 presents the concentration (q_e) of fuel oil (average ± standard deviation) on VMP after contact with synthetic seawater with different initial concentrations of fuel oil. The mixture of fuel oil and synthetic seawater remained emulsified during the experiment. This and the continuous mixing of the emulsion due to aeration allowed a similar interaction between fuel oil and different types of plastics, independently of the density of plastics. In natural environments, the density of plastics affects their position in the water column, plastics with a smaller density than seawater float, and plastics with a higher density sink. Biofilm formation on the MP's surface affects the MP's density and its interaction with the chemical substances present in seawater. This research did not measure or control the droplet size of fuel oil emulsified in synthetic seawater.

Table 2. Concentration of fuel oil (measured as organic carbon in mg/kg) sorbed on different types of virgin microplastics.

Plastic Type	Initial Concentration of Fuel Oil in Emulsions			
	126.35 ± 10.12 mg/L	**14.38 ± 2.20 mg/L**	**1.93 ± 0.39 mg/L**	**0.23 ± 0.10 mg/L**
PS	2894.48 ± 484.76	1935.44 ± 2.28	224.55 ± 3.9	62.80 ± 0.08
HDPE	1744.11 ± 119.36	1970.01 ± 13.15	207.90 ± 25.43	59.08 ± 1.97
PET	1844.93 ± 502.15	1892.01 ± 19.0	186.13 ± 9.05	50.78 ± 4.13
PP	2255.91 ± 264.34	1964.56 ± 17.55	239.52 ± 21.74	63.38 ± 1.23
LDPE	2914.39 ± 359.49	2014.64 ± 34.67	223.71 ± 1.29	64.89 ± 0.40
PVC	2543.01 ± 422.39	1704.27 ± 3.22	198.57 ± 13.11	55.90 ± 3.24

PS, PP, and LDPE sorbed the highest amounts of fuel oil. At the highest initial fuel oil concentration evaluated (126.35 ± 10.12 mg/L), the highest sorption was obtained with LDPE (2914.39 ± 359.49 mg/kg), while the lowest was with HDPE (1744.11 ± 119.36). In addition, at the lowest initial fuel oil concentration (0.23 ± 0.10 mg/L), the lowest sorption was obtained with PET (50.78 ± 4.13 mg/kg). VMP presented a significant change in color after exposition to fuel oil.

The toxicity of TPH depends not only on the concentration of petroleum hydrocarbons, but also on the compounds present in the mix [35]. The composition of TPH mixtures varies with sources of petroleum hydrocarbons or changes with time due to weathering and transport phenomena, among other factors. Hence, mixtures with the same TPH concentration may have different risks to the environment and human beings.

TPH are non-polar, hydrophobic, and lipophilic compounds with low solubility. TPH solubility decreases with the number of rings. Solubility influences the migration of pollutants in soil, superficial water, and groundwater. It has been demonstrated that PS and PE microplastics sorb low molecular weight PAHs with two or three rings [20].

It has been proposed that PS and PE microplastics can sorb high quantities of PAHs and other organic compounds [36,37]. In 2018, PAH concentrations between 6.9 and 77 mg/kg MP in PS and PE pellets were reported in China [20]. Other studies have reported the high sorption of various compounds on PS and PE. A study in fresh and seawater found that MP sorbed hydrophobic organic compounds [38]. The sorption capacity of MP increased with aging and salinity and decreased with size; sorption was between 10 and 100 times higher in nanoplastics than in microplastics (due to the surface area to volume ratio) [38].

The concentrations of organic compounds sorbed on MP reported in the literature are significantly lower than those obtained in this study. This may be explained by the relatively high concentration of emulsified fuel oil used in this study (126.35 ± 10.12 mg/L). Weathering of MP could also explain the differences in the values reported in the literature; in natural environments, MPs are subjected to weathering, which modifies their affinity for hydrophobic pollutants due to the introduction of oxygen-containing groups [39]. Finally, the density of MPs can affect their position in the water column. In this research, fuel oil sorption experiments were carried out with aeration. The mix of fuel oil, synthetic seawater, and MP was constantly agitated by air. Aeration did not allow the MPs with a low density to float or the MPs with a high density to sink. Other factors that affect MP's position in the water column are biofilm formation on MP and marine currents. In natural environments, these factors affect the interaction of chemical substances (as pollutants) and MP.

3.3. Surface Area of Virgin Microplastics and Sorption of Fuel Oil

The surface area, average pore diameter, and pore volume of VMP are presented in Table 3; these properties were used for the kinetics study. Since the pore size for all the VMPs evaluated was between 2 and 50 nm, the materials were classified as mesoporous [31]. On the other hand, the sorption data presented in Table 2 indicate that for an initial concentration of fuel oil of 126.35 ± 10.12 mg/L, sorption decreased in the following order: LDPE (2914.39 ± 359.49 mg/kg) > PS (2894.48 ± 484.76 mg/kg) > PVC (2543.01 ± 422.39 mg/kg) > PP (2255.91 ± 264.34 mg/kg) > PET (1844.93 ± 502.15 mg/kg) > HDPE (1744.11 ± 119.36 mg/kg). Other initial concentrations of fuel oil presented a similar trend (Table 2). MP from LDPE (from 64.89 ± 0.40 mg/kg to 2914.39 ± 359.49 mg/kg), PS (from 62.80 ± 0.08 mg/kg to 2894.48 ± 484.76 mg/kg) and PP (from 63.38 ± 1.23 mg/kg to 2255.91 ± 264.34 mg/kg) presented the highest sorption capacities in the range of the fuel oil concentrations studied; these VMPs also presented the highest surface areas (PS: 17.45 m^2/g, LDPE: 13.25 m^2/g, PP: 11.23 m^2/g). Several studies have found that surface area affects the sorption of organic compounds on MP. Furthermore, MP properties could influence sorption behavior.

Table 3. Surface area and pore size and volume of virgin microplastics.

Plastic Type	Surface Area (m^2/g)	Average Pore Diameter (nm)	Pore Volume (10^{-2} cm^3/g)
PS	17.45	5.57	1.42
HDPE	2.85	11.66	2.43
PET	6.55	4.91	0.83
PP	11.23	5.05	0.96
LDPE	13.25	4.25	0.80
PVC	7.73	4.94	1.41

In the following paragraphs, the effect of surface area on the sorption of organic compounds on MP is presented. Goedecke et al. found that the sorption of metformin, a type-2 diabetes drug, increased when PP was cryo-milled (particle size: 0.2–0.6 mm) [40,41]. Furthermore, a study of 3,6-dibromocarbazole and 1,3,6,8-tetrabromocarbazole sorption on MP found the highest sorption capacity at a small particle size (0.45–5 mm) [42]. Wang and Wang studied pyrene sorption on MP, with sorption decreasing in the order PE > PS> PVC [33]. This behavior was attributed to the physicochemical properties of plastics. For instance, PE presented the highest surface area (6.91 m^2/g), followed by PS (2.35 m^2/g), and PVC (1.87 m^2/g). In our study, the highest sorption of fuel oil at an initial concentration of 126.35 ± 10.12 mg/L was obtained for LDPE, PS, and PVC, in agreement with previous reports [33].

The porosity of materials could also affect the sorption of pollutants. Materials have pores of different sizes, and organic pollutants may enter those pores and be trapped [39]. It has been found that oxytetracycline sorption in weathered PS microplastics was enhanced by the high specific and micropore area and the degree of oxidation of MP [43]. The sorption of 19 different compounds (pesticides, pharmaceuticals, and personal care products) on PE and PS microplastics was studied, finding that the high porosity of PS enhanced its sorption capacity [44]. Other researchers found that the high sorption capacity of nitrobenzene and naphthalene on PP microplastics was due to their porosity and high surface area (3.327 m^2/g) [45].

The type of polymer may also influence the sorption of some pollutants on MP [46,47]; it was found that non-polar compounds (perfluoro octane sulfonate and perfluoro octane sulfonamide) present higher affinity for PE (polar) than for PVC (nonpolar) [48]. In several studies, PE and PS have presented high values of sorption for pollutants [20,36,38,45].

Hüffer and Hofmann investigated the sorption of seven aliphatic and aromatic organic probe sorbates on four types of MP. Sorption increased in the order polyamide < PE < PVC < PS, suggesting hydrophobic interactions are highly important [49]. When changes in energy due to interactions between compound plastic are more favorable than for interactions between compound aqueous phases, the highest concentration of the compound at equilibrium will be on the plastic [40].

In conclusion, the order and rate of sorption depend on the hydrophobicity of compounds, type of plastic, aging of plastics (virgin materials or aged), and properties of the aqueous phase (temperature, pH, salinity, and composition) [50]. Wang et al. published a detailed discussion on the factors affecting the sorption of organic pollutants on MP [39].

3.4. Sorption Kinetics of Fuel Oil on Virgin Microplastics

Figure 3 shows experimental data for sorption experiments and experimental data fitted to pseudo first-, pseudo second-order kinetic models and intraparticle diffusion. Table 4 presents the sorption parameters for the kinetic models evaluated in this research. Kinetics were studied only for fuel oil sorption on PS, PP, and LDPE VPM because this type of plastic presented the highest sorption in previous experiments.

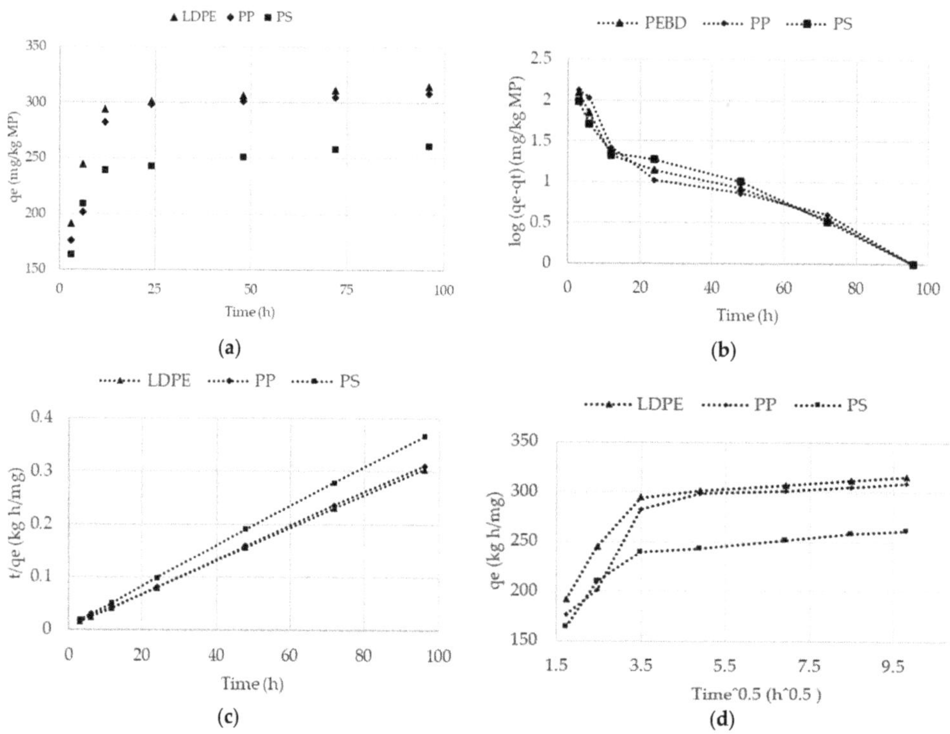

Figure 3. Sorption kinetics. (**a**) Experimental data for sorption of fuel oil on virgin microplastics, (**b**) pseudo first-order kinetic model, (**c**) pseudo second-order kinetic model, and (**d**) intraparticle diffusion model.

Table 4. Kinetic parameters for the pseudo first-order, pseudo second-order, and ID models.

Plastic Type	q_e (mg/kg)	Pseudo First-Order		Pseudo Second-Order		Intraparticle Diffusion (ID)	
		k (1/h)	R^2	k (kg/mg h)	R^2	k (mg/ kg h)	R^2
PS	261.42	0.02	0.95	0.0038	0.99	12.00	0.64
PP	308.73	0.02	0.88	0.0032	0.99	15.00	0.67
LDPE	315.52	0.02	0.91	0.0031	0.99	9.00	0.69

The sorption capacity (q_e) of TPH on VMP followed the order LDPE (315.52 mg/kg) > PP (308.73 mg/kg) > PS (315.52 mg/kg); this can be observed in Figure 3a. The kinetic model that presents the lowest linear correlation, between VMP and fuel oil, was that of intraparticle diffusion (ID). In this model, the principle is that if the straight line passes through the origin, intraparticle diffusion is a rate-limiting step that controls the adsorption process, whereas if the line does not pass through the origin, both intraparticle diffusion and external diffusion play an important role in sorption [34]. Factors such as sorbent particle size, temperature, solute concentration, to mention a few also have an effect on sorption [34]. The VMP evaluated presented a very high linear correlation with the pseudo second-order kinetic model, suggesting that sorption may involve electron sharing or transfer between the sorbent and the sorbate [42]. The very high correlation with the pseudo second-order model indicates that the sorption of fuel oil on MP depends on sorption capacity and that the rate-limiting step is chemisorption. Several studies have found similar results [51,52]. For instance, these include fipronil on non-degradable and biodegradable MP (R^2 = 0.953–0.998) [53]; pyrene on PE, PS, and PVC MP (R^2 > 0.99) [33]; phenanthrene on PE and nylon fibers from a mariculture farm in Xiangshan Bay, China (R^2 > 0.99); and fipronil on PBS, polylactic acid (PLA), PP, PE, PS, and PVC (R^2 = 0.953–0.998) [53]. However, the results of the kinetic models may have been influenced by fuel oil uptake on the MP surface due to the high oleophobicity of MP.

The sorption capacity of organic compounds on MP has been recently studied through partitioning [33,48,54]. The octanol–water coefficient has been effectively used to study hydrophobic partitioning interactions between organic pollutants and MP.

Regarding contact time, in this research, 93–97% of sorption occurred mainly during the first 24 h. After this period, the sorption rate decreased significantly. In the literature, equilibrium sorption times vary widely. Bakir et al. studied the sorption of phenanthrene and DDT on PVC and PE (particle size 200–250 μm); these compounds reached equilibrium in 24 h, except for the sorption of DDT on PE that required 48 h [55]. Wang et al. found that the phenanthrene sorption on nylon MP achieved equilibrium at 24 h [40]. Guo et al. studied the sorption of tylosin on PE, PP, PS, and PVC, with an equilibrium time of 36 h [56]. The sorption of pyrene on PE, PS, and PVC reached equilibrium in 48 h [33]. Zhao et al. found an equilibrium sorption time of 5 days for PAH on polar (PBS, PCL; and PU) and non-polar (PS) microplastics [57].

3.5. Sampling and Sorption Evaluation of Organic Compound on Real Microplastics

Table 5 presents the concentration of organic compounds on RMP sampled at four beaches in Veracruz, Mexico. The concentration reported here was calculated by subtracting the concentration of organic compounds in the blanks from the concentration in the samples (to account for the contribution of biofilm in the samples).

Table 5. Concentration of organic compounds in real microplastic samples from four Mexican beaches.

Beach	Tamiahua	Tuxpan	La Barra de Sontecomapan	Coatzacoalcos
q_e * (mg/kg)	35,258.81	16,133.76	1660.85	8776.99
Standard deviation (mg/kg)	12,921.23	2980.46	1114.73	3088.20

* Units of q_e and standard deviation are mg of organic compound per kg of microplastic.

The concentration of organic compounds on RMP ranged from 1660.85 to 32,258.81 mg/kg. In the literature, the concentration of organic pollutants varies widely, especially in samples from natural environments. Variation in the concentration of organic compounds in RMP could be due to the type of organic compound studied, the type of MP studied, the chemistry of the medium in which sorption takes place, weathering of samples of MP, and the method used to extract the compounds from MP. In a study where MP samples were cleaned with a 30% hydrogen peroxide (H_2O_2) solution before extraction, concentrations of PAH from 3.4 to 119 mg/kg were found [20]. Bouhroum et al. reported average PAH concentrations of 5.8×10^{-2} mg/kg on PE and 0.142 mg/kg on PET in samples collected in the North Atlantic gyre. The extraction of PAH was carried out with a mixture of solvents (dichloromethane and heptane) [58]. Furthermore, concentrations of organic compounds of 1–10,000 mg/kg in microplastics from the Pacific Ocean have been reported before, with extraction with dichloromethane [59].

4. Conclusions

The interactions between microplastics and other pollutants found in the marine environment are a very relevant topic due to the potential of plastic particles to act as carriers of contaminants in the food chain. Different researchers have suggested that specific organic and inorganic compounds can be sorbed on the surface of plastics.

In this study, the approach was to analyze a family of contaminants found in the Gulf of Mexico due to leaks occurring during petroleum extraction and processing. Analyzing samples artificially polluted with fuel oil in the lab allowed us to define a methodology that efficiently recovered the studied TPH fraction. This extraction technique could be used with other organic pollutants, especially when they are quantified together.

The interactions between plastics and organic pollutants are complex because different factors can influence the sorption process, such as the hydrophobicity of the compounds, the composition and physical characteristics of the plastics, the polarity of the species involved, and the properties of the aqueous phase. On the other hand, the possible sorption–desorption process and kinetics could significantly influence the ecotoxicity associated with the ingestion of microplastics. The presence in the marine environment of different pollutants could simultaneously result in competition or synergistic effects; more research is needed to understand these dynamics. Measuring families of compounds together can be a useful tool to achieve this.

The high concentration of OC found in the field samples on beaches polluted by hydrocarbons proves the high level of interaction between the studied contaminants. These findings are relevant, as both are common on the beaches of the Gulf of Mexico, where the petrochemical industry has a high presence. Further research could help us to better understand the magnitude of the problem and its potential effects on natural ecosystems.

Author Contributions: Conceptualization, M.B.-V. and A.V.-M.; methodology, J.C.A.-Z. and M.B.-V.; formal analysis, S.O.-B. and M.V.-P.; investigation, N.M.-M., J.C.A.-Z. and A.A.C.-S.; resources, A.V.-M.; data curation, M.V.-P.; writing—original draft preparation, M.B.-V. and A.A.C.-S.; writing—review and editing, M.B.-V. and S.O.-B.; project administration, A.V.-M. All authors have read and agreed to the published version of the manuscript.

Funding: This research was partially funded by the National Council of Science and Technology of Mexico (CONACYT), through the scholarships No. 291137 and No. 492495 and registration No. 637179, provided to two of the authors.

Institutional Review Board Statement: Not applicable.

Informed Consent Statement: Not applicable.

Data Availability Statement: Not applicable.

Conflicts of Interest: The authors declare no conflict of interest.

References

1. Plastics Europe. *Plastics—The Facts 2021 An Analysis of European Plastics Production, Demand and Waste Data*; Plastics Europe: Brussels, Belgium, 2021.
2. Aguirrezabal, I.U. Plásticos en México; Oficina Económica y Comercial de España en Ciudad de México. In *Fecha: 01/07/2019 NIPO: 114-19-040-2. Fichas Sector México*; ICEX España Exportación e Inversiones: Madrid, Spain, 2019.
3. SEMARNAT. *Diagnóstico Básico Para La Gestión Integral de Los Residuos 2020*; SEMARNAT: Mexicali, Mexico, 2020.
4. Bigdeli, M.; Mohammadian, A.; Pilechi, A.; Taheri, M. Lagrangian Modeling of Marine Microplastics Fate and Transport: The State of the Science. *J. Mar. Sci. Eng.* **2022**, *10*, 481. [CrossRef]
5. Andrady, A.L. Microplastics in the Marine Environment. *Mar. Pollut. Bull.* **2011**, *62*, 1596–1605. [CrossRef] [PubMed]

6. Hidalgo-Ruz, V.; Gutow, L.; Thompson, R.C.; Thiel, M. Microplastics in the Marine Environment: A Review of the Methods Used for Identification and Quantification. *Environ. Sci. Technol.* **2012**, *46*, 3060–3075. [CrossRef] [PubMed]
7. Lagarde, F.; Olivier, O.; Zanella, M.; Daniel, P.; Hiard, S.; Caruso, A. Microplastic Interactions with Freshwater Microalgae: Hetero-Aggregation and Changes in Plastic Density Appear Strongly Dependent on Polymer Type. *Environ. Pollut.* **2016**, *215*, 331–339. [CrossRef]
8. Frias, J.P.G.L.; Sobral, P.; Ferreira, A.M. Organic Pollutants in Microplastics from Two Beaches of the Portuguese Coast. *Mar. Pollut. Bull.* **2010**, *60*, 1988–1992. [CrossRef]
9. Camacho, M.; Herrera, A.; Gómez, M.; Acosta-Dacal, A.; Martínez, I.; Henríquez-Hernández, L.A.; Luzardo, O.P. Organic Pollutants in Marine Plastic Debris from Canary Islands Beaches. *Sci. Total Environ.* **2019**, *662*, 22–31. [CrossRef]
10. Salman, M.S.; Sheikh, M.C.; Hasan, M.M.; Hasan, M.N.; Kubra, K.T.; Rehan, A.I.; Awual, M.E.; Rasee, A.I.; Waliullah, R.M.; Hossain, M.S.; et al. Chitosan-coated cotton fiber composite for efficient toxic dye encapsulation from aqueous media. *Appl. Surf. Sci.* **2023**, *622*, 157008. [CrossRef]
11. Salman, M.S.; Hasan, M.N.; Hasan, M.M.; Kubra, K.T.; Sheikh, M.C.; Rehan, A.I.; Awual, M.E.; Hossain, M.S.; Awual, M.R. Improving copper(II) ion detection and adsorption from wastewater by the ligand-functionalized composite adsorbent. *J. Mol. Struct.* **2023**, *1282*, 135259. [CrossRef]
12. Hasan, M.M.; Kubra, K.T.; Hasan, M.N.; Awual, M.E.; Salman, M.S.; Sheikh, M.C.; Rehan, A.I.; Rasee, A.I.; Waliullah, R.M.; Islam, M.S.; et al. Sustainable ligand-modified based composite material for the selective and effective cadmium(II) capturing from wastewater. *J. Mol. Liq.* **2023**, *371*, 121125. [CrossRef]
13. Rochman, C.M.; Browne, M.A.; Underwood, A.J.; van Franeker, J.A.; Thompson, R.C.; Amaral-Zettler, L.A. The Ecological Impacts of Marine Debris: Unraveling the Demonstrated Evidence from What Is Perceived. *Ecology* **2016**, *97*, 302–312. [CrossRef]
14. van Raamsdonk, L.W.D.; van der Zande, M.; Koelmans, A.A.; Hoogenboom, P.L.A.; Peters, R.J.B.; Groot, M.J.; Peijnenburg, M.A.C.; Weesepoel, Y.J.A. Current Insights into Monitoring, Bioaccumulation, and Potential Health Effects of Microplastics Present in the Food Chain. *Foods* **2020**, *9*, 72. [CrossRef] [PubMed]
15. O'Donovan, S.; Mestre, N.C.; Abel, S.; Fonseca, T.G.; Carteny, C.C.; Cormier, B.; Keiter, S.H.; Bebianno, M.J. Ecotoxicological Effects of Chemical Contaminants Adsorbed to Microplastics in the Clam Scrobicularia Plana. *Front. Mar. Sci.* **2018**, *5*, 143. [CrossRef]
16. le Bihanic, F.; Clérandeau, C.; Cormier, B.; Crebassa, J.C.; Keiter, S.H.; Beiras, R.; Morin, B.; Bégout, M.L.; Cousin, X.; Cachot, J. Organic Contaminants Sorbed to Microplastics Affect Marine Medaka Fish Early Life Stages Development. *Mar. Pollut. Bull.* **2020**, *154*, 111059. [CrossRef] [PubMed]
17. Kleinteich, J.; Seidensticker, S.; Marggrander, N.; Zarf, C. Microplastics Reduce Short-Term Effects of Environmental Contaminants. Part II: Polyethylene Particles Decrease the Effect of Polycyclic Aromatic Hydrocarbons on Microorganisms. *Int. J. Environ. Res. Public. Health* **2018**, *15*, 287. [CrossRef] [PubMed]
18. Gewert, B.; Plassmann, M.M.; Macleod, M. Pathways for Degradation of Plastic Polymers Floating in the Marine Environment. *Environ. Sci. Process. Impacts* **2015**, *17*, 1513–1521. [CrossRef]
19. Rios, L.M.; Moore, C.; Jones, P.R. Persistent Organic Pollutants Carried by Synthetic Polymers in the Ocean Environment. *Mar. Pollut. Bull.* **2007**, *54*, 1230–1237. [CrossRef]
20. Mai, L.; Bao, L.J.; Shi, L.; Liu, L.Y.; Zeng, E.Y. Polycyclic Aromatic Hydrocarbons Affiliated with Microplastics in Surface Waters of Bohai and Huanghai Seas, China. *Environ. Pollut.* **2018**, *241*, 834–840. [CrossRef]
21. PEMEX. *Informe de Sustentabilidad 2020*; PEMEX: Mexico City, Mexico, 2020.
22. ATSDR. *Toxicological Profile for Total Petroleum Hydrocarbons*; ATSDR: Atlanta, GA, USA, 1999.
23. Kim, K.H.; Jahan, S.A.; Kabir, E.; Brown, R.J.C. A Review of Airborne Polycyclic Aromatic Hydrocarbons (PAHs) and Their Human Health Effects. *Environ. Int.* **2013**, *60*, 71–80. [CrossRef]
24. Lee, S.Y.; Kang, H.J.; Kwon, J.H. Toxicity Cutoff of Aromatic Hydrocarbons for Luminescence Inhibition of Vibrio Fischeri. *Ecotoxicol. Environ. Saf.* **2013**, *94*, 116–122. [CrossRef]
25. González-Soto, N.; Campos, L.; Navarro, E.; Bilbao, E.; Guilhermino, L.; Cajaraville, M.P. Effects of microplastics alone or with sorbed oil compounds from the water accommodated fraction of a North Sea crude oil on marine mussels (Mytilus galloprovincialis). *Sci. Total Environ.* **2022**, *851*, 157999. [CrossRef]
26. Potter, B.B.; Wimsatt, J. *Method 415.3, Rev. 1.2: Determination of Total Organic Carbon and Specific UV Absorbance at 254 Nm in Source Water and Drinking Water*; EPA: Washington, DC, USA, 2009.
27. Rios, L.M.; Jones, P.R.; Moore, C.; Narayan, U.V. Quantitation of Persistent Organic Pollutants Adsorbed on Plastic Debris from the Northern Pacific Gyre's "Eastern Garbage Patch". *J. Environ. Monit.* **2010**, *12*, 2226–2236. [CrossRef] [PubMed]
28. Siddique, T.; Rutherford, P.M.; Arocena, J.M.; Thring, R.W. A Proposed Method for Rapid and Economical Extraction of Petroleum Hydrocarbons from Contaminated Soils. *Can. J. Soil Sci.* **2011**, *86*, 725–728. [CrossRef]
29. Fisner, M.; Majer, A.P.; Balthazar-Silva, D.; Gorman, D.; Turra, A. Quantifying Microplastic Pollution on Sandy Beaches: The Conundrum of Large Sample Variability and Spatial Heterogeneity. *Environ. Sci. Pollut. Res.* **2017**, *24*, 13732–13740. [CrossRef] [PubMed]
30. Pons-Jiménez, M.; Guerrero-Peña, A.; Zavala-Cruz, J.; Alarcón, A. Extracción de Hidrocarburos y Compuestos Derivados del Petróleo en Suelos Con Características Físicas y Químicas Diferentes. *Univ. Cienc.* **2011**, *27*, 1–15.

31. Naderi, M. Chapter Fourteen Surface Area Brunauer–Emmett–Teller (BET). In *Progress in Filtration and Separation*; Elsevier Ltd.: Amsterdam, The Netherlands, 2015; pp. 585–608. ISBN 9780123983077.
32. Cifuentes Lemus, J.L.; Torres García, M.d.P.; Frías, M.M. El Océano y Sus Recursos II. Las Ciencias Del Mar: Oceanografía Geológica y Oceanografía Química. Available online: http://bibliotecadigital.ilce.edu.mx/sites/ciencia/volumen1/ciencia2/12/htm/oceano2.html (accessed on 26 August 2022).
33. Wang, W.; Wang, J. Comparative Evaluation of Sorption Kinetics and Isotherms of Pyrene onto Microplastics. *Chemosphere* **2018**, *193*, 567–573. [CrossRef]
34. Chen, Z.; Yang, J.; Huang, D.; Wang, S.; Jiang, K.; Sun, W.; Chen, Z.; Cao, Z.; Ren, Y.; Wang, Q.; et al. Adsorption behavior of aniline pollutant on polystyrene microplastics. *Chemosphere* **2023**, *323*, 138187. [CrossRef]
35. Heath, J.S.; Koblis, K.; Sager, S.L. Review of Chemical, Physical, and Toxicologic Properties of Components of Total Petroleum Hydrocarbons. *J. Soil Contam.* **2008**, *2*, 1–25. [CrossRef]
36. Tang, G.; Liu, M.; Zhou, Q.; He, H.; Chen, K.; Zhang, H.; Hu, J.; Huang, Q.; Luo, Y.; Ke, H.; et al. Microplastics and Polycyclic Aromatic Hydrocarbons (PAHs) in Xiamen Coastal Areas: Implications for Anthropogenic Impacts. *Sci. Total Environ.* **2018**, *634*, 811–820. [CrossRef]
37. Liu, L.; Fokkink, R.; Koelmans, A.A. Sorption of Polycyclic Aromatic Hydrocarbons to Polystyrene Nanoplastic. *Environ. Toxicol. Chem.* **2016**, *35*, 1650–1655. [CrossRef]
38. Velzeboer, I.; Kwadijk, C.J.A.F.; Koelmans, A.A. Strong Sorption of PCBs to Nanoplastics, Microplastics, Carbon Nanotubes, and Fullerenes. *Environ. Sci. Technol.* **2014**, *48*, 4869–4876. [CrossRef]
39. Wang, F.; Zhang, M.; Sha, W.; Wang, Y.; Hao, H.; Dou, Y.; Li, Y. Sorption Behavior and Mechanisms of Organic Contaminants to Nano and Microplastics. *Molecules* **2020**, *25*, 1827. [CrossRef]
40. Endo, S.; Koelmans, A.A. Sorption of Hydrophobic Organic Compounds to Plastics in the Marine Environment: Equilibrium. *Handb. Environ. Chem.* **2019**, *78*, 185–204. [CrossRef]
41. Goedecke, C.; Stollin, U.M.; Hering, S.; Richter, J.; Piechotta, C.; Paul, A.; Braun, U. A First Pilot Study on the Sorption of Environmental Pollutants on Various Microplastic Materials. *J. Environ. Anal. Chem.* **2017**, *4*, 1000191. [CrossRef]
42. Zhang, X.; Zheng, M.; Yin, X.; Wang, L.; Lou, Y.; Qu, L.; Liu, X.; Zhu, H.; Qiu, Y. Sorption of 3,6-Dibromocarbazole and 1,3,6,8-Tetrabromocarbazole by Microplastics. *Mar. Pollut. Bull.* **2019**, *138*, 458–463. [CrossRef]
43. Zhang, H.; Wang, J.; Zhou, B.; Zhou, Y.; Dai, Z.; Zhou, Q.; Chriestie, P.; Luo, Y. Enhanced Adsorption of Oxytetracycline to Weathered Microplastic Polystyrene: Kinetics, Isotherms and Influencing Factors. *Environ. Pollut.* **2018**, *243*, 1550–1557. [CrossRef] [PubMed]
44. Seidensticker, S.; Grathwohl, P.; Lamprecht, J.; Zarfl, C. A Combined Experimental and Modeling Study to Evaluate PH-Dependent Sorption of Polar and Non-Polar Compounds to Polyethylene and Polystyrene Microplastics. *Environ. Sci. Eur.* **2018**, *30*, 30. [CrossRef]
45. Wang, J.; Liu, X.; Liu, G. Sorption Behaviors of Phenanthrene, Nitrobenzene, and Naphthalene on Mesoplastics and Microplastics. *Environ. Sci. Pollut. Res. Int.* **2019**, *26*, 12563–12573. [CrossRef] [PubMed]
46. Teuten, E.L.; Rowland, S.J.; Galloway, T.S.; Thompson, R.C. Potential for Plastics to Transport Hydrophobic Contaminants. *Environ. Sci. Technol.* **2007**, *41*, 7759–7764. [CrossRef]
47. Teuten, E.L.; Saquing, J.M.; Knappe, D.R.U.; Barlaz, M.A.; Jonsson, S.; Björn, A.; Rowland, S.J.; Thompson, R.C.; Galloway, T.S.; Yamashita, R.; et al. Transport and Release of Chemicals from Plastics to the Environment and to Wildlife. *Philos. Trans. R. Soc. B* **2009**, *364*, 2027–2045. [CrossRef]
48. Wang, F.; Shih, K.M.; Li, X.Y. The Partition Behavior of Perfluorooctanesulfonate (PFOS) and Perfluorooctanesulfonamide (FOSA) on Microplastics. *Chemosphere* **2015**, *119*, 841–847. [CrossRef]
49. Hüffer, T.; Hofmann, T. Sorption of Non-Polar Organic Compounds by Micro-Sized Plastic Particles in Aqueous Solution. *Environ. Pollut.* **2016**, *214*, 194–201. [CrossRef] [PubMed]
50. Seltenrich, N. New Link in the Food Chain? Marine Plastic Pollution and Seafood Safety. *Environ. Health Perspect.* **2015**, *123*, A34–A41. [CrossRef]
51. Ho, Y.S.; McKay, G. Kinetic Models for the Sorption of Dye from Aqueous Solution by Wood. *Process Saf. Environ.* **1998**, *76*, 183–191. [CrossRef]
52. Ho, Y.S.; McKay, G. Pseudo-Second Order Model for Sorption Processes. *Process Biochem.* **1999**, *34*, 451–465. [CrossRef]
53. Gong, W.; Jiang, M.; Han, P.; Liang, G.; Zhang, T.; Liu, G. Comparative Analysis on the Sorption Kinetics and Isotherms of Fipronil on Nondegradable and Biodegradable Microplastics. *Environ. Pollut.* **2019**, *254*, 112927. [CrossRef]
54. Kook, H.; Cha, M.; Park, C. Transport of Emerging Organic Ultraviolet (UV) Filters in Ceramic Membranes: Role of Polyethylene (PE) Microplastics. *Chemosphere* **2022**, *309*, 136570. [CrossRef]
55. Bakir, A.; Rowland, S.J.; Thompson, R.C. Competitive Sorption of Persistent Organic Pollutants onto Microplastics in the Marine Environment. *Mar. Pollut. Bull.* **2012**, *64*, 2782–2789. [CrossRef]
56. Guo, X.; Pang, J.; Chen, S.; Jia, H. Sorption Properties of Tylosin on Four Different Microplastics. *Chemosphere* **2018**, *209*, 240–245. [CrossRef]
57. Zhao, L.; Rong, L.; Xu, J.; Lian, J.; Wang, L.; Sun, H. Sorption of Five Organic Compounds by Polar and Nonpolar Microplastics. *Chemosphere* **2020**, *257*, 127206. [CrossRef]

58. Bouhroum, R.; Boulkamh, A.; Asia, L.; Lebarillier, S.; ter Halle, A.; Syakti, A.D.; Doumenq, P.; Malleret, L.; Wong-Wah-chung, P. Concentrations and Fingerprints of PAHs and PCBs Adsorbed onto Marine Plastic Debris from the Indonesian Cilacap Coast and The North Atlantic Gyre. *Reg. Stud. Mar. Sci.* **2019**, *29*, 100611. [CrossRef]
59. Hirai, H.; Takada, H.; Ogata, Y.; Yamashita, R.; Mizukawa, K.; Saha, M.; Kwan, C.; Moore, C.; Gray, H.; Laursen, D.; et al. Organic Micropollutants in Marine Plastics Debris from the Open Ocean and Remote and Urban Beaches. *Mar. Pollut. Bull.* **2011**, *62*, 1683–1692. [CrossRef] [PubMed]

polymers

MDPI

Article

Synthesis of PMMA Microspheres with Tunable Diameters: Evaluation as a Template in the Synthesis of Tin Oxide Coatings

José L. Mendoza-Castellanos, Juan C. Pantoja-Espinoza *, Luis C. Rodríguez-Pacheco and Francisco Paraguay-Delgado *

Centro de Investigación en Materiales Avanzados S. C. (CIMAV), Laboratorio de Síntesis de Óxidos Semiconductores, Departamento de Física de Materiales, Miguel de Cervantes 120, Complejo Industrial Chihuahua, Chihuahua 31136, Mexico; luis.mendoza@cimav.edu.mx (J.L.M.-C.); luis.rodriguez@cimav.edu.mx (L.C.R.-P.)

* Correspondence: juan.pantoja@cimav.edu.mx (J.C.P.-E.); francisco.paraguay@cimav.edu.mx (F.P.-D.); Tel.: +52-(614)-649-4801 (J.C.P.-E.); +52-(614)-649-1107 (F.P.-D.)

Abstract: The synthesis of polymethyl methacrylate (PMMA) spheres with different sizes has been a challenge. PMMA has promise for future applications, e.g., as a template for preparing porous oxide coatings by thermal decomposition. Different amounts of SDS as a surfactant are used as an alternative to control PMMA microsphere size through the formation of micelles. The objectives of the study were twofold: firstly, to determine the mathematical relationship between SDS concentration and PMMA sphere diameter, and secondly, to assess the efficacy of PMMA spheres as templates for SnO_2 coating synthesis and their impact on porosity. The study used FTIR, TGA, and SEM techniques to analyze the PMMA samples, and SEM and TEM techniques were used for SnO_2 coatings. The results showed that PMMA sphere diameter could be adjusted by varying the SDS concentration, with sizes ranging from 120 to 360 nm. The mathematical relationship between PMMA sphere diameter and SDS concentration was determined with a $y = ax^b$ type equation. The porosity of SnO_2 coatings was found to be dependent on the PMMA sphere diameter used as a template. The research concludes that PMMA can be used as a template to produce oxide coatings, such as SnO_2, with tunable porosities.

Keywords: PMMA spheres; macroporous; coatings; tin oxide

1. Introduction

Polymethyl methacrylate (PMMA) is widely used in electronic equipment, medical technology, and polymer membrane areas [1]. In addition, some authors have described PMMA as a template for metallic oxide synthesis to improve properties such as pore size distribution and superficial area [2]. Hyodo et al. used PMMA microspheres at 150, 250, 400, and 800 nm to synthesize macroporous SnO_2 films to improve NO_x and hydrogen (H_2) detection [3]. Furthermore, Kamitani et al. used 98.6 μm microspheres to make high porosity alumina films for possible uses as filters, insulators, and catalytic support [4]. Another interesting work is that of Khan et al. who, with spheres of approximately 300 nm diameter, fabricated a porous film of TiO_2 in heterojunction with ZnO for potential use as an efficient photoanode in a photoelectrochemical cell [5]. It is also important to mention the research of Sordello and Minero, who synthesized microspheres with 250 nm diameter porous TiO_2 powders with Pt to improve the H_2 production of this semiconductor [6].

Tin oxide (SnO_2) coatings have gained significant attention due to their remarkable properties and potential applications in various research fields [7]. The properties of SnO_2, such as high transparency, high refractive index, high chemical stability, and excellent electrical conductivity, make it an attractive material for use in a wide range of technology applications. These applications include gas sensors [8], solar cells [9], photocatalysts [10], electrochromic devices [11], and antistatic coatings [12]. The versatility and flexibility of

SnO_2 coatings have made them a popular research topic, with numerous studies being conducted to explore their various applications [13]. The future of SnO_2 coatings is bright, with potential for further research and development to expand their applications and improve their properties [14].

The polymer PMMA is a transparent thermoplastic synthesized by emulsion polymerization, solution polymerization, and bulk polymerization from the MMA (methyl methacrylate) monomer [15,16]. Emulsion polymerization is one of the most commonly used methods to synthesize PMMA owing to its high reactivity and property of simultaneous increase in the polymerization rate and molecular weight [1]. This method uses a surfactant, a monomer, and an initiator [1,2].

Some initiators as tertiary amines, such as 2-(N, N-dialkylamino)ethanol and N, N, N′, N′-tetramethyl ethylenediamine, exhibit a higher initiation activity during PMMA polymerization [17]. In addition, H_2O_2 [10] and potassium peroxy-disulfate [18] have been reported as initiators. However, the most common initiators used for PMMA synthesis are ammonium persulfate (APS) [19–22] or APS/ascorbic acid mixtures [23] that act as redox initiators.

Various factors impact the PMMA synthesis process. Among these, the utilization of surfactants is often explored in existing literature due to their impact on PMMA's morphology, particle size variation, and particle shape. Parra et al. used anionic (sodium lauryl sulfate, SLS) and nonionic (nonylphenol ethoxylated) surfactants and reported a high dependence of the nature and concentration of the surfactant on the morphology and particle size, with the anionic surfactant SLS being better than the non-ionic surfactant [24]. Baissac et al. prepared PMMA under different time and power parameters and small amounts of the surfactants Triton X-100 and Tween 20 and studied their effect on particle diameters between 78 and 310 nm [25]. Mahmoudian et al. used Triton X-100 as a surfactant for PMMA synthesis and reported sphere diameters between 100 and 300 nm [26]. Vargas-Salazar et al. synthesized PMMA by ultrasound at 50 °C with Hitenol BC10 as a surfactant and reported average particle diameters between 39 and 63 nm [27]. However, the most frequently used surfactant for PMMA synthesis is sodium dodecyl sulfate (SDS), for which some reports have found particle diameters between 20 and 97 nm [28] and between 100 and 1000 nm [29].

Sodium dodecyl sulfate can form micelles in water, which can adsorb onto the surface of the growing PMMA particles and prevent them from coagulating or aggregating [30]. SDS can also lower the interfacial tension between the water and the monomer phases, facilitating the formation of smaller and more uniform droplets [31].

The role of SDS in the formation of PMMA microspheres depends on several factors, such as the concentration of SDS, the monomer concentration, the initiator concentration, the reaction temperature, and the stirring speed [30–32]. The optimal conditions for obtaining monodisperse PMMA microspheres with high yield and stability vary depending on the desired size and morphology of the microspheres [30–32].

Few works report PMMA sphere synthesis with different diameter sizes depending on the variation of the amount of surfactant. Kamras et al. prepared PMMA by emulsion polymerization using microwave radiation as a heating source. They measured the Z-average particle size using the dynamic light scattering (DLS) technique, reporting particle sizes between 97.36 and 23.19 nm using 0.008 and 0.2 % SDS [28]. There has been no report about PMMA sphere synthesis by emulsion polymerization and conventional heating as a function of the variation of the amount of SDS as surfactant. In addition, the effect of PMMA sphere size as a template on SnO_2 porosity synthesis has not been reported.

Therefore, the objectives of this research were twofold: first, find a mathematical function that expresses the change in the size of PMMA sphere diameter as a function of the surfactant SDS concentration. Second, evaluate PMMA spheres as a template and their effect on porosity for SnO_2 coating synthesis.

2. Materials and Methods

2.1. Synthesis of PMMA

The synthesis method for all PMMA spheres is by batch emulsion polymerization [19,33–35]. The followed reagents were used: sodium dodecyl sulfate (SDS) (Sigma-Aldrich, 99%, Tokyo, Japan) as a surfactant; methyl methacrylate (MMA) (Aldrich, 99%, St. Louis, MI, USA) as a monomer, and ammonium persulfate (APS) (Sigma-Aldrich, 98%, Japan) was used as initiator.

Figure 1a shows the experimental arrangement system used in the polymerization reaction to obtain PMMA spheres. First, 150 mL tri-distilled water (J. T. Baker, México) was placed with stirring at 400 rpm in a 3-hole flask. An inert argon atmosphere was maintained at a flow rate of 100 mL/min (Figure 1a) to maintain the inert atmosphere. Under these conditions, the tri-distilled water was heated using a heating mantle (Figure 1b) up to 65 °C and maintained by the temperature controller (Figure 1c) with ±1 °C. Following this, 0.0055 g of SDS was dissolved in 10 mL of tri-distilled water and pre-heated at 65 °C on the heater (hot plate); this last solution was added to a 3-hole flask and maintained with stirring for 20 min. Then, 42 mL of the MMA monomer, previously heated on the hot plate at 65 °C, was added to the solution (3-hole flask); the resultant dissolution was maintained with stirring for 30 min, obtaining a clear solution (Figure 1b). Following this, the polymerization initiator was prepared by dissolving 0.3434 g of APS in 10 mL of tri-distilled water.

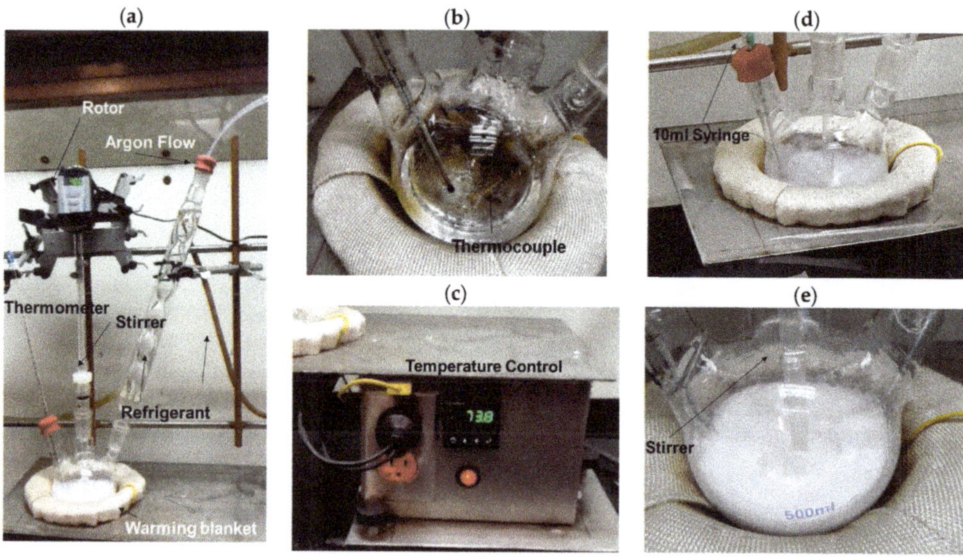

Figure 1. PMMA sphere synthesis by emulsion polymerization: (**a**) overview, (**b**) temperature controller, (**c**) thermometer and thermocouple, (**d**) stirrer, and (**e**) PMMA.

Finally, the APS solution was slowly dosed to initiate the polymerization reaction. At this time, the polymerization process began; after 1 h, the transparent solution changed to a white color as observed in Figure 1d,e. Then, this PMMA-spheres dispersion was cooled to room temperature, and the rest of the reagents were separated by centrifuge at 5000 RPM for 1 h; this washing process was made in tri-distilled water 5 times continuously. The resulting product was dried at 70 °C for 12 h until a white powder was obtained. The sample was identified as 0.0139, which corresponds to the SDS surfactant percentage concentration concerning the MMA monomer. The above procedure was repeated in triplicate with the same synthesis method using the following weights of SDS 0.0055, 0.0083, 0.011, 0.0138, 0.0165, 0.0275, and 0.55 g of surfactant; then the obtained samples were identified as 0.0139,

0.0210, 0.0279, 0.0349, 0.0418, 0.0696, and 0.1391; these values represent the different SDS % value concentrations.

2.2. Synthesis of Porous Tin Oxide Coatings

PMMA spheres as a template were used to synthesize SnO_2 coatings with different pore sizes according to the previously reported procedure [36]. In brief, the precursor dispersion was made using PMMA microspheres as templates; the precursor salt of SnO_2 was tin (IV) tetrachloride pentahydrate with 98% purity purchased from the Sigma-Aldrich company. Then, the dispersion was deposited on a glass substrate by the doctor blade method and dried in a closed system at room temperature. Then, the sample was underwent heat treatment at 400 °C, and it was applied to decompose the microspheres of PMMA to obtain porous SnO_2 coatings (Figure 2). Finally, the coatings' structure and morphology were characterized with scanning and transmission electron microscopy (SEM and TEM, JEOL) techniques for all obtained coatings with different porous diameters.

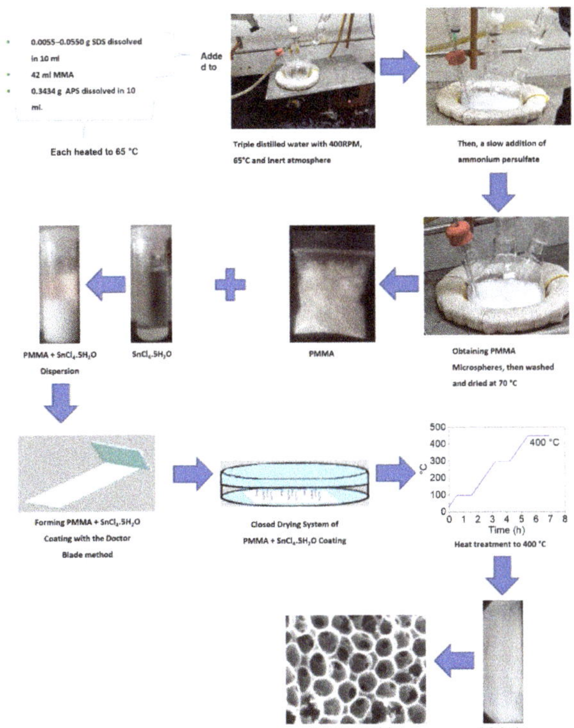

Figure 2. General scheme, starting with the PMMA sphere synthesis as a template and ending with the synthesis of porous SnO_2 coatings.

2.3. Characterization

The PMMA microspheres were characterized to determine their purity. First, Fourier Transform Infrared Spectroscopy (FT-IR, Schimadzu IRAffinity-1S, Kyoto, Japan) in transmittance mode with an ATR accessory was used to acquire spectra at room temperature from 4000 to 400 cm^{-1} to identify the characteristic functional groups of the obtained samples. From the thermogram obtained by TGA was determined the decomposition temperatures (TGA-SDT Q600, New Castle, Delaware, USA). The morphology of PMMA microspheres was observed with a scanning electron microscope (SEM, JEOL Ltd., JSM7401F,

Tokyo, Japan). The structure of SnO_2 obtained was determined by X-ray diffraction and electron diffraction (Panalytical, Almelo, Holland).

The morphology of the SnO_2 coatings was studied by transmission electron microscope (TEM, JEOL Ltd., JEM 2200FS+Cs, Tokyo, Japan) and SEM (Hitachi SU3500, Tokyo, Japan) techniques. The selected area electron diffraction (SAED) technique confirmed the crystalline structure. In addition, from the SEM micrographs were measured the pore diameters using Gatan Digital Micrograph(r) Software (version 3.7.0., Pleasanton, CA, USA).

3. Results and Discussion

3.1. FTIR Analysis

The FTIR spectrum of the PMMA spheres recorded by ATR mode is shown in Figure 3. Three characteristic peaks of PMMA appear at 1062, 985, and 845 cm^{-1} [37]. The bands observed at 1387 and 750 cm^{-1} correspond to the α-CH_3 group [38]. The absorption bands centered at 1483, 1445, and 1435 cm^{-1} are associated with symmetrical deformation vibrations of the CH_3 group [19]. The CH_2/CH_3 stretching modes are at 2994 cm^{-1} and 2950 cm^{-1}; the C=O stretch is at 1725 cm^{-1} from the C-COO group, and the asymmetric C–O–C stretch is at 1243, 1194 cm^{-1}, and 1144 cm^{-1}. These vibrations are visible for synthesized PMMA spheres [39]. Different quantities of SDS as a surfactant did not affect the chemical composition of PMMA; the SDS surfactant used during the emulsion polymerization process was removed through multiple washings and centrifugations. Therefore, it is unlikely that there would be any significant SDS peaks present in the FTIR spectra of the resulting PMMA product.

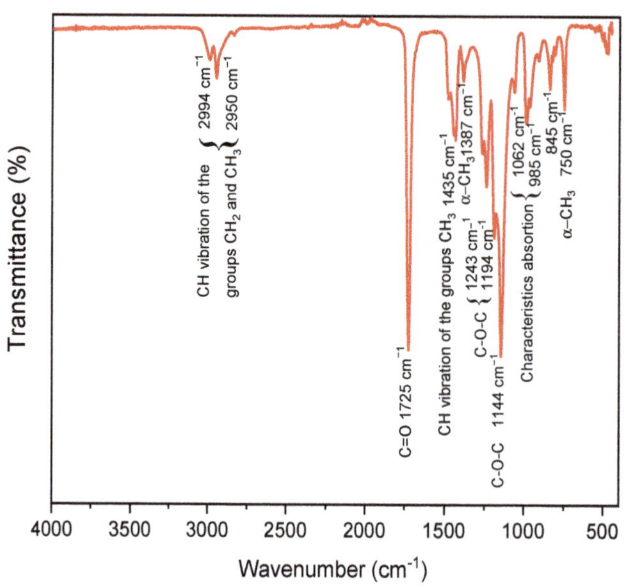

Figure 3. FTIR spectrum acquired from the synthesized PMMA spheres.

3.2. TGA Analysis

The TGA thermogram of the synthesized PMMA material by emulsion polymerization is shown in Figure 4.

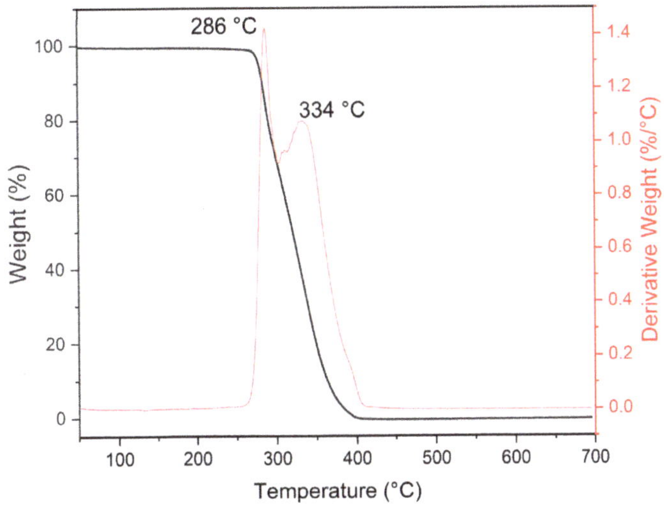

Figure 4. TGA thermogram showing decomposition of PMMA.

It reveals a thermal decomposition process in a single step. According to the TGA sample behavior, it was separated into two regions, as follows. (i) First step: the mass loss for the PMMA samples remains unchanged up to 286 °C, where it starts to decompose. (ii) Second step: the PMMA decomposes into monomers between 286 and 400 °C [40,41], ending with the CO_2 and CO formation [42]. It can be deduced that the PMMA obtained is a thermally stable polymer [19], and the temperature needed to decompose the spheres to form the tin oxide coating must be set above 400 °C. The amount of SDS used in the emulsion polymerization process does not affect the decomposition temperatures observed in the TGA analysis because the SDS is removed from the final product through washing and centrifugation. The TGA analysis measures the thermal stability of the polymer, and the SDS does not contribute to this property. Therefore, the amount of SDS used in the emulsion polymerization process does not affect the TGA results.

3.3. *Morphology by SEM*

Figure 5 shows SEM images of PMMA nanoparticles for all synthesized samples. It can be seen that the aspect for all synthesis conditions is completely spheric.

The sphere diameter decreases as the surfactant quantity concentration increases, with uniform size. These images confirm the formation of spherical micelles for this range of surfactant concentrations. This means the micelles are arranged with small sizes at higher concentrations, remaining spherical. In addition, it is possible to observe some particle agglomerates with different sizes.

Figure 6 shows the statistical distribution of the diameter measurement values of PMMA spheres.

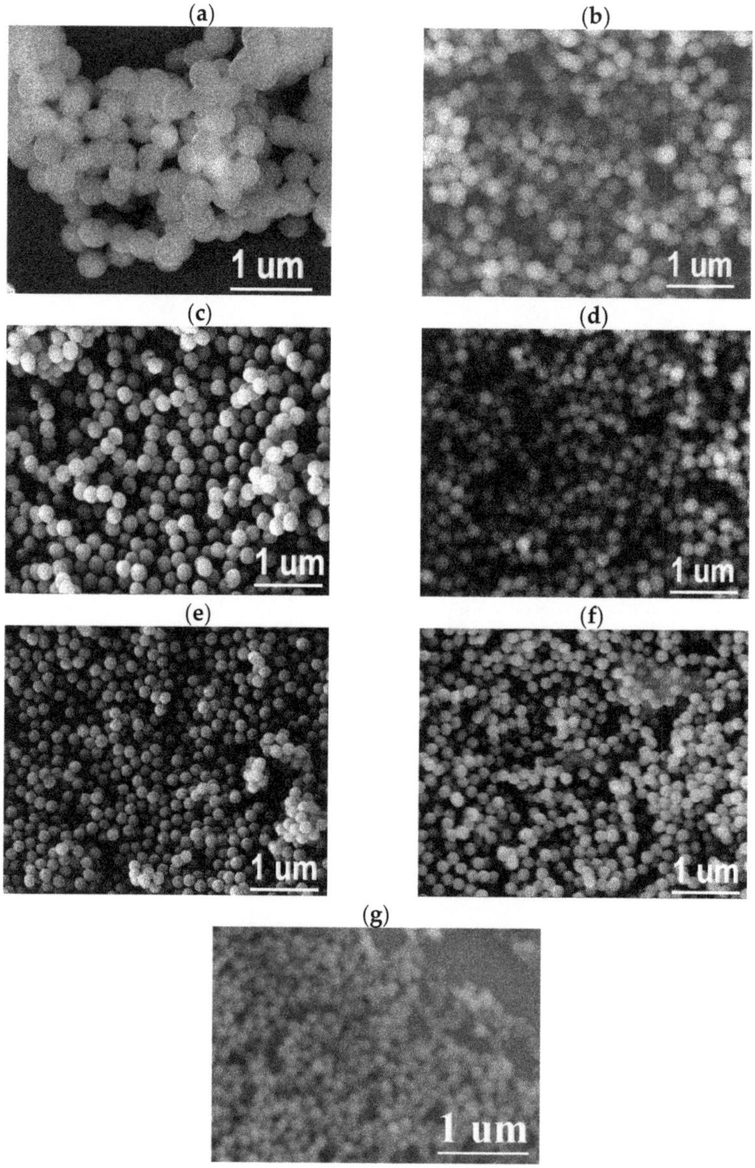

Figure 5. SEM images of PMMA spheres for (**a**) 0.0139, (**b**) 0.0210, (**c**) 0.0279, (**d**) 0.0349, (**e**) 0.0418, (**f**) 0.0696, and (**g**) 0.1391% SDS content.

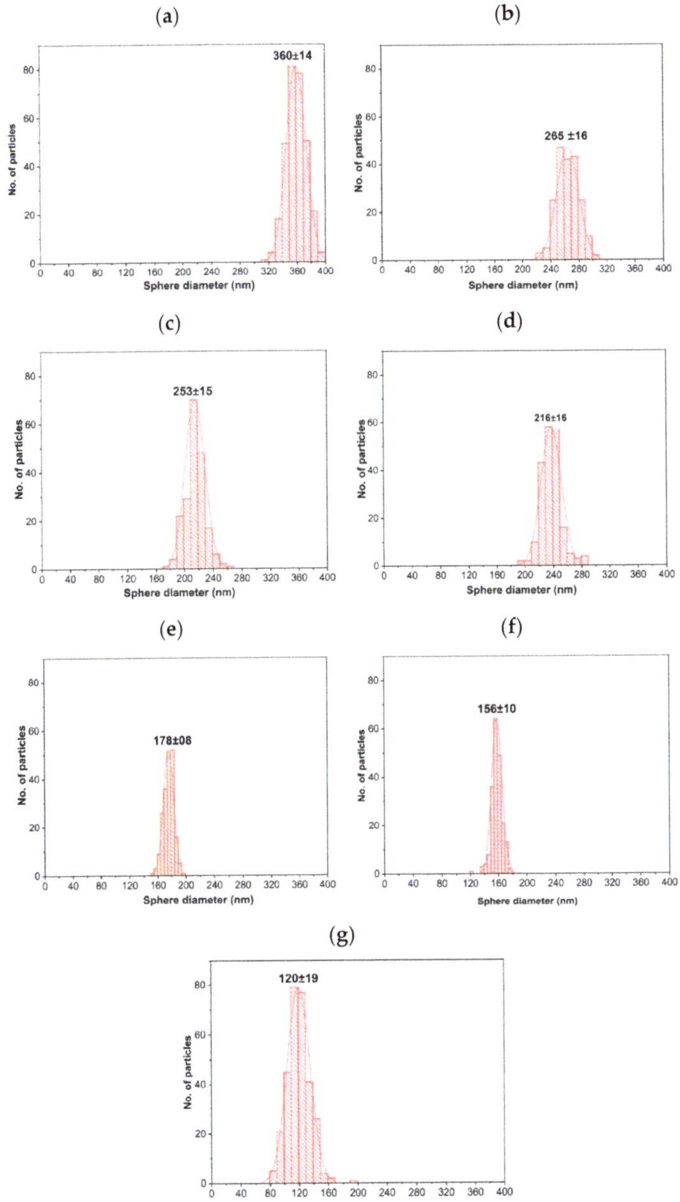

Figure 6. Statistics distribution of PMMA sphere diameter for (**a**) 0.0139, (**b**) 0.0210, (**c**) 0.0279, (**d**) 0.0349, (**e**) 0.0418, (**f**) 0.0696, and (**g**) 0.1391% SDS content.

The quantitative data corresponding to the sphere sizes were obtained from acquired SEM images (Figure 5). Figure 6a shows the distribution of PMMA sphere sizes, starting with the lowest surfactant concentration up to the higher SDS content. These average diameters of PMMA sphere sizes decreased systematically as the SDS concentration increased. These values were 360 ± 14 265 ± 16, 245 ± 15, 216 ± 16, 178 ± 8, 156 ± 10, and 120 ± 8 nm for the concentrations of 0.0139, 0.0210, 0.0279, 0.0349, 0.0418, 0.0696, and

0.1391% of SDS content, respectively. The lowest standard deviation of measurement values confirm their uniformity, as can be noticed in the SEM images. These diameter values are bigger than those obtained by [43]; they find diameter values between 20 and 100 nm at a similar surfactant concentration. This confirms the small spherical micelle formation in the synthesis procedure. Our procedure aims to find a spheres diameter between 120 and 360 nm, and this interval size is helpful for different technological applications.

Table 1 shows the relationship between surfactant concentration and sphere diameter in previous studies and in our own. The data shows that increased surfactant concentration leads to decreased sphere diameter. This trend is consistent with our results, where the size of the spheres decreased as the amount of SDS increased. These findings demonstrate the importance of surfactant concentration in controlling the size of the spheres and, ultimately, the pore size in coatings, which has important implications for various applications.

Table 1. Summary reported activities for PMMA sphere synthesis using SDS as surfactant.

Surfactant	Surfactant (%)	Sphere Diameter (nm)	Synthesis Method	Reference
SDS	0.008–0.2	97–23	Microwave heating	[28]
SDS	0.09–0.017	60–20	Magnetic stirring and ultrasound	[43]
SDS	0.0139–0.1391	360–120	Emulsion polymerization	This work

Some studies have shown that increasing the concentration of SDS can lead to smaller and more spherical PMMA microspheres and lower conversion and higher polydispersity [30,32]. Other studies have reported that increasing the concentration of SDS can increase the conversion and decrease the polydispersity but also increase the size and reduce the sphericity of the PMMA microspheres [31]. These discrepancies may be due to different experimental conditions and methods researchers use [30–32]. This is the case of the information in Table 1, where different synthesis techniques led to a different PMMA sphere size.

The interaction between SDS and PMMA microspheres or micelles can be explained by two models: (1) the electrostatic stabilization model, which assumes that SDS micelles carry a negative charge that repels other micelles or particles; and (2) the steric stabilization model, which assumes that SDS micelles form a layer around the particles that prevents them from coming into contact with each other [30]. Both models may contribute to the phenomenon observed in the works presented in Table 1, as well as the results of this article, where an increase in SDS causes a decrease in sphere diameter size.

3.4. *Fit a Function*

Figure 7 shows the PMMA sphere diameter size changes as a function of SDS concentration. The behavior between these variables showed an exponential relationship, which was fitted using the Levenberg–Marquardt algorithm, given by the general Equation:

$$y = ax^{b}, \quad (1)$$

The fit of the function relating the diameter of PMMA spheres to the SDS surfactant concentration is shown as follows:

$$\text{PMMA diameter (nm)} = 37.99x^{-0.5168}, \quad (2)$$

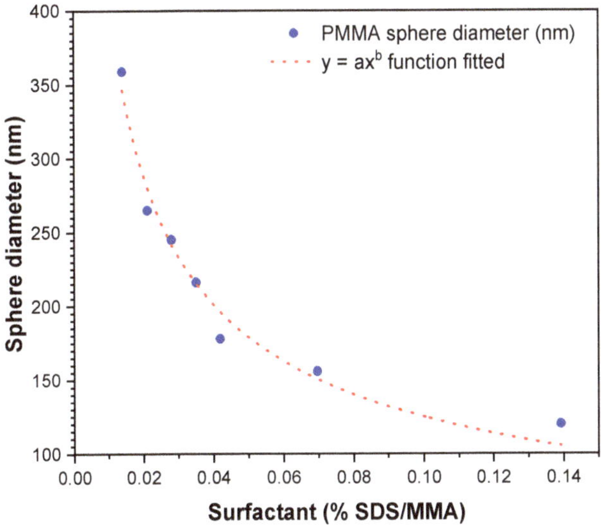

Figure 7. Diameter size fitting as a function of surfactant shows tunability of PMMA sphere diameter.

Table 2 shows the parameters of the fitted function to relate the change in PMMA sphere diameter as a function of SDS surfactant concentration.

Table 2. Parameters model in fitted function.

Parameter	Value
Equation	$y = ax^b$
a	37.99 ± 5.75
b	−0.5168 ± 0.0406
R^2	0.9746

The results of this investigation are similar to those found by other authors on the change of PMMA sphere size as a function of SDS surfactant concentration, in which they report a nonlinear behavior, although with a second-order exponential decay function [28]. Although surfactant concentration was similar, the particle size intervals show considerable differences in nanoparticle sizes (20–97 nm) relative to this work. Possible causes of the differences in particle size ranges may be factors such as the type of conventional or microwave heating, the polymerization temperatures at 65 or 90 °C, or the different initiators used, such as APS or potassium persulfate (KPS).

The table shows that the calculated PMMA diameters are within the minimum and maximum ranges of each mean value obtained for each concentration of the SDS surfactant. Regarding 0.0418% SDS, the particle diameter calculated falls outside the minimum and maximum value of the measured value. The equation modeling the diameter size of PMMA spheres as a function of % SDS is valid between 120 to 360 nm. Kamras et al. reported the validity of a second-order exponential decay equation between 20 and 97 nm [28]. Other works reported a relationship with particle size between 300 and 1000 [28,44]. This fitting shows a good mathematical function that is predictable for the diameter parameters tunability synthesis.

Table 3 shows the PMMA sphere diameters values obtained from the statistical distribution measurements on the micrographs and the diameters calculated from the nonlinear function.

Table 3. SEM-analysis vs. formula-calculated sphere diameter of PMMA.

SDS (%)	PMMA Sphere Diameter (nm)	
	Measured SEM Analyses	Calculated
0.0139	360 ± 14	346.31
0.0210	265 ± 16	279.80
0.0279	245± 15	241.59
0.0349	216 ± 14	215.19
0.0418	178 ± 08	196.03
0.0696	156 ± 10	150.62
0.1391	120 ± 19	105.30

3.5. Synthesis of SnO_2 Coatings

Figure 8 shows SAED patterns of the SnO_2 particles (Figure 8a) and the brightfield TEM images of the SnO_2 coating under different magnifications (Figure 8b–d). Figure 8a shows the selected area electron diffraction (SAED) patterns of SnO_2 nanoparticles prepared with a PMMA microspheres template. The rings correspond to the (1 1 0), (1 0 1), (2 0 0), and (2 1 1) d-spaces; they belong to the tetragonal phase of SnO_2, similar to the XRD pattern.

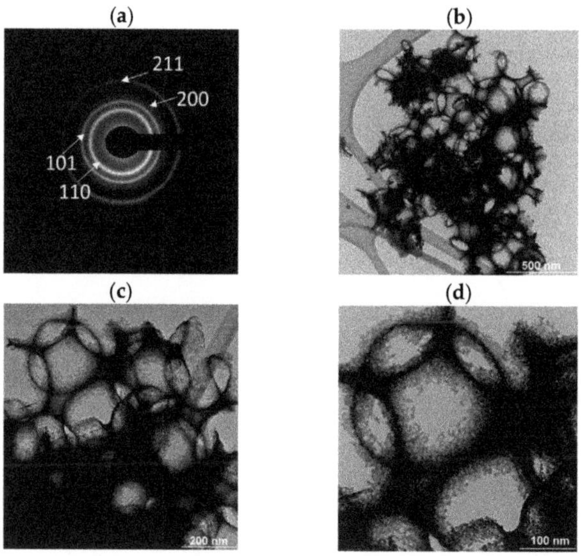

Figure 8. (**a**) SAED pattern and brightfield TEM images with (**b**) 500 nm, (**c**) 200 nm, and (**d**) 100 nm scale bars.

These rings agreed with the XRD intensities previously reported, and the SAED patterns pattern indexed to SnO_2 [41] too. Micrographs with different scale sizes show a macroporous morphology and pore diameter of 262 nm. The macroporous size of SnO_2 coating correlated with the average microsphere size of PMMA used for synthesis. This porosity is due to the decomposition of the PMMA microspheres template by heat treatment at 400 °C which was previously synthesized using the surfactant under the concentration of 0.0279 % SDS, obtaining an average diameter of 245 nm.

Figure 9 shows SEM images of SnO_2 coatings morphology with different-sized porous, generated using different PMMA spheres as a template.

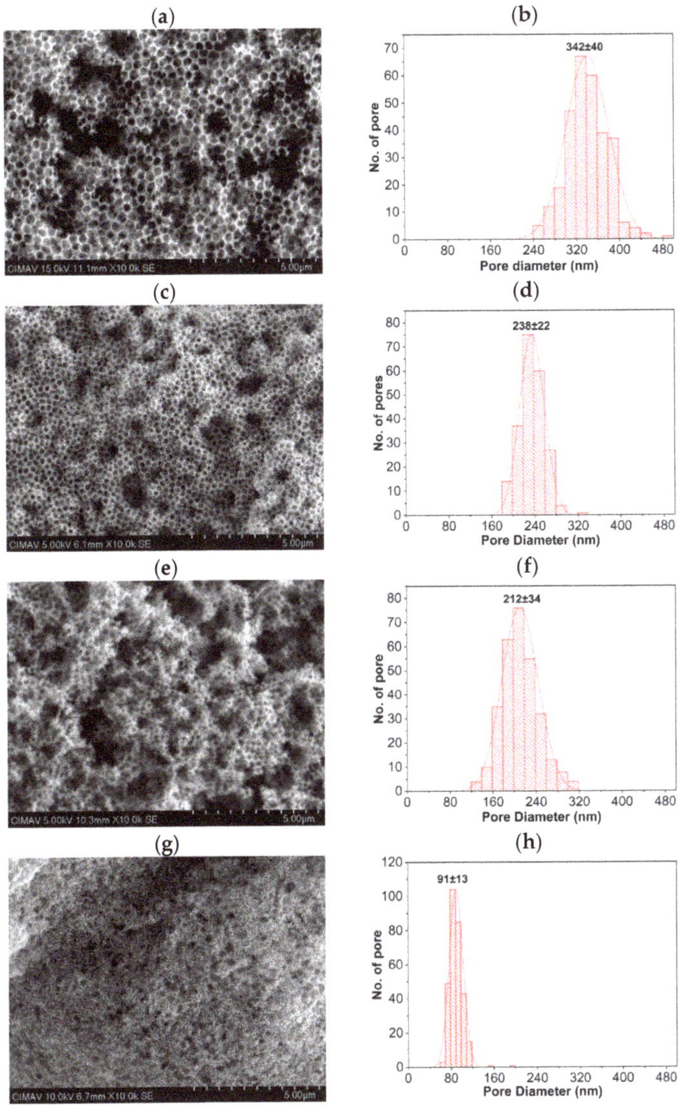

Figure 9. SEM images of SnO_2 coatings (**left**) and statistical distribution of pore sizes (**right**) prepared with PMMA spheres as a template obtained with different % of SDS content: 0.0139 (**a**,**b**), 0.0279 (**c**,**d**), 0.0349 (**e**,**f**), and 0.1391 (**g**,**h**).

The micrographs were obtained at 10,000 × times magnifications with 5 μm scales. Figure 9a corresponds to the SnO_2 coating synthesized using the PMMA spheres diameter of 360 nm (0.0139% SDS), in which the porous morphology can be clearly seen. The statistical distribution shown in Figure 9b corresponds to the pore size measurements giving an average of 342 nm, which is a pore size caused by the spaces generated due to the decomposition of PMMA with a sphere diameter size of 360 nm. The pore size started to decrease, as shown in the micrographs in Figure 9c,e,g, maintaining the expected behavior since PMMA spheres were used as templates with sphere diameters of 245, 216, and 120 nm synthesized using the concentrations of 0.0279, 0.0349, and 0.1391% SDS as a surfactant, respectively. The statistical pore size distributions corresponding to these micrographs are shown in Figure 9d,f,h, in which the frequency shift to the left can indicate the decrease in pore size consistent with the porosity observed in the micrographs. The average pore sizes obtained were 238, 212, and 91 nm, and likewise, it can be affirmed that the porosity size in the SnO_2 coatings was due to the thermal decomposition of the PMMA spheres formed by the different diameter sizes.

Table 4 shows a summary of pore sizes of different materials synthesized using PMMA as a template; the pore sizes of the SnO_2 coatings obtained in this work are comparable with the sphere size of PMMA used as a template reported in the literature, regardless of the synthesis conditions for their preparation. This work highlights, in addition to the synthesis of PMMA with controllable sphere size, the synthesis of SnO_2 coatings with controllable pore size.

Table 4. Summary values of porous materials synthesized using PMMA as a template.

Material	Sphere Diameter (nm)	Pore Size (nm)	Ref.
TiO_2	87	87	[45]
$ZnFe_2O_4$	291	203	[46]
TiO_2	250 420	125 220	[6]
CeO_2	325	240	[47]
WC	490 180	371 149	[48]
$KTiOPO_4/SiO_2$	250	200	[49]
$LaFeO_3$	270	191	[50]
SnO_2	360	342	[36]

Figure 10 shows the relationship between the porosity of SnO_2 coatings with the diameters of PMMA spheres obtained experimentally under the effect of surfactant concentration. In addition, the relationship of the pore size of SnO_2 coatings with the PMMA sphere diameter obtained from the mathematical fitted function to the experimental data is shown. It can be seen in Figure 10 that the pore diameter of the coatings decreases as the PMMA sphere size decreases, and that this has a relationship with the SDS surfactant concentration quantity, whose diameter variation has been studied in this work. It is again demonstrated that the function is also exponential.

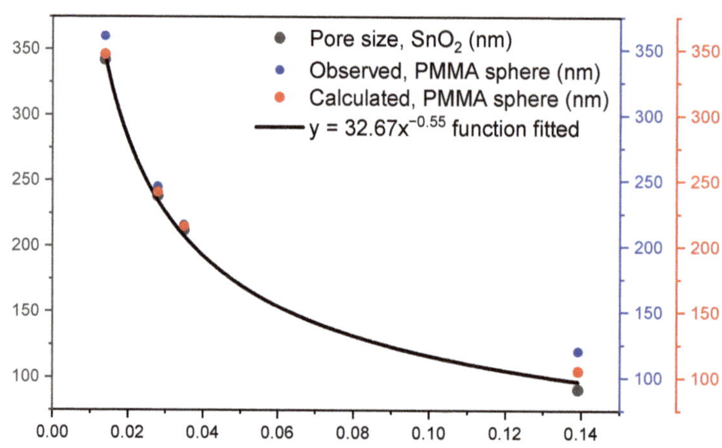

Figure 10. Comparison of the pore size of SnO_2 coatings (black) with the diameter of observed (blue) and calculated (red) PMMA spheres.

4. Conclusions

The effect of SDS surfactant on the synthesis of PMMA with different sphere diameters was studied and evaluated as a template for obtaining SnO_2 coatings with different porosities. The PMMA sphere diameters between 120 and 360 nm were synthesized using different amounts of surfactant in the range of 0.0139 and 0.139 % SDS with a conventional radical polymerization technique. For size-tuning of PMMA sphere diameter as a function of % SDS was fitted to the mathematical equation $y = ax^b$. The pore size of SnO_2 coatings varies as a function of the PMMA-sphere diameter used as a template. This research demonstrates that it is possible to obtain oxide coatings such as SnO_2 with size-tunable porous using PMMA as a template. The wall of these macropores is built with nanometric-size SnO_2 particles. Similar coating porosities could be synthesized using PMMA spheres with different oxides for different technological applications.

Author Contributions: Conceptualization, J.L.M.-C., J.C.P.-E. and F.P.-D.; methodology, J.L.M.-C., L.C.R.-P. and F.P.-D.; validation, J.L.M.-C., L.C.R.-P. and F.P.-D.; formal analysis, J.C.P.-E. and F.P.-D.; investigation, J.L.M.-C., L.C.R.-P. and F.P.-D.; resources, F.P.-D.; data curation, J.L.M.-C. and L.C.R.-P.; writing—original draft preparation, J.L.M.-C.; writing—review and editing, J.C.P.-E. and F.P.-D.; visualization, J.C.P.-E. and F.P.-D.; supervision, F.P.-D.; project administration, F.P.-D.; funding acquisition, J.C.P.-E. and F.P.-D. All authors have read and agreed to the published version of the manuscript.

Funding: The APC was funded by Instituto de Innovación y Competitividad de la Secretaría de Innovación y Desarrollo Económico del Estado de Chihuahua, México.

Institutional Review Board Statement: Not applicable.

Informed Consent Statement: Not applicable.

Data Availability Statement: The data presented in this study are available on request from the corresponding authors.

Acknowledgments: Thanks to C. Leyva-Porras, K. Campos, and D. Lardizabal for technical support at the Labs of CIMAV Chihuahua, Mexico. In addition, J.L. Mendoza-Castellanos thanks CONACYT for scholarship support.

Conflicts of Interest: The authors declare no conflict of interest.

References

1. Yuan, M.; Huang, D.; Zhao, Y. Development of Synthesis and Application of High Molecular Weight Poly(Methyl Methacrylate). *Polymers* **2022**, *14*, 2632. [CrossRef] [PubMed]
2. Duan, G.; Zhang, C.; Li, A.; Yang, X.; Lu, L.; Wang, X. Preparation and Characterization of Mesoporous Zirconia Made by Using a Poly (Methyl Methacrylate) Template. *Nanoscale Res. Lett.* **2008**, *3*, 118. [CrossRef] [PubMed]
3. Hyodo, T.; Sasahara, K.; Shimizu, Y.; Egashira, M. Preparation of Macroporous SnO_2 Films Using PMMA Microspheres and Their Sensing Properties to NO_x and H_2. *Sens. Actuators B Chem.* **2005**, *106*, 580–590. [CrossRef]
4. Kamitani, K.; Hyodo, T.; Shimizu, Y.; Egashira, M. Fabrication of Porous Alumina Ceramics Having Cell Windows with Controlled Size by PMMA Template Method. *J. Mater. Sci.* **2010**, *45*, 3602–3609. [CrossRef]
5. Khan, H.; Samanta, S.; Seth, M.; Jana, S. Fabrication and Photoelectrochemical Activity of Hierarchically Porous TiO_2–ZnO Heterojunction Film. *J. Mater. Sci.* **2020**, *55*, 11907–11918. [CrossRef]
6. Sordello, F.; Minero, C. Photocatalytic Hydrogen Production on Pt-Loaded TiO_2 Inverse Opals. *Appl. Catal. B Environ.* **2015**, *163*, 452–458. [CrossRef]
7. Masuda, Y. Recent Advances in SnO_2 Nanostructure Based Gas Sensors. *Sens. Actuators B Chem.* **2022**, *364*, 131876. [CrossRef]
8. Tan, Y.; Zhang, J. Highly Sensitive Ethanol Gas Sensors Based on Co-Doped SnO_2 Nanobelts and Pure SnO_2 Nanobelts. *Phys. E Low-Dimens. Syst. Nanostructures* **2023**, *147*, 115604. [CrossRef]
9. Rameshkumar, C.; Ananth, D.; Divyalakshmi, V.; Balakrishnan, M.; Senthilkumar, G.; Subalakshmi, R. An Investigation of SnO_2 Nanofilm for Solar Cell Application by Spin Coating Technique. *AIP Conf. Proc.* **2021**, *2341*, 040024. [CrossRef]
10. Alfaro Cruz, M.R.; Saldaña-Ramírez, A.; Juárez-Ramírez, I.; Torres-Martínez, L.M. Development of SnO_2–ZnO Thin Films as a Photocatalyst for Obtaining Alternative Fuels through Photocatalytic Reactions. *Solid State Sci.* **2023**, *137*, 107112. [CrossRef]
11. Jo, M.-H.; Koo, B.-R.; Ahn, H.-J. Accelerating F-Doping in Transparent Conducting F-Doped SnO_2 Films for Electrochromic Energy Storage Devices. *Ceram. Int.* **2020**, *46*, 25066–25072. [CrossRef]
12. Wu, S.; Zhao, Q.; Miao, D.; Dong, Y. Synthesis and Characterization of Sb-Doped SnO_2-(CeO_2-TiO_2) Composite Thin Films Deposited on Glass Substrates for Antistatic Electricity and UV-Shielding. *J. Rare Earths* **2010**, *28*, 189–193. [CrossRef]

13. Bathula, B.; Gurugubelli, T.R.; Yoo, J.; Yoo, K. Recent Progress in the Use of SnO_2 Quantum Dots: From Synthesis to Photocatalytic Applications. *Catalysts* **2023**, *13*, 765. [CrossRef]
14. Zhang, K.; Tan, X.; Xiang, P.; Li, B.; Li, J.; Ren, Y.; Zhu, Y.; Liu, Y.; Yan, W.; Chen, X.; et al. Modifying the Photoelectric Performance of SnO_2 via D-Arginine Monohydrochloride for High-Performance Perovskite Solar Cells. *J. Alloys Compd.* **2023**, *946*, 169361. [CrossRef]
15. Goseki, R.; Ishizone, T. Poly(Methyl Methacrylate) (PMMA). In *Encyclopedia of Polymeric Nanomaterials*; Kobayashi, S., Müllen, K., Eds.; Springer: Berlin/Heidelberg, Germany, 2021; pp. 1–11. ISBN 978-3-642-36199-9.
16. Forte, M.A.; Silva, R.M.; Tavares, C.J.; Silva, R.F. e Is Poly(Methyl Methacrylate) (PMMA) a Suitable Substrate for ALD?: A Review. *Polymers* **2021**, *13*, 1346. [CrossRef]
17. Gao, J.; Jiang, F.; Zhai, G. Ultra-High Molecular Weight Alpha-Amino Poly(Methyl Methacrylate) with High T g through Emulsion Polymerization by Using Transition Metal Cation-Tertiary Amine Pairs as a Mono-Centered Initiator. *Macromol. React. Eng.* **2016**, *10*, 269–279. [CrossRef]
18. Arora, P.; Jain, R.; Mathur, K.; Sharma, A.; Gupta, A. Synthesis of Polymethyl Methacrylate (PMMA) by Batch Emulsion Polymerization. *Afr. J. Pure Appl. Chem.* **2010**, *4*, 152–157.
19. Hamlaoui, F.Z.; Naar, N. Improvement of the Structural and Electrical Properties of PMMA/PANI-MA Blends Synthesized by Interfacial in Situ Polymerization in a Continuous Organic Phase. *Polym. Bull.* **2022**, *79*, 37–63. [CrossRef]
20. Chatterjee, A.; Mishra, S. Novel Synthesis with an Atomized Microemulsion Technique and Characterization of Nano-Calcium Carbonate ($CaCO_3$)/Poly(Methyl Methacrylate) Core-Shell Nanoparticles. *Particuology* **2013**, *11*, 760–767. [CrossRef]
21. Yang, Y.; He, J.; Zhang, Y.; Hong, Y.; Wang, X. Understanding the Interface Structures of Water-Based and Solvent-Based Poly(Methyl Methacrylate) Coatings at the Molecular Level. *Appl. Surf. Sci.* **2022**, *579*, 152239. [CrossRef]
22. Yang, M.; Wang, L.; Xia, Y. Ammonium Persulphate Induced Synthesis of Polymethyl Methacrylate Grafted Sodium Alginate Composite Films with High Strength for Food Packaging. *Int. J. Biol. Macromol.* **2019**, *124*, 1238–1245. [CrossRef] [PubMed]
23. Li, X.; Huang, Y.; Dan, Y. Synthesis of Sub-100 Nm PMMA Nanoparticles Initiated by Ammonium Persulfate/Ascorbic Acid in Acetone-Water Mixture. *Colloid Polym. Sci.* **2020**, *298*, 225–232. [CrossRef]
24. Parra, C.; Albano, C.; González, G. Effect of Surfactant Type on the Synthesis of PMMA Using Redox Initiation and High Frequency Ultrasound. *Polym. Eng. Sci.* **2008**, *48*, 2066–2073. [CrossRef]
25. Baissac, L.; Buron, C.C.; Hallez, L.; Berçot, P.; Hihn, J.-Y.; Chantegrel, L.; Gosse, G. Synthesis of Sub-Micronic and Nanometric PMMA Particles via Emulsion Polymerization Assisted by Ultrasound: Process Flow Sheet and Characterization. *Ultrason. Sonochem.* **2018**, *40*, 183–192. [CrossRef]
26. Mahmoudian, M.; Torbati, S.; AliMirzayi, N.; Nozad, E.; Kochameshki, M.G.; Shokri, A. Preparation and Investigation of Poly(Methylmethacrylate) Nano-Capsules Containing Haloxyfop-R-Methyl and Their Release Behavior. *J. Environ. Sci. Health -Part B Pestic. Food Contam. Agric. Wastes* **2020**, *55*, 301–309. [CrossRef]
27. Vargas-Salazar, C.Y.; Ovando-Medina, V.M.; Ledezma-Rodríguez, R.; Peralta, R.D.; Martínez-Gutiérrez, H. Ultrasound-Assisted Polymerization of Methyl Methacrylate Using the Reactive Surfactant Hitenol BC10 in a Semicontinuous Heterophase Process. *Iran. Polym. J. Engl. Ed.* **2015**, *24*, 41–50. [CrossRef]
28. Kamras, B.L.; Mirzanasiri, N.; Korir, D.K.; Mandal, S.; Hariharakumar, S.L.; Petros, R.A.; Marpu, S.B.; Simmons, D.P.; Omary, M.A. Formula-Driven, Size-Tunable Synthesis of PMMA Nanoparticles by Varying Surfactant Concentration. *Materials* **2020**, *13*, 1834. [CrossRef]
29. Bao, J.; Zhang, A. Poly(methyl methacrylate) nanoparticles prepared through microwave emulsion polymerization. *J. Appl. Polym. Sci.* **2004**, *93*, 2815–2820. Available online: https://onlinelibrary.wiley.com/doi/abs/10.1002/app.20758 (accessed on 10 March 2023). [CrossRef]
30. Xie, C.; Leng, K.; Sheng, J.; Wang, X.; Li, Q.; Song, L.; Liu, L.; Sun, H.; Huang, X.; Wang, Z.; et al. Preparation of Poly(Methyl Methacrylate) Microspheres via Photopolymerization Initiated by LED Light Source. *Colloid Polym. Sci.* **2020**, *298*, 1285–1291. [CrossRef]
31. Yoshida, S.; Kikuchi, S.; Kanehashi, S.; Okamoto, K.; Ogino, K. Microfluidic Fabrication of Morphology-Controlled Polymeric Microspheres of Blends of Poly(4-Butyltriphenylamine) and Poly(Methyl Methacrylate). *Materials* **2018**, *11*, 582. [CrossRef]
32. Gao, Y.; Zhang, J.; Liang, J.; Yuan, D.; Zhao, W. Research Progress of Poly(Methyl Methacrylate) Microspheres: Preparation, Functionalization and Application. *Eur. Polym. J.* **2022**, *175*, 111379. [CrossRef]
33. Ortiz-Landeros, J.; Pfeiffer, H. Métodos De Síntesis De Microesferas Poliméricas Y Su Uso En El Proceso De Síntesis De Materiales Cerámicos Macroporosos. *Tip Rev. Espec. En Cienc. Quím.-Biológicas* **2010**, *13*, 113–120.
34. Chern, C.S. Emulsion Polymerization Mechanisms and Kinetics. *Prog. Polym. Sci.* **2006**, *31*, 443–486. [CrossRef]
35. Czajka, A.; Lovell, P.A.; Armes, S.P. Time-Resolved Small-Angle X-Ray Scattering Studies during the Aqueous Emulsion Polymerization of Methyl Methacrylate. *Macromolecules* **2022**, *55*, 10188–10196. [CrossRef]
36. Mendoza-Castellanos, J.L.; Morales-Mendoza, J.E.; Paraguay-Delgado, F. Synthesis and Characterization of Macroporous Tin Oxide Coatings. *J. Mater. Res. Technol.* **2022**, *19*, 4092–4102. [CrossRef]
37. Abdelrazek, E.M.; Hezma, A.M.; El-khodary, A.; Elzayat, A.M. Spectroscopic Studies and Thermal Properties of PCL/PMMA Biopolymer Blend. *Egypt. J. Basic Appl. Sci.* **2016**, *3*, 10–15. [CrossRef]

38. Sayyah, S.M.; Khaliel, A.B.; El-Shafiey, Z.A.; Barsoum, B.N. Infrared Studies on Polymethyl Methacrylate Doped with a Sulphur-Containing Ligand and Its Cobalt (II) Complex During Gamma Radiolysis. *Int. J. Polym. Mater. Polym. Biomater.* **2005**, *54*, 445–466. [CrossRef]
39. Wypych-Puszkarz, A.; Cetinkaya, O.; Yan, J.; Udovytska, R.; Jung, J.; Jenczyk, J.; Nowaczyk, G.; Jurga, S.; Ulański, J.; Matyjaszewski, K.; et al. Molecular Dynamics and Structure of Poly(Methyl Methacrylate) Chains Grafted from Barium Titanate Nanoparticles. *Molecules* **2022**, *27*, 6372. [CrossRef]
40. Ali, U.; Karim, K.J.B.A.; Buang, N.A. A Review of the Properties and Applications of Poly (Methyl Methacrylate) (PMMA). *Polym. Rev.* **2015**, *55*, 678–705. [CrossRef]
41. Novaković, K.; Katsikas, L.; Popović, I.G. The Thermal Degradation of Poly(Iso-Butyl Methacrylate) and Poly(Sec-Butyl Methacrylate). *J. Serbian Chem. Soc.* **2000**, *65*, 867–875. [CrossRef]
42. Rojas, A.F.; Aranzazu, L.M.; Gaviria, G.H.; Carrero, J.I. Degradación Térmica de Polimetíl Metacrilato En Múltiples Extrusiones. *Ingeniería y Competitividad* **2014**, *16*, 131–142. Available online: http://www.scielo.org.co/scielo.php?script=sci_arttext&pid=S0123-30332014000200012&lng=es&tlng= (accessed on 27 February 2023). [CrossRef]
43. Guo, Z.; Liu, J.; Li, Y.; Lin, H.; Wang, H.; Tam, K.C.; Liu, G. Effects of Dispersion Techniques on the Emulsion Polymerization of Methyl Methacrylate. *Colloid Polym. Sci.* **2021**, *299*, 1147–1159. [CrossRef]
44. Shang, L.; Nienhaus, K.; Nienhaus, G.U. Engineered nanoparticles interacting with cells: Size matters. *J. Nanobiotechnol.* **2014**, *12*, 5. Available online: https://jnanobiotechnology.biomedcentral.com/articles/10.1186/1477-3155-12-5 (accessed on 7 March 2023). [CrossRef] [PubMed]
45. Zhang, H.; Dong, F.; Fang, S.; Ye, C.; Wang, M.; Cheng, H.; Han, Z.; Zhai, S. Fabrication of Macroporous Titanium Dioxide Film Using PMMA Microspheres as Template. *J. Colloid Interface Sci.* **2012**, *386*, 73–79. [CrossRef] [PubMed]
46. Sadakane, M.; Takahashi, C.; Kato, N.; Ogihara, H.; Nodasaka, Y.; Doi, Y.; Hinatsu, Y.; Ueda, W. Three-Dimensionally Ordered Macroporous (3DOM) Materials of Spinel-Type Mixed Iron Oxides. Synthesis, Structural Characterization, and Formation Mechanism of Inverse Opals with a Skeleton Structure. *Bull. Chem. Soc. Jpn.* **2007**, *80*, 677–685. [CrossRef]
47. Waterhouse, G.I.N.; Metson, J.B.; Idriss, H.; Sun-Waterhouse, D. Physical and Optical Properties of Inverse Opal CeO_2 Photonic Crystals. *Chem. Mater.* **2008**, *20*, 1183–1190. [CrossRef]
48. Bosco, J.P.; Sasaki, K.; Sadakane, M.; Ueda, W.; Chen, J.G. Synthesis and Characterization of Three-Dimensionally Ordered Macroporous (3DOM) Tungsten Carbide: Application to Direct Methanol Fuel Cells. *Chem. Mater.* **2010**, *22*, 966–973. [CrossRef]
49. Li, D.; Zhao, Z.; Kong, X.; Tian, M.; Wang, Y.; Cheng, Y.; Cao, T.; Zhou, J.; Fan, X.; Gou, L. Solution-Processable Three-Dimensionally Macroporous $KTiOPO_4/SiO_2$ Inverse Opal Powders with Enhanced Second Harmonic Emission. *J. Alloys Compd.* **2018**, *746*, 256–261. [CrossRef]
50. Qin, J.; Cui, Z.; Yang, X.; Zhu, S.; Li, Z.; Liang, Y. Synthesis of Three-Dimensionally Ordered Macroporous LaFeO3 with Enhanced Methanol Gas Sensing Properties. *Sens. Actuators B Chem.* **2015**, *209*, 706–713. [CrossRef]

polymers

MDPI

Article

Crosslinked Chitosan Films Supplemented with *Randia* sp. Fruit Extract

Felipe López-Saucedo [1,2], Leticia Buendía-González [1], Héctor Magaña [3], Guadalupe Gabriel Flores-Rojas [2] and Emilio Bucio [2,*]

[1] Facultad de Ciencias, Campus El Cerrillo Piedras Blancas, Universidad Autónoma del Estado de México, Carretera Toluca-Ixtlahuaca Km 15.5, Toluca 50200, Mexico
[2] Departamento de Química de Radiaciones y Radioquímica, Instituto de Ciencias Nucleares, Universidad Nacional Autónoma de México, Circuito Exterior, Ciudad Universitaria, Mexico City 04510, Mexico
[3] Facultad de Ciencias Químicas e Ingeniería, Universidad Autónoma de Baja California, Calzada Universidad 14418, Parque Industrial Internacional Tijuana, Tijuana 22390, Mexico
* Correspondence: ebucio@nucleares.unam.mx

Abstract: This work proposes the development of a polymer film made up of affordable components for its use as a healthcare material. Chitosan, itaconic acid, and *Randia capitata* fruit extract (Mexican variation) are the unique ingredients of this biomaterial prospect. Chitosan (from crustacean chitin) is crosslinked with itaconic acid, and in situ added *R. capitata* fruit extract in a one-pot reaction carried out in water as the sole solvent. Structurally, the film formed is an ionically crosslinked composite characterized by IR spectroscopy and thermal analysis (DSC and TGA); cell viability was also performed in vitro using fibroblasts BALB/3T3. Dry and swollen films were analyzed to determine affinity and stability in water. This chitosan-based hydrogel is designed as a wound dressing due to the combined properties of the chitosan with *R. capitata* fruit extract, which has potential as bioactive material due to its properties in epithelial regeneration.

Keywords: biomaterial; chitosan; crosslinking; hydrogel; natural product; *Randia capitata*

1. Introduction

In recent years, the trend of using natural materials has become popular. Preference for materials obtained from plants, animals, or microorganisms over synthetics is diverse; one benefit is that once the life cycle is closed, the residues can be easily reincorporated into the environment as new mineral nutrients [1], unlike synthetic polymer materials that require careful residue management once disposed [2]; unfortunately, however, synthetic polymers usually turn into pollutant agents due to poor treatment. For example, waste plastics cause serious damage to soils [3] and effluents such as rivers, lakes, and seas [4]. A way to help stop the disproportionate increase of human waste is promoting the use of biodegradable materials to reincorporate simpler compounds rich in C, N, and O into the environment to be used as nutrients for plants and microorganisms [5].

Chitosan is a polymer of natural origin obtained by deacetylation of chitin, which is abundantly found in the exoskeletons of arthropods and the cell wall of some fungi [6,7]. Chitosan contains −OH groups that allow the formation of intermolecular hydrogen bonds; it also has a $-NH_2$ group, which can react with electrophiles such as carboxylic acids −COOH to form covalent bonds in a condensation (forming amides), or ionic bonds (forming carboxylate-ammonium ions) [8]. This polysaccharide is quite thermally stable and is structured of linear chains, but can be easily crosslinked [9], so it is possible to obtain higher molecular weight polymers to form films or hydrogel networks [10]. The use of chitosan is convenient for sanitary dressing, nanoparticle loading [11], and drug loading/delivery systems [7,12] due to its high hydrophilicity. Traditionally, the crosslinking of chitosan hass been carried out in acetic acid [13,14], with different crosslinking agents [15–17], and even using ionic liquids as solvent [18]. Recently, the crosslinking of chitosan and itaconic acid has been reported with promising results [8,19], where the versatility of this diacid functioning as a crosslinking agent is highlighted.

Randia sp. is a fruity arboreal plant belonging to the Rubiaceae family that is found around the world [20], including Mexico, where it is known by different names, including "crucecita", "cruceta", and "crucillo". In the past, the fruit of this plant was used in traditional medicine, although recently in this century some varieties of *Randia*, such as *R. aculeata* [21], *R. echinocarpa* [22,23], and *R. dumetorum* [24,25] have attracted the attention of organic chemistry and pharmacology, since there is evidence of their bioactivity. Thus, incorporating *Randia* sp. extracts into biomedical devices and disposables could be an eco-friendly alternative for healing purposes.

This paper puts forward a film containing only three natural-based ingredients chitosan (polymer), itaconic acid (crosslinking agent), and *R. capitata* fruit extract (bioactive compound). This film was synthesized in water as the sole solvent and under catalyst-free conditions, which is our proposed method. The goal is to get an absorbable, biodegradable, and biocompatible film as a potential wound dressing [26].

2. Materials and Methods

2.1. Reagents and Solvents

Chitosan (75–85% deacetylated, viscosity 200–800 cP, 1 wt.%, in acetic acid (1% vol.), at 25 °C, CAS 9012-76-4) and itaconic acid (99%) were obtained from Sigma Aldrich (South 2nd, Saint Louis, MO, USA). Dry fruits of *Randia capitata* (*R. capitata*) were harvested from trees located in Veracruz, Mexico. Double distilled water was used for the reaction and analysis, and ethanol reagent grade from Sigma Aldrich (South 2nd, Saint Louis, MO, USA) was used to isolate the *Randia* extract.

2.2. Randia Extract

Complete dry fruit *R. capitata* was pulverized with a mortar. Then, 1 g of the powder was placed into a flask (100 mL capacity), dissolved in 20 mL of a 1:1 water/ethanol mixture, and sonicated for 30 min at room temperature. Separation of solid and liquid phases was performed in a vacuum line system with a funnel. Whatman filter paper (number 3) and fiberglass were used to retain most of solid particles and to allow a faster filtration. The liquid phase obtained was reserved inside a ball flask, and subsequently the solution was concentrated under reduced pressure to isolate the *R. capitata* extract (a semi-solid), which was stored at 4 °C.

2.3. Synthesis of Chitosan/Itaconic Acid Film

Chitosan/itaconic acid film (Ch/It) was prepared mixing 100 mg of chitosan and 42 mg of itaconic acid in 5 mL of water inside a beaker of 100 mL capacity. Composition was based on stoichiometry of chitosan units and molecular weight of itaconic acid (Figure 1a). This mixture was kept under stirring for 60 min at 70 °C. Once the reaction time was over, the magnetic stirrer was removed, and the solution (of viscous appearance) remained inside the beaker at room temperature for 72 h, to allow slow evaporation of water. The dry film was carefully taken out the bottom of beaker and stored into a vacuum desiccator chamber at room temperature until use (Figure 1b left).

2.4. Synthesis of the Chitosan/Itaconic Acid/Randia Film

Chitosan/itaconic acid/*Randia* film (Ch/It/Ran) was synthesized, weighing 100 mg of chitosan and 42 mg of itaconic acid in 5 mL of water. This mixture was stirred for 60 min at 70 °C and cooled until room temperature. Then, 3 mL of an aqueous solution containing 40 mg of *Randia* extract was added. This amount of fruit extract was chosen because a homogeneous film was obtained, as well as the high cell viability observed (see Section 3.6).

The film was dried at room temperature for 72 h and stored inside a desiccator at room temperature until use (Figure 1b right).

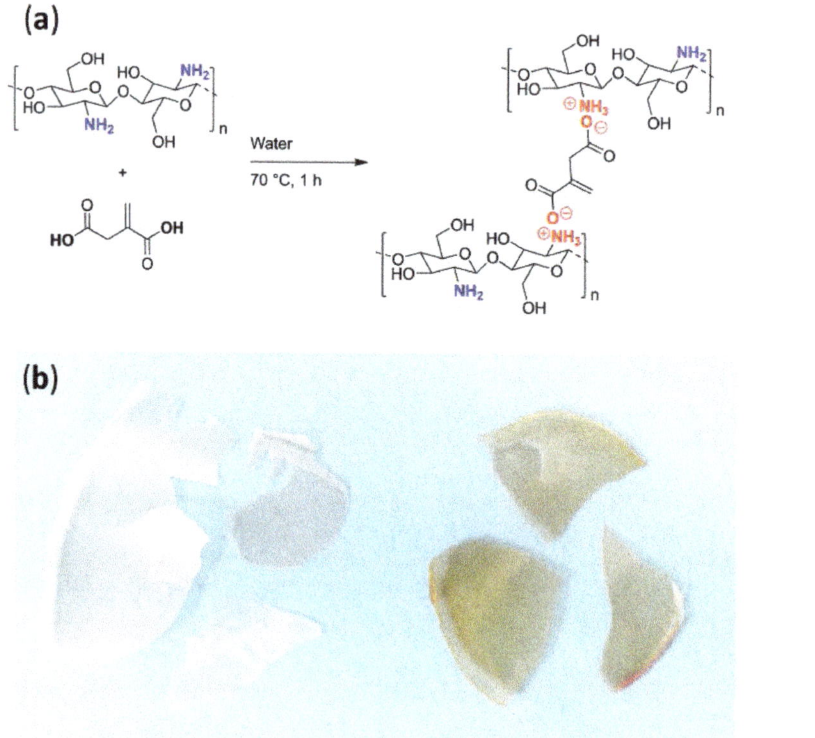

Figure 1. (**a**) Scheme of synthesis of ionic crosslinking of chitosan and itaconic acid. (**b**) Side by side chitosan/itaconic acid film (**left**) and chitosan/itaconic acid/*Randia* film (**right**).

2.5. Limit Swelling

Small pieces of Ch/It or Ch/It/Ran films (between 2 and 4 mg) were weighed and placed inside small aluminum pans. Later, distilled water at room temperature was added until the sample got submerged. Swelling measurements were chronometered and monitored for up to 6 h. The swollen film was taken out of the water in between each weighing, the excess water was absorbed with a paper wipe to weigh the sample in the balance, and according to Equation (1) the *Swelling (%)* was calculated.

$$Swelling\ (\%) = [ws - wd/wd] \times 100 \tag{1}$$

where *ws* is the weight of swollen film (hydrogel) and *wd* is the weight of dry film.

2.6. Cell Viability

The fibroblast cell line BALB/3T3 (mouse) was used to test the cell viability of different samples Ch/It, Ch/It/Ran, *Randia* complete fruit, and *Randia* fruit extract in in vitro conditions. First, an indirect method was performed, in which 20 mg of each sample was suspended in 2 mL of DMEM (Dulbecco's Modified Eagle Medium) for 24 h. Subsequently, 50 μL of this medium was placed in contact with cell medium in 96-well plates. Each well contained 30,000 cells mL^{-1} in DMEM medium complemented with fetal bovine serum medium, streptomycin solution, and gentamicin. Incubation was for 24 h at 5% CO_2 and 37 °C. Subsequently, 3-(4,5-dimethylthiazol-2-yl)-2,5-diphenyltetrazolium bromide (MTT) kit reagent (Roche, Switzerland) was placed in each well to determine the colorimetric cell viability. The absorbance of samples and controls was recorded at a wavelength of 620 nm. Cell viability was determined using Equation (2). Results were performed in triplicate, collected, and statistically analyzed.

$$Cytocompatibility\ (\%) = [A\ Sam)/[A\ Con] \times 100 \quad (2)$$

where *A Sam* is the absorbance of the sample, and *A Con* corresponds to absorbance of the control recorded at 620 nm.

2.7. Instrumental

2.7.1. UV-Vis Spectrophotometry

A Multiskan FC spectrophotometer (Thermo Scientific, Waltham, MA, USA) was used to determine the absorbance at a wavelength of 620 nm. Cell viability was performed in 96-well plates. The experiments were performed in triplicate.

2.7.2. Infrared Spectroscopy

A Perkin-Elmer Spectrum 100 (PerkinElmer, Inc., Waltham, MA, USA) equipped with ATR was used to analyze all samples. First, a background scan was run, and then the dry sample was analyzed with 16 scans in the wavelength range from 4000 to 650 cm^{-1}.

2.7.3. Differential Scanning Calorimetry (DSC)

Differential scanning calorimetry was carried out in a Differential Scanning Calorimeter 2010 (TA Instruments, New Castle, DE, USA). Around 10 mg of a dry sample was weighed, encapsulated in an aluminum pan, and analyzed from 25 to 250 °C in a nitrogen atmosphere at a heating rate of 10 °C min^{-1}.

2.7.4. Thermogravimetric Analysis (TGA)

Thermogravimetric analysis was performed using a TGA Q50 (TA Instruments, New Castle, DE, USA). Then, 5 to 10 mg of a sample previously dried into a vacuum oven at 60 °C was weighed. Each sample was placed in a platinum tray, and the analysis was performed at a heating rate of 10 °C min^{-1} up to 800 °C under nitrogen atmosphere.

2.7.5. Contact Angle

Samples of Ch/It and Ch/It/Ran were selected and placed between two glasses and pressed for 2 days at room temperature to get flattened surfaces. For this technique, a Kruss DSA 100 droplet analyzer (Hamburg, Germany) was used. The native software was used to determine the angles formed between the film surface and the water microdroplet. Angles were recorded at 0, 1, and 2 min; each experiment was performed in triplicate.

2.7.6. Direct Analysis in Real-Time Mass Spectrometry (DART-MS)

The mass spectrometer instrument was a JEOL AccuTOF: JMS-T100LC with ionization mode $DART^{+}$.

3. Results and Analysis

3.1. Mass Spectrometry of Randia Extract

Regarding the composition of the ethanol/water *Randia* extract, it was found in the literature that alcoholic *Randia* extracts contain mainly nitrogen-free compounds, particularly phenols and flavonoids, characterized by ferric chloride and Constantinescu assays, respectively. In said investigation, the total composition of *Randia laevigata* fruit alcoholic extract was 57% humidity, 3% ashes, 3.31% protein, and 37.75% nitrogen-free molecules [27].

Following the reports by Santos-Cervantes et al. [28] and Cruz-Silva et al. [29], who reported the abundance of flavonoids in alcohol *Randia* extracts, in this investigation, with the ethanol/water *Randia capitata* extract, the existence of flavonoids was confirmed, where the presence of flavanone and quercetin is likely. Since in the DART$^+$ spectrum (Figure 2), the most intense peaks correspond to these compounds; firstly, the base peak m/z = 183 may be attributed to the fragmentation of flavanone in the alpha position to the carbonyl, while the second most intense peak at m/z = 165 possibly corresponds to the quercetin fragment formed by a rearrangement during ionization [30].

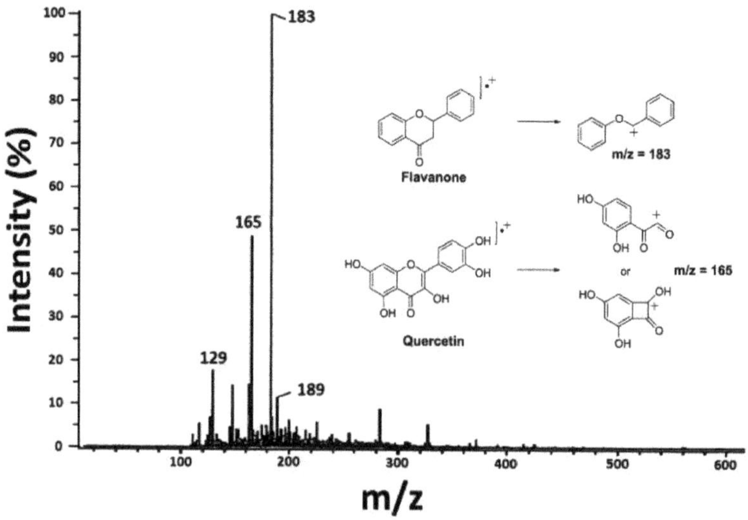

Figure 2. DART+ mass spectrometry of *Randia* extract and proposed fragmentation of two flavonoids corresponding to the main signals detected.

3.2. Spectroscopic Analysis

IR-ATR spectroscopy was carried out in dry samples to avoid signals of absorbed water from moisture. The starting compounds chitosan, itaconic acid, and *Randia* extract share functional groups in common, but there are certain differences that allow their characterization (Figure 3) before and after crosslinking reaction (in the composite films). First, in the chitosan spectrum, the O-H stretching band is distinguished, which is wide and appears above 3000 cm^{-1}; in this zone, two small overlapped peaks at 3354 and 3296 cm^{-1} may correspond to N-H stretching of amines; and at 2868 cm^{-1}, the C-H stretching band is observed; at 1575 cm^{-1} appears the N-H bending; while the symmetric methyl bending signal is observed at 1373 cm^{-1}; finally, at 1145 and 1025 cm^{-1}, the asymmetric and symmetric stretching of C-O are observed [31]. In the spectrum of itaconic acid, besides the bands of O-H and C-H stretching, the C=O of carbonyl from the diacid is observed at 1682 cm^{-1}; the alkene band of C=C appears at 1623 cm^{-1}, which is particularly strong; methyl bending at 1393 cm^{-1} is distinguished; and finally, the C-O stretching bands appear at 1215 and 1164 cm^{-1}. The third spectrum of *Randia* extract, due to its natural origin, displayed similar bands to those found in natural products (plants and fruits) [32]; despite that *Randia* extract is a mixture, it was possible to identify some bands, for example, at 3291 cm^{-1} the stretching of O-H, which is observed in carbohydrates, polysaccharides, aliphatic alcohols, and also in phenols [23]. Additionally, in the zone of 1600–1000 cm^{-1}, the secondary signals of C-H, O-H, and C-C confirmed the bonds found in the mentioned natural compounds (Table 1).

Regarding FTIR of films, in addition to functional groups already described, there was evidence of the crosslinking reaction, with forming the ionic bond R-COO$^-$ R-NH3$^+$ from carboxyl (from itaconic acid) and the amine (from chitosan). In both infrared spectra of Ch/It and Ch/It/Ran, the carboxylates are observed, although in the Ch/It film it is easier to identify the carboxylate anion R-(O=C-O$^-$), since the band corresponding to the asymmetric stretching is observed at 1528 cm^{-1} and the symmetric stretching band is also observed at 1375 cm^{-1} [33], which overlaps with the methyl bending and is intensified for this reason. Although the O-H stretching band should decrease, it was not the case, since the band appeared more intense and broad, due to humidity and/or residual water from the reaction. Moreover, there was no evidence of carbonyl of amides. The carbonyl band broadened, and stronger bands are a consequence of a higher molecular weight; in other words, the chitosan/itaconic acid is crosslinked through the formation of ionic bonds of ammonium-carboxylate [9]. On the other hand, in the Ch/It/Ran film, the O-H bands are also observed, and the band at 1524 cm^{-1} of carboxylate anion is observed; while the intense band at 1375 cm^{-1} is attributed to two types of vibrations, these are carboxylate symmetric stretching and methyl bending; finally, the group of bands at 1151, 1061, and 1013 cm^{-1} confirmed the C-O stretching of alcohols and phenols from flavonoids [34], as well as complex carbohydrates [22] and polyglucosamine [35].

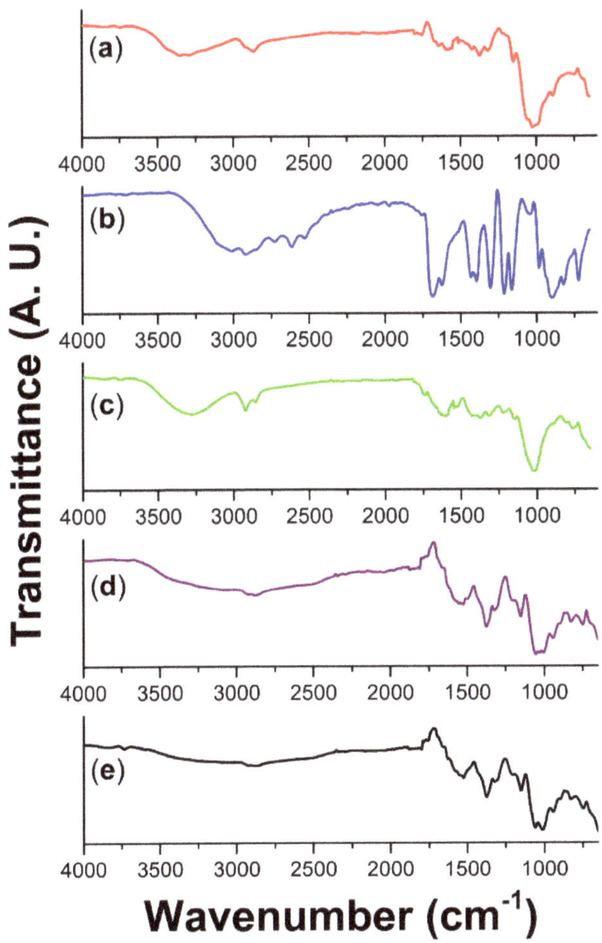

Figure 3. FTIR–ATR spectra of (**a**) chitosan, (**b**) itaconic acid, (**c**) *Randia* extract, (**d**) Ch/It, and (**e**) Ch/It/Ran.

Table 1. Most important bands of FTIR assigned to starting materials and products [33].

Sample	Wavenumber (cm^{-1})	Type
Chitosan	3354, 3296 (wide band)	O-H st and N-H st
	2868	C-H st
	1575	N-H bd
	1373	C-H (methyl) bd
	1145, 1025	C-O st as and sym
Itaconic acid	3000 (wide band)	O-H st
	3013, 2917	C-H st
	1682	C=O st
	1623	C=C
	1395	C-H bd
	1304, 1213	C-O st
Randia extract	3291	O-H st
	2920	C-H st
	1699	C=O st
	1594	C=C
	1403	C-N
	1021	C-O st
Ch/It	3030 (wide band)	O-H st
	2885	C-H st
	1528	C=C
	1375	C-H methyl bd and O-C=O as st
	1066, 1022	C-O st
Ch/It/Ran	3000 (wide band)	O-H st
	2878	C-H st
	1524	C=C
	1375	C-H methyl bd and O-C=O as st
	1151, 1061, 1013	C-O st, as and sym

Abbreviations: stretching (st), bending (bd), asymmetric (as), symmetric (sym).

Summarizing, FTIR spectra indicated that the Ch/It and Ch/It/Ran films formed ionic crosslinked structures through the carboxylic and amino groups, and conformed intermolecular hydrogen bonds; this is confirmed, since the film bands become more intense and they broaden when the molecular weight of chitosan is increased, as in hydrogels [36].

3.3. Mechanism of Crosslinking

Regarding the crosslinking reaction mechanism, Sirviö et al. found that during the crosslinking of itaconic acid and chitosan in the first step, the addition between the alkene group of itaconic acid and the amine of chitosan takes place. In the second step, the amine intermediate subsequently reacts with the carboxyl group of itaconic acid, resulting in a 5-membered cyclic amide (or substituted γ-lactam), where reaction conditions at reflux temperature and reaction time of 24 h were reported [19].

Other works also shed some light concerning this type of reaction. For example, Qu et al. reported the polymerization of ethanolamine with itaconic acid in N_2 atmosphere, heating at 180 °C, and using a catalyst [37], which are extreme reaction conditions compared to those reported in this manuscript. On the other hand, Milosavljevic et al. (like us) proposed that ionic crosslinking is the primary mechanism, since the formation of carboxylate and ammonium species is favored in an aqueous reaction medium [8,38]. While in an independent research, a similar ionic crosslinking mechanism was reported for chitosan with oxalic acid [39]. Finally, according to Gabriele et al., it is indicated that during crosslinking of chitosan and succinic acid to form membranes, both, the formation of covalent and ionic bonds occurred [40]. Once again, the reaction conditions determine the path of bonding during the reaction.

3.4. Hydrophilicity: Contact Angle and Swelling

Both Ch/It and Ch/It/Ran films were subjected to two different tests to determine the degree of hydrophilicity; these tests were the contact angle per drop of water and swelling by immersion in water. The experiments were carried out at room temperature with double distilled water, with the following results.

The contact angle was measured at short times because at longer times the surface of the film is deformed, and recording is not correct. Both films showed contact angles under 90° at time = 0, which means that surfaces wet quickly and are hydrophilic; in addition, the films increase in size when the experiment is over, which is a first indication of physisorption phenomena [41]. Quantitatively, the Ch/It angles range from 56 ± 3.9° to 35 ± 12.5°, while for the Ch/It/Ran film, the angles found were between 53.4 ± 8.3° and 51.3 ± 10.6° (Figure 4), which means that the Ch/It film is more hydrophilic and the drop of water adheres to the surface faster; it is important to notice that the error bars are larger at 2 min, which is attributed to the difficulty of measuring the angle due to deformations on the surface caused by the water droplet absorbed.

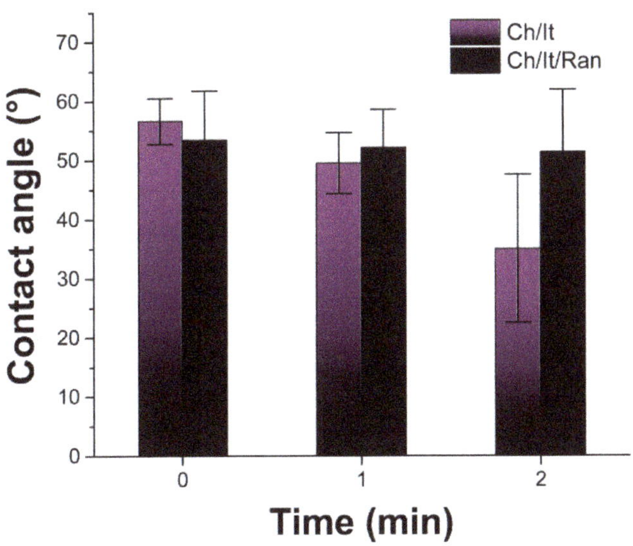

Figure 4. Contact angle of Ch/It and Ch/It/Ran measures between 0–2 min.

Swelling capacity of films was measured by gravimetry (Figure 5). Ch/It film swelled the most (1582 ± 415%), and even reached the fastest maximum swelling in the first minutes, while the Ch/It/Ran film showed a curve with a less pronounced slope in the first minutes and with a lower swelling percentage (1069 ± 372%). Despite the differences in the swelling percentages, in both films this physicochemical behavior is like that found in hydrogels [7,42,43]. Because of this affinity for water, it is assumed that the interactions between the water in the medium and the films is not limited to the surface; water molecules most likely penetrated between the crosslinked polymer chains forming hydrogen bonds, being that this water retained and facilitated a fast saturation (swelling limit) [44]. The wettability and subsequent swelling of the films did not cause their destruction by infinite swelling; this is possible because the crosslinking of the films stops the swelling, since the ionic bond between the carboxylate and the amino groups must be stronger [45] than the hydrogen bond with the surrounding water [46]. However, at a macroscopic level, the film can be broken by rude mechanical action, so films must be carefully handled to keep its integrity.

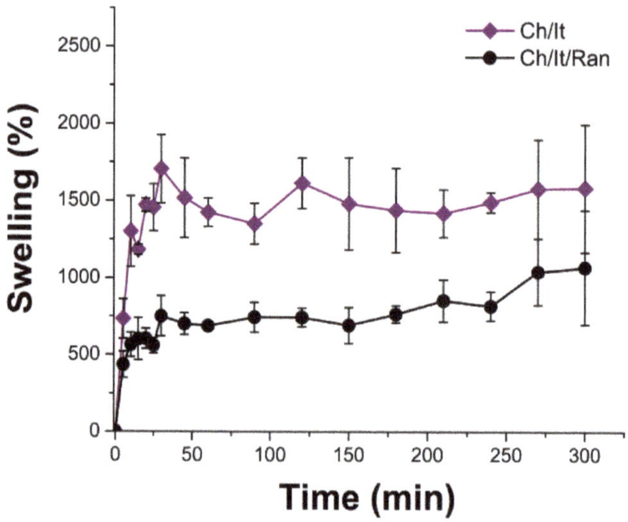

Figure 5. Percentage of water uptake of samples in distilled water.

3.5. Thermal Analysis

Reagents (raw materials) and products were subjected to two types of thermal analysis, DSC and TGA. In general, it was noticed that the thermal properties of the analyzed films versus reagents have a certain relationship, although it is evident that the films are not just a mixture; the formation of ionic bonds between chitosan and itaconic acid through a crosslinking reaction is inferred, since the transitions found in the films correspond to a material with new characteristics.

In the analyzed DSC thermograms, mainly endothermic transitions are observed (Figure 6); in the DSC of chitosan, the largest signal is a very broad and undefined endothermic transition that reaches the maximum energy change at 101.48 °C and is attributed to water loss [6]. For itaconic acid thermogram, a narrow signal that corresponds to the melting temperature (Tm) is observed; this type of signal appears in crystalline organic compounds of high purity. In the thermogram of the *Randia* extract, an endothermic transition of intermediate intensity is observed at 135 °C; in addition, several peaks are observed between 172 and 208 °C (with a midpoint at 191 °C) (Table 2), and these signals are attributed to the large number of organic compounds present in the extract with Tm in this temperature range. Finally, in the DSC thermograms of the Ch/It film, it is not possible to clearly observe transitions, and only a small endothermic transition is observed at 128 °C, which can be attributed to water loss; this lack of signals is indicative of films with a mainly amorphous structure. In the thermogram of the Ch/It/Ran film, two small (but recognizable) endothermic transitions appear at 167 and 210 °C that can be attributed to domains of itaconic acid and *Randia* extract, due to the proximity of the Tm of these compounds.

The information collected from TGA thermograms completed the thermal analysis (Figure 7). It was found that 10% weight loss temperature was higher for chitosan, followed by Ch/It/ and Ch/It/Ran, while the lowest temperatures were for *Randia* extract and itaconic acid, respectively. This trend was also observed in the decomposition temperatures (Td). It should be noted that chitosan and itaconic acid show a single Td, while the Td of the *Randia* extract and the Ch/It/ and Ch/It/Ran films show multi-stage decompositions, due to the components of these mixtures having different resistance to heating. Respecting the analysis at 800 °C, the Ch/It/Ran film was the one that most resisted heating with 54.2% of residue mass, due to the crosslinking with itaconic acid and a structural reinforcement with the *Randia* extract; although chitosan also has a residue of 35.4%, this value is smaller than in the composite. Finally, in both itaconic acid and the *Randia* extract, residues at 800

°C showed very low residual percentages, which corresponds to an organic compound with volatile thermal decomposition, these parameters are indicated in Table 1.

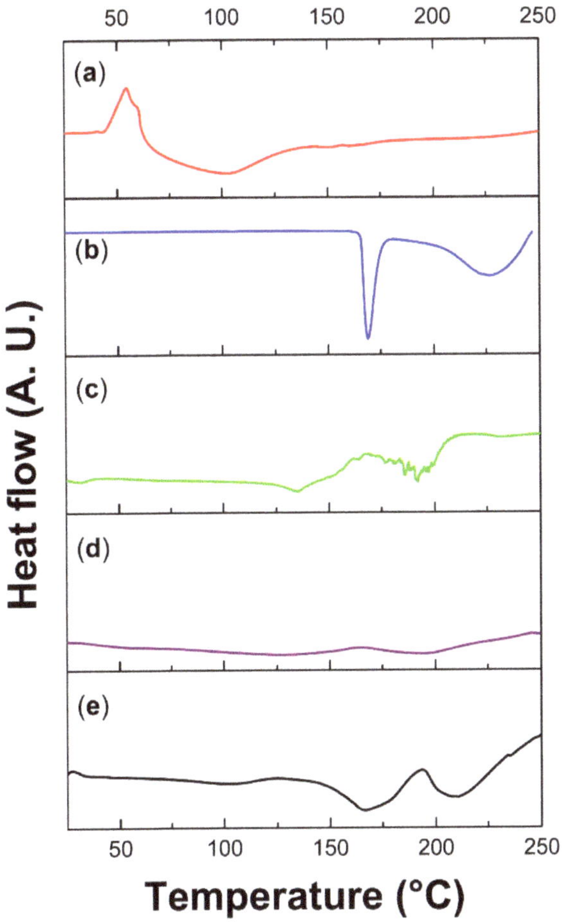

Figure 6. DSC thermograms of samples run under nitrogen atmosphere, up to 250 °C, and heating rate at 10 °C min^{-1}. **(a)** chitosan; **(b)** itaconic acid; **(c)** *Randia* extract; **(d)** Ch/It; **(e)** Ch/It/Ran.

Table 2. Condensed information of DSC and TGA analysis.

Sample	DSC		TGA		
	Transition (°C)	Thermal Transition	10 wt% Loss (°C)	Decomposition Temperature Td (°C)	% Char Yield (800 °C)
Chitosan	101	Endo	275	297	35.4
Itaconic acid	168	Endo	176	203	0.6
Randia extract	135, 191	Endo	176	192, 258	12.5
Ch/It	128	Endo	224	196, 302	42.1
Ch/It/Ran	170, 210	Endo	200.3	136, 185,311	54.2

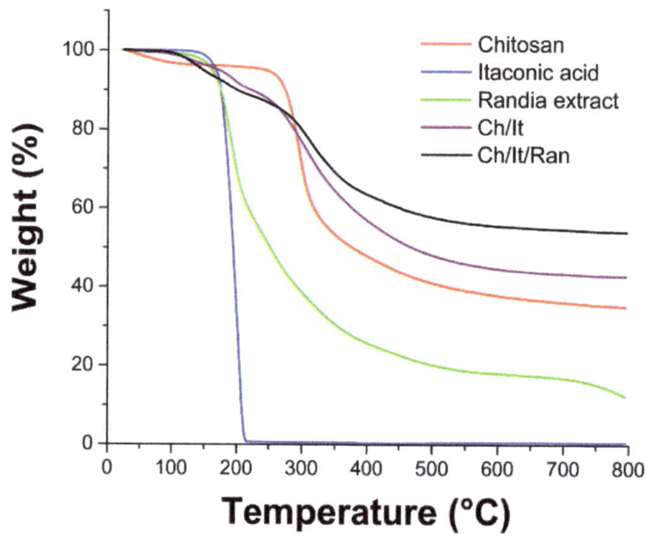

Figure 7. TGA thermograms of samples run at conditions of nitrogen atmosphere, up to 800 °C, and heating rate at 10 °C min^{-1}.

3.6. Cell Viability

Cell viability was performed using the MTT kit by formazan formation. Samples were tested with the healthy cell strain BALB/3T3. In addition to Ch/It and Ch/It/Ran films, the pulverized complete fruit and the *Randia* extract were analyzed. All experimental details can be checked in the corresponding sections of "Materials and Methods". As additional information, the weight percentage (%wt.) of fractions are shown in Table 3.

Table 3. Weight percentage of films Chitosan/Itaconic acid and Chitosan/Itaconic acid/*Randia* extract.

	Chitosan (%wt.)	Itaconic Acid (%wt.)	*Randia* Extract (%wt.)
Ch/It	70	30	-
Ch/It/Ran	55	23	22

Cell viability values obtained in the controls (the medium without the sample) are adjusted to 100% for a proper comparison among samples regarding the effect on cell survival. As expected, Ch/It film was the one that showed the highest cell survival, 84% ± 1.4, since both biomaterials chitosan and itaconic acid are extensively used in the industry [16,47]. The second sample of higher cell survival 77% ± 2.3 was the film with the three components Ch/It/Ran. This slight decrease in cell viability is attributed to the bioactive compounds in the *Randia* extract (22% wt.), even though this sample is also profiled as a biocompatible material since cell survival was sufficient [48]. Lastly, the cell viability percentages for *Randia* complete fruit 12% ± 3.4 and *Randia* extract 9% ± 2.1 were similarly very low. Therefore, it is assumed that 10 mg mL^{-1} exceeds the concentration at which cells can survive in vitro; moreover, both the fruit extract and the complete fruit decrease cell survival in a statistically equivalent percentage, despite the difference in solubility between the extract and the complete fruit (higher solubility and higher amount of bioactive principle in the *Randia* extract).

Based on these preliminary results, it is concluded that the composition of the Ch/It/Ran film produces a hydrogel biocompatible with BALB/3T3, which allows continuing in subsequent stages of analysis (Figure 8). The concentration of bioactive compounds such as flavonoids in the Ch/It/Ran film is acceptable for a biomaterial. As final note, the potential use of this material as a wound dressing can be studied on epithelial regeneration [49], and in the case of satisfactory results, an in vivo study can be proceeded to [50].

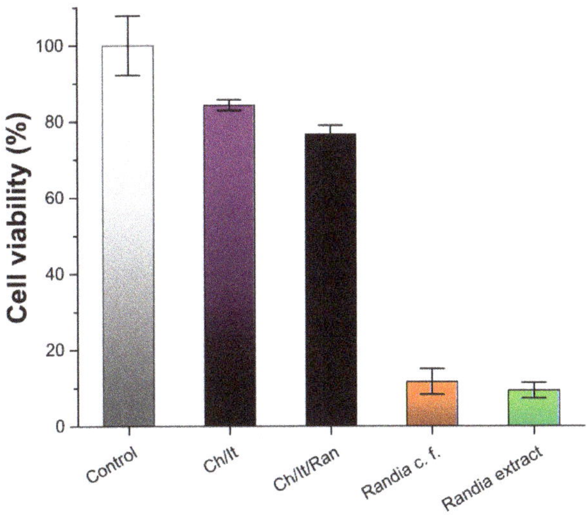

Figure 8. Cell viability (BALB/3T3) of chitosan and itaconic acid film (Ch/It); chitosan, itaconic acid, and *Randia* extract film (Ch/It/Ran); *Randia* complete fruit (*Randia* c. f.); and *Randia* extract.

4. Conclusions

The proposal of ionic crosslinking chitosan provides advantages over other crosslinking methods; firstly, the crosslinking was achieved under reaction conditions free of acetic acid and halogenated crosslinking agents, since itaconic acid enhances crosslinking chitosan, forming a hydrosoluble network. A second benefit is that the reaction system only required water as solvent to form the hydrogel in situ, which is positive from a green approach. Additionally, this methodology can be easily adapted in an industrial process to obtain an interesting material for sanitary dressings because this chitosan-based hydrogel allowed the incorporation of *R. capitata* extract into the network, forming a biomaterial with high cell viability. As a final annotation, we sowed the annals to continue in the study of a Chitosan/Itaconic acid/*Randia* extract system as a biomaterial for medical applications due to the promising bioactive properties of *R. capitata* extract for epithelial wound healing.

Author Contributions: Conceptualization, F.L.-S. and L.B.-G.; methodology, F.L.-S., H.M. and G.G.F.-R.; software, F.L.-S. and G.G.F.-R.; validation, L.B.-G., G.G.F.-R. and E.B.; formal analysis, F.L.-S., L.B.-G. and H.M.; investigation, F.L.-S. and L.B.-G.; resources, L.B.-G. and E.B.; data curation, L.B.-G. and G.G.F.-R.; writing—original draft preparation, F.L.-S.; writing—review and editing, F.L.-S. and L.B.-G.; visualization, F.L.-S. and L.B.-G.; supervision, L.B.-G.; project administration, L.B.-G.; funding acquisition, L.B.-G. and E.B. All authors have read and agreed to the published version of the manuscript.

Funding: This research was partially funded by the Universidad Autónoma del Estado de México grant number 6187/2020CI, and by the Dirección General de Asuntos del Personal Académico (DGAPA), Universidad Nacional Autónoma de México under Grant IN204223.

Institutional Review Board Statement: Not applicable.

Data Availability Statement: The data presented in this study are available on request from the corresponding author.

Acknowledgments: The authors would like to thank Y. Gamboa-Calderón, V. Sánchez-Correa, and E. Palacios from ICN-UNAM for technical assistance. FLS (CVU 409872) thanks to CONAHCyT for scholarship support.

Conflicts of Interest: The authors declare no conflict of interest.

References

1. Birajdar, M.S.; Joo, H.; Koh, W.-G.; Park, H. Natural bio-based monomers for biomedical applications: A review. *Biomater. Res.* **2021**, *25*, 8. [CrossRef] [PubMed]
2. Huang, S.J. Polymer waste management–biodegradation, incineration, and recycling. *J. Macromol. Sci. Part A* **1995**, *32*, 593–597. [CrossRef]
3. Tudor, V.C.; Mocuta, D.N.; Teodorescu, R.F.; Smedescu, D.I. The issue of plastic and microplastic pollution in soil. *Mater. Plast.* **2019**, *56*, 484–487. [CrossRef]
4. Elliott, J.E.; Elliott, K.H. Tracking marine pollution. *Science* **2013**, *340*, 556–558. [CrossRef]
5. Kharrazi, S.M.; Younesi, H.; Abedini-Torghabeh, J. Microbial biodegradation of waste materials for nutrients enrichment and heavy metals removal: An integrated composting-vermicomposting process. *Int. Biodeterior. Biodegrad.* **2014**, *92*, 41–48. [CrossRef]
6. Ferrero, F.; Periolatto, M. Antimicrobial finish of textiles by chitosan UV-curing. *J. Nanosci. Nanotechnol.* **2012**, *12*, 4803–4810. [CrossRef]
7. Peers, S.; Montembault, A.; Ladavière, C. Chitosan hydrogels for sustained drug delivery. *J. Control. Release* **2020**, *326*, 150–163. [CrossRef] [PubMed]
8. Milosavljević, N.B.; Ristić, M.Đ.; Perić-Grujić, A.A.; Filipović, J.M.; Štrbac, S.B.; Rakočević, Z.L.; Kalagasidis Krušić, M.T. Hydrogel based on chitosan, itaconic acid and methacrylic acid as adsorbent of Cd2+ ions from aqueous solution. *Chem. Eng. J.* **2010**, *165*, 554–562. [CrossRef]
9. Yang, Y.; Chen, G.; Murray, P.; Zhang, H. Porous chitosan by crosslinking with tricarboxylic acid and tuneable release. *SN Appl. Sci.* **2020**, *2*, 435. [CrossRef]
10. Noel, S.P.; Courtney, H.; Bumgardner, J.D.; Haggard, W.O. Chitosan films: A potential local drug delivery system for antibiotics. *Clin. Orthop. Relat. Res.* **2008**, *466*, 1377–1382. [CrossRef] [PubMed]
11. Flores-Rojas, G.G.; López-Saucedo, F.; Vera-Graziano, R.; Magaña, H.; Mendizábal, E.; Bucio, E. Silver Nanoparticles Loaded on Polyethylene Terephthalate Films Grafted with Chitosan. *Polymers* **2023**, *15*, 125. [CrossRef] [PubMed]
12. Ortega, A.; Sánchez, A.; Burillo, G. Binary graft of poly(N-vinylcaprolactam) and poly(acrylic acid) onto chitosan hydrogels using ionizing radiation for the retention and controlled release of therapeutic compounds. *Polymers* **2021**, *13*, 2641. [CrossRef]
13. Lukman Hekiem, N.L.; Md Ralib, A.A.; Mohd Hatta, M.A.; Ahmad, F.B.; Nordin, A.N.; Ab Rahim, R.; Za'bah, N.F. Effect of chitosan dissolved in different acetic acid concentration towards VOC sensing performance of quartz crystal microbalance overlay with chitosan. *Mater. Lett.* **2021**, *291*, 129524. [CrossRef]
14. Zhang, H.; Kong, M.; Jiang, Q.; Hu, K.; Ouyang, M.; Zhong, F.; Qin, M.; Zhuang, L.; Wang, G. Chitosan membranes from acetic acid and imidazolium ionic liquids: Effect of imidazolium structure on membrane properties. *J. Mol. Liq.* **2021**, *340*, 117209. [CrossRef]
15. Yang, S.; Liu, L.; Chen, H.; Wei, Y.; Dai, L.; Liu, J.; Yuan, F.; Mao, L.; Li, Z.; Chen, F.; et al. Impact of different crosslinking agents on functional properties of curcumin-loaded gliadin-chitosan composite nanoparticles. *Food Hydrocoll.* **2021**, *112*, 106258. [CrossRef]
16. Wang, J.; Zhuang, S. Chitosan-based materials: Preparation, modification and application. *J. Clean. Prod.* **2022**, *355*, 131825. [CrossRef]
17. Kyzas, G.Z.; Siafaka, P.I.; Lambropoulou, D.A.; Lazaridis, N.K.; Bikiaris, D.N. Poly(itaconic acid)-grafted chitosan adsorbents with different cross-linking for Pb(II) and Cd(II) uptake. *Langmuir* **2014**, *30*, 120–131. [CrossRef] [PubMed]
18. Ko, E.; Kim, H. Preparation of chitosan aerogel crosslinked in chemical and ionical ways by non-acid condition for wound dressing. *Int. J. Biol. Macromol.* **2020**, *164*, 2177–2185. [CrossRef]
19. Sirviö, J.A.; Kantola, A.M.; Komulainen, S.; Filonenko, S. Aqueous modification of chitosan with itaconic acid to produce strong oxygen barrier film. *Biomacromolecules* **2021**, *22*, 2119–2128. [CrossRef]
20. WFO Randia Capitata DC. Available online: http://www.worldfloraonline.org/taxon/wfo-0000294365 (accessed on 27 January 2023).
21. Gallardo-Casas, C.; Guevara-Balcázar, G.; Morales-Ramos, E.; Tadeo-Jiménez, Y.; Gutiérrez-Flores, O.; Jiménez-Sánchez, N.; Valadez-Omaña, M.; Valenzuela-Vargas, M.; Castillo-Hernández, M. Ethnobotanic study of Randia aculeata (Rubiaceae) in Jamapa, Veracruz, Mexico, and its anti-snake venom effects on mouse tissue. *J. Venom. Anim. Toxins Incl. Trop. Dis.* **2012**, *18*, 287–294. [CrossRef]
22. Cuevas-Juárez, E.; Yuriar-Arredondo, K.Y.; Pío-León, J.F.; Montes-Avila, J.; López-Angulo, G.; Páz Díaz-Camacho, S.; Delgado-Vargas, F. Antioxidant and α-glucosidase inhibitory properties of soluble melanins from the fruits of Vitex mollis Kunth, Randia echinocarpa Sessé et Mociño and Crescentia alata Kunth. *J. Funct. Foods* **2014**, *9*, 78–88. [CrossRef]
23. Montes-Avila, J.; Ojeda-Ayala, M.; López-Angulo, G.; Pío-León, J.F.; Díaz-Camacho, S.P.; Ochoa-Terán, A.; Delgado-Vargas, F. Physicochemical properties and biological activities of melanins from the black-edible fruits Vitex mollis and Randia echinocarpa. *J. Food Meas. Charact.* **2018**, *12*, 1972–1980. [CrossRef]
24. Kumar, D.; Mudgade, S.C.; Bhat, Z.A.; Bhujbal, S.S.; Rub, R. Anti allergic and anti-inflammatory effects of the fruits of Randia dumetorum Lamk. *Orient. Pharm. Exp. Med.* **2011**, *11*, 161–167. [CrossRef]
25. Kandimalla, R.; Kalita, S.; Saikia, B.; Choudhury, B.; Singh, Y.P.; Kalita, K.; Dash, S.; Kotoky, J. Antioxidant and hepatoprotective potentiality of Randia dumetorum Lam. Leaf and bark via inhibition of oxidative stress and inflammatory cytokines. *Front. Pharmacol.* **2016**, *7*, 205. [CrossRef] [PubMed]

26. Martínez-Barbosa, M.E.; Moreno-Corral, R.A. Washable, reusable and disposable medical textiles. In *Medical Textiles from Natural Resources*; Elsevier: Amsterdam, The Netherlands, 2022; pp. 717–765.
27. Jiménez Ortega, L.A.; Barrientos Ramírez, L.; Tena Meza, M.P. Evaluación Fisicoquímica y fitoquímica de frutos de sapuche (*Randia laevigata* Standl.). *e-CUCBA* **2020**, *13*, 30–39. [CrossRef]
28. Santos-Cervantes, M.E.; Ibarra-Zazueta, M.E.; Loarca-Piña, G.; Paredes-López, O.; Delgado-Vargas, F. Antioxidant and antimutagenic activities of Randia echinocarpa fruit. *Plant Foods Hum. Nutr.* **2007**, *62*, 71–77. [CrossRef]
29. Bezerra da Cruz-Silva, S.C.; Matias, R.; Bono, J.A.M.; Santos, K.S.; Ludwig, J. Antifungal potential of extracts and fractions of Randia nitida leaves on soybean pathogens and their phytochemistry. *Rev. Caatinga* **2016**, *29*, 594–602. [CrossRef]
30. Scigelova, M.; Hornshaw, M.; Giannakopulos, A.; Makarov, A. Fourier transform mass spectrometry. *Mol. Cell. Proteom.* **2011**, *10*, M111.009431. [CrossRef]
31. Fernandes Queiroz, M.; Melo, K.; Sabry, D.; Sassaki, G.; Rocha, H. Does the use of chitosan contribute to oxalate kidney stone formation? *Mar. Drugs* **2014**, *13*, 141–158. [CrossRef]
32. Cozzolino, D. Near infrared spectroscopy in natural products analysis. *Planta Med.* **2009**, *75*, 746–756. [CrossRef]
33. Silverstein, R.M.; Webster, F.X.; Kiemle, D.J. *Spectrometric Identification of Organic Compounds*, 7th ed.; John Wiley & Sons: Hoboken, NJ, USA, 2005; ISBN 0471393622.
34. Baranović, G.; Šegota, S. Infrared spectroscopy of flavones and flavonols. Reexamination of the hydroxyl and carbonyl vibrations in relation to the interactions of flavonoids with membrane lipids. *Spectrochim. Acta Part A Mol. Biomol. Spectrosc.* **2018**, *192*, 473–486. [CrossRef]
35. Kim, L.-S.; Hong, S.-J.; Son, M.-K.; Lee, Y.-H. Polymeric and compositional properties of novel extracellular microbial polyglucosamine biopolymer from new strain of *citrobacter* sp. BL-4. *Biotechnol. Lett.* **2006**, *28*, 241–245. [CrossRef] [PubMed]
36. Farahani, B.V.; Ghasemzadeh, H.; Afraz, S. Thermodynamic studies of insulin loading into a glucose responsive hydrogel based on chitosan-polyacrylamide-polyethylene glycol. *J. Chin. Chem. Soc.* **2016**, *63*, 438–444. [CrossRef]
37. Qi, P.; Chen, H.-L.; Nguyen, H.T.H.; Lin, C.-C.; Miller, S.A. Synthesis of biorenewable and water-degradable polylactam esters from itaconic acid. *Green Chem.* **2016**, *18*, 4170–4175. [CrossRef]
38. Milosavljević, N.B.; Kljajević, L.M.; Popović, I.G.; Filipović, J.M.; Kalagasidis Krušić, M.T. Chitosan, itaconic acid and poly(vinyl alcohol) hybrid polymer networks of high degree of swelling and good mechanical strength. *Polym. Int.* **2010**, *59*, 686–694. [CrossRef]
39. Pérez-Calderón, J.; Santos, M.V.; Zaritzky, N. Synthesis, characterization and application of cross-linked chitosan/oxalic acid hydrogels to improve azo dye (Reactive Red 195) adsorption. *React. Funct. Polym.* **2020**, *155*, 104699. [CrossRef]
40. Gabriele, F.; Donnadio, A.; Casciola, M.; Germani, R.; Spreti, N. Ionic and covalent crosslinking in chitosan-succinic acid membranes: Effect on physicochemical properties. *Carbohydr. Polym.* **2021**, *251*, 117106. [CrossRef] [PubMed]
41. Lee, L.-H. Roles of molecular interactions in adhesion, adsorption, contact angle and wettability. *J. Adhes. Sci. Technol.* **1993**, *7*, 583–634. [CrossRef]
42. Malkin, A.; Ilyin, S.; Roumyantseva, T.; Kulichikhin, V. Rheological evidence of gel formation in dilute poly(acrylonitrile) solutions. *Macromolecules* **2013**, *46*, 257–266. [CrossRef]
43. Kulichikhin, V.G.; Ilyin, S.O.; Mironova, M.V.; Berkovich, A.K.; Nifant'ev, I.E.; Malkin, A.Y. From polyacrylonitrile, its solutions, and filaments to carbon fibers: I. Phase state and rheology of basic polymers and their solutions. *Adv. Polym. Technol.* **2018**, *37*, 1076–1084. [CrossRef]
44. Rizwan, M.; Yahya, R.; Hassan, A.; Yar, M.; Azzahari, A.D.; Selvanathan, V.; Sonsudin, F.; Abouloula, C.N. pH sensitive hydrogels in drug delivery: Brief history, properties, swelling, and release mechanism, material selection and applications. *Polymers* **2017**, *9*, 137. [CrossRef]
45. Blanco, Y.S.; Topel, Ö.; Bajnóczi, É.G.; Werner, J.; Björneholm, O.; Persson, I. Chemical equilibria of aqueous ammonium–carboxylate systems in aqueous bulk, close to and at the water–air interface. *Phys. Chem. Chem. Phys.* **2019**, *21*, 12434–12445. [CrossRef] [PubMed]
46. Shan, S.; Herschlag, D. The change in hydrogen bond strength accompanying charge rearrangement: Implications for enzymatic catalysis. *Proc. Natl. Acad. Sci. USA* **1996**, *93*, 14474–14479. [CrossRef]
47. Cunha da Cruz, J.; Machado de Castro, A.; Camporese Sérvulo, E.F. World market and biotechnological production of itaconic acid. *3 Biotech.* **2018**, *8*, 138. [CrossRef] [PubMed]
48. Damiani, E.; Solorio, J.A.; Doyle, A.P.; Wallace, H.M. How reliable are in vitro IC50 values? Values vary with cytotoxicity assays in human glioblastoma cells. *Toxicol. Lett.* **2019**, *302*, 28–34. [CrossRef]
49. Leoni, G.; Neumann, P.-A.; Sumagin, R.; Denning, T.L.; Nusrat, A. Wound repair: Role of immune–epithelial interactions. *Mucosal Immunol.* **2015**, *8*, 959–968. [CrossRef]
50. Coraux, C.; Hajj, R.; Lesimple, P.; Puchelle, E. In vivo models of human airway epithelium repair and regeneration. *Eur. Respir. Rev.* **2005**, *14*, 131–136. [CrossRef]

polymers

MDPI

Article

Biodegradability Assessment of Prickly Pear Waste–Polymer Fibers under Soil Composting

Zormy Nacary Correa-Pacheco [1,*], Silvia Bautista-Baños [1], José Jesús Benítez-Jiménez [2], Pedro Ortega-Gudiño [3,*], Erick Omar Cisneros-López [3] and Mónica Hernández-López [1]

[1] Centro de Desarrollo de Productos Bióticos, Instituto Politécnico Nacional, Carretera Yautepec-Jojutla, Km. 6, Calle CEPROBI, No. 8, San Isidro, Yautepec 62731, Morelos, Mexico; sbautis@ipn.mx (S.B.-B.); mohernandezl@ipn.mx (M.H.-L.)

[2] Instituto de Ciencia de Materiales de Sevilla, CSIC-Universidad de Sevilla, Avda. Américo Vespucio 49, Isla de la Cartuja, 41092 Sevilla, Spain; benitez@icmse.csic.es

[3] Departamento de Ingeniería Química, Centro Universitario de Ciencias Exactas e Ingenierías, Universidad de Guadalajara, Blvd. Gral. Marcelino García Barragán #1451, Guadalajara 44430, Jalisco, Mexico; erick.cisneros@academicos.udg.mx

* Correspondence: zcorreap@ipn.mx (Z.N.C.-P.); pedro.ogudino@academicos.udg.mx (P.O.-G.)

Abstract: Nowadays, solving the problems associated with environmental pollution is of special interest. Therefore, in this work, the morphology and thermal and mechanical properties of extruded fibers based on polylactic acid (PLA) and poly(butylene adipate-co-terephthalate) (PBAT) added to prickly pear flour (PPF) under composting for 3 and 6 months were evaluated. The highest weight loss percentage (92 ± 7%) was obtained after 6-month degradation of the PLA/PBAT/PPF/CO/AA blend, in which PPF, canola oil (CO), and adipic acid (AA) were added. Optical and scanning electron microscopy (SEM) revealed structural changes in the fibers as composting time increased. The main changes in the absorption bands observed by Fourier transform infrared spectroscopy (FTIR) were related to the decrease in -C=O (1740 cm^{-1}) and -C-O (1100 cm^{-1}) groups and at 1269 cm^{-1}, associated with hemicellulose in the blends with PPF. Differential scanning calorimetry (DSC) showed an increase in the cold crystallization and melting point with degradation time, being more evident in the fibers with PPF, as well as a decrease in the mechanical properties, especially Young's modulus. The obtained results suggest that PPF residues could promote the biodegradability of PLA/PBAT-based fiber composites.

Keywords: degradation time; morphology; thermal properties; mechanical properties; weight loss

1. Introduction

Currently, compatibilized polymer blends based on biodegradable polymers added to recycled agriculture crop residues have gained special attention in polymer applications [1,2]. In packaging applications, PBAT poly (butylene adipate-co-terephthalate) is considered a good candidate to improve the toughness of PLA (polylactic acid) because the later has a low heat-resistance capacity, which limits its usefulness in food packaging. Therefore, PLA and PBAT complement each other. However, the PLA/PBAT blend is considered immiscible; therefore, having good interface compatibility between PLA and PBAT is essential to obtaining a blend with satisfactory properties. Both are biodegradable polymers that have recently become popular due to their potential application in the packaging field [3,4].

Opuntia ficus-indica, commonly known as prickly pear, belongs to the Cactaceae family and consists of four main parts: cladodes (succulent stems), flowers, fruit, and seeds. This is a plant that easily adapts to the environment. The cladodes contain mucilage and cellulosic fibers (80–95%), carbohydrates (3–7%), fibers (1–2%), and small amounts of protein (0. 5–1%) [5,6].

Scaffaro et al. [7] obtained a lignocellulosic flour by grinding *O. ficus-indica* cladodes to prepare extruded biocomposites with PLA. They found a notable increase in stiffness, up to 135% of the polymer matrix, and, at the same time, a decrease in tensile strength, elongation at break, and toughness. Regarding the size of *O. ficus-indica,* it was found that the larger the size of the filling, the greater the stiffness and fragility due to the contact area between the filler and the polymer matrix. On the other hand, Spiridon et al. [8] studied the effect of the incorporation of lignin microparticles into PLA by extrusion. From the morphology, a homogeneous distribution of lignin into the PLA matrix without holes, cracks, and aggregations was observed by scanning electron microscopy (SEM). Young's modulus increased and the tensile strength decreased with the addition of 7% lignin compared to the PLA polymer matrix. When 15% lignin was incorporated into the PLA matrix, there was an increase in the tensile strength; therefore, the lignin content was an important factor in controlling the mechanical properties of the composites.

The use of compatibilizers such as adipic acid (AA) incorporated into blown-film polymer blends has been evaluated by Kim et al. [9]. They found that AA improves the flexibility of PLA and its compatibility with PBAT, resulting in enhanced mechanical properties. Differential scanning calorimetry (DSC) revealed that AA exerted a plasticizing effect, lowering the transition peaks with the presence of one melting peak. SEM revealed phase-separated structures and voids. However, when the AA content increased, a continuous phase morphology was seen due to the improvement in the interfacial adhesion. The mechanical properties showed a decrease in the tensile strength and an increase in the elongation at break with the addition of AA.

Studying the biodegradability of polymer blends combined with additives and agricultural waste is necessary due to their application as compostable materials. Borelbach et al. [10] studied the degradation of biodegradable polymer fibers from PLA under a soil environment at 23 °C for 12 weeks. After this time, no degradation was observed. After 2 weeks, the tensile strength decreased. However, after 8 weeks it increased. Fourier transform infrared spectroscopy (FTIR) and DSC showed no changes over the degradation time. Ramle et al. [11] employed cellulose from bamboo as a reinforced filler in PLA/PBAT blends, studying their degradation in soil. The cellulose content in the prepared films was 0%, 3%, 6%, and 9%. The weight loss was evaluated, and the results showed a value of 12.39% for the 9% PLA/cellulose film, followed by the PLA/PBAT/cellulose blend at the same concentration (9.69%). Moreover, a higher cellulose content enhanced the biodegradability after a period of 15 days. Rebelo et al. [12] simulated an outdoor accelerated weathering environment for a commercial PLA/PBAT blend (BioFlex®) with different organic acids (acetic and its derivatives). They reported that the organic acids accelerated the blend´s degradation. The results of FTIR before and after degradation, after 200 h by UV, did not show significant differences. Only small changes in the band at 1741 cm^{-1} corresponding to C-O stretching were observed. SEM revealed cracks and holes after degradation. DSC before and after degradation showed similar events. However, in the presence of the acetic and its derivatives, some chain scission occurred in the polymer blend with small crystalline peaks due to oligomer formation. In terms of mechanical performance, the tensile strength and elongation at break decreased in the PLA/PBAT blend without additives.

The aim of this work was to study the biodegradation process in soil of PLA/PBAT polymer fibers with the addition of prickly pear flour as agro-industrial waste, adipic acid as a compatibilizer, and canola oil as a plasticizer.

2. Materials and Methods

2.1. Materials

To make the fibers, the PLA, Biopolymer 7001D (Ingeo™, NatureWorks® LLC, Minnesota, MN, USA), with a specific gravity of 1.24 g cm^{-3} and a melt flow index of 6 g/10 min (210 °C, 2.16 kg), and the PBAT, Ecoflex® F Blend C1200 (BASF SE, Ludwigshafen, Germany), with a specific gravity of 1.25–1.27 g cm^{-3} and a melt flow index of 2.7–4.9 g/10 min (190 °C,

2.16 kg), were used. The neat polymer pellets were dried at 60 °C for 24 h in a conventional oven before use. A prickly pear flour (PPF) from waste cladodes collected in the municipality of Tlalnepantla, Morelos, Mexico, was incorporated into the extruder by an emulsion that contained canola oil (CO) (Valley Foods®, Monterrey, Mexico) and Tween 80 (Hycel®, Zapopan, Jal., Zapopan, Mexico). Adipic acid (AA) (Meyer®, Mexico City, Mexico) was used as a compatibilizer.

2.2. Methods

2.2.1. PPF Preparation

To obtain the PPF, the methodology of Scaffaro et al. [7] was followed with some modifications. First, the cladodes were rinsed with potable water to reduce surface contamination and cut into 2 × 2 cm pieces. Then, they were oven-dried (Binder® FD23ULE2, Tuttlingen, Germany) at 60 °C for 24 h. After that, the sample was ground with a blender (Osterizer® 465-43, Mexico city, Mexico) for 1 min. Then, the flour was passed through a 50-mesh sieve (100 μm) and sealed and stored at room temperature.

2.2.2. Fiber Preparation

Fibers were prepared using a twin-screw extruder (Process 11, Thermo Scientific™, Waltham, MA, USA). The temperature profile was 160/170/170/180/180/190/190/160 °C from the feeding zone to the die. The outlet nozzle was 2.5 mm. The fibers were cooled in water. The composition of the polymer blend was PLA/PBAT 60/40 based on previous studies [13]. To add the compatibilizer and the plasticizers to the polymer blend in the extruder, an emulsion based on Tween 80 (0.5%), PPF (3%), CO (4%), and AA (7%) was prepared considering two temperatures, 27 ± 2 °C and 60 ± 2 °C, which was added to the extruder using a peristaltic pump (MasterFlex C/L, Cole-Parmer, Vernon Hills, IL, USA). The feeding speed was 0.15 mL/min. In total, six fibers were extruded: neat PLA and PBAT, PLA/PBAT, PLA/PBAT/CO, PLA/PBAT/PPF/CO/AA without emulsion heating or temperature (PLA/PBAT/PPF/CO/AA) and without emulsion heating (PLA/PBAT/PPF/CO/AA_T). The fiber composition is explained in Table 1.

Table 1. Composition of the biodegradable fibers.

Fiber	Polymer (%)	CO (%)	PPF (%)	AA (%)
PLA	100	0	0	0
PBAT	100	0	0	0
PLA/PBAT	100	0	0	0
PLA/PBAT/CO	96	4	0	0
PLA/PBAT/PPF/CO/AA	86	4	3	7
PLA/PBAT/PPF/CO/AA_T	86	4	3	7

2.2.3. Biodegradability Tests

The biodegradability of the fibers knitted as nets in soil was assessed using the methodology of López et al. [14] with some modifications. A blend of litter/soil/donkey manure in a proportion of 6/2/1 (v/v) was allowed to settle for 30 days with constant watering. Then, the nets were buried at a depth of 5 cm from the surface. The composting time was 3 and 6 months. After that, the nets were removed from the soil, cleaned, and weighed using a portable digital scale (OHAUS, Parsippany, NJ, USA) to determine the percentage of weight loss. This variable was calculated between the initial and the final net weight after composting.

2.2.4. Optical and SEM Observations

The morphology of the fibers from the nets was observed throughout the composting process using a zoom lens (Edmund Optics, VZM 200i, York, UK) and an optical microscope (Nikon, Optiphot2, Tokyo, Japan), both coupled to a digital camera. Higher-resolution imaging was performed using a SEM (Hitachi, S-4800, Tokyo, Japan) operated at 15 kV. In this latter case, the fibers were previously coated with a thin layer of gold.

2.2.5. Fourier Transform Infrared Spectroscopy–Attenuated Total Reflection (FTIR-ATR)

The ATR spectra of the neat fiber polymers and blends before and after composting were recorded using an FT-IR spectrometer (Thermo Nicolet, IS50, Minneapolis, MN, USA) equipped with an attenuated total reflection accessory (Thermo Nicolet, Smart Performer, USA) with a ZnSe single reflection crystal. A resolution of 4 cm^{-1} was used and 32 scans within the range of 4000–650 cm^{-1} were accumulated.

2.2.6. Differential Scanning Calorimetry (DSC)

The DSC measurements were carried out using a DSC-Q20 (TA Instruments, Wakefield, MA, USA) before and after composting. The instrument was calibrated with an indium standard and non-hermetic aluminum crucibles were used.

2.2.7. Mechanical Properties

The uniaxial tensile tests of the fibers were conducted with a universal machine (MTS, Criterion C42, Huntsville, AL, USA) equipped with a 50 N load cell (MTS, LSB.501, USA) using an initial crosshead distance of 10 mm and a deformation rate of 5 mm/min.

2.2.8. Statistical Analysis

To establish the differences between the formulations, a one-way analysis of variance with Tukey's mean comparison test ($p \leq 0.05$) was performed. InfoStat software, version 2020 (National University of Córdoba, Córdoba, Argentina) was used.

3. Results and Discussion

3.1. Biodegradability Tests

According to the results, the highest weight loss was obtained by the PLA/PBAT/PPF/CO/AA with and without temperature after 6 months. For neat polymers, after 3 months, PLA showed a higher weight loss compared to PBAT. This could be attributed to the ease of degradation of the aliphatic backbone of PLA compared to the aromatic component in the backbone of PBAT [15]. This behavior was reported by Carbonell-Verdu et al. [16] for PLA in PLA/PBAT films. However, the final value (6 months) was almost the same for both polymers. PLA and PBAT degradation efficiency is related to the difference in polymer structure and the action of microorganisms in the soil [17]. Some authors have demonstrated for PLA/PBAT blends that, during composting, there is a selectivity to degrade the PLA phase with the formation of a porous 3D network [18] as will be discussed in the Section 3.2. In the literature, it has been reported that the lower weight loss of PBAT can be attributed to the presence of benzene rings [19]. During degradation, microorganisms attack the ester bond in PLA and PBAT, with the generation of small molecules or oligomers, which are subsequently metabolized by these microorganisms [17]. The 60/40 PLA/PBAT blend has a higher composition of PLA than PBAT and more ester bonds are available, favoring microorganism attacks, as observed in the higher weight loss of the blend compared to the neat polymers. On the other hand, weight loss was higher for the blends with and without canola oil compared to the neat polymers after the 3 and 6 months. Significant statistical differences ($p < 0.05$) were found between the fibers. The weight loss of the fibers after 3 and 6 months of composting is shown in Table 2.

According to ASTM 6400 [20] for an aerobically composted plastic, 90% of film degradation must be achieved in 180 days (6 months). According to the results, the PLA/PBAT/PPF/CO/AA with and without temperature reached this percentage (92 $\pm$ 7 and 87 $\pm$ 11%, respectively). In a previous study with PLA/PBAT nets with and without cinnamon essential oil (CEO), weight loss of 15, 7, 3, and 5% was observed for PLA, PBAT, PLA/PBAT 60/40, and PLA/PBAT/CEO, respectively, with 6.1% of CEO after 3 months of composting. At the end of the composting time (6 months) these percentages were 15,

5, 4, and 6% for PLA, PBAT, PLA/PBAT 60/40, and PLA/PBAT/CEO, respectively. After 3 months, the values were similar for neat polymers (PLA and PBAT) [21]. However, the weight loss values for PLA/PBAT were higher in this work. Among the factors that affect composting are aeration, moisture, nutrients, temperature, lignin content, polyphenols, and pH [22]. These factors could influence the kinetics of aerobic composting [23].

Table 2. Weight loss (%) of the fibers after 3 and 6 months of composting in soil.

	Weight Loss Percentage (%) *	
Fiber	**3 Months**	**6 Months**
PLA	20 ± 1^{B}	45 ± 13^{A}
PBAT	8 ± 1^{A}	49 ± 12^{A}
PLA/PBAT	44 ± 7^{C}	65 ± 11^{B}
PLA/PBAT/CO	25 ± 1^{B}	64 ± 3^{A}
PLA/PBAT/PPF/CO/AA	62 ± 4^{D}	92 ± 7^{C}
PLA/PBAT/PPF/CO/AA_T	56 ± 9^{D}	87 ± 11^{BC}

* Different letters represent statistical differences between the fibers by Tukey ´s test ($p < 0.05$).

Ruggero et al. [24] reported a weight loss of 8% and 3% for pure PBAT and PLA films, respectively, under soil and composting conditions for 60 days. The PBAT value was similar to that found in this work after 3 months. However, the authors mentioned that the degradation of PLA increased depending on the thermophilic and mesophilic conditions, and on film thickness.

Moreover, depending on the biopolymer, lower or faster degradation can be found in soil. For example, starch degradation is faster than PLA because PLA maintains its weight for a period of 3 weeks [25]. On the other hand, a higher biodegradation rate between 80–100% has been found for cellulose-based biopolymers after 3 months [26,27], similar to the values obtained in this work for the PPF that contained the fibers. The size and structure of the lignocellulosic filler, such as PPF, influenced the degradation process. It has been reported that for pure cellulose, more accessibility points and microorganism attacks were created due to moisture and adhesion to the polymer [28].

3.2. Optical and SEM Observations

According to the results, the initial fibers present a smooth surface. However, after 3 months of degradation in soil, the PLA surface became rough. Dark spots could be due to microorganism colonies or residual soil/compost. As composting time increased, a color change to yellowish-brown was observed in the fibers. After 6 months, cracks and a broken fiber surface were observed as a result of biodegradation [29]. Higher fiber disintegration was observed in the PLA/PBAT/CO fiber after 3 months of decomposition. As the degradation progressed, a loss of transparency of the fibers was observed due to possible loss of crystallinity associated with PLA component hydrolysis and sample biodeterioration, as reported by Brdlík et al. [30] and Carbonell-Verdu et al. [15]. PLA/PBAT/PPF/CO/AA fibers with and without temperature showed similar degradation behavior. In Table 3, the optical surface micrographs of the fibers before and after 3 and 6 months of degradation are shown.

The SEM results (Tables 4 and 5) revealed differences between the initial and degraded fibers. Before degradation, the neat PLA and PBAT fibers showed a smooth and homogeneous surface. As for the blends, a rougher surface can be seen. The appearance of pores or cavities is observed after degradation in all of the fibers due to physicochemical degradation and biological attack by microorganisms [10]. In PLA it has been found that during chain scission in degradation, lactic acid oligomers are formed, thereby increasing the concentration and catalytic action of carboxylic acid end groups [31]. For PBAT, chain scission of ester groups and the reaction of carboxylic groups with water in the surrounding medium occurs [32].

Table 3. Optical surface micrographs of the fibers initially and after 3 and 6 months of composting in soil.

Fiber	Initial	3 Months	6 Months
PLA	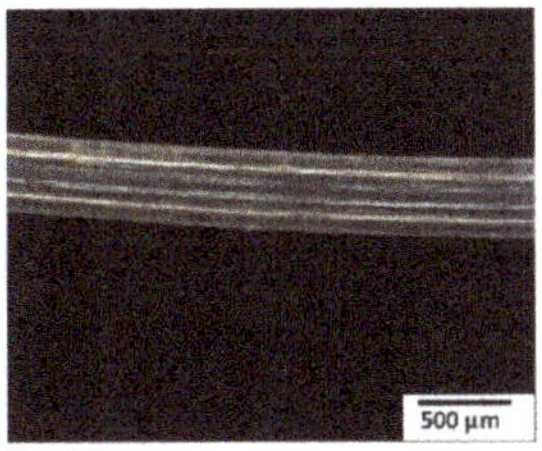	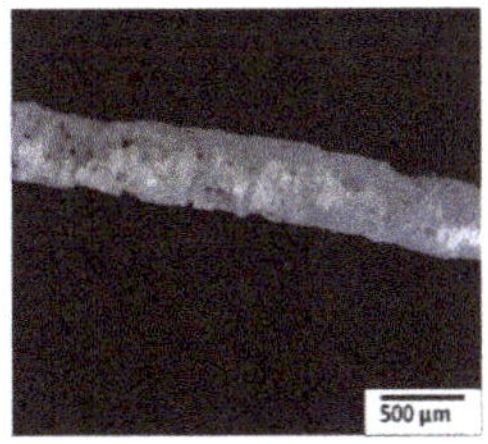	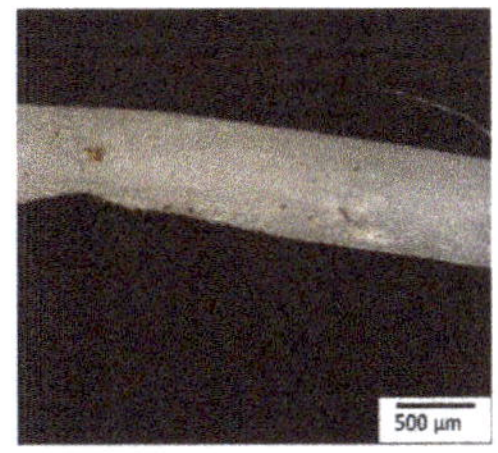
PBAT	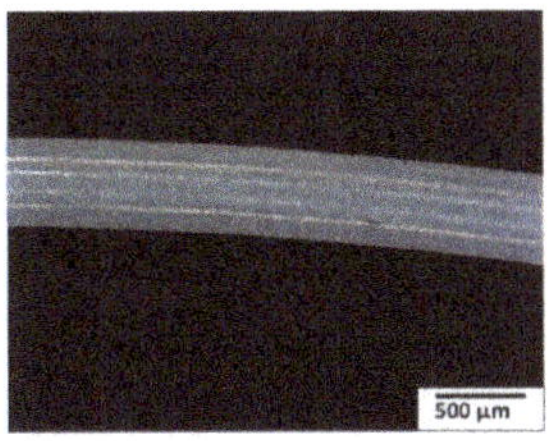	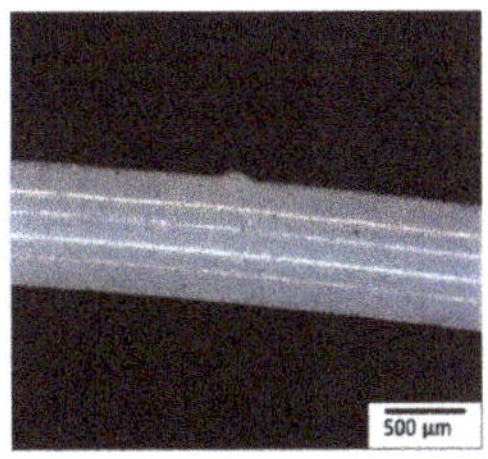	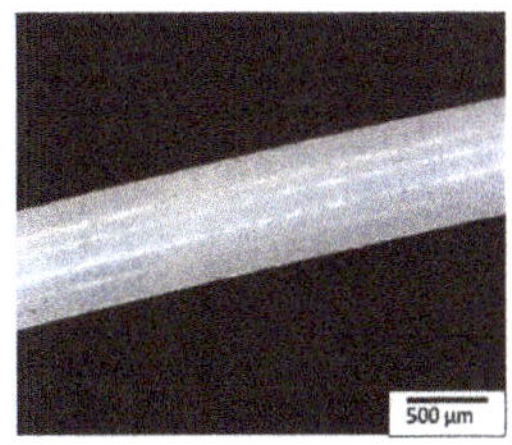
PLA/PBAT	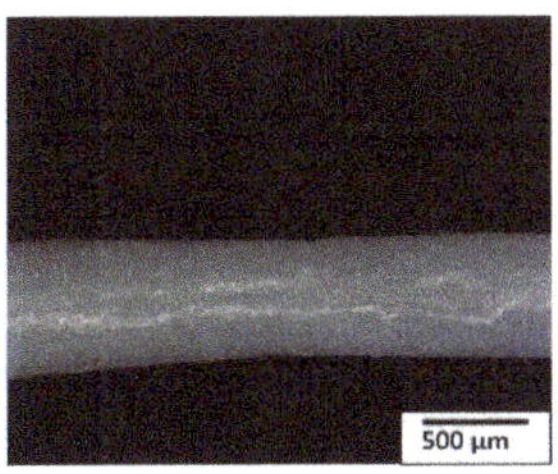	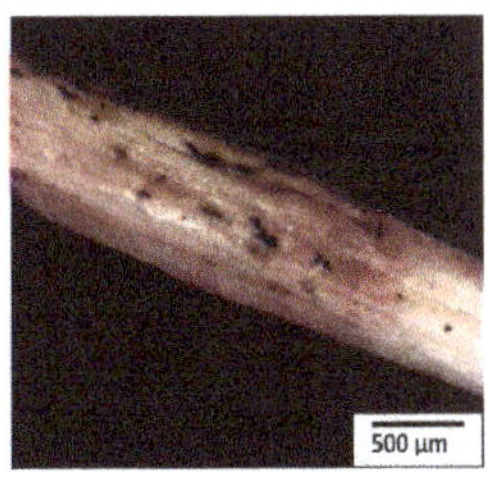	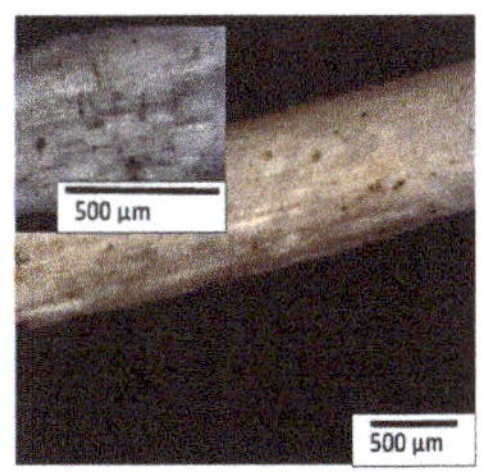
PLA/PBAT/CO	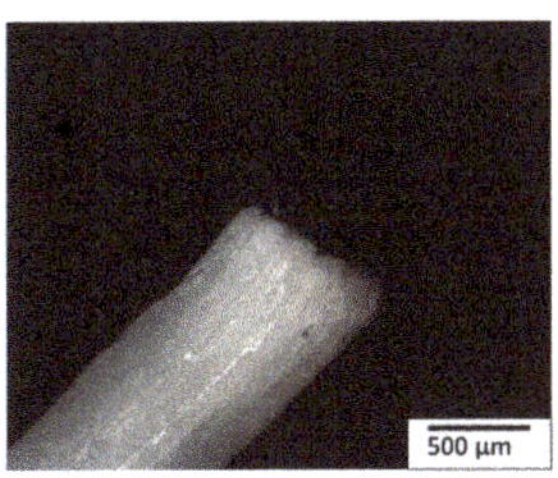	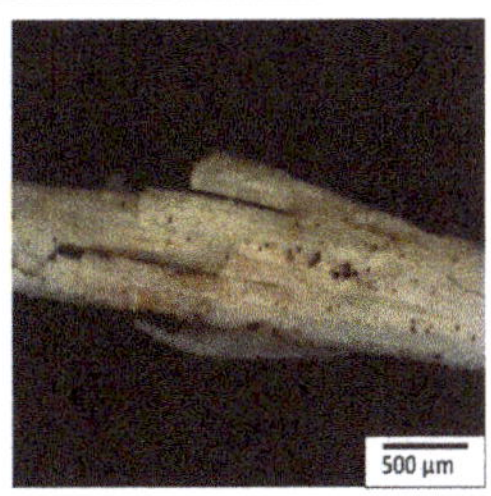	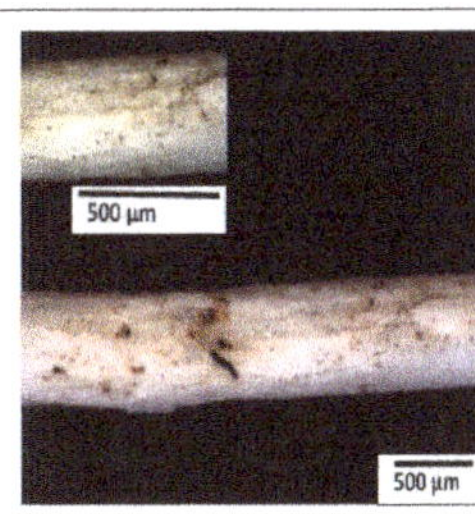
PLA/PBAT/PPF/CO/AA		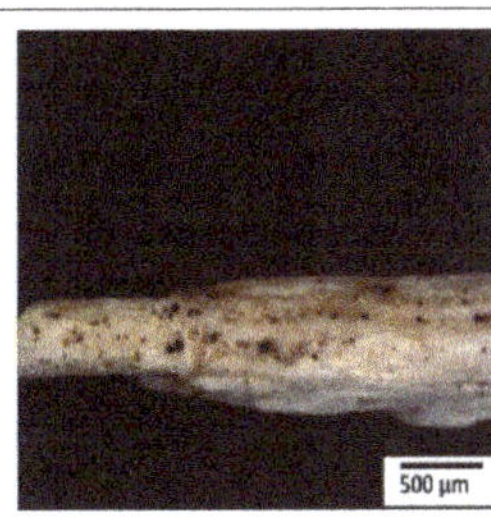	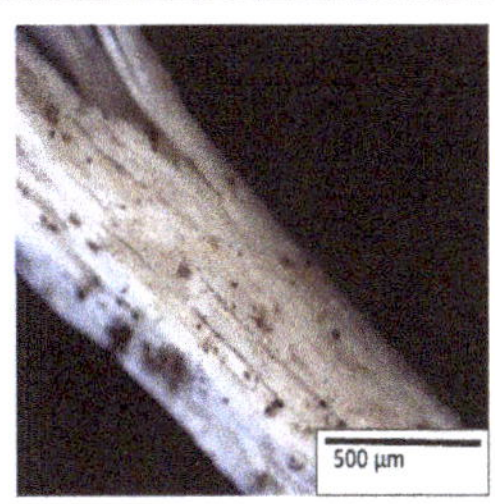

Table 3. *Cont.*

Fiber	Initial	3 Months	6 Months
PLA/PBAT/PPF/CO/AA_T	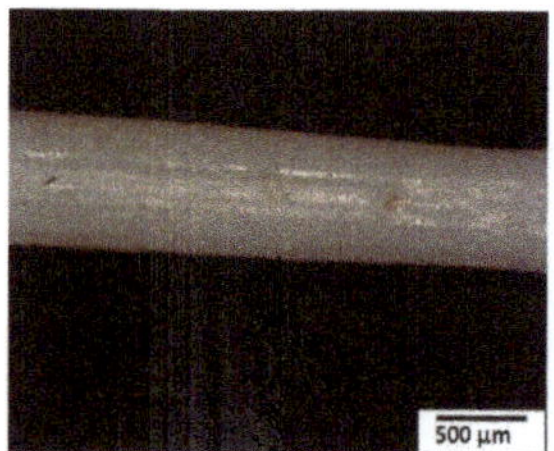	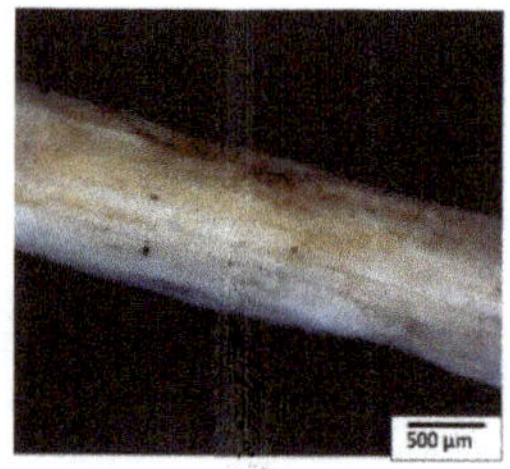	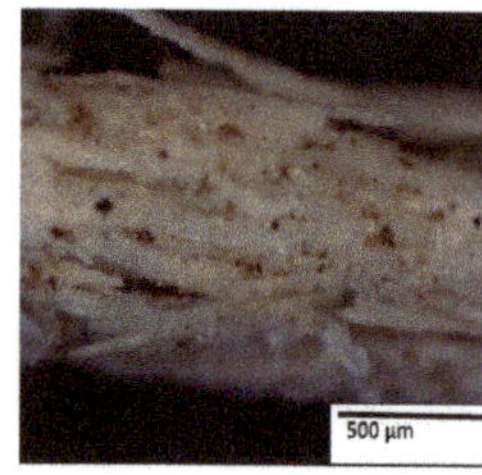

Table 4. SEM micrographs of the fiber's surface morphology initially and after 3 and 6 months of composting in soil (magnification ×100).

Fiber	Initial	3 Months	6 Months
PLA	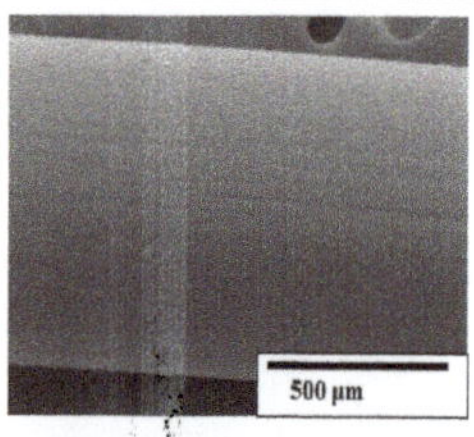		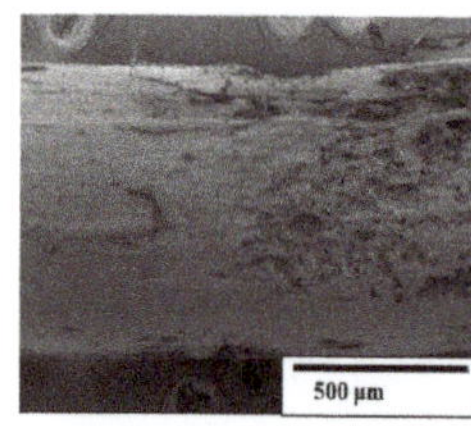
PBAT	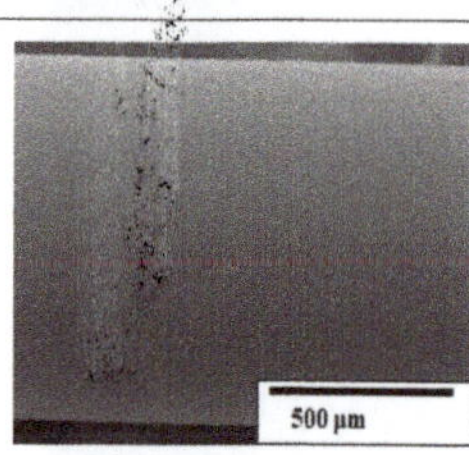	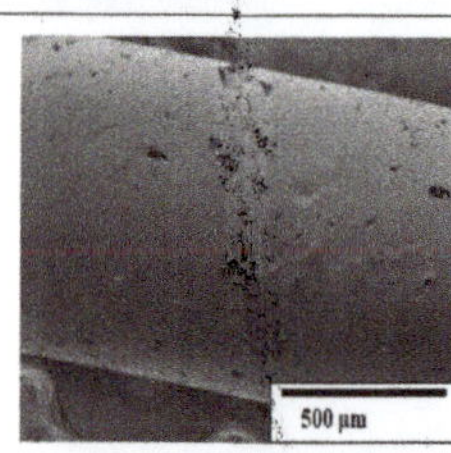	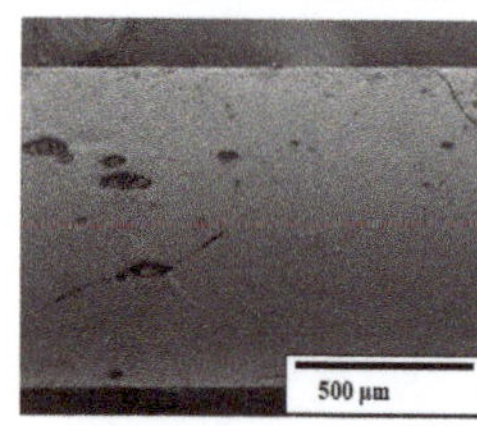
PLA/PBAT	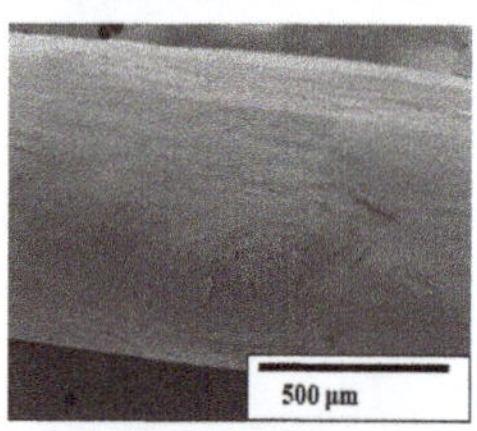		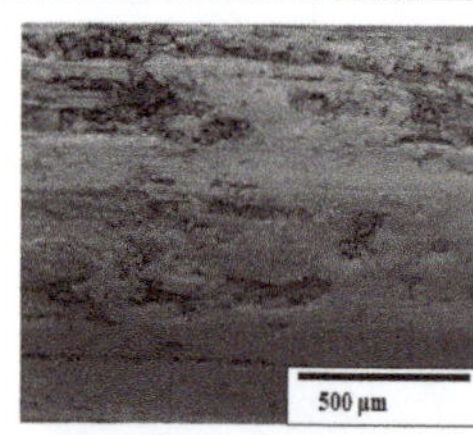
PLA/PBAT/CO	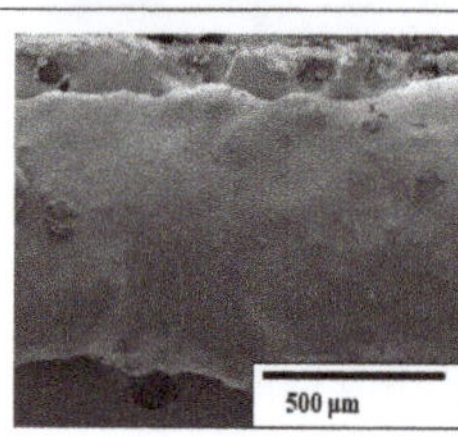	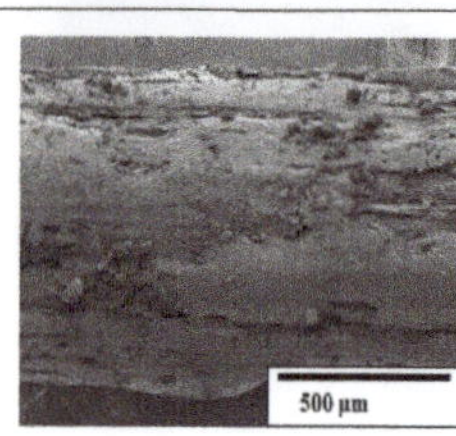	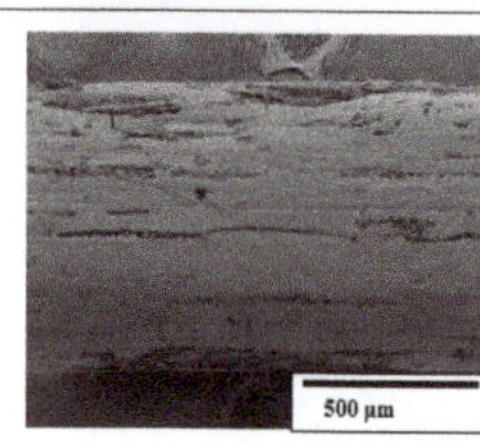

Table 4. *Cont.*

Fiber	Initial	3 Months	6 Months
PLA/PBAT/PPF/CO/AA	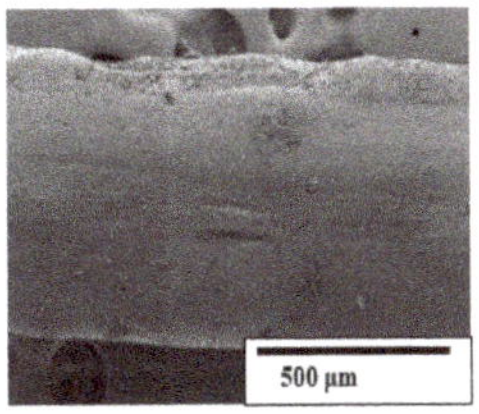		
PLA/PBAT/PPF/CO/AA_T			

Table 5. SEM micrographs of the fiber's surface morphology initially and after 3 and 6 months of composting in soil (magnification ×5k).

Fiber	Initial	3 Months	6 Months
PLA	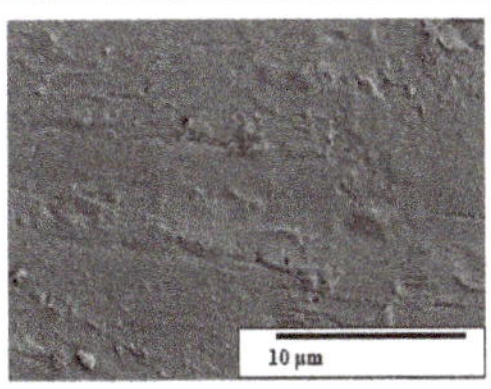	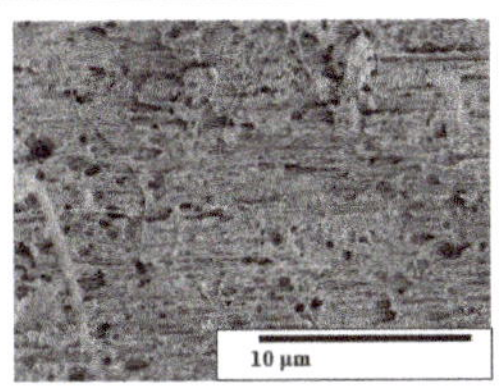	
PBAT			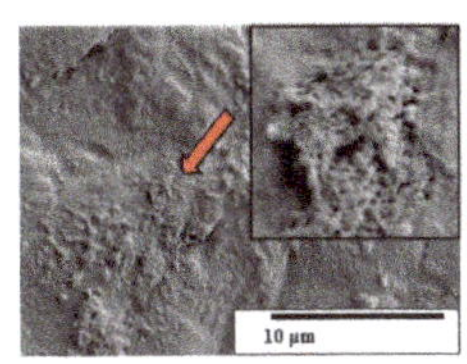
PLA/PBAT	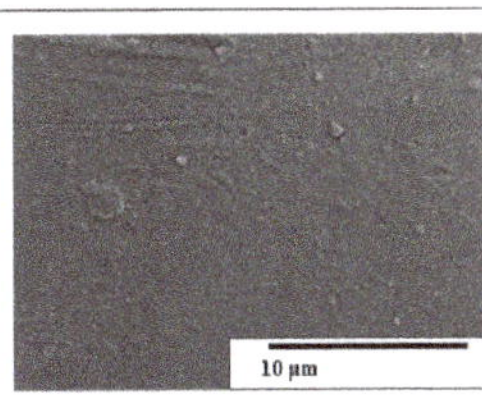	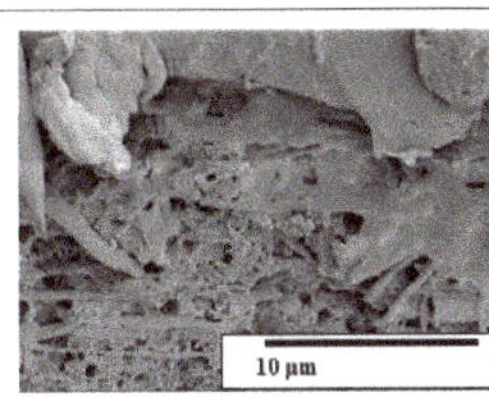	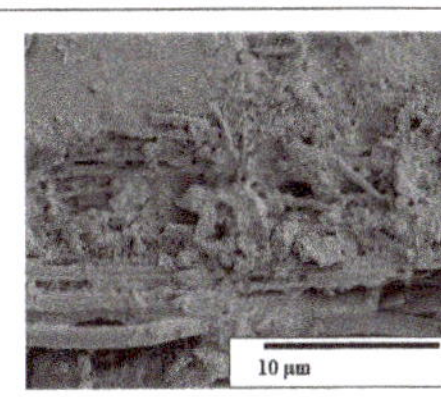

Table 5. *Cont.*

Fiber	Initial	3 Months	6 Months
PLA/PBAT/CO		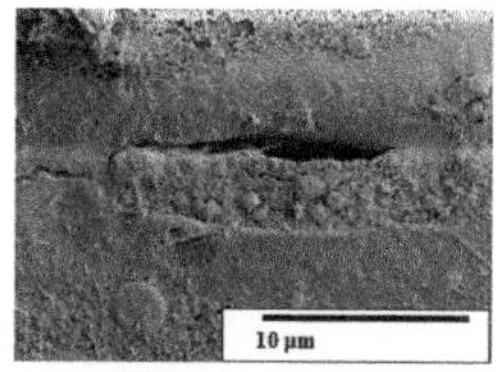	
PLA/PBAT/PPF/CO/AA	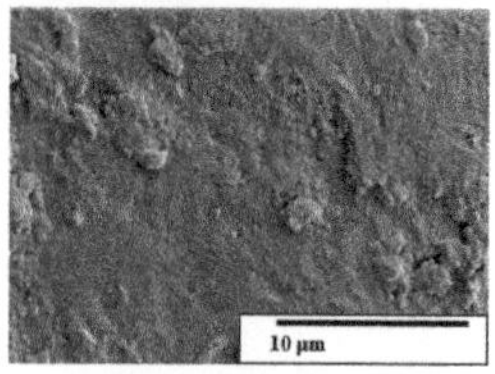	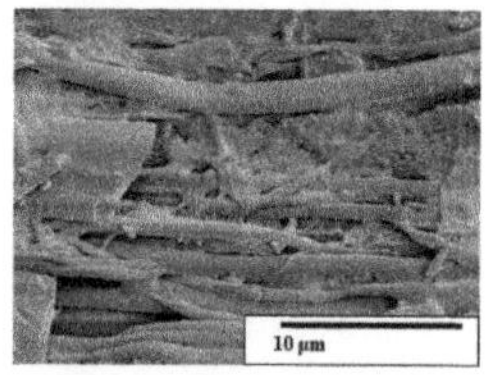	
PLA/PBAT/PPF/CO/AA_T	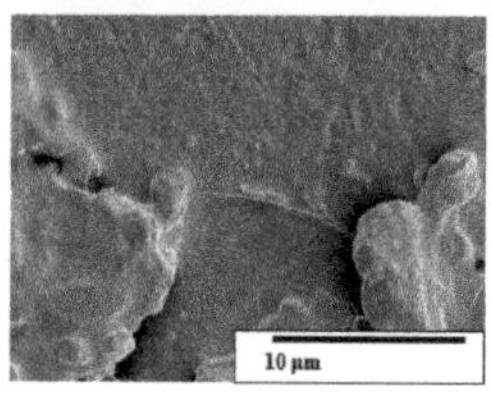	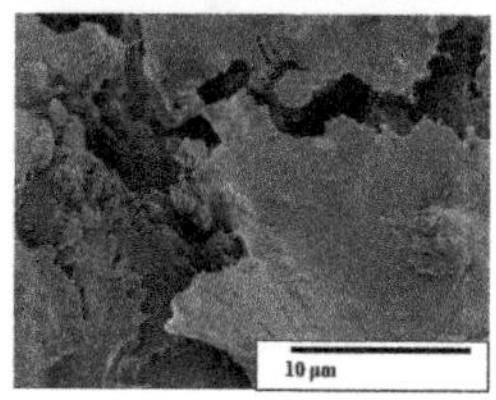	

A porous structure with nanometric holes can be seen more clearly for PLA and PLA/PBAT blends after 3 and 6 months of degradation in Table 5. This was also observed by Ruggero et al. [24] for PLA/PBAT/starch blends. On the other hand, a "lunar surface" morphology is observed in the PLA/PBAT/CO fibers after 3 months. This can be related to the viscosity of the oil and the miscibility between the CO and the PLA/PBAT matrix [30].

In general, on the blends, pores and surface erosion are observed during the composting process due to the environmental aging of the fibers. Fiber delamination can be seen due to the progressive erosion of the material, especially in the PLA/PBAT/CO after 6 months of degradation.

In the PLA/PBAT/PPF/CO/AA fibers with and without temperature, the PPF particles are not seen in the micrographs, which is indicative of good adhesion between the PPF and the polymer matrix. Spiridon et al. [8] found a similar observation for lignin particles in a PLA matrix. Moreover, the delamination of the fiber due to erosion is not observed compared to PLA/PBAT/CO. This could be due to the AA enhancing the phase compatibility between polymers.

Comparing the PLA/PBAT/PPF/CO/AA fibers with and without temperature, higher degradation can be seen in PLA/PBAT/PPF/CO/AA. The temperature in emulsions promotes the kinetic energy and mobility of the particles, thereby increasing the collisions between them with a subsequent improvement in emulsion stability [33,34].

The detachment of fragments from the fibers and the formation of pores or cavities contributes to higher weight loss, being more evident after 3 months for the neat polymers. In Table 5, more cavities can be seen in PLA compared to PBAT, with a weight loss of 19 and 7%, respectively (see Table 2). The same behavior was found for PLA fibers under composting by Feijoo et al. [28].

Tables 4 and 5 show the SEM micrographs of the surface morphology of the fibers before and after 3 and 6 months of degradation for 100 and 5 k magnification, respectively.

3.3. Fourier Transform Infrared Spectroscopy–Attenuated Total Reflection (FTIR-ATR)

In Figure 1, the FTIR-ATR spectra of the fibers show the changes that occurred on their surface due to the degradation process.

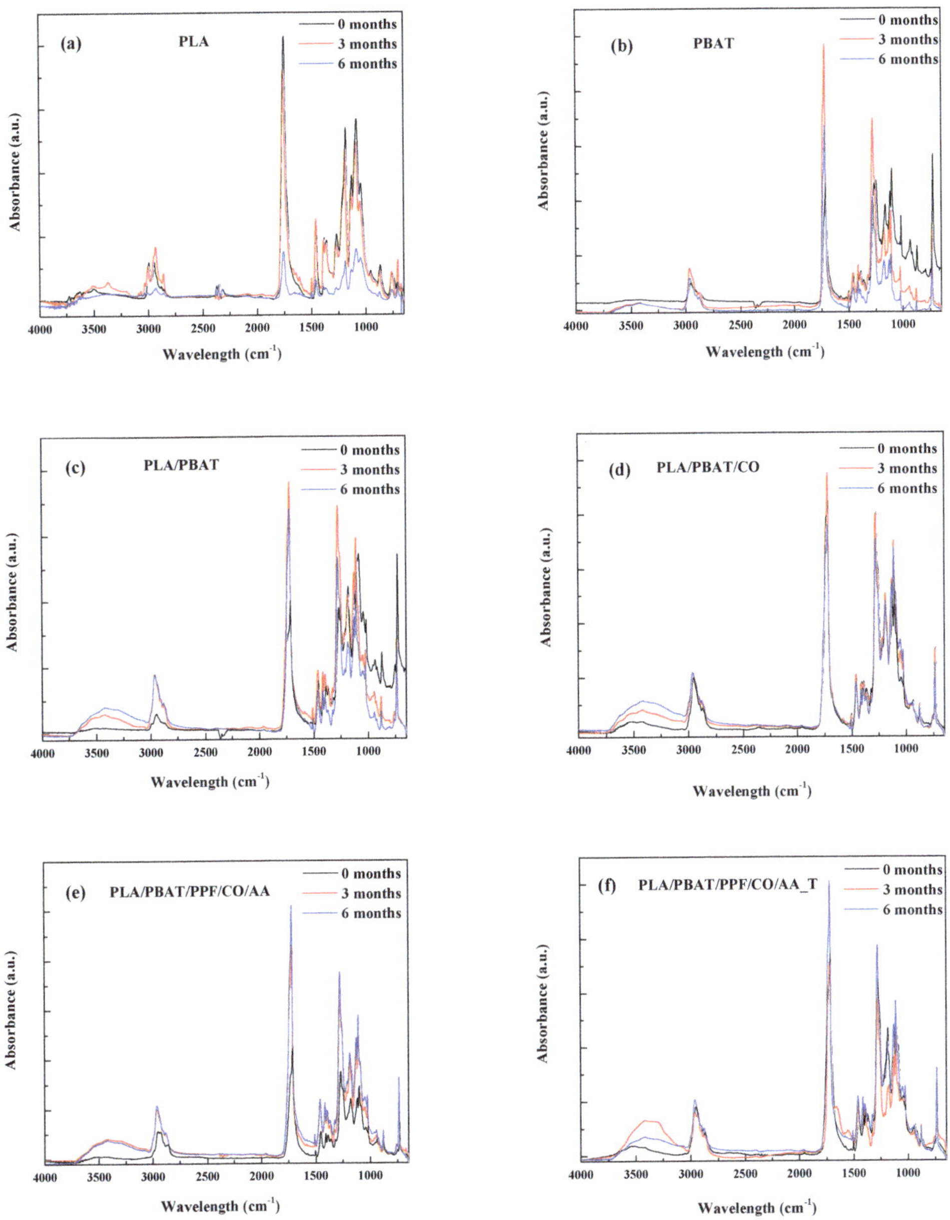

Figure 1. FTIR spectra of (**a**) PLA, (**b**) PBAT, (**c**) PLA/PBAT, (**d**) PLA/PBAT/CO, (**e**) PLA/PBAT/PPF/CO/AA, (**f**) PLA/PBAT/PPF/CO/AA_T.

At the initial time, the main characteristic absorption peaks for PLA, PBAT, and PLA/PBAT have already been reported [13,24,30,35,36]. The main PLA bands and peaks (see Figure 1a) are between 3760 and 3115 cm^{-1}, assigned to the -OH groups, and between 3000 and 2775 cm^{-1}, corresponding to -CH deformations (symmetric and asymmetric bends). The sharp peak at 1740 cm^{-1} is attributed to C=O from the ester groups; at 1450 cm^{-1} to -CH in CH_3; at 1272 cm^{-1} it is related to C-O stretching; at 1180 for C-O stretching in the -CH-O group; and at 1079 cm^{-1} it is related to C-O-C stretching. For PBAT, the main peaks observed in Figure 1b are similar to those observed for PLA, due to the similar chemical structure except for the band between 1210 and 1000 cm^{-1}, which is related to the benzene rings, and the peak at 725 cm^{-1}, for the bending vibration of CH-plane in the benzene ring [37]. No differences were observed on the bands for the blends after PBAT incorporation into PLA and the addition of CO, PPF, and the influence of temperature.

After biodegradation, the following changes were observed. In PBAT and the polymer blends, an increase in the band between 3760 and 3115 cm^{-1} corresponding to the -OH groups with degradation time was observed (Figure 1b–f). This is related to bond scission and -OH formation, characteristic of the hydrolysis process that occurs in the fibers during composting [10,30].

A detail of the FTIR spectra of the carboxyl groups (-C=O) in the region between 1800 and 1600 cm^{-1} to trace the degradation process is shown in Figure 2.

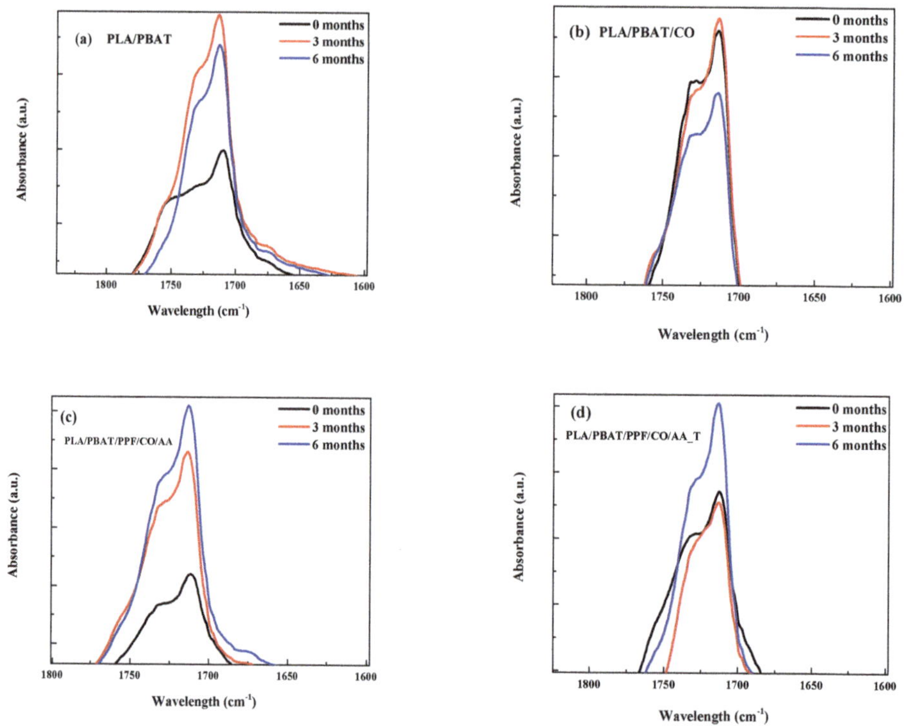

Figure 2. Detailed FTIR spectra of (**a**) PLA/PBAT, (**b**) PLA/PBAT/CO, (**c**) PLA/PBAT/PPF/CO/AA, (**d**) PLA/PBAT/PPF/CO/AA_T at 1740 cm^{-1}.

A decrease in the peak at 1740 cm^{-1} related to the stretching vibrations of -C=O can be seen as a result of chain scission during degradation [36,38] as well as in the band at around 1100 cm^{-1}, correlated to -C-O groups as shown in Figure 3 [36].

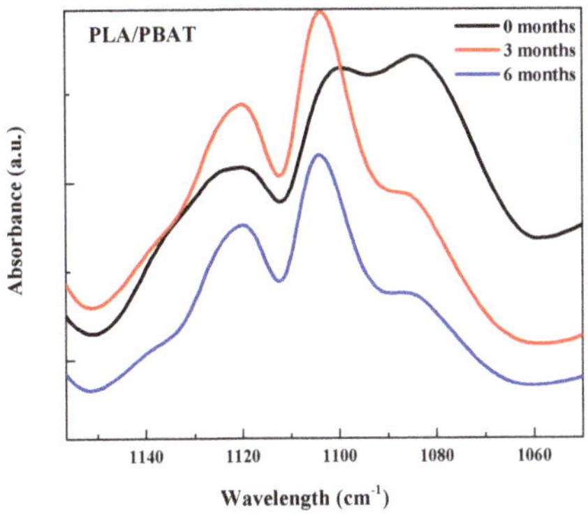

Figure 3. Detailed FTIR spectra of PLA/PBAT at 1110 cm^{-1}.

Moreover, the absorption band at 1269 cm^{-1} showed a decrease for the PLA/PBAT/PPF/CO/AA with and without temperature as observed in Figure 4a,b. This band is related to hemicelluloses biodegradation from acetyl and carboxylic vibration of the carbohydrate component of plant cell walls. It has been reported that lignin absorbs at a wavelength near 1269 cm^{-1} [39]. The FTIR-ATR spectrum of PPF is shown in Figure 4c. The decrease in this peak was only observed in the PLA/PBAT/PPF/CO/AA with and without temperature fibers. The decrease in this absorption band could be related to the higher weight loss observed in Table 2.

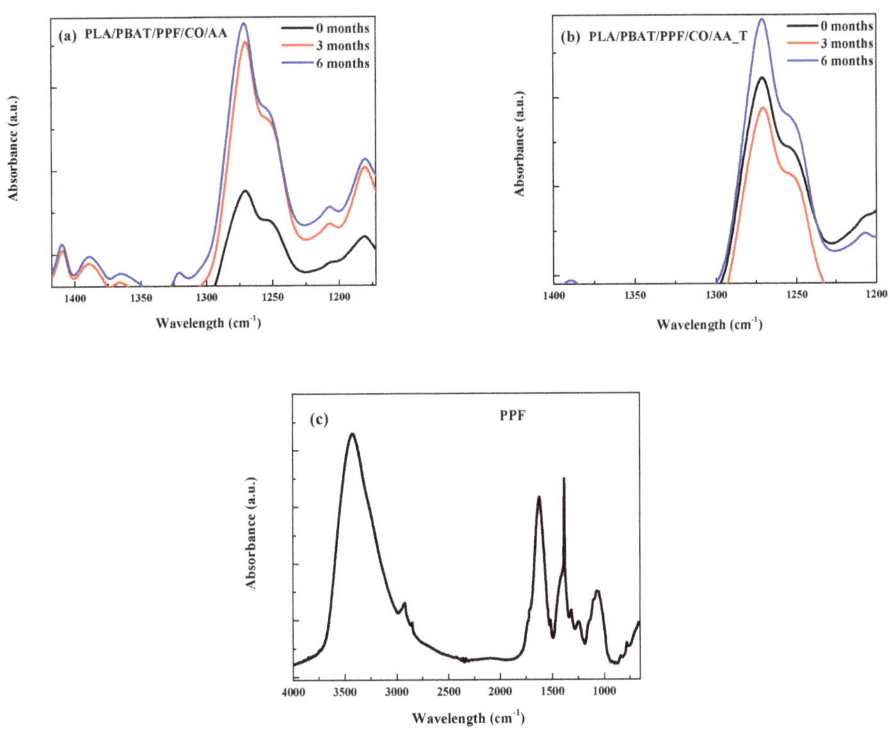

Figure 4. Detailed FTIR spectra of (**a**) PLA/PBAT/PPF/CO/AA, (**b**) PLA/PBAT/PPF/CO/AA_T, and (**c**) PPF at 1269 cm^{-1}.

3.4. *Differential Scanning Calorimetry (DSC)*

The results of thermal analysis by DSC for the fibers before and after degradation are shown in Figures 5 and 6. The cooling and heating thermograms are shown in these two figures, also.

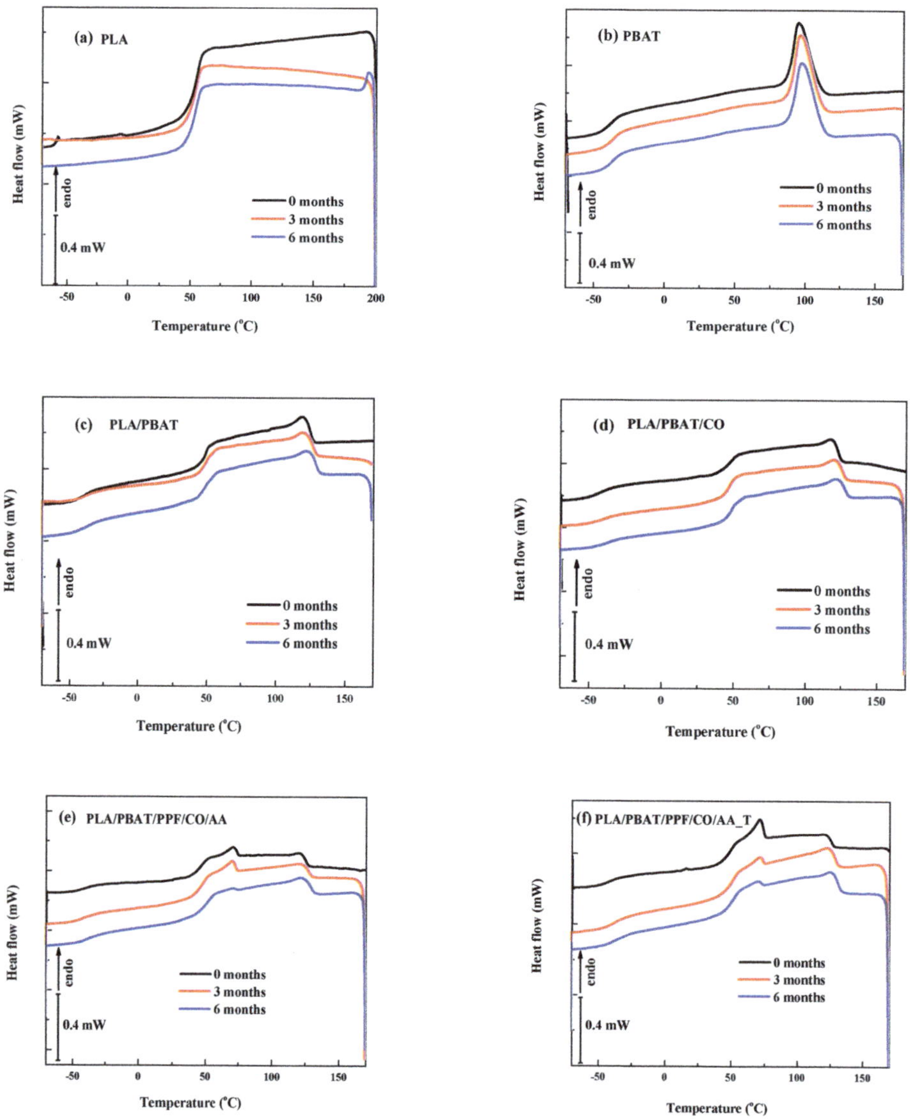

Figure 5. Cooling DSC thermogram of neat polymers and blend fibers before and after degradation.

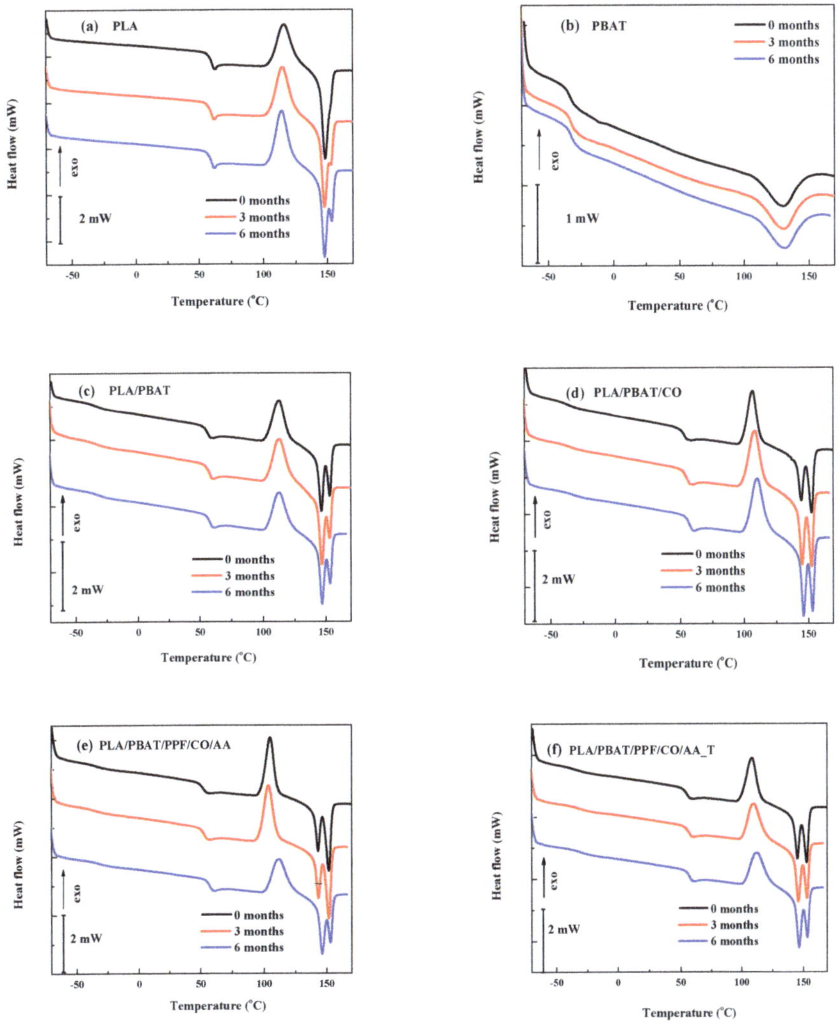

Figure 6. Second heating DSC thermogram of neat polymers and blend fibers before and after degradation.

In Figure 5a, the glass transition temperature (T_g) of PLA can be seen at around 61–62 °C. For PBAT in Figure 5b, the crystallization temperature (T_c) increased for the fibers as the degradation time progressed with values of 95.6 °C (initial), 97.4 °C (3 months), and 98.1 °C (6 months). For the blends, the same behavior was observed. As the degradation time increased, the T_c increased as well as the PLA T_g (Figure 5c,d). On the other hand, when PPF and AA were added to the blends, the T_c value shifted to higher values, being highest for the PLA/PBAT/PPF/CO/AA _T fiber. For example, after 6 months of degradation the T_c value of the PLA/PBAT/CO fiber was 122.4 °C and of the PLA/PBAT/PPF/CO/AA _T fiber it was 125.8 °C with a difference of almost 3 °C. For the PLA/PBAT/PPF/CO/AA fiber, a similar value to that of PLA/PBAT/CO was found (122.5 °C). The increase in T_c is associated with interaction between phases in the polymer blends. For the PLA/PBAT/PPF/CO/AA and PLA/PBAT/PPF/CO/AA fibers with and without temperature, the stiffening effect of the PPF and the enhancement of blend compatibility due to AA influenced the crystallization process of the blend. The increase in T_c is indicative of an increase in crystal size [26].

From Figure 6, the second heating of the fibers before and after degradation can be observed.

The DSC curves corresponding to the second heating can be seen in Figure 6. In Figure 6a, the T_g and T_m of neat PLA were observed at 60.5 and 148.4 °C, respectively. Some authors have reported T_g values between 59.1 °C and 62.4 °C and T_m values between 147.8 °C and 154 °C [13,16,37]. While the T_m of PLA before and after degradation remained constant (60.4–60.6 °C), the melting showed some differences. In the detail of Figure 7a, before degradation, only one peak at 148.4 °C can be seen. After degradation the curve splits into two peaks showing T_m at 148.5 °C and 153.5 °C after 3 months and 148.4 °C and 154.3 °C after 6 months. While the main melting peak (148.4–148.5 °C) remained steady, the area and T_m of the new one increased progressively. This higher temperature component is attributed to new crystalline phases formed by smaller chains resulting from the degradation process [31,37].

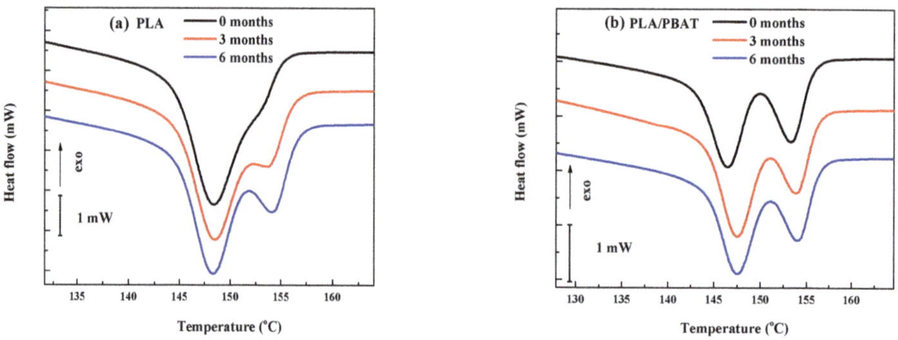

Figure 7. Details of the second heating DSC thermogram for (**a**) PLA and (**b**) the PLA/PBAT blend before and after degradation.

The T_g value for PBAT was found at −33.1 °C (Figure 6b), close to the −34 °C reported by Turek et al. [40]. On the other hand, the T_m (129.8 °C) was slightly above those reported between 120 °C and 124.1 °C [13,16,34]. As for PLA, the T_m of PBAT increased with degradation time (129.8 °C neat, 130.5 °C after 3 months, and 131.2 °C after 6 months).

The thermal behavior of the PLA component in the PLA/PBAT blend (Figure 6c) is slightly different when compared to that of neat PLA, as shown in Figure 7a,b. Both T_g and T_m decreased by 4 and 1.9 °C, respectively, and a higher melting temperature (153.2 °C) developed. These findings suggest the interaction between PLA and PBAT in the blend. Additionally, both the T_g and T_m values increased with degradation time: 56.5, 58.7 and 59.1 °C and 146.5, 147.5 and 147.5 °C initially and after 3 and 6 months, respectively, and approached those of neat PLA. This result can be interpreted as the selective degradation of either or both PBAT or PBAT-altered PLA regions.

For the PLA/PBAT/CO, PLA/PBAT/PPF/CO/AA, and PLA/PBAT/PPF/CO/AA_T fibers, a similar behavior was observed, as shown in Figure 6d–f. First, there was a reduction in T_g for both the PBAT and PLA phases with respect to the PLA/PBAT blend, which is attributable to the plasticizing effect of canola oil. Second, there was a more noticeable increment in T_g values with degradation time, which points to the leaching of the additives along the degradation.

Between the T_g and T_m, a cold crystallization (endothermic peak) is observed for PLA and the blends. The cold crystallization peak for the PLA (Figure 6a) was at around 115 °C and for the PLA/PBAT blend (Figure 6c) at around 113 °C. For the other blends, the cold crystallization temperature (T_{cc}) was lower compared to PLA and, as degradation time increased, the T_{cc} increased. Values between 117 °C and 120 °C have been reported in the literature [15,30]. Nomadolo et al. [41] and Vasile et al. [38] found higher values of T_{cc} for PLA compared to the blends, as shown in this work. This is correlated with the presence of a nucleating agent in the blend composition accelerating the crystallization and decreasing the T_{cc} [37,41].

Comparing the fibers as the degradation process progresses, it has been reported that the crystalline domains increased during degradation. These domains act as networks or cross-linking points, limiting the mobility of PLA. Only amorphous chains with high molecular mobility undergo rearrangement and crystallize at low temperature resulting in cold crystallization. On the other hand, the chains with low mobility also crystallize, but at higher temperatures [3,38,42].

The detail of Figure 8a shows that for the PLA/PBAT/PPF/CO/AA blend, a similar T_{cc} value was obtained for the initial fiber and after 3 months of degradation (105.0 °C and 104.3 °C, respectively). After 6 months of degradation, this value increased by 8.5 °C (112.6 °C). For the PLA/PBAT/PPF/CO/AA_T fiber, there was a displacement of T_{cc} to higher values from 108.5 °C (initial) to 110.5 °C (3 months) and 113.2 °C (6 months). As seen in Figure 8b, after degradation, a similar T_{cc} value was reached after both 3 and 6 months. This could be related to the adhesion or uniform dispersion of the PPF with the use of temperature in the emulsion limiting the blend's segmental motion [43,44].

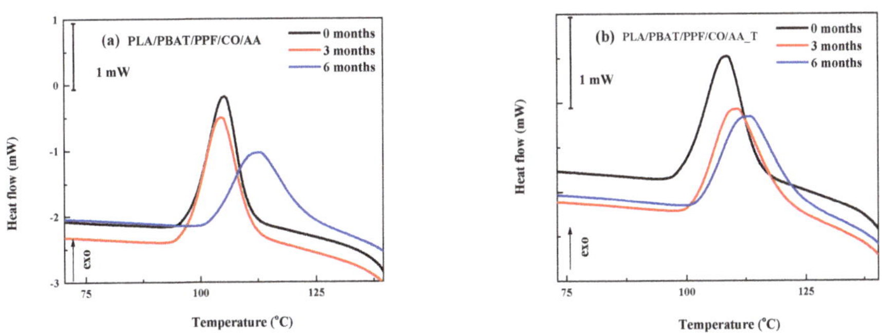

Figure 8. Details of the second heating DSC thermogram for (**a**) PLA/PBAT/PPF/CO/AA and (**b**) PLA/PBAT/PPF/CO/AA_T fibers before and after degradation.

DSC showed that the crystallization temperature (T_c) increased for the fibers as the degradation time progressed, as well as when PPF and AA were incorporated into the blend. For the melting, new crystalline phases resulting from the degradation process were found. Regarding cold crystallization, the most important change was observed for the PLA/PBAT/PPF/CO/AA fiber with an increase in T_{cc} after 6 months of degradation.

3.5. Mechanical Properties

The mechanical properties of the fibers are reported in Table 6 and also illustrated in Figure 9. For the initial fibers, PLA showed a higher Young´s modulus and tensile strength but lower elongation at break than PBAT, as shown in Table 6. A decrease in the tensile strength of PLA fibers in compost over time has been reported by Borelbach et al. [10]. In the case of the PLA/PBAT blend, the modulus and tensile strength were close to averages of both components. However, the deformation was similar to that of neat PLA; therefore, the incorporation of PBAT into PLA did not increase the flexibility of PLA. By adding CO, Young´s modulus and tensile strength decreased compared to PLA/PBAT. On the other hand, the elongation at break was improved. It has been reported in the literature that CO improves the elongation at break [45]. On the other hand, when AA (PLA/PBAT/PPF/CO/AA) and temperature (PLA/PBAT/PPF/CO/AA_T) were used, an increase in Young´s modulus, tensile strength, and a decrease in elongation at break was observed compared to PLA/PBAT/CO. This can be attributed to the stiffness of the PPF and the improvement of the interfacial adhesion and stress transfer between the polymer matrix and the PPF [44]; therefore, ductility and toughness were enhanced with the addition of PPF [43]. However, after 3 and 6 months of degradation, those properties decreased, indicating a restriction in segmental mobility [8] and an increase in crystalline regions with degradation time [38], as clearly seen in Figure 9. This was also observed via the increase in T_{cc} and T_m in the DSC thermograms (Figure 6).

Table 6. Mechanical properties of PLA, PBAT, PLA/PBAT, PLA/PBAT/CO, PLA/PBAT/PPF/CO/AA, and PLA/PBAT/PPF/CO/AA_T before and after degradation.

	Mechanical Properties		
Fiber Composition and Time of Composting	**Young's Modulus (MPa) ***	**Tensile Strength (MPa) ***	**Elongation at Break (%) ***
Initial			
PLA	1109 ± 47^{Cg}	55 ± 2^{Ce}	6 ± 1^{Aa}
PBAT	115 ± 10^{Aa}	11 ± 1^{Abc}	224 ± 31^{Bc}
PLA/PBAT	507 ± 89^{Bcdef}	21 ± 4^{Bd}	5 ± 1^{Aa}
PLA/PBAT/CO	412 ± 131^{Bbcd}	8 ± 2^{Ab}	40.87 ± 5^{ABc}
PLA/PBAT/PPF/CO/AA	435 ± 83^{Bbcde}	9 ± 1^{Ab}	3 ± 1^{Aa}
PLA/PBAT/PPF/CO/AA_T	552 ± 89^{Bdef}	18 ± 3^{Bcd}	5 ± 1^{Aa}
3 months			
PLA	605 ± 61^{Cf}	25 ± 2^{Cde}	5 ± 1^{Aa}
PBAT	98 ± 5^{Aa}	11 ± 1^{Abc}	130 ± 34^{Bb}
PLA/PBAT	430 ± 52^{Bbcde}	16 ± 4^{Bcd}	5 ± 1^{Aa}
PLA/PBAT/CO	497 ± 88^{Bcdef}	14 ± 3^{ABc}	3 ± 1^{Aa}
PLA/PBAT/PPF/CO/AA	463 ± 87^{Bbcdef}	11 ± 3^{ABbc}	3 ± 1^{Aa}
PLA/PBAT/PPF/CO/AA_T	385 ± 55^{Bbc}	11 ± 3^{Abc}	3 ± 1^{Aa}
6 months			
PLA	562 ± 71^{Def}	27 ± 6^{Cde}	6 ± 1^{Aa}
PBAT	98 ± 5^{Aa}	11 ± 1^{ABbc}	5 ± 1^{Aa}
PLA/PBAT	397 ± 31^{BCbc}	13 ± 3^{ABbc}	4 ± 1^{Aa}
PLA/PBAT/CO	455 ± 117^{Cbcde}	13 ± 5^{Bbc}	4 ± 1^{Aa}
PLA/PBAT/PPF/CO/AA	413 ± 39^{BCbcd}	7 ± 1^{Aa}	3 ± 1^{Aa}
PLA/PBAT/PPF/CO/AA_T	340 ± 29^{Bb}	7 ± 2^{ABb}	3 ± 1^{Aa}

* Different letters represent statistical differences between compositions (upper case letters) and months (lower case letters) by Tukey test ($p < 0.05$).

After 3 and 6 months of degradation, the trend was the same for the initial neat polymers and the PLA/PBAT blend. However, for the PLA/PBAT/CO fibers, Young´s modulus was higher compared to the PLA/PBAT fibers, the opposite of the initial ones. The tensile strength and elongation at break values were similar. After AA incorporation (PLA/PBAT/PPF/CO/AA), Young´s modulus decreased as well as the tensile strength and elongation at break. Finally, for the PLA/PBAT/PPF/CO/AA_T fibers, the modulus decreased and the tensile strength and elongation at break remained almost the same compared to the PLA/PBAT/PPF/CO/AA fibers.

As the degradation time increased, there was a decrease in the mechanical properties of the PLA, PBAT, and PLA/PBAT/PPF/CO/AA_T fibers. This difference was more noteworthy when comparing the initial and degraded fibers, being similar for the 3- and 6-month degraded ones. This decrease in mechanical properties can be seen in SEM micrographs (Table 5), in which the fiber delamination, holes, and cracks could be observed in the blend fibers, associated with the weight loss (Table 2). It has been reported by several authors [46–48] that the mechanical properties decreased after degradation in PLA, PBAT, and their blends.

The incorporation of PPF and AA into the blends modifies interactions between the components. Moreover, when temperature was used for the emulsion preparation, lower weight loss was observed compared to the PLA/PBAT/PPF/CO/AA fibers, so perhaps the PPF loss was hindered by these interactions [38]. As can be observed from the results, degradation is a complex process that depends on the environmental conditions [49–51].

Statistically, the main significant differences ($p \leq 0.05$) between compositions and degradation time were obtained for the mechanical properties of PLA and PBAT.

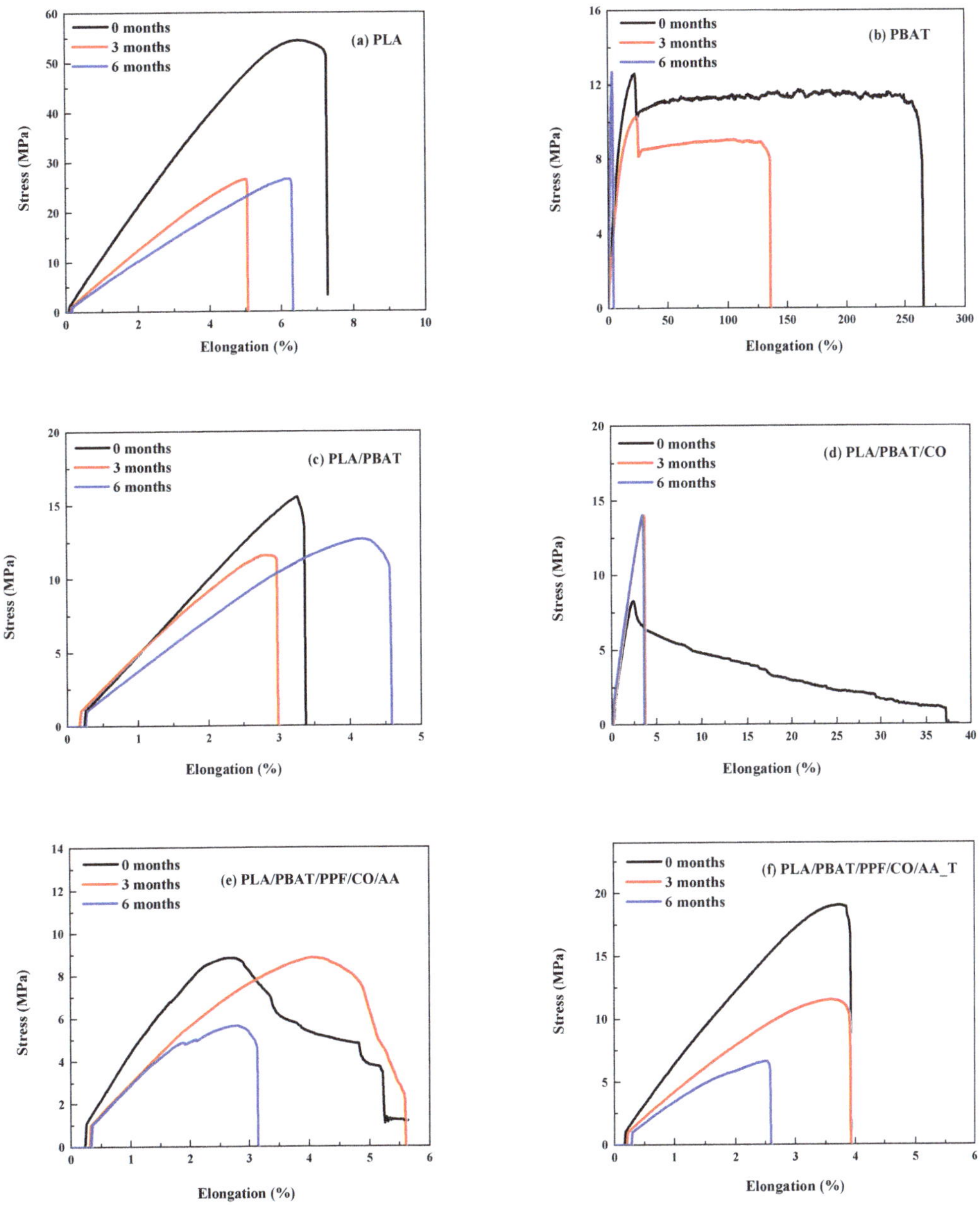

Figure 9. Stress–strain curves for (**a**) PLA, (**b**) PBAT, (**c**) PLA/PBAT, (**d**) PLA/PBAT/CO, (**e**) PLA/PBAT/PPF/CO/AA, and (**f**) PLA/PBAT/PPF/CO/AA_T before and after degradation.

4. Conclusions

This study investigated the effect of adding PPF to a biodegradable PLA/PBAT polymer blend. After 6 months of composting, the studied fibers containing PF were degraded by 87 ± 11%. Morphological changes were observed by optical microscopy and SEM. The analysis of the FTIR-ATR spectra showed a decrease in the absorption bands associated with polymer matrix degradation. The thermal stability and mechanical properties decreased as degradation progressed. The PLA/PBAT/PPF/CO/AA composites resulted in accelerated biodegradation, useful for the disposal of any material.

Author Contributions: Conceptualization, Z.N.C.-P. and J.J.B.-J.; methodology, Z.N.C.-P., E.O.C.-L. and J.J.B.-J.; validation, J.J.B.-J., P.O.-G., E.O.C.-L. and M.H.-L.; formal analysis, Z.N.C.-P., J.J.B.-J., P.O.-G. and M.H.-L.; investigation, Z.N.C.-P. and J.J.B.-J.; data curation, M.H.-L. and E.O.C.-L.; writing—original draft preparation, Z.N.C.-P.; writing—review and editing, Z.N.C.-P., J.J.B.-J., P.O.-G. and S.B.-B.; visualization, S.B.-B. and Z.N.C.-P.; supervision, Z.N.C.-P. All authors have read and agreed to the published version of the manuscript.

Funding: This research received no external funding.

Institutional Review Board Statement: Not applicable.

Data Availability Statement: Data available on request from the authors.

Conflicts of Interest: The authors declare no conflict of interest.

References

1. Muthuraj, R.; Misra, M.; Mohanty, A.K. Biodegradable Compatibilized Polymer Blends for Packaging Applications: A Literature Review. *J. Appl. Polym. Sci.* **2018**, *135*, 45726. [CrossRef]
2. Mohite, A.S.; Jagtap, A.R.; Avhad, M.S.; More, A.P. Recycling of Major Agriculture Crop Residues and Its Application in Polymer Industry: A Review in the Context of Waste to Energy Nexus. *Energy Nexus* **2022**, *7*, 100134. [CrossRef]
3. Bianchi, M.; Dorigato, A.; Morreale, M.; Pegoretti, A. Evaluation of the Physical and Shape Memory Properties of Fully Biodegradable Poly(Lactic Acid) (PLA)/Poly(Butylene Adipate Terephthalate) (PBAT) Blends. *Polymers* **2023**, *15*, 881. [CrossRef] [PubMed]
4. Meng, M.; Wang, S.; Xiao, M.; Meng, Y. Recent Progress in Modification and Preparations of the Promising Biodegradable Plastics: Polylactide and Poly(Butylene Adipate-Co-Terephthalate). *Sustain. Polym. Energy* **2023**, *1*, 10006. [CrossRef]
5. De Farias, P.M.; de Vasconcelos, L.B.; Ferreira, M.E.S.; Pascall, M.; Tapia-Blácido, D.R. Nopal Cladode (Opuntia Ficus-Indica) Flour: Production, Characterization, and Evaluation for Producing Bioactive Film. *Food Packag. Shelf Life* **2021**, *29*, 100703. [CrossRef]
6. Rives-Castillo, S.C.H.; Correa-Pacheco, Z.N.; Bautista-Baños, S.; Ventura-Aguilar, R.I.; Schettino-Bermudez, B.S.; Ortega-Gudiño, P.; Barajas-Cervantes, A. Chemical Proximate and Thermal Analyses on Flour from Residual Products of Pear-Cactus (*Opuntia ficus-indica*) towards Obtaining Biobased Polymeric Compounds. *Agrociencia* **2021**, *55*, 403–416. [CrossRef]
7. Scaffaro, R.; Maio, A.; Gulino, E.F.; Megna, B. Structure-Property Relationship of PLA-Opuntia Ficus Indica Biocomposites. *Compos. B Eng.* **2019**, *167*, 199–206. [CrossRef]
8. Spiridon, I.; Tanase, C.E. Design, Characterization and Preliminary Biological Evaluation of New Lignin-PLA Biocomposites. *Int. J. Biol. Macromol.* **2018**, *114*, 855–863. [CrossRef] [PubMed]
9. Kim, D.Y.; Lee, J.B.; Lee, D.Y.; Seo, K.H. Plasticization Effect of Poly(Lactic Acid) in the Poly(Butylene Adipate–Co–Terephthalate) Blown Film for Tear Resistance Improvement. *Polymers* **2020**, *12*, 1904. [CrossRef]
10. Borelbach, P.; Kopitzky, R.; Dahringer, J.; Gutmann, P. Degradation Behavior of Biodegradable Man-Made Fibers in Natural Soil and in Compost. *Polymers* **2023**, *15*, 2959. [CrossRef]
11. Mhd Ramle, S.F.; Ahmad, N.A.; Mohammad Rawi, N.F.; Zahidan, N.S.; Geng, B.J. Physical Properties and Soil Degradation of PLA/PBAT Blends Film Reinforced with Bamboo Cellulose. *IOP Conf. Ser. Earth Environ. Sci.* **2020**, *596*, 012021. [CrossRef]
12. Rebelo, R.C.; Gonçalves, L.P.C.; Fonseca, A.C.; Fonseca, J.; Rola, M.; Coelho, J.F.J.; Rola, F.; Serra, A.C. Increased Degradation of PLA/PBAT Blends with Organic Acids and Derivatives in Outdoor Weathering and Marine Environment. *Polymer* **2022**, *256*, 125223. [CrossRef]
13. Correa-Pacheco, Z.N.; Black-Solís, J.D.; Ortega-Gudiño, P.; Sabino-Gutiérrez, M.A.; Benítez-Jiménez, J.J.; Barajas-Cervantes, A.; Bautista-Baños, S.; Hurtado-Colmenares, L.B. Preparation and Characterization of Bio-Based PLA/PBAT and Cinnamon Essential Oil Polymer Fibers and Life-Cycle Assessment from Hydrolytic Degradation. *Polymers* **2019**, *12*, 38. [CrossRef] [PubMed]
14. López, Á.; Rivas, J.; Loaiza, M.; Sabino, M. Degradación de Películas Plastificadas de Quitosano Obtenidas a Partir de Conchas de Camarón (L.Vannamei). *Rev. Fac. Ing. Univ. Cent. Venez.* **2010**, *25*, 133–143.
15. Wang, J.H.; Tian, Y.; Zhou, B. Degradation and Stabilization of Poly(Butylene Adipate-Co-Terephthalate)/Polyhydroxyalkanoate Biodegradable Mulch Films Under Different Aging Tests. *J. Polym. Environ.* **2022**, *30*, 1366–1379. [CrossRef]

16. Carbonell-Verdu, A.; Ferri, J.M.; Dominici, F.; Boronat, T.; Sanchez-Nacher, L.; Balart, R.; Torre, L. Manufacturing and Compatibilization of PLA/PBAT Binary Blends by Cottonseed Oil-Based Derivatives. *Express Polym. Lett.* **2018**, *12*, 808–823. [CrossRef]
17. Zhang, Y.; Gao, W.; Mo, A.; Jiang, J.; He, D. Degradation of Polylactic Acid/Polybutylene Adipate Films in Different Ratios and the Response of Bacterial Community in Soil Environments. *Environ. Pollut.* **2022**, *313*, 120–167. [CrossRef]
18. Tabasi, R.Y.; Ajji, A. Selective Degradation of Biodegradable Blends in Simulated Laboratory Composting. *Polym. Degrad. Stab.* **2015**, *120*, 435–442. [CrossRef]
19. Liu, B.; Guan, T.; Wu, G.; Fu, Y.; Weng, Y. Behavior of Degradable Mulch with Poly (Butylene Adipate-coTerephthalate) (PBAT) and Poly (Butylene Succinate) (PBS) in Simulation Marine Environment. *Polymers* **2022**, *14*, 1515. [CrossRef]
20. *ASTM D6400-19*; Standard Specification for Labeling of Plastics Designed to Be Aerobically Composted in Municipal or Industrial Facilities. ASTM International: West Conshohocken, PA, USA, 2018.
21. Black-Solis, J.; Ventura-Aguilar, R.I.; Correa-Pacheco, Z.; Corona-Rangel, M.L.; Bautista-Baños, S. Preharvest Use of Biodegradable Polyester Nets Added with Cinnamon Essential Oil and the Effect on the Storage Life of Tomatoes and the Development of Alternaria Alternata. *Sci. Hortic.* **2019**, *245*, 65–73. [CrossRef]
22. Misra, R.V.; Roy, R.N.; Hiraoka, H. *On-Farm Composting Methods*; FAO Land and Water Discussion Papers: Rome, Italy, 2003.
23. Jain, M.S.; Paul, S.; Kalamdhad, A.S. Kinetics and Physics during Composting of Various Organic Wastes: Statistical Approach to Interpret Compost Application Feasibility. *J. Clean. Prod.* **2020**, *255*, 120324. [CrossRef]
24. Ruggero, F.; Onderwater, R.C.A.; Carretti, E.; Roosa, S.; Benali, S.; Raquez, J.-M.; Gori, R.; Lubello, C.; Wattiez, R. Degradation of Film and Rigid Bioplastics During the Thermophilic Phase and the Maturation Phase of Simulated Composting. *J. Polym. Environ.* **2021**, *29*, 3015–3028. [CrossRef]
25. Kawashima, N.; Yagi, T.; Kojima, K. How Do Bioplastics and Fossil-Based Plastics Play in a Circular Economy? *Macromol. Mater. Eng.* **2019**, *304*, 1900383. [CrossRef]
26. Bilo, F.; Pandini, S.; Sartore, L.; Depero, L.E.; Gargiulo, G.; Bonassi, A.; Federici, S.; Bontempi, E. A Sustainable Bioplastic Obtained from Rice Straw. *J. Clean. Prod.* **2018**, *200*, 357–368. [CrossRef]
27. Hashimoto, K.; Sudo, M.; Ohta, K.; Sugimura, T.; Yamada, H.; Aoki, T. Biodegradation of Nylon4 and Its Blend with Nylon6. *J. Appl. Polym. Sci.* **2002**, *86*, 2307–2311. [CrossRef]
28. Feijoo, P.; Marín, A.; Samaniego-Aguilar, K.; Sánchez-Safont, E.; Lagarón, J.M.; Gámez-Pérez, J.; Cabedo, L. Effect of the Presence of Lignin from Woodflour on the Compostability of PHA-Based Biocomposites: Disintegration, Biodegradation and Microbial Dynamics. *Polymers* **2023**, *15*, 2481. [CrossRef]
29. Ebadi-Dehaghani, H.; Barikani, M.; Borhani, S.; Bolvardi, B.; Khonakdar, H.A.; Jafari, S.H.; Aarabi, A. Biodegradation and Hydrolysis Studies on Polypropylene/Polylactide/Organo-Clay Nanocomposites. *Polym. Bull.* **2016**, *73*, 3287–3304. [CrossRef]
30. Brdlík, P.; Novák, J.; Borůvka, M.; Běhálek, L.; Lenfeld, P. The Influence of Plasticizers and Accelerated Ageing on Biodegradation of PLA under Controlled Composting Conditions. *Polymers* **2022**, *15*, 140. [CrossRef]
31. Gorrasi, G.; Pantani, R. Hydrolysis and Biodegradation of Poly(Lactic Acid). In *Synthesis, Structure and Properties of Poly(lactic acid)*; Di Lorenzo, M.L., Androsch, R., Eds.; Springer International Publishing: Cham, Switzerland, 2018; pp. 119–151. [CrossRef]
32. Fu, Y.; Wu, G.; Bian, X.; Zeng, J.; Weng, Y. Biodegradation Behavior of Poly(Butylene Adipate-Co-Terephthalate) (PBAT), Poly(Lactic Acid) (PLA), and Their Blend in Freshwater with Sediment. *Molecules* **2020**, *25*, 3946. [CrossRef]
33. Ng, S.-W.; Chong, W.-T.; Soo, Y.-T.; Tang, T.-K.; Ab Karim, N.A.; Phuah, E.-T.; Lee, Y.-Y. Pickering Emulsion Stabilized by Palm-Pressed Fiber Cellulose Nanocrystal Extracted by Acid Hydrolysis-Assisted High Pressure Homogenization. *PLoS ONE* **2022**, *17*, e0271512. [CrossRef]
34. Soo, Y.; Ng, S.; Tang, T.; Ab Karim, N.A.; Phuah, E.; Lee, Y. Preparation of Palm (*Elaeis oleifera*) Pressed Fibre Cellulose Nanocrystals via Cation Exchange Resin: Characterisation and Evaluation as Pickering Emulsifier. *J. Sci. Food Agric.* **2021**, *101*, 4161–4172. [CrossRef] [PubMed]
35. Hernández-López, M.; Correa-Pacheco, Z.N.; Bautista-Baños, S.; Zavaleta-Avejar, L.; Benítez-Jiménez, J.J.; Sabino-Gutiérrez, M.A.; Ortega-Gudiño, P. Bio-Based Composite Fibers from Pine Essential Oil and PLA/PBAT Polymer Blend. Morphological, Physicochemical, Thermal and Mechanical Characterization. *Mater. Chem. Phys.* **2019**, *234*, 345–353. [CrossRef]
36. Lamparelli, R.; Montagna, L.; da Silva, A.; Montanheiro, T.; Lemes, A. Study of the Biodegradation of PLA/PBAT Films after Biodegradation Tests in Soil and the Aqueous Medium. *Biointerface Res. Appl. Chem.* **2022**, *12*, 833–846. [CrossRef]
37. Weng, Y.-X.; Jin, Y.-J.; Meng, Q.-Y.; Wang, L.; Zhang, M.; Wang, Y.-Z. Biodegradation Behavior of Poly(Butylene Adipate-Co-Terephthalate) (PBAT), Poly(Lactic Acid) (PLA), and Their Blend under Soil Conditions. *Polym. Test.* **2013**, *32*, 918–926. [CrossRef]
38. Vasile, C.; Pamfil, D.; Râpă, M.; Darie-Niţă, R.N.; Mitelut, A.C.; Popa, E.E.; Popescu, P.A.; Draghici, M.C.; Popa, M.E. Study of the Soil Burial Degradation of Some PLA/CS Biocomposites. *Compos. B Eng.* **2018**, *142*, 251–262. [CrossRef]
39. Pozo, C.; Diaz-Visurraga, J.; Contreras, D.; Freer, J.; Rodriguez, J. Characterization of Temporal Biodegradation of Radiata Pine by *Gloeophyllum trabeum* through Principal Component Analysis-Based Two-Dimensional Correlation FTIR Spectroscopy. *J. Chil. Chem. Soc.* **2016**, *61*, 2878–2883. [CrossRef]
40. Turek, C.; Stintzing, F.C. Stability of Essential Oils: A Review. *Compr. Rev. Food Sci. Food Saf.* **2013**, *12*, 40–53. [CrossRef]
41. Nomadolo, N.; Dada, O.E.; Swanepoel, A.; Mokhena, T.; Muniyasamy, S. A Comparative Study on the Aerobic Biodegradation of the Biopolymer Blends of Poly(Butylene Succinate), Poly(Butylene Adipate Terephthalate) and Poly(Lactic Acid). *Polymers* **2022**, *14*, 1894. [CrossRef]

42. Arrieta, M.; Samper, M.; Aldas, M.; López, J. On the Use of PLA-PHB Blends for Sustainable Food Packaging Applications. *Materials* **2017**, *10*, 1008. [CrossRef]
43. Mu, C.; Xue, L.; Zhu, J.; Jiang, M.; Zhou, Z. Mechanical and Thermal Properties of Toughened Poly (L-Lactic) Acid and Lignin Blends. *Bioresources* **2014**, *9*, 5557–5566. [CrossRef]
44. Lugoloobi, I.; Li, X.; Zhang, Y.; Mao, Z.; Wang, B.; Sui, X.; Feng, X. Fabrication of Lignin/Poly(3-Hydroxybutyrate) Nanocomposites with Enhanced Properties via a Pickering Emulsion Approach. *Int. J. Biol. Macromol.* **2020**, *165*, 3078–3087. [CrossRef]
45. Lopera-Valle, A.; Caputo, J.V.; Leão, R.; Sauvageau, D.; Luz, S.M.; Elias, A. Influence of Epoxidized Canola Oil (eCO) and Cellulose Nanocrystals (CNCs) on the Mechanical and Thermal Properties of Polyhydroxybutyrate (PHB)—Poly(lactic acid) (PLA) Blends. *Polymers* **2019**, *11*, 933. [CrossRef] [PubMed]
46. Myalenko, D.; Fedotova, O. Physical, Mechanical, and Structural Properties of the Polylactide and Polybutylene Adipate Terephthalate (PBAT)-based Biodegradable Polymer during Compost Storage. *Polymers* **2023**, *15*, 1619. [CrossRef]
47. Dammak, M.; Fourati, Y.; Tarrés, Q.; Delgado-Aguilar, M.; Mutjé, P.; Boufi, S. Blends of PBAT with Plasticized Starch for Packaging Applications: Mechanical Properties, Rheological Behaviour and Biodegradability. *Ind. Crops Prod.* **2020**, *144*, 112061. [CrossRef]
48. Aversa, C.; Barletta, M.; Cappiello, G.; Gisario, A. Compatibilization Strategies and Analysis of Morphological Features of Poly(Butylene Adipate-Co-Terephthalate) (PBAT)/Poly(Lactic Acid) PLA Blends: A State-of-Art Review. *Eur. Polym. J.* **2022**, *173*, 111304. [CrossRef]
49. Palsikowski, P.A.; Kuchnier, C.N.; Pinheiro, I.F.; Morales, A.R. Biodegradation in Soil of PLA/PBAT Blends Compatibilized with Chain Extender. *J. Polym. Environ.* **2018**, *26*, 330–341. [CrossRef]
50. Syuhada, D.N.; Azura, A.R. Waste Natural Polymers as Potential Fillers for Biodegradable Latex-Based Composites: A Review. *Polymers* **2021**, *13*, 3600. [CrossRef] [PubMed]
51. Tolga, S.; Kabasci, S.; Duhme, M. Progress of Disintegration of Polylactide (PLA)/Poly(Butylene Succinate) (PBS) Blends Containing Talc and Chalk Inorganic Fillers under Industrial Composting Conditions. *Polymers* **2020**, *13*, 10. [CrossRef]

polymers

MDPI

Article

Dielectric and Viscoelastic Behavior of Polyvinyl Butyral Films

Jesús G. Puente-Córdova [1,*], Flor Y. Rentería-Baltiérrez [2,*], Beatriz López-Walle [1] and Juan A. Aguilar-Garib [1]

1 Facultad de Ingeniería Mecánica y Eléctrica, Universidad Autónoma de Nuevo León, Av. Universidad s/n, Cd. Universitaria, San Nicolás de los Garza 66455, Mexico
2 Facultad de Ciencias Químicas, Universidad Autónoma de Nuevo León, Av. Universidad s/n, Cd. Universitaria, San Nicolás de los Garza 66455, Mexico
* Correspondence: jesus.puentecr@uanl.edu.mx (J.G.P.-C.); flor.renteriabltz@uanl.edu.mx (F.Y.R.-B.)

Abstract: Dielectric and thermal properties of polyvinyl butyral (PVB) were studied in this work, using dynamic electrical analysis (DEA) at frequencies from 100 Hz to 1 MHz and temperatures from 293 K to 473 K. Two electrical relaxation processes were investigated: glass transition and interfacial polarization. Above the glass transition temperature (~343 K), interfacial polarization dominates conductive behavior in polyvinyl butyral. The framework of the complex electric modulus was used to obtain information about interfacial polarization. The viscoelastic behavior was analyzed through dynamic mechanical analysis (DMA), where only the mechanical manifestation of the glass transition is observed. The experimental results from dielectric measurements were analyzed with fractional calculus, using a fractional Debye model with one cap-resistor. We were successful in applying the complex electric modulus because we had a good correlation between data and theoretical predictions. The fractional order derivative is an indicator of the energy dissipated in terms of molecular mobility, and the calculated values close to 1 suggest a conductive behavior at temperatures above the glass transition temperature of PVB.

Keywords: dielectric properties; polyvinyl butyral; electric modulus; fractional calculus; relaxation phenomena

1. Introduction

Electrical conductivity, permittivity, and dielectric strength are among the most important properties of materials for the electrical industry. In the case of polymers, they are widely applied as electrical insulators or dielectrics, and seldom as semiconductors. However, there are new synthesized polymers with a specific molecular structure that can be converted into electrically conductive materials [1–3]. Some structures are far from thermodynamic equilibrium, thus physical aging affects the electrical properties of polymers [4,5]. Physical aging is a phenomenon associated with structural recovery or relaxation of the glassy state towards the metastable equilibrium of the amorphous state, presenting important changes in physical properties as a function of temperature and time. The accumulation of electrical charges on the surface (electrodes) or bulk (traps) contributes to the physical aging of polymers under electrical fields. This generates a scenario that produces space charge relaxation, also known as interfacial polarization or Maxwell–Wagner–Sillars polarization [6,7]. Some experiments have been performed on space charge relaxation above the glass transition temperature (Tg). Among them, there are some works that use dynamic electrical analysis (DEA) to study the space charge relaxation or interfacial polarization in polymers [8–11]. It is established that above the Tg and at low frequencies (typically 10^{-3}–10^{2} Hz), extrinsic and intrinsic carriers contribute to the generation of interfacial polarization. Since these observations are not easy to obtain in the spectrum of complex relative permittivity ε_r^* using DEA, the use of the complex electric modulus M^* has been proposed as an alternative [12,13].

The concept of electric modulus was first proposed by McCrum et al. [13,14] and intensively used for the investigation of the electrical relaxation phenomena of organic dielectric materials [8,9,12–14]. This is analogous to the complex elastic modulus used in linear viscoelasticity. M^* represents the electric field relaxation in the dielectric material when the electric displacement remains constant. Richert & Wagner [15] present experimental results about M^* considering it as a true dielectric relaxation, while ε_r^* refers to the dielectric retardation process. M^* is calculated as the inverse of ε_r^*, Equation (1):

$$M^* = M' + M'' = \frac{1}{\varepsilon_r^*} = \frac{1}{\varepsilon_r' - i\varepsilon_r''} = \frac{\varepsilon_r'}{(\varepsilon_r')^2 + (\varepsilon_r'')^2} + i\frac{\varepsilon_r''}{(\varepsilon_r')^2 + (\varepsilon_r'')^2} \quad (1)$$

where M' is the real part and M'' the imaginary part of the complex electric modulus, and ε_r' is the real part and ε_r'' the imaginary part of the complex relative permittivity.

Considering this knowledge, in this study we investigate the thermal and dielectric response of polyvinyl butyral (PVB). We are interested in this polymer due to its potential applications, such as for hybrid magnetic materials and nanocomposites, another topic of our research group [16–18]. PVB is a polyacetal considered as an amorphous random copolymer with viscoelastic properties and optical clarity, suitable for use in the manufacture of safety glass laminates, especially in automotive and architectural glass [19,20]. PVB is obtained by modifying polyvinyl alcohol by condensation with butyraldehyde in an acid media. The result of this chemical modification produces polymeric chains whose structure is formed by three structural units along the PVB chains: butyral, alcohol, and acetate units [21,22]. The butyral unit is hydrophobic, and the alcohol unit is considered hydrophilic, giving in the adhesion property to be used as coating on wood, glass, or metals.

Regarding studies of electrical properties of PVB, most were carried out below Tg (~343 K). For instance, Jain et al. carried out the study of the electrical properties of PVB using the thermally stimulated discharge currents (TSDC) technique [23]. They observed two relaxation phenomena, the first one at a temperature of 347 K and the second one at 423 K. They attributed the first phenomenon to the disorientation of electric dipoles, and the second phenomenon to the release of electric charges trapped in the PVB. Some authors [24–26] worked on the dielectric properties of PVB, around the glass transition temperature, by measuring ε_r^*. As a general conclusion of these studies, the main relaxation consists of a thermally activated process, where electric dipoles of various chemical groups of the polymer chains participate. Carine et al. [27] studied the secondary and principal relaxations of PVB using dielectric spectroscopy, for ε_r' and ε_r''. It was observed that when the temperature increased from 120 to 250 K, ε_r'' increases and passes through the secondary β-relaxation peak at about 190 K. Above this region, ε_r'' shows the α-relaxation peak at approximately 340 K, which is associated with the glass transition phenomenon.

The objective of this work is to evaluate the thermal, viscoelastic, and dielectric properties and the interfacial polarization of PVB films at temperatures above Tg. The novelty of this study lies in applying the fractional calculus approach under the theoretical framework of the electric modulus. The main feature of PVB is that it is a good alternative to perform functions as organic insulator or dielectric material in technological devices such as solar cells, electronic components, and supercapacitors, among others.

2. Materials and Methods

2.1. Preparation of Polyvinyl Butyral Films

Polymer samples were manufactured as thin films from solutions of polyvinyl butyral (PVB-BM, Mw = 53,000 g/mol, Sekisui, Osaka, Japan) in tetrahydrofuran (THF, 99%, Sigma-Aldrich, St. Louis, MI, USA) as a solvent. A concentration of 10% by weight was used, which ensures an appropriate viscosity for the manufacture of the films. To guarantee the dissolution process, the PVB-THF mixture was stirred at 700 rpm for 1 h at 313 K. The solutions obtained were tape-casted on a PTFE surface, after which the solvent was removed by natural convection at room temperature for 24 h. The thickness of the obtained films was ~20 μm.

2.2. Thermal, Viscoelastic, and Dielectric Characterization

To determine the thermal response of the PVB, thermogravimetric analysis (TGA) was carried out in an air atmosphere at a heating rate of 10 K/min in a temperature range between 295 and 873 K. The viscoelastic response of the PVB was analyzed by dynamic mechanical analysis (DMA) using a Perkin Elmer DMA8000 (Waltham, MA, USA), in tension mode, for three frequency values (0.1 Hz, 1 Hz and 10 Hz). The sample was subjected to a controlled strain with an amplitude of 0.01 mm, in a temperature range from 300 to 443 K, using a 2 K/min heating rate. By this isochronal condition, the storage modulus E' and loss modulus E'' were registered. Tangent delta was calculated as $\tan\delta_m = E''/E'$. The dimensions of the analyzed films were 20 mm in length, 7 mm in width, and a thickness of 20 μm.

Dynamic electrical analysis (DEA) measurements were taken in a frequency interval from 100 Hz to 1 MHz using an Agilent E4980A electrometer (Santa Clara, CA, USA). The PVB sample was placed between two electrodes of copper. The applied voltage that defines the electric field oscillates between −1 and 1 V, and the measurements were carried out at different temperatures, from 293 to 473 K in steps of 10 K. To corroborate that space charge relaxation manifests at high temperatures, electrical current measurements were fulfilled using the technique identified as thermally stimulated discharge current (TSDC). For this purpose, a polarization voltage Vp = 500 V in direct current was applied, for a polarization temperature Tp = 363 K in a time interval of 1200 s. The sample was cooled rapidly (~20 K/min). Then it was short-circuited between the electrodes, and the recording of the electric current was started from 303 K. A heating rate of 5 K/min was used, and a Keithley 6517B electrometer (Beaverton, OR, USA) with a precision of 10^{-14} A was used for the TSDC measurements.

2.3. The Fractional Debye Model and the Cap-Resistor

Nowadays, fractional calculus is a branch of mathematics that deals with the derivatives and integrals of arbitrary order [28] and offers diverse applications in science and engineering. In the field of rheology, fractional operators are used to model the mechanical behavior of polymeric systems, where there is a partial storage or dissipation of energy. Regarding the study of the dielectric behavior of materials, the empirical Cole-Cole, Cole–Davidson, and Havriliak–Negami equations were used for the interpretation and analysis of experimental data. However, full precision has not yet been achieved. In previous studies it has been shown [29–31] that by using fractional calculus, the dielectric behavior of organic materials can be described with a higher degree of accuracy than by using integer order calculus. In this sense, we use an electrical element, the "cap-resistor" [32,33], which is analogous to the spring-pot or Scott Blair element used in linear viscoelasticity. The cap-resistor, Equation (2), intimately combines the electrical response of a capacitor (energy storage) with the response of a resistor (energy dissipation) by a differential operator of fractional order γ:

$$V = \frac{\tau^{\gamma}}{C} D_t^{\gamma} Q = \frac{(RC)^{\gamma}}{C} D_t^{\gamma} Q \tag{2}$$

where V is the voltage, Q is the electric charge, C is the capacitance, and R is the resistance. $D_t^{\gamma} Q$ is the fractional derivative operator of the Q, with $0 \leq \gamma \leq 1$. The electric relaxation time $\tau = RC$ can be associated with the time required by electric charge carriers in movement for a complete reorganization to a new equilibrium state. When $\gamma = 0$ in Equation (2), the capacitance or dielectric response is obtained; otherwise, when $\gamma = 1$, the electrical resistance response is obtained. The fractional derivative operator, used in Equation (2), can be expressed by the Riemann-Liouville definition, Equation (3):

$$D_t^{\gamma}Q = \frac{1}{\Gamma(1-\gamma)}\frac{d}{dt}\int_0^t \frac{Q(\xi)}{(t-\xi)^{\gamma}}d\xi \quad (3)$$

with $0 < \gamma < 1$, and $\Gamma(\cdot)$ is the gamma function. The fractional order derivative can be related to the partial energy dissipated by the system; several studies have shown that fractional orders are related to the molecular mobility of organic dielectric materials [17,29]. Therefore, the equivalent electrical circuit of the classical Debye model was modified by replacing the resistor with one cap-resistor, γ. Figure 1 shows the components that describe this fractional model. The cap-resistor, γ, characterizes times (τ) associated with dielectric response in the region around dielectric relaxation, and the two capacitor elements represent the storage dielectric response. C_s is the capacitance at low frequencies or high temperatures, and C_∞ is the capacitance at high frequencies or low temperatures.

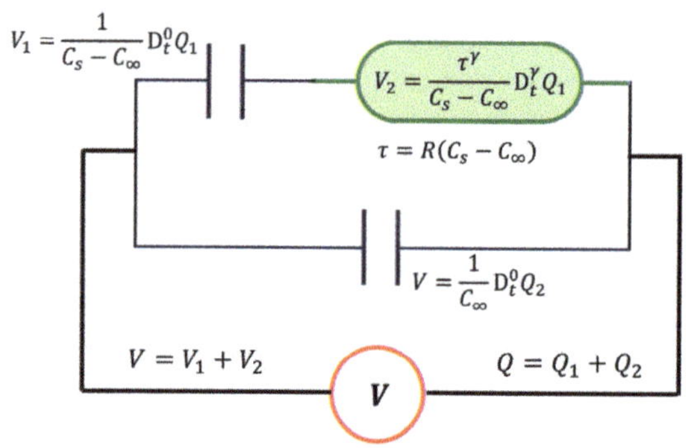

Figure 1. The fractional Debye model with one cap-resistor.

Then, from the constitutive equations of the capacitors and the cap-resistor, the fractional differential equation can be obtained, Equation (4), with $0 < \gamma \leq 1$:

$$V = \frac{Q - C_\infty V}{C_s - C_\infty} + \frac{\tau^\gamma}{C_s - C_\infty} D_t^\gamma [Q - C_\infty V] \quad (4)$$

From Equation (4), the complex dielectric permittivity ε_r^* is calculated as a function of the angular frequency ω, under isothermal conditions, see Equation (5). The Fourier transform of the fractional derivative D_t^γ is obtained as the product of $(i\omega)^\gamma$ by the Fourier transform of the function [29,31].

$$\varepsilon_r^* = \frac{\varepsilon_{rs} + \varepsilon_{r\infty}(i\omega\tau)^\gamma}{1+(i\omega\tau)^\gamma} \quad (5)$$

where ε_{rs} is the permittivity at low frequencies or high temperatures and $\varepsilon_{r\infty}$ is the permittivity at high frequencies or low temperatures. It is interesting that Equation (5) also corresponds to the empirical Cole-Cole function used to fit the dielectric data. Furthermore, it is possible to obtain a mathematical expression for the complex electric modulus M^* from Equation (5), as pointed out in Equation (1). The result is presented in Equation (6).

$$M^* = M' + iM'' = \frac{1}{\varepsilon_r^*} = M_\infty M_s \frac{1+(i\omega\tau)^\gamma}{M_\infty + M_s(i\omega\tau)^\gamma} \quad (6)$$

Here, $M_\infty = 1/\varepsilon_{r\infty}$ is the electric modulus at high frequencies and $M_s = 1/\varepsilon_{rs}$ is the electric modulus at low frequencies. Equations (7) and (8) show the mathematical expressions for M' and M'', respectively, which were obtained from Equation (6).

$$M' = M_\infty M_s \frac{M_\infty + (M_s + M_\infty)\left[\cos\left(\gamma\frac{\pi}{2}\right)(\omega\tau)^\gamma\right] + M_s(\omega\tau)^{2\gamma}}{M_\infty^2 + 2M_sM_\infty\left[\cos\left(\gamma\frac{\pi}{2}\right)(\omega\tau)^\gamma\right] + M_s^2(\omega\tau)^{2\gamma}} \tag{7}$$

$$M'' = M_\infty M_s \frac{(M_\infty - M_s)\left[\sin\left(\gamma\frac{\pi}{2}\right)(\omega\tau)^\gamma\right]}{M_\infty^2 + 2M_sM_\infty\left[\cos\left(\gamma\frac{\pi}{2}\right)(\omega\tau)^\gamma\right] + M_s^2(\omega\tau)^{2\gamma}} \tag{8}$$

It is important to highlight that the fractional order derivative γ can only take values between 0 and 1. When $\gamma = 1$, we recover the classical equations for the Debye model (single relaxation process). Equations (7) and (8) correspond to a symmetric response, which is typical of interfacial polarization or space charge relaxation [11,12]. From imaginary modulus M'', Equation (8), it is possible to establish a relationship from the maximum peak to estimate the relaxation time $\tau_M = \tau(M_s/M_\infty)^{1/\gamma}$, where τ is defined as mentioned above. The loss tangent is calculated as $\tan\delta_e = M''/M'$.

3. Results and Discussion

3.1. Thermal and Viscoelastic Analysis

Thermal stability and degradation of PVB due to the heating effect were determined by means of TGA. Figure 2 shows the thermogravimetric curve obtained. The degradation temperature is identified at 641 K, whose decomposition products correspond mainly to butyraldehyde and acetic acid, due to the degradation of butyral and acetate units of polymer chains, respectively. At a temperature above 741 K, the sample degrades completely [34,35]. The presence of water was not detected, since no significant decrease in weight loss at the boiling temperature was observed. These results are important because they impact the electrical response and must be considered in the dielectric measurements above the Tg of PVB.

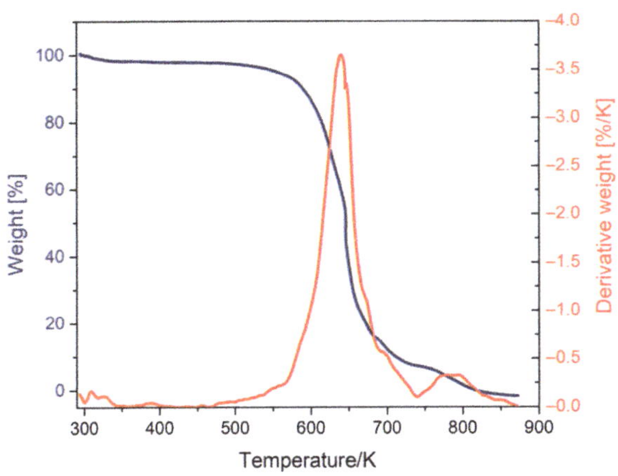

Figure 2. Thermogravimetric curves of PVB.

The mechanical manifestation of the relaxation phenomena of the PVB was analyzed using DMA, in a temperature range from 303 to 393 K. Figure 3a presents the graph of the storage modulus E' as a function of temperature. A decrease in E' was observed when the temperature increased; this is associated with the α-relaxation, which corresponds to the mechanical manifestation of the glass transition, and is correlated with the cooperative mobility of the chain segments. This process corresponds with a notable increase in molecular mobility when the temperature increases and with long-range molecular movements at temperatures around the glass transition temperature Tg. At temperatures above 353 K, a region in which E′ remains constant, but sensitive to frequency, was identified. This region was found to be elastomeric, occurs at temperatures above Tg, and is a function of the number of physical crosslinks generated between the polymer chains and chemical groups. At higher temperatures, the polymer begins to flow. In Figure 3b, tan δ_m is shown as a function of temperature for each of the analyzed frequencies. In these plots, tan δ_m is the ratio between the energy dissipated and the energy stored by the PVB. The maximum or peak of tan δ_m corresponds to the maximum energy dissipation of the main PVB relaxation process. The temperature at which these peaks occur is estimated as the Tg; and, when the frequency increases, the tan δ_m peaks shift toward higher temperatures. The estimated Tg values were 340 K, 346 K, and 353 K for the frequencies of 0.1 Hz, 1 Hz, and 10 Hz, respectively. Additionally, a differential scanning calorimetry (DSC) measurement was carried out to calculate the Tg of the PVB, whose value was 333 K. These values are consistent with those reported by Calucci et al. [36] and by Kirchberg et al. [37]. In DMA, the stimulus comes from a surface force, which generates significant mobility in the polymer chains of PVB. Through DEA, it is possible to analyze the molecular mobility in a more selective way, since a body force (electric field) is applied and acts on the electric charge carriers, electrons, ions, and electric dipoles.

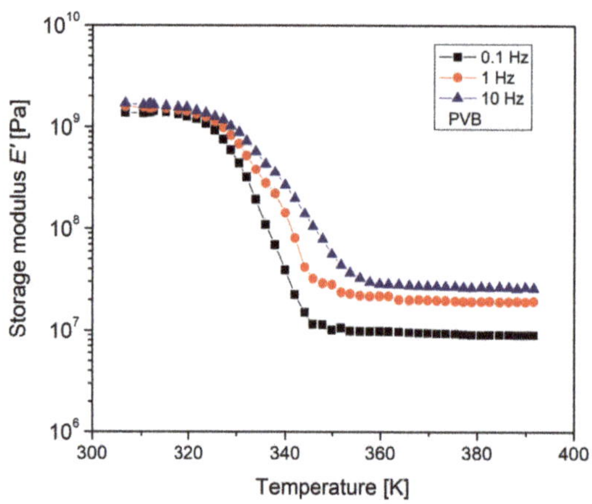

(**a**)

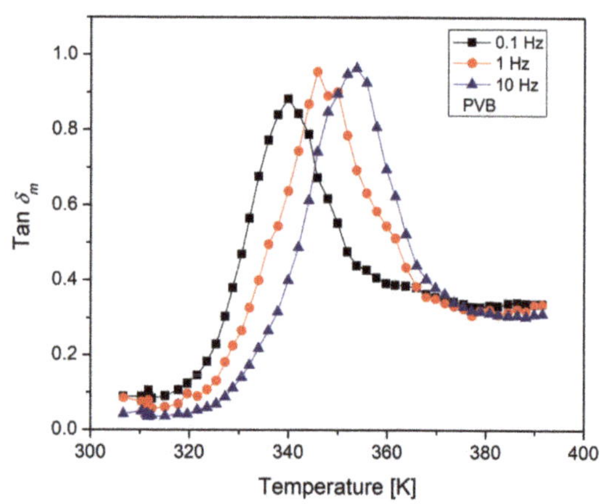

(**b**)

Figure 3. (**a**) storage modulus and (**b**) tan δ_m, as a function of temperature and frequency, for PVB.

3.2. Dynamic Dielectric Analysis DEA

The measurements of real ε_r' and imaginary relative permittivity ε_r'' were obtained as a function of frequency in an interval from 100 Hz to 1 MHz, for different constant temperatures. Figure 4 shows the isothermal curves obtained. Regarding the isothermal curves of ε_r' (Figure 4a), at temperatures above 353 K, as the frequency increases, the magnitude of ε_r' decreases. This decrement is more pronounced as the temperature increases. Among the electrical phenomena that must define the shape of these curves, there is, in addition to the α-relaxation (electrical manifestation of the glass transition), the interfacial polarization and the conductivity process. These phenomena are strongly influenced by the polymer structure and morphology [5,31]. Conductive processes, ionic conduction, and interfacial polarization must occur at temperatures above the Tg [8,10,12]. However, it is a complex process to separately identify the presence of any electrical relaxation phenomena.

Figure 4. (**a**) real permittivity ε_r' and (**b**) imaginary permittivity ε_r'', as a function of frequency and temperature, for PVB.

Components of the complex permittivity, ε_r' and ε_r'', depend on the electric dipoles of the PVB. The value of ε_r' depends directly on the number of them that are oriented according to the intensity and direction of the applied electric field. The dipoles that present significant mobility are the hydroxyl groups (~1.6 D), since acetals have greater mass and volume, presenting restricted mobility [22,38]. Instead, as the frequency of the applied electric field increases, there is not enough time to be aligned by the field so that the number of oriented dipoles decreases, which is reflected in the magnitude of ε_r'. At low frequencies and above the Tg, an increase in the value of ε_r' was observed, which represents a greater electrical charge storage capacity. In the curves of ε_r'' (Figure 4b), as the temperature increases, a broad relaxation peak was observed. This peak shifts towards high frequencies as the temperature increases, corroborating the presence of electrical phenomena that are constituent by thermally activated movements. The relaxation peaks are associated with a distribution of relaxation times of the electric dipoles of the PVB and correspond mainly to the electrical manifestation of the glass transition. In Figure 4b, when the temperature increases above the Tg, a behavior of type $\varepsilon_r'' \propto 1/f$ at low frequencies is observed, resulting [27,36] in ionic conductivity due to the rubbery state and the activation of the micro-Brownian motion of the polymer segments. This observed behavior is consistent with various reports for polymers like PMMA, PS, PET, and PEEK [8,10–12]. However, in this temperature region, interfacial polarization can also manifest itself and thus overlap with the ionic conductivity. Therefore, the data are approached using the theoretical framework of the electric modulus M^*.

3.3. Electric Modulus Approach

Using the Equation (1), M' and M'' were calculated from the Figure 4a,b. In Figure 5a the isothermal curves for M' are presented, and in Figure 5b those obtained for M'' are presented. These curves exhibit shapes close to the typical curves of electrical relaxation phenomena in polymeric materials [13,16]. In Figure 5a, M' increases as the frequency increases and becomes more pronounced as the temperature increases, which is the opposite to what was observed for ε_r'. The increment of M' with a frequency at temperatures above Tg corresponds to a maximum or peak in the curves of M''.

In the M'' curves (Figure 5b), two types of relaxation are distinguished. Some peaks located in a temperature interval of 293–373 K are identified and correspond to the electrical manifestation of the glass transition of the PVB (α-relaxation). The peaks identified in the temperature interval 393–473 K correspond to the interfacial polarization, which is consistent with the results obtained by Jain et al. [23]. The interfacial polarization peak is only detectable in electrical measurements and not from DMA, DSC, or TGA. The interfacial polarization phenomenon occurs through the accumulation of electrical charge in regions of the electrode-polymer interface or within the polymer bulk [8,39,40]. This process is due to the complex structure of the PVB, an amorphous copolymer, which presents traps with different energy levels (shallow and deep traps). For PMMA and PEI it was reported that the physical model of Coelho and the macrodipole concept are utilized to explain the space charge relaxation or interfacial polarization that occurs at temperatures above the Tg [8,12]. In Figure 5b, the relaxation peaks shift toward high frequencies as the temperature increases [8,9,41].

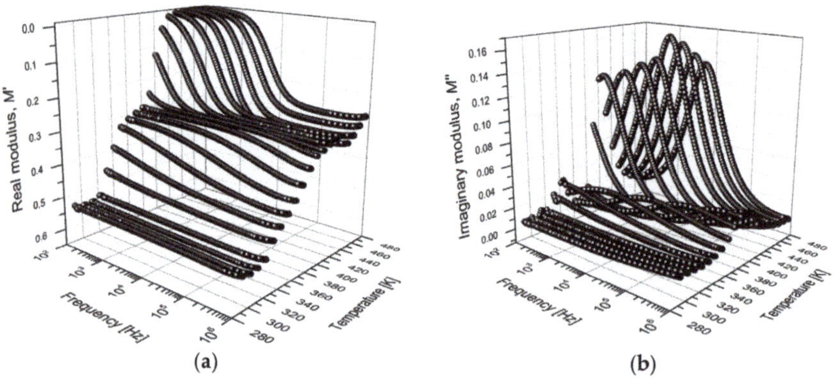

Figure 5. (**a**) real modulus M′ and (**b**) imaginary modulus M″, as a function of frequency and temperature, for PVB. The axis of M′ has been inverted for better visualization.

From Figure 5b, the activation energy (E_a) for interfacial polarization was calculated, using the relation $\tau = \exp(-E_a/kT)$, by plotting $\ln(\tau)$ against 1000/T (Figure 6). The value obtained for $E_a = 0.75$ eV, consistent with others results reported for PMMA, epoxy resin, EPDM, among others [12,14,42]. This is higher compared to the activation energy calculated from α-relaxation ($E_a = 0.45$ eV). The contribution of conductivity σ at low frequencies was estimated from Figure 4b using the relation $\sigma = \omega\varepsilon_0\varepsilon_r''$. The Arrhenius plot for conductivity is also shown in Figure 6. The calculated activation energy E_a is 0.87 eV, suggesting that conductive processes are associated and originated due to ionic conduction by segmental motions of polymer chains of PVB.

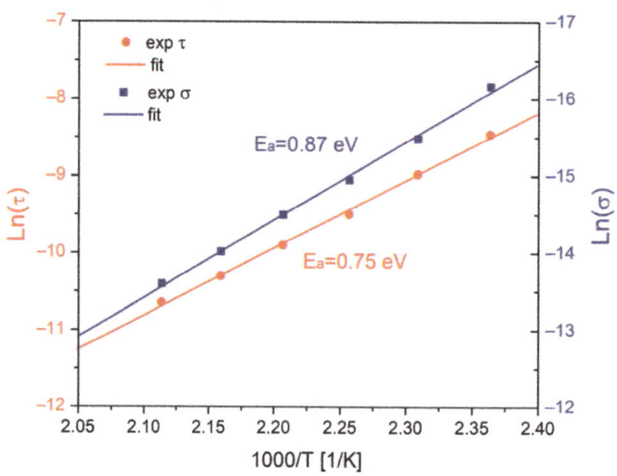

Figure 6. Arrhenius plot of relaxation time and conductivity for PVB.

An isochronous graph was constructed at 100 Hz for ε_r'' and M″. The results obtained are presented in Figure 7. In this figure, in both isochronal curves at a temperature around 343 K, the peak associated with the α-relaxation is clearly identified. For a temperature around 393 K, only in the curve M″ is it possible to identify the relaxation peak associated with the phenomenon of interfacial polarization. This finding corroborates that by calculating the M*, it is possible to differentiate the interfacial polarization from the α-relaxation and conductive processes. The analysis of ε_r'' and M″ presents identical information, but with different emphasis. The temperature at which the peaks associated with the interfacial polarization appear is an indicator of a transition from short-range to long-range mobility, corresponding to charge carriers trapped inside the PVB, thus requiring more energy to be released, in comparison with the cooperative motions of dipoles on the α-relaxation. At low temperatures, for the peaks associated with the α-relaxation, the electric charge carriers move short distances. At higher temperatures, such carriers can travel long distances, in the order of magnitude of the size of a repeating unit of PVB, which results in segmental movements. Moreover, the mobility of the charge carriers corresponding to the α-relaxation exhibits more localized mobility than the movements corresponding to the interfacial polarization.

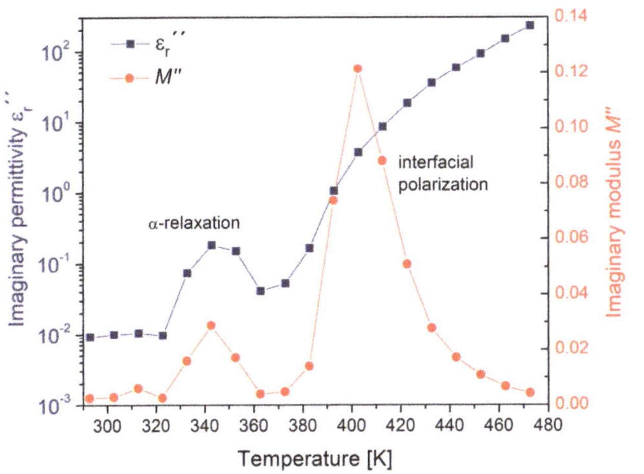

Figure 7. Comparison for ε_r'' and M″ at 100 Hz, as a function of temperature.

3.4. Comparison between Experimental Data and Theoretical Predictions from the Fractional Debye Model

The experimental data of interfacial polarization in the frequency domain can be interpreted theoretically based on the fractional Debye model, Equations (7) and (8). The parameters that describe this model (M_s, M_∞, τ_M, γ) were obtained from the experimental curves for the PVB and are presented in Table 1. Figure 8 shows a good agreement between the theoretical predictions and experimental data obtained for M′, M″, tan δ_e and the Cole-Cole plot, at a temperature of 433 K. Only in the tan δ_e plot did there occur a deviation observed at low frequencies, related to the contribution of the DC conductivity, which is not considered in the proposed model.

Table 1. Parameters obtained experimentally for the fractional Debye model.

	423 K	433 K	443 K	453 K	463 K	473 K
M_s	5×10^{-4}	1×10^{-4}	2×10^{-4}	4×10^{-4}	3×10^{-4}	3×10^{-4}
M_∞	0.304	0.303	0.333	0.317	0.295	0.274
τ_M (s)	2.11×10^{-4}	1.26×10^{-4}	7.51×10^{-5}	5.02×10^{-5}	3.35×10^{-5}	2.37×10^{-5}
γ	0.932	0.958	0.985	0.989	0.992	0.994

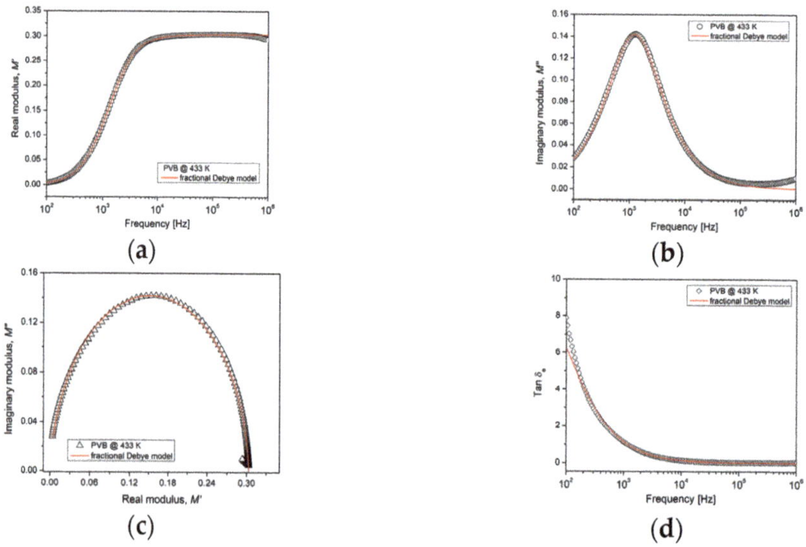

Figure 8. Comparison of the fractional Debye model with experimental data at 433 K. (**a**) real modulus, (**b**) imaginary modulus, (**c**) Cole-Cole plot, and (**d**) Tanδ_e curve.

From Table 1, there is observed a strong dependence of the parameters on the temperature. Relaxation time τ_M decreases as the temperature increases, due to an increment in conductivity. Regarding the fractional order, this increases as the temperature increases, meaning that the PVB presents a conductive response. The fractional-order derivative is an indicator of the energy dissipated in the polymer. When the value of γ is close to 0, the mobility reflects a capacitor-like behavior, and when the value of γ is close to 1, the mobility indicates a resistor-like behavior. Therefore, interfacial polarization is strongly influenced by the conductive properties of PVB at temperatures above Tg.

3.5. Thermally Stimulated Discharge Current (TSDC)

Electrical current measurements were carried out using TSDC to corroborate that space charge or interfacial polarization manifests at temperatures above Tg. The current as a function of temperature was obtained and is presented in Figure 9. As a reference, in this figure the results of M″ are added. On the TSDC curve, two peaks were recognized at different temperatures. The first one, with the lowest amplitude (~70 pA), is located around 342 K and corresponds to the α-relaxation of the PVB. This corresponds to several electric dipoles aligned with the applied electric field. It can be stated that this peak is due to the stored electric charge from the elastic orientation of the electric dipoles. The second peak is observed at a 423 K, and its amplitude (~545 pA) is greater than that obtained for the α-relaxation. This peak can be associated with the space charge relaxation. When PVB is polarized above Tg, charge carriers are trapped. The thermal energy produces a release of trapped carriers from localized energy states. The trapping sites in amorphous polymers are favored by surface states, chain ends, and molecular disorder. Also, polar groups in polymer chains can act as traps. Consequently, this creates a scenario for the generation, transport, and diffusion of space charge. The differences between the results by M″ and TSDC are mainly due to the experimental nature of each analysis technique.

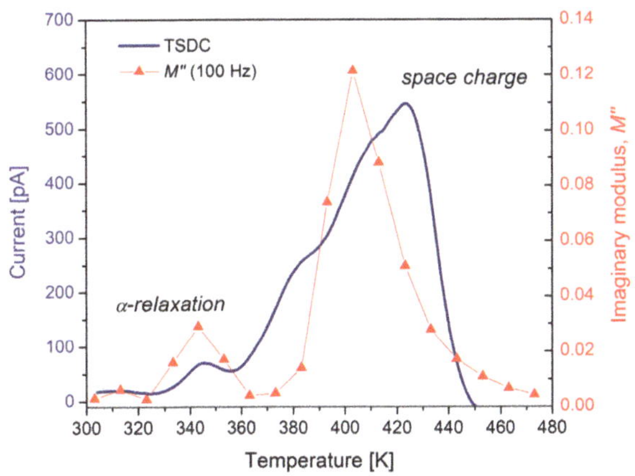

Figure 9. Comparison between TSDC and M″ (100 Hz) for the PVB.

The area under the curve for each of the peaks is proportional to the electric charge that the PVB can store; it is evident that at high temperatures the storage capacity increases. Taking the ratio between the maxima of space charge and α-relaxation, a value of 7.78 is obtained. For M″ at 100 Hz, from Figure 5b, a value of 4.75 is obtained. This can be considered an indicator of the energy dissipated in these relaxation mechanisms. These results are important for potential applications of PVB such as thermo-electret, or sensor in the areas of electronic, mechatronic, and electrical engineering. To accomplish this, several authors suggest that PVB must have the form of a thin film, with an appropriate relative permittivity, low loss factor, facility of fabrication, high transparency, and good mechanical stability [43,44].

4. Conclusions

The properties that polyvinyl butyral PVB exhibits at temperatures around Tg are in concordance with the occurrence of interfacial polarization or space charge relaxation. At low frequencies and high temperatures, the conductivity of the polymer overlaps with the dielectric characteristics. We achieve the application of the complex electric modulus approach to separate the contribution of conductivity from relaxation phenomena. At temperatures higher than Tg, a relaxation phenomenon was obtained which is called space charge. The activation energy associated with this relaxation is 0.75 eV, which is influenced by the segmental movements of the PVB. In this study, it was proposed to use a fractional Debye model under the framework of the electric modulus. The theoretical predictions compared with the experimental results are good enough to demonstrate the capability of fractional derivatives for the interpretation of dielectric data under the electric modulus formalism.

Author Contributions: Conceptualization, J.G.P.-C. and F.Y.R.-B.; methodology, J.G.P.-C., B.L.-W. and F.Y.R.-B.; validation, F.Y.R.-B. and B.L.-W.; formal analysis, J.G.P.-C. and F.Y.R.-B.; investigation, J.G.P.-C. and F.Y.R.-B.; resources, J.A.A.-G. and F.Y.R.-B.; writing—original draft preparation, J.G.P.-C.; writing—review and editing, J.A.A.-G., F.Y.R.-B. and B.L.-W.; supervision, J.A.A.-G.; project administration, B.L.-W.; funding acquisition, F.Y.R.-B. All authors have read and agreed to the published version of the manuscript.

Funding: This research received no external funding.

Institutional Review Board Statement: Not applicable.

Data Availability Statement: The data generated during the current study are available from the corresponding author upon reasonable request.

Acknowledgments: In memory of Edgar Reyes-Melo (31 March 1969–1 March 2022), who dedicated his career to the application of fractional calculus to the relaxation phenomena of polymers. The authors acknowledge the support provided by SNII-CONAHCYT.

Conflicts of Interest: The authors declare no conflict of interest.

References

1. Chen, J.; Zhu, Y.; Huang, J.; Zhang, J.; Pan, D.; Zhou, J.; Ryu, J.E.; Umar, A.; Guo, Z. Advances in Responsively Conductive Polymer Composites and Sensing Applications. *Polym. Rev.* **2021**, *61*, 157–193. [CrossRef]
2. Wang, Y.; Liu, A.; Han, Y.; Li, T. Sensors Based on Conductive Polymers and Their Composites: A Review. *Polym. Int.* **2020**, *69*, 7–17. [CrossRef]
3. Kumar, D.; Sharma, R.C. Advances in Conductive Polymers. *Eur. Polym. J.* **1998**, *34*, 1053–1060. [CrossRef]
4. Merrick, M.M.; Sujanani, R.; Freeman, B.D. Glassy Polymers: Historical Findings, Membrane Applications, and Unresolved Questions Regarding Physical Aging. *Polymer* **2020**, *211*, 123176. [CrossRef]
5. Mandal, S.; Dasmahapatra, A.K. Effect of Aging on the Microstructure and Physical Properties of Poly(Vinyl Alcohol) Hydrogel. *J. Polym. Res.* **2021**, *28*, 269. [CrossRef]
6. McKenzie, R.; Zurawsky, W.; Mijovic, J. A Molecular Interpretation of Maxwell-Wagner-Sillars Processes. *J. Non. Cryst. Solids* **2014**, *406*, 11–21. [CrossRef]
7. Samet, M.; Levchenko, V.; Boiteux, G.; Seytre, G.; Kallel, A.; Serghei, A. Electrode Polarization vs. Maxwell-Wagner-Sillars Interfacial Polarization in Dielectric Spectra of Materials: Characteristic Frequencies and Scaling Laws. *J. Chem. Phys.* **2015**, *142*, 194703. [CrossRef]
8. Mudarra, M.; Belana, J.; Cañadas, J.C.; Diego, J.A.; Sellarès, J.; Díaz-Calleja, R.; Sanchís, M.J. Space Charge Relaxation in Polyetherimides by the Electric Modulus Formalism. *J. Appl. Phys.* **2000**, *88*, 4807–4812. [CrossRef]
9. Mudarra, M.; Sellarès, J.; Canãdas, J.C.; Diego, J.A. Sublinear Dispersive Conductivity in Polyetherimides by the Electric Modulus Formalism. *IEEE Trans. Dielectr. Electr. Insul.* **2015**, *22*, 3327–3333. [CrossRef]
10. Arous, M.; Karray, F.; Hammami, H.; Perrier, G. Study of the Space Charge Relaxation in Poly Ether Ether Ketone (PEEK). *Phys. Chem. News* **2003**, *10*, 5–7.
11. Hasegawa, Y.; Ohki, Y.; Fukunaga, K.; Mizuno, M.; Sasaki, K. Complex Permittivity Spectra of Various Insulating Polymers at Ultrawide-Band Frequencies. *Electr. Eng. Jpn.* **2017**, *198*, 11–18. [CrossRef]
12. Mudarra, M.; Díaz-Calleja, R.; Belana, J.; Cañadas, J.C.; Diego, J.A.; Sellarès, J.; Sanchís, M.J. Study of Space Charge Relaxation in PMMA at High Temperatures by Dynamic Electrical Analysis. *Polymer* **2001**, *42*, 1647–1651. [CrossRef]
13. Tian, F.; Ohki, Y. Electric Modulus Powerful Tool for Analyzing Dielectric Behavior. *IEEE Trans. Dielectr. Electr. Insul.* **2014**, *21*, 929–931. [CrossRef]
14. Tsangaris, G.M.; Psarras, G.C.; Kouloumbi, N. Electric Modulus and Interfacial Polarization in Composite Polymeric Systems. *J. Mater. Sci.* **1998**, *33*, 2027–2037. [CrossRef]
15. Richert, R.; Wagner, H. The Dielectric Modulus: Relaxation versus Retardation. *Solid State Ion.* **1998**, *105*, 167–173. [CrossRef]
16. Reyes-Melo, M.E.; Miranda-Valdez, I.Y.; Puente-Córdova, J.G.; Camarillo-Hernández, C.A.; López-Walle, B. Fabrication and Characterization of a Biocompatible Hybrid Film Based on Silver Nanoparticle/Ethyl Cellulose Polymer. *Cellulose* **2021**, *28*, 9227–9240. [CrossRef]
17. Rentería-Baltiérrez, F.Y.; Reyes-Melo, M.E.; Puente-Córdova, J.G.; López-Walle, B. Application of Fractional Calculus in the Mechanical and Dielectric Correlation Model of Hybrid Polymer Films with Different Average Molecular Weight Matrices. *Polym. Bull.* **2023**, *80*, 6327–6347. [CrossRef]
18. Puente-Córdova, J.G.; Reyes-Melo, M.E.; López-Walle, B.; Miranda-Valdez, I.Y.; Torres-Castro, A. Characterization of a Magnetic Hybrid Film Fabricated by the In-Situ Synthesis of Iron Oxide Nanoparticles into Ethyl Cellulose Polymer. *Cellulose* **2022**, *29*, 3845–3857. [CrossRef]
19. Vedrtnam, A.; Pawar, S.J. Experimental and Simulation Studies on Fatigue Behavior of Laminated Glass Having Polyvinyl Butyral and Ethyl Vinyl Acetate Interlayers. *Fatigue Fract. Eng. Mater. Struct.* **2018**, *41*, 1437–1446. [CrossRef]
20. Iwasaki, R.; Sato, C.; Latailladeand, J.L.; Viot, P. Experimental Study on the Interface Fracture Toughness of PVB (Polyvinyl Butyral)/Glass at High Strain Rates. *Int. J. Crashworthiness* **2007**, *12*, 293–298. [CrossRef]
21. Hoepfner, J.C.; Loos, M.R.; Pezzin, S.H. Role of the Degree of Acetalization on Dynamic Mechanical Properties of Polyvinyl Butyral/Carbon Nanotube Composites. *J. Appl. Polym. Sci.* **2019**, *136*, 48146. [CrossRef]
22. Bartolotta, A.; Marco, G.D.I.; Federico, M.; Carini, G. Rising the Molecular Mobility of Polyvinyl Butyral by Plasticizers: Towards Energy Storage Applications. *Atti Della Accad. Peloritana Dei Pericolanti-Cl. Di Sci. Fis. Mat. E Nat.* **2020**, *98*, 1–11. [CrossRef]
23. Jain, K.; Kumar, N.; Mehendru, P.C. Electrical and Dielectric Properties of Polyvinyl Butyral: I. Studies of Charge Storage Mechanism. *J. Electrochem. Soc.* **1979**, *126*, 1958–1963. [CrossRef]
24. Funt, B.L. Dielectric Dispersion in Solid Polyvinyl Butyral. *Can. J. Chem.* **1952**, *30*, 84–91. [CrossRef]
25. Saad, G.R.; El-Shafee, E.; Sabaa, M.W. Dielectric and Mechanical Properties in the Photodegradation of Poly(Vinyl Butyral) Films. *Polym. Degrad. Stab.* **1995**, *47*, 209–215. [CrossRef]

26. Mehendru, P.C.; Kumar, N.; Arora, V.P.; Gupta, N.P. Dielectric Relaxation Studies in Polyvinyl Butyral. *J. Chem. Phys.* **1982**, *77*, 4232–4235. [CrossRef]
27. Carini, G.; Bartolotta, A.; Carini, G.; D'Angelo, G.; Federico, M.; Di Marco, G. Water-Driven Segmental Cooperativity in Polyvinyl Butyral. *Eur. Polym. J.* **2018**, *98*, 172–176. [CrossRef]
28. Ortigueira, M.; Machado, J. Which Derivative? *Fractal Fract.* **2017**, *1*, 3. [CrossRef]
29. Rentería-Baltiérrez, F.Y.; Reyes-Melo, M.E.; Puente-Córdova, J.G.; López-Walle, B. Correlation between the Mechanical and Dielectric Responses in Polymer Films by a Fractional Calculus Approach. *J. Appl. Polym. Sci.* **2021**, *138*, 49853. [CrossRef]
30. Reyes-Melo, E.; Martinez-Vega, J.; Guerrero-Salazar, C.; Ortiz-Mendez, U. Application of Fractional Calculus to the Modeling of Dielectric Relaxation Phenomena in Polymeric Materials. *J. Appl. Polym. Sci.* **2005**, *98*, 923–935. [CrossRef]
31. Reyes-Melo, M.E.; Martínez-Vega, J.J.; Guerrero-Salazar, C.A.; Ortiz-Méndez, U. Mechanical and Dielectric Relaxation Phenomena of Poly(Ethylene-2,6- Napthalene Dicarboxylate) by Fractional Calculus Approach. *J. Appl. Polym. Sci.* **2006**, *102*, 3354–3368. [CrossRef]
32. Meng, R. Application of Fractional Calculus to Modeling the Nonlinear Behaviors of Ferroelectric Polymer Composites: Viscoelasticity and Dielectricity. *Membranes* **2021**, *11*, 409. [CrossRef] [PubMed]
33. Luo, D.; Chen, H.S. A New Generalized Fractional Maxwell Model of Dielectric Relaxation. *Chin. J. Phys.* **2017**, *55*, 1998–2004. [CrossRef]
34. Ivanov, V.B.; Zavodchikova, A.A.; Popova, E.I.; Lazareva, O.L.; Belova, O.A.; Kryuchkov, I.A.; Bykov, E.V. Accelerated Testing of Thermo-Oxidative Degradation of Polyvinyl Butyral. *Thermochim. Acta* **2014**, *589*, 70–75. [CrossRef]
35. Hajian, M.; Reisi, M.R.; Koohmareh, G.A.; Jam, A.R.Z. Preparation and Characterization of Polyvinylbutyral/Graphene Nanocomposite. *J. Polym. Res.* **2012**, *19*, 409. [CrossRef]
36. Calucci, L.; Pizzanelli, S.; Mandoli, A.; Birczyński, A.; Lalowicz, Z.T.; De Monte, C.; Ricci, L.; Bronco, S. Unravelling Main- and Side-Chain Motions in Polymers with NMR Spectroscopy and Relaxometry: The Case of Polyvinyl Butyral. *Polymers* **2021**, *13*, 2686. [CrossRef]
37. Kirchberg, S.; Rudolph, M.; Ziegmann, G.; Peuker, U.A. Nanocomposites Based on Technical Polymers and Sterically Functionalized Soft Magnetic Magnetite Nanoparticles: Synthesis, Processing, and Characterization. *J. Nanomater.* **2012**, *2012*, 670531. [CrossRef]
38. Carini, G.; Carini, G.; D'Angelo, G.; Federico, M.; Di Marco, G.; Bartolotta, A. Enhancing the Molecular Cooperativity of Polyvinyl Butyral Using Liquid Additives. *J. Polym. Sci. Part B Polym. Phys.* **2018**, *56*, 340–346. [CrossRef]
39. Hoang, M.Q.; Boudou, L.; Le Roy, S.; Teyssedre, G. Dissociating Space Charge Processes from Orientation Polarization in Poly(Ethylene Naphthalate) Films. *J. Phys. D. Appl. Phys.* **2014**, *47*, 455306. [CrossRef]
40. Mazzanti, G.; Montanari, G.C.; Dissado, L.A. Electrical Aging and Life Models: The Role of Space Charge. *IEEE Trans. Dielectr. Electr. Insul.* **2005**, *12*, 876–890. [CrossRef]
41. Montanari, G.C.; Seri, P.; Dissado, L.A. Aging Mechanisms of Polymeric Materials under DC Electrical Stress: A New Approach and Similarities to Mechanical Aging. *IEEE Trans. Dielectr. Electr. Insul.* **2019**, *26*, 634–641. [CrossRef]
42. Marín-Genescà, M.; Mujal-Rosas, R.; García-Amorós, J.; Mudarra, M.; Juan, X.R.; Fajula, X.C. Study Analysis of Thermal, Dielectric and Functional Characteristics of an Ethylene Polyethylene Diene Monomer Blended with End-of-Life Tire Microparticles Amounts. *Polymers* **2021**, *13*, 509. [CrossRef] [PubMed]
43. Pan, J.; Liu, S.; Zhang, H.; Lu, J. A Flexible Temperature Sensor Array with Polyaniline/Graphene–Polyvinyl Butyral Thin Film. *Sensors* **2019**, *19*, 4105. [CrossRef] [PubMed]
44. Ruan, Y.; Shi, P.; Lei, Y.; Weng, S.; Li, S.; Huang, L.; Lin, X.; Yao, H. Polyvinyl butyral/graphene oxide nanocomposite modified electrode for the integrate determination of terminal metabolites of catecholamines in human urine. *J. Electroanal. Chem.* **2019**, *848*, 113267. [CrossRef]

polymers

MDPI

Article

Influence of Reduced Graphene Oxide and Carbon Nanotubes on the Structural, Electrical, and Photoluminescent Properties of Chitosan Films

Jesús R. González-Martínez [1], Ana B. López-Oyama [2,3,*], Deyanira Del Ángel-López [2], Crescencio García-Guendulain [4,*], Eugenio Rodríguez-González [2], Eder U. Pulido-Barragan [2,3], Felipe Barffuson-Domínguez [5], Aurora G. Magallanes-Vallejo [2] and Pablo J. Mogica-Cantú [2]

1 Departamento de Investigación en Física (DIFUS), Universidad de Sonora, Blvd. Transversal S/N., Hermosillo 83000, Sonora, Mexico; jrgm18@hotmail.com

2 Centro de Investigación en Ciencia Aplicada y Tecnología Avanzada, Unidad Altamira del Instituto Politécnico Nacional, Km. 14.5 Carr. Puerto Industrial, Altamira 89600, Tamaulipas, Mexico; ddelangel@ipn.mx (D.D.Á.-L.); eugenior62@gmail.com (E.R.-G.); eder.pulido.bar@gmail.com (E.U.P.-B.); aumagallanes@gmail.com (A.G.M.-V.); pmogicac2300@alumno.ipn.mx (P.J.M.-C.)

3 Conahcyt-Cicata Unidad Altamira, IPN. Km. 14.5 Carretera Puerto Industrial, Altamira 89600, Tamaulipas, Mexico

4 Tecnologico de Monterrey, Escuela de Ingeniería y Ciencias, Blvd. Petrocel Km. 1.3, Altamira 89603, Tamaulipas, Mexico

5 Departamento de Física, Universidad de Sonora, Blvd. Transversal S/N., Hermosillo 83000, Sonora, Mexico; felipe.barffuson@unison.mx

* Correspondence: ablopezoy@conahcyt.mx (A.B.L.-O.); crescencio.garcia@tec.mx (C.G.-G.)

Abstract: Chitosan is a biopolymer with unique properties that have attracted considerable attention in various scientific fields in recent decades. Although chitosan is known for its poor electrical and mechanical properties, there is interest in producing chitosan-based materials reinforced with carbon-based materials to impart exceptional properties such as high electrical conductivity and high Young's modulus. This study describes the synergistic effect of carbon-based materials, such as reduced graphene oxide and carbon nanotubes, in improving the electrical, optical, and mechanical properties of chitosan-based films. Our findings demonstrate that the incorporation of reduced graphene oxide influences the crystallinity of chitosan, which considerably impacts the mechanical properties of the films. However, the incorporation of a reduced graphene oxide–carbon nanotube complex not only significantly improves the mechanical properties but also significantly improves the optical and electrical properties, as was demonstrated from the photoluminescence studies and resistivity measurements employing the four-probe technique. This is a promising prospect for the synthesis of new materials, such as biopolymer films, with potential applications in optical, electrical, and biomedical bioengineering applications.

Keywords: carbon nanotubes; reduced graphene oxide; chitosan; film-forming properties; electrical properties; photoluminescence

1. Introduction

Chitosan (Cs) is a naturally occurring material produced from chitin, usually found in crustaceans. Cs is a polysaccharide and, after cellulose, is the second largest renewable biopolymer. It is constituted of β-(1-4)-D-glucosamine units and β-(1-4)-N-acetyl-glucosamine according to the deacetylation degree of chitin. The chemical and functional properties of chitosan allow it to form gels and films, opening up new possibilities for industrial, optical, electrical, and biomedical applications. In recent years, chitosan films have attracted wide attention and have been applied in many fields, such as water treatment, dye removal, tissue engineering, wound healing [1,2], photocatalytic activity [3], and flexible electronics. Despite the numerous advantages and exceptional properties of Cs, its poor intrinsic mechanical and electrical properties impede its use in a wider range of applications. An effective approach for improving chitosan's physical and mechanical properties is to develop films through the incorporation of fillers, such as clays, metal nanoparticles, carbon nanotubes, and/or reduced graphene oxide.

Dinakaran et al. [4] developed a composite nanostructure of nickel oxide (NiO) and NiO-embedded graphene hybrid nanostructure loaded into a PVDF (polyvinylidene difluoride) matrix. The nanocomposites were synthesized by hydrothermal reactions at 80 °C and fabricated through the solvent-casting technique. Buinov et al. fabricated conductive tissue engineering nanocomposite films based on chitosan and surfactant-stabilized graphene dispersions obtained by solution casting. Conductivity was linearly enhanced to 0.43 S/cm by increasing the concentration of rGO to 5 wt% [5]. Talebi et al. [6] prepared a polycaprolactone/chitosan/polypyrrole conductive composite film and reported that polypyrrole addition resulted in enhanced electrical conductivity. Although Cs has been widely used in many fields, there is scarce information on studying its electrical properties by embedding reduced graphene oxide and/or CNTs (carbon nanotubes). Investigations on the properties of nanocellulose [7–11], photoluminescent materials such as boron nitride [12], as well as carbon-based materials such as CNTs [13] and reduced graphene oxide (rGO) [14,15] have been developed for understanding the interesting properties of the distinct materials, which, when embedded in Cs, can integrate their individual properties for an enhanced synergistic effect in the development of electrically conductive films.

Reduced graphene oxide, a two-dimensional honeycomb lattice material, has attracted considerable attention due to its exceptional mechanical, thermal, electronic, optical, and electrical properties. Furthermore, rGO exhibits a large surface area, oxygen-containing chemical groups, strong ion exchangeability, and high electrical conductivity. These characteristics provide rGO/polymer composites with high strength, high modulus, and superior electrical conductivity. Barra et al. employed a green method for the preparation of electrically conductive bionanocomposites from chitosan and reduced graphene oxide subjected to a hydrothermal reduction using caffeic acid and then dispersed in chitosan. In the final bionanocomposites, the electrical conductivity was achieved at 0.7 S/m. Since increased tensile strength was accompanied by decreased water solubility, the reduced graphene oxide led to a significant improvement in the mechanical performance of the chitosan matrix [16].

Reduced graphene oxide–carbon nanotubes (rGO-CNTs) nanocomposites have been developed for enhancing electrochemical energy storage [17], supercapacitors [18,19], electromagnetic shielding [20], and water treatment [21]. Surprisingly, the development of chitosan-based films loaded with rGO-CNTs mixture has been employed only for a few applications, as shown below. Water-soluble rGO sheets grafted with chitosan effectively disperse multiwalled carbon nanotubes that efficiently enhance the well-known poor mechanical properties of Cs by the effective non-covalent π-π interactions between graphene sheets and CNTs through hydrogen bonding between grafted Cs were synthesized by Pan et al. [22], Mergen et al. [23] reported the study of electrical, optical, and mechanical properties of Cs/graphene nanoplatelets and Cs/MWCNT (multiwalled carbon nanotube) biocomposites films with promising properties for food packaging, UV protection, and biomedical applications. Their findings demonstrate that with the increase in graphene nanoplatelets or MWCNTs, the surface conductivity and mechanical and optical properties significantly increased.

In this paper, an rGO/Cs and rGO-CNTs/Cs composite film with enhanced electrical properties and photoluminescent properties was developed using the solvent-cast technique. When reduced graphene oxide was added, the mechanical properties of Cs films were improved, but not the electrical values. However, it is worth noting that the crystallinity of chitosan films was also improved. This indicates that rGO not only improves the mechanical properties of Cs but also improves the crystalline arrangement. In addition, we explore the synergistic effect of a mixture of rGO-CNTs into Cs. When the mixture was added to chitosan, the electrical conductivity was enhanced, demonstrating that CNTs considerably influence the carrier transport into Cs, improving the crystallinity, electrical conductivity, and photoluminescent (PL) properties of chitosan. This demonstrates that CNTs not only improve the mechanical properties of chitosan by physical enhancement but also enhance the electrical and optical properties of chitosan. Under the condition of maintaining high flexibility, the mechanical properties of chitosan films were improved as well, thus promoting the application of films in many fields, such as tissue engineering, biosensors, and biomedical applications.

2. Materials and Methods

2.1. Materials

Chitosan powder having a deacetylation degree of 75% (50 kDa, 190 kDa, and 310 kDa), multiwalled carbon nanotubes, HOPG (8000 mesh, purity 99.99%), HCl, NaCl, and acetic acid were acquired from Sigma-Aldrich, St. Louis, MO, USA. All the reagents were analytically graded and used as received without further purification. Reduced graphene oxide (rGO) was provided by the Center of Investigation in Applied Science and Advanced Technology, Altamira, México, previously synthesized as described in reference [14] and was employed without further purification.

2.2. Films Preparation

Films containing rGO and a combination of CNTs and rGO were made by dissolving chitosan of three molecular weights (low, medium, and high) in a 2.0% wt. aqueous solution of acetic acid. The solvent was then evaporated for 72 h at 30 °C under controlled conditions within a ventilated oven. Then, the films were characterized at room conditions of 35 °C and a relative humidity of ≈30%, characteristic of Hermosillo, Sonora, México city [coordinates 29°04′30″ N 110°57′30″ W].

For more details of labels assigned to the films produced in this work, see Supplementary Information (Figure S1).

2.3. UV–Vis Spectroscopy

The UV–visible spectra of the aqueous solutions were obtained using a PerkinElmer Lambda 45 dual-beam spectrophotometer at a scan rate of 280 nm/min. Measurements were made in the 200–900 nm range using a 1 cm quartz cell. Samples in acetic acid were allowed to equilibrate in the cell for one hour before being transferred to the quartz cell. The final volume was 3 mL.

2.4. Films Transparency

Following the procedure described by Vega et al. [24], the transparency of the film was calculated. Each sample was cut into a square shape measuring 1.25 × 1.25 cm and placed on the interior section of a spectrophotometer cell (Agilent Technologies Cary 5000 UV-vis NIR) that scanned within the 200–900 nm range. The measured absorbance at 800 nm (A_{800}) was employed for transparency calculation, according to Han et al. [25].

$$Transparency = \frac{A_{800}}{T} \quad (1)$$

A_{800} is the measure of absorbance at a wavelength of 800 nm, whereas T stands for the thickness of the film in millimeters. To obtain an average value, each film was measured three times.

2.5. Fourier Transform Infrared Spectroscopy

FTIR spectra were obtained by employing a Thermo Scientific FTIR instrument, the Nicolet 5 s model (Madison, WI, USA), equipped with an ATR iD3 accessory. A total of 64 scans were taken from 600 to 4000 cm^{-1} in range, with a resolution of 4 cm^{-1}.

2.6. Films Thickness

We employed a digital micrometer (Mitutoyo, MDC-1 PX, Kawasaki-Shi, Kanawaga, Japan) to measure the film thickness. Four measurements were taken at different positions on the films [26].

2.7. Color Analysis

The film color analyses were conducted using a Minolta Chroma Meter colorimeter CR 300 (Osaka, Japan). The use of the CIELAB scale was employed in determining color parameters, specifically the lightness scale from black to white (L), and the chromaticity parameters, a* (redness–greenness) and b* (yellowness–blueness). All measures were carried out on a standard white plate. Six replicates for each sample were measured. The calculation of color differences (ΔE) was obtained as follows:

$$\Delta E = \left[(\Delta L^*)^2 + (\Delta a^*)^2 + (\Delta b^*)^2\right]^{1/2} \quad (2)$$

$$\Delta L^* = L^* - L_0, \ \ \Delta a = a^* - a_0 \text{ and } \Delta b^* = b^* - b_0 \quad (3)$$

L^*, a^*, and b^*, represent the standard color parameters while L_0, a_0, and b_0 represent the color parameter of the sample.

2.8. Ionic Exchange Capacity (IEC)

We measured the IEC of three samples of each film using the titration method. The films were subjected to a prior IEC activation process, in which the films were alternately converted to the H^+ and OH^- forms. The process involved exposing the films to a solution of 0.1 M HCl and 0.1 M NaCl for 48 h.

2.9. Swelling and Solubility

Swelling degree (DS) and solubility in water (WS) were analyzed. The state of equilibrium is reached by films (1.25 cm × 1.25 cm) immersed in distilled water at room temperature (25 °C) within an hour. The percentage of swelling degree (% DS) was determined through the following equation:

$$\%DS = \frac{w_e - w_0}{w_0} \quad (4)$$

w_0 is the initial dry weight of each film. In contrast, w_e represents the weight of the absorbed films. After 24 h, the swollen films were dried at 60 °C. The percentage of solubility in water ($\%SW$) is calculated:

$$\%SW = \frac{w_0 - w_d}{w_0} \times 100 \quad (5)$$

where w_0 is the initial weight of the dry films and w_d is the dry weight of the films after drying. The DS and SW values were measured in triplicate.

2.10. XRD Analysis

A Bruker D8 Advance X-ray diffractometer using Cu Ka radiation was used to obtain the X-ray diffraction patterns of the films. The X-ray wavelength was 0.15406 nm, and data were collected at a scan rate of 2°/min over a scattering angle range (2θ) of 5 to 90°.

2.11. Mechanical Characterization

A total of six specimens were tested for tensile strength (TS), elongation at break (EB), and Young's modulus using a Texture Technologies Corp. texturometer under ASTM D882 [27]. The test was conducted at room temperature and ambient conditions (35 °C and ≈30% relative humidity). Each strip in all samples (10 mm wide, 40 mm long, 23 mm longitudinal slit) was cut with a razor blade. A crosshead speed of 10 mm/min and a gauge length of 30 mm were used as experimental parameters. Each film was subjected to five replicates of the tensile test.

2.12. X-ray Photoelectron Spectroscopy

The X-ray photoelectron spectroscopy (XPS) technique (PHI 5000 Versaprobe instrument, MN, USA) was utilized to conduct a surface analysis of four films selected for their use of MCs (medium molecular weight) chitosan (190 kDa) and the addition of carbon nanotubes and reduced graphene oxide. The high-resolution spectra for the detected and analyzed elements were obtained through the utilization of a monochromatic Mg Ka radiation as an X-ray source.

2.13. SEM Analysis

The high-magnification SEM images of medium-molecular-weight chitosan with rGO and rGO-CNTs were acquired using a JEOL JSM-7800F (Tokyo, Japan) with an operating energy of 5.0 eV.

2.14. Resistivity Measurements

The resistivity measurements of the films were carried out in ambient conditions using a source measuring unit (Keithley Model 4200-SCS system, Solon, OH, USA) equipped with a 0.40-inch equidistant tips configuration. The data were collected for 10 s with a gate voltage in the range of −0.0001 to +0.0001 V. The resistivity (ρ) measurement, using the four-probe method, considers the thickness and shape of the films and was calculated as follows:

$$Resistivity\ (\rho) = (V/I)2\pi S \tag{6}$$

where S is the distance of the tips, V is the measured voltage, and I is the current applied.

Finally, the electrical conductivity (σ) was calculated as follows:

$$\sigma = \frac{1}{\rho} \tag{7}$$

2.15. Photoluminescence

The photoluminescence technique was used to investigate the optical properties of the films. The films were excited by a 375 nm CW laser (CNI, MDL-III-375). A 45° focused laser beam was directed at the sample surface, and perpendicular photoluminescence spectra were obtained using an Ocean Optics HR4000 spectrometer, Dunedin, FL, USA.

3. Results

3.1. UV–Vis Analysis

To characterize the absorption features of homogeneous dispersions, the solubilized rGO/Cs and rGO-CNTs/CS solutions were subjected to UV–vis spectroscopic analysis in acetic acid (0.001 mg mL^{-1}) in the 200–900 nm wavelength range. Figure 1a depicts UV–vis absorption for pure chitosan of low, medium, and high molecular weight. Spectra reveal the absorption peaks at 228 nm and 280 nm corresponding to the $\eta \rightarrow \pi$ and $\pi \rightarrow \pi$ transitions, respectively. Figure 1b exhibits the absorption spectra of pure rGO and rGO/Cs solutions from low-, medium-, and high-molecular-weight chitosan. The typical electronic transitions related to rGO were observed in the UV–vis spectrum with a maximum absorbance of around 227 nm, which agrees with previous results [14,28,29]. Mixtures of rGO/Cs from low and medium molecular weights exhibit the absorption bands of both the CS and rGO; however, the sample containing reduced graphene oxide/high-molecular-weight chitosan (rGO/HCs) solution shows an intense absorption at around 228 nm, which masks the CS transitions. This high absorption can be attributed to the absorption capacity of chitosan chromophores in the blue region than rGO. Figure 1c shows the UV–vis absorption of rGO-CNTs/Cs solution. We observe that this sample displays a higher capacity to disperse carbon nanotubes due to the repulsion of the charged groups of Cs when dispersed in an acidic medium. The hydrophobic character of rGO/Cs is enhanced by the area of π-conjugated aromatic domains, which facilitates the attachment of CNTs to the rGO/Cs surface [30].

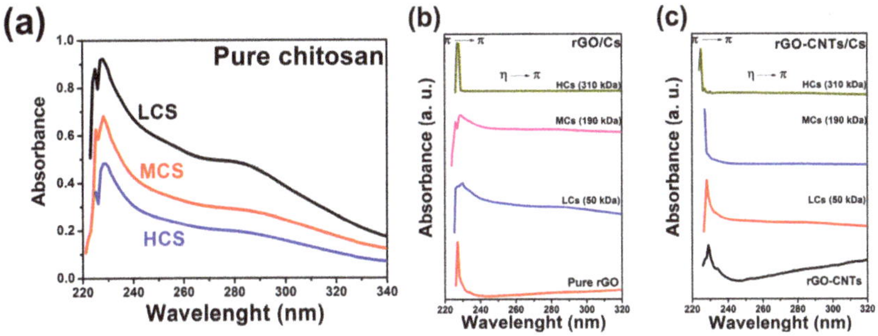

Figure 1. UV–vis absorption spectra of (**a**) pure chitosan, (**b**) rGO/chitosan and (**c**) rGO-CNTs/chitosan solutions from 50, 190, and 310 kDa of molecular weight.

3.2. FTIR Spectroscopy

The Fourier Transform Infrared (FTIR) spectroscopy proved to be sufficient in analyzing the functional groups and structural modifications of films. rGO/Cs films FTIR spectra are depicted in Figure 2a. The typical superimposed -OH and NH_2 groups in chitosan are observed. In both reduced graphene oxide/medium-molecular-weight chitosan (rGO/MCs) and reduced graphene oxide/high-molecular-weight chitosan (rGO/HCs), the high wavenumber of 2883 and 2866 cm^{-1} was observed for the antisymmetric and symmetric stretching vibrations of the methyl ($-CH_3$) group. Notably, the FTIR spectra of the rGO/LCs film did not exhibit the methyl group peaks, which suggests an interaction between rGO and Cs. The peak related to the acetyl units exhibiting stretching vibration of C=O has undergone a shift toward lower wavenumber values (1640 cm^{-1}). A shift toward 1537 cm^{-1} of the protonated amide group's N-H bending vibration on the positively charged Cs suggests that it interacts with the oxygen-containing groups of the rGO surface. The β-1,4-glycosidic bond of Cs exhibited a shift toward a lower wavenumber (1066 cm^{-1}) while the C-O-C stretching vibration shifted from 1091 to 1066 cm^{-1}, maybe due to the interaction between -OH groups of Cs and the hexagonal lattice of rGO. Among all the films, the rGO/MCs films exhibited the most intense bands in the FTIR spectrum. The peak at 1637 cm^{-1} is attributed to the -C=C- asymmetric stretching of the rGO. Also, the peaks observed at 3391, 3060, 1642, 1433, and 1091 cm^{-1} correspond to C-OH, CH stretching, -C=C- stretching, CH bending, and C-O stretching, respectively. The centered band at 1637 cm^{-1} could be ascribed to the conjugation of -C=C- with a carbonyl group, displaying reduced intensities [31].

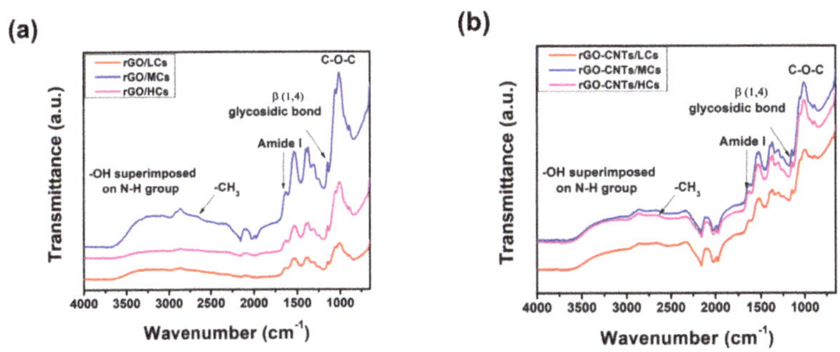

Figure 2. FTIR spectra of (**a**) rGO-Cs and (**b**) rGO-CNTs/Cs films.

The rGO-CNTs/Cs films FTIR spectra are depicted in Figure 2b. The typical -OH superimposed NH_2 of chitosan is observed. Antisymmetric and symmetric stretching vibrations of the methyl ($-CH_3$) did not appear in the spectra, indicating that the rGO-CNTs and Cs could interact. The peak associated with the acetyl units with stretching vibration of C=O shifted to 1632 cm^{-1}. The N-H bending vibration of the protonated amide group of N-H slightly shifted to 1543 cm^{-1}, implying that positively Cs interacts with the oxygen groups on the rGO-CNTs surface. β-1,4-glycosydic bond of chitosan shifted to 1151 cm^{-1} and the C-O-C stretching vibration shifted from 1091 to 1057 cm^{-1}. All rGO-CNTs/Cs films displayed almost the same intensity in the FTIR, indicating that, by embedding a mixture of CNTs and rGO into Cs, a highly condensed structure is formed. Typical vibrations of carbon-based materials also were observed. The probable interaction between CNTs and rGO is ascribed to the peak at 1637 cm^{-1}, which is attributed to the -C=C- asymmetric stretching. The peaks at 3391, 1433, and 1091 cm^{-1} are ascribed to C-OH, -C=C- stretching, and C-O stretching, respectively. The low-intensity band observed at 1637 cm^{-1} can be attributed to the conjugation of -C=C- with a carbonyl group [32]. The interaction of -OH from Cs within rGO-CNTs rings is associated with the stretching of C-O-C at 1070–1010 cm^{-1}.

The bands attributed to the amine groups in chitosan appear more intense in reduced graphene oxide–carbon nanotubes/medium-molecular-weight chitosan (rGO-CNTs/MCs) and reduced graphene oxide–carbon nanotubes/high-molecular-weight chitosan (rGO-CNTs/HCs) films. This suggests that positively charged Cs and oxygen-containing groups of carbon-based materials could effectively interact among them. Extensive hydrogen bonding has become broader in reduced graphene oxide–carbon nanotubes/low-molecular-weight chitosan (rGO-CNTs/LCs) films in suitable agreement with previous reports [33]; both -OH and NH^{+} groups could contribute to the ability of chitosan to effectively wrap along CNTs [34] and to interact with the oxygen-containing groups of the rGO. After incorporating CNTs and rGO, the peaks attributed to β (1,4) glycosidic bonds and CH_3 angular deformation of Cs appear more intense in rGO-CNTs/LCs films, suggesting that the skeleton of chitosan prevails after the addition of carbon-based materials. By increasing the chitosan long chain, a more condensed structure is observed. The slight variations in the stretching of $C-CH_3$ groups could be attributed to the π-bonds of chitosan and carbon-based materials. Embedding rGO and rGO-CNTs to chitosan influences hydrogen bonds that contribute to the interactions within the entire system. Interestingly, the antisymmetric and symmetric $-CH_3$ groups in chitosan did not appear in rGO-CNTs/Cs films, suggesting that chitosan could interact with the rGO-CNTs structure due to the formation of platelet-like structures observed in SEM micrographs elsewhere in this paper. Chitosan interacting with carbon-based materials could decrease the availability of protonated group concentration in Cs, leading to the formation of condensed structures that have great implications for the optical, mechanical, and electrical properties of films. The synthesis of CNTs and rGO could incorporate oxygen functionalities that could fall under the classification of structural defects, such as edges and basal planes, responsible for lattice vacancies that promote the synergistic effect between carbon-based materials and Cs.

3.3. Films Thickness

The thickness of the films is presented in Table 1. From our results, we report that rGO/Cs film thickness ranged around ~50–70 μm, and the rGO-CNTs/Cs thickness films ranged from ~70 to 110 μm. However, the rGO addition increases the thickness of films as chitosan molecular weight also increases, but the incorporation of rGO-CNTs complex decreases the thickness of film as the chitosan molecular weight increases, which agrees with previous reports. A thicker film was formed by adding CNTs and rGO into chitosan, which suggests that carbon-based materials could interact with the long chains of the polymer, reducing the interplanar distance and approaching crystalline structures, as was detailed in the FTIR analysis. Aryaei et al. reported that the modification of the mechanical properties of polymeric films is feasible through distribution, filler content, and orientation, which affects the stress transfer between the polymeric matrix and the filler. Their findings of dried uniform thin films demonstrate that the chitosan/MWCNT composite displayed an average thickness of 0.06 mm and 60 mm [35]. According to SEM images, the rGO-CNTs/Cs films showed more homogeneous and condensed films. It is suggested that thicker films arise from hydrogen bond formation between oxygen-containing groups of rGO and the NH_3^+ of chitosan, leading to a more compact structure.

Table 1. Thickness, opacity, and color attributes of films.

Films	Color Attributes			ΔE	Opacity ($A \cdot mm^{-1}$)	Thickness (μm)
	L*	a*	b*			
rGO/LCs	34.82	−0.07	1.39	62.84	3.31	48
rGO-CNTs/LCs	34.53	0.05	−0.11	63.16	0.85	106
rGO/MCs	38.54	0.08	0.88	59.12	2.4	52
rGO-CNTs/MCs	31.14	0.02	0.23	66.53	1.18	70
rGO/HCs	36.41	0.26	0.94	61.26	1.9	76
rGO-CNTs/HCs	32.88	−0.11	−0.09	64.80	1.04	60

3.4. Color and Opacity

The film color values are shown in Table 1. The brightness of the film is indicated by the L* value. Attribute a* denotes greenness/redness, while b* represents blueness/yellowness in the test films. Based on the analysis of the L* parameter, it is evident that rGO has played a role in the decrease in opacity as chitosan molecular weight increases, leading to a decrease in the luminosity of the films. This increase in opacity is attributed to a possible interaction between the oxygen-containing groups or rGO and the positive Cs, which leads to a reduction in the intrinsic yellow color of the pure chitosan. The rGO films appear darker due to their higher L-values, showing less browning. Similar values of *L*-attribute were shown in rGO-CNTs/Cs films. Since film transparency depends on the crystallinity of the polymeric lattice, the correlation between XRD and color results revealed that an increase in crystallinity leads to a decrease in transparency (or an increase in opacity) of the polymer due to a corresponding increase in density. The addition of rGO and rGO-CNTs was found to increase the crystallite size of films. At this point, it is important to mention that non-functionalized CNTs display a high aspect ratio and strong aggregation due to the van der Waals attraction, which leads to the formation of complex networks that contribute to the crystallite size increase [36].

3.5. Ion Exchange Capacity and Swelling Degree

Ion exchange capacity (IEC) is a parameter that quantifies the capability of functional groups to bind protons in an aqueous solution, serving as an indirect indication of the proton conductivity of polymers. This parameter is correlated with the amount of interchangeable ionic groups that impact the water absorption properties. The IEC values are shown in Table 2. In pure chitosan film, the ionic groups that provide support to the ion exchangers are -NH^{3+}. An increased proportion of protonated groups leads to a rise in the IEC. The role of water content in films is crucial for ionic conductivity, as water molecules are involved in OH- transportation. The high water absorption capacity in the films is a result of the abundant hydroxyl and amino groups of chitosan. The increase in the percentage of swelling degree (DS) in films resulted in high IEC values. Nhung et al. reported that the increase in water absorption is predominantly attributed to the increase in the IEC value. The highly hydrophilic polar groups in Cs are responsible for the increase in swelling ratio and IEC values, thereby augmenting the water absorption ability of films. The transfer of ions relies on the absorption of water. An increased uptake of water could lead to better ionic conductivity [37]. IEC values in rGO/Cs films dropped because of scarce functional groups able to transport electrons. Positively chitosan and highly charged density of oxygen-containing groups in rGO could promote the formation of the condensed structure of films from LCs to HCs molecular weight via hydrogen bonds. Interestingly, the IEC values in rGO-CNTs/Cs films remain almost constant as chitosan molecular weight increases attributed to the highly condensed structures.

Table 2. Swelling, solubility, and IEC values of films.

Films	DS (%)	SW (%)	*IEC* (meq g^{-1})
rGO/LCs	270.1	21.1	0.0077
rGO-CNTs/LCs	324.7	34.6	0.0073
rGO/MCs	323.9	29.1	0.0074
rGO-CNTs/MCs	208.7	21.4	0.0076
rGO/HCs	208.7	21.4	0.0063
rGO-CNTs/HCs	193.2	15.0	0.0076

The swelling of chitosan can be influenced by chemical composition (i.e., chain flexibility, molecular weight, chemical cross-linking, and solvent effect) and the porous structure of the polymer. The highest percentage of swelling is referred to as a more rigid and condensed structure as chitosan molecular weight increases (Table 2).

The results of this research are consistent with the typical values of three-dimensional polymeric networks, which are appropriate for swelling without disrupting their structure for up to 48 h, at which point absorption equilibrium was established [38]. The rGO/Cs and rGO-CNTs/Cs films exhibited a decrease in the percentage of swelling degree, maybe due to the formation of a more condensed structure, which leads to reduced swelling. The addition of rGO and rGO-CNTs into Cs led to a less hydrophilic network structure, primarily due to functional groups like amide, amine, hydroxyl, and a non-porous network. Carbon-based materials can influence the crystalline structure of chitosan, diminishing the availability of hydroxyl groups. Cohesive forces are involved in CNTs and rGO assembly, resulting in enhanced mechanical properties and reduced water swelling. The FTIR spectrum of films showed a less intense OH band, which could suggest the interaction between positive Cs and oxygen-containing groups of CNTs and rGO. This interaction causes a decrease in the accessibility of polar groups, leading to a reduction in the percentage of DS.

Chitosan exhibits hydrophilicity, and the rapid diffusion of water into its structure results in the swelling of films prior to their degradation. During the initial stages of the hydration process, bond breaking occurs prior to chitosan degradation. Nevertheless, swelling suppresses degradation. When an equilibrium state is reached due to the presence of salts, residual acetic acid, and other factors, the swelling profile of chitosan films reaches a peak and thereafter decreases with time. The influence of rGO on the decrease in DS of CS-rGO films is related to the presence of hydrophilic groups, cross-linking, and polar ionic groups present in the chitosan structure. This allows it to have a high interaction with an ionic solution [38,39]. We suggest that the low DS value observed due to the incorporation of rGO into chitosan may be attributed to a possible repulsion of water molecules by the presence of negative charges on the film surface. The pores of chitosan can be filled with nano-sized graphene, thereby substituting free water without any swelling effect, thereby complicating the rehydration process. This phenomenon may be attributed to the size of rGO nanosheets, which prevent water molecules from entering the chitosan during the rehydration process. Additionally, the material undergoes dehydration, resulting in a decrease in size. This may be accompanied by a rearrangement of the chitosan molecules coating the rGO, thereby limiting the hydration process. From Figure 3, it is observed that the film rGO/MCS shows a maximum DS, attributed to the availability of amino groups interacting with the hydrophilic groups of water. According to the FTIR analysis, the superimposed OH-NH_2 band is more intense in the rGO/MCS film than in the rGO/LCS and rGO/HCS films. The percentage of DS observed in films containing a mixture of rGO-CNTs exhibits a distinct profile as the molecular weight increases. For all three molecular weights of chitosan with rGO-CNTs, the intensity of the superimposed OH-NH_2 band was similar. The decrease in DS can be attributed to the presence of CNTs, which confer higher negative charges, which in turn cause higher repulsion of water molecules [40].

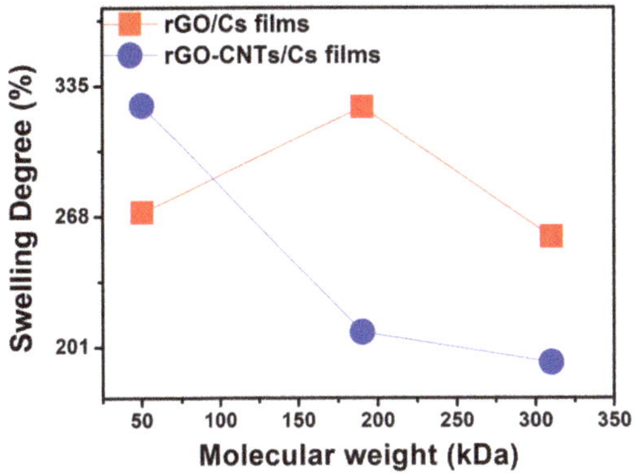

Figure 3. Swelling degree percentage of rGO/Cs and rGO-CNTs/Cs films.

3.6. XRD Diffraction Analysis

The changes in the crystallinity of films were identified using X-ray diffraction analysis. Figure 4a–c display the XRD patterns of Cs (LCs, MCs, and HCs), rGO/Cs, and rGO-CNTs/Cs films. The diffraction peaks observed at around $2\theta = 11°$ and $2\theta = 20°$ are consistent with prior research [41]. The XRD patterns of films exhibit a significant reduction in peak sharpness at $2\theta = 10°$ upon the addition of rGO-CNTs. The reflections at $2\theta = 11.61°$ and $2\theta = 16.91°$ seen in both LCs and HCs chitosan films correspond to form I and form II. As shown in Figure 4c, the rGO-CNTs/Cs film's XRD pattern exhibits a decrease in the $2\theta = 11.5°$ peak, which is absent in rGO films, indicating significant changes in the Cs form I, as displayed in Table 3.

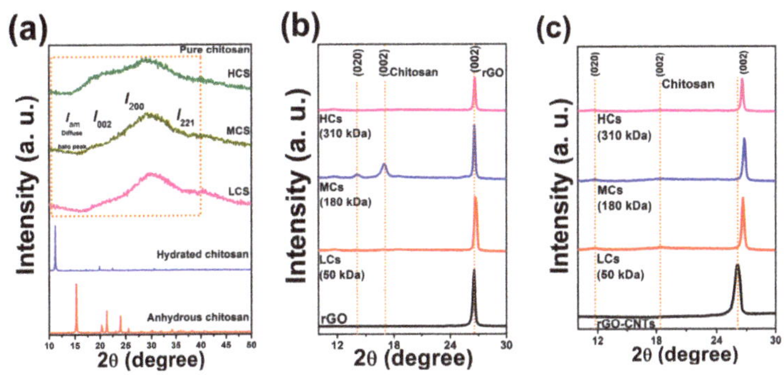

Figure 4. XRD patterns of (**a**) low-, medium-, and high-molecular-weight chitosan, (**b**) rGO/Cs films, and (**c**) rGO-CNTs/Cs films.

Table 3. Crystallinity index and crystallite size of the rGO/Cs and rGO-CNTs/Cs films.

Cs Molecular Weight	CrI (%)		Crystallite Size (nm)	
	rGO	rGO/CNTs	rGO	rGO/CNTs
LCs	95	94	31.57	42.08
MCs	56	93	42.08	31.58
HCs	92	97	42.08	42.08

According to Barbosa et al. [42], the chitosan crystallinity is highly influenced by molecular weight. The XRD patterns have shown that both CNTs and rGO exert a considerable influence on the crystalline structure of Cs, as evidenced in Figure 4b,c. Graphemic structures could influence the crystalline structure of MCs chitosan since the peaks at 14.1° and 16.95° still appear. The incorporation of rGO was found to have a significant impact on the crystalline structure of chitosan, as indicated by the XRD pattern of the rGO/Cs film. The diffractograms of rGO/LCs and rGO/HCs films depict low-intensity XRD peaks of chitosan at 16.95°. The observation suggests that the composite structure formation is significantly influenced by the intercalation between rGO and chitosan, which may be ascribed to the interaction between the oxygen-containing groups of rGO and the Cs amine groups. The peak at 2θ = 26.1° typical of carbon-based materials shifted to higher angles is due to the strain in both inter- and intra-layer directions. FWHM (Full Width at Half Maximum) values from rGO and rGO-CNTs films (002) peaks suggest that, after incorporating carbon materials into LCs and MCs, a more crystalline structure was obtained, demonstrating that carbon-based materials are responsible for the changes in the crystallinity of the films. The method to calculate the CrI is based on the formula showing the relationship between the intensity of a crystalline reflection and the intensity of the minimum peak at 2θ = 12–16°, which conditionally describes the diffuse halo peak (I_{am}). Recently, Podgorbunskikh et al. [41] emphasized the applicability of the method for calculating the crystallinity index (CrI) of chitosan. As is well known, to calculate the CrI, the method that shows the ratio between the intensities of the peaks of the crystalline and amorphous region of chitosan, centered at 2θ = 19–20° and 2θ = 12–16°, respectively, is used. However, the applicability of the method decreases substantially if we have a mixture of anhydrous and hydrated polymorphs. The method based on the ratio of the crystalline peak area to the total area of the XRD pattern has been proposed to calculate the crystallinity index of samples consisting of a mixture of polymorphs more accurately. In our work, we analyze the amorphous and crystalline region of films by the deconvolution method, as can be observed in the Supplementary Information (Figure S2): the XRD pattern was deconvoluted into internal peaks (deconvolution into five or six peaks was carried out for a more accurate description) using the Voight approximation. The crystallinity index was calculated from the ratio of the sum of the areas of the crystalline regions and the sum of the areas of the amorphous component peaks.

$$CrI = S_{cryst} / S_{total}$$

For the chitosan used in this work, the reflections corresponding to the hydrated crystalline form were detected at ~2θ = 10° (020) and 20° (200). The single reflection (220) is not observed in the XRD pattern for any of the films fabricated with the three chitosan molecular weights studied; however, by applying the deconvolution method, it was possible to observe the component of this reflection. The peaks corresponding to (100), (200), and (001) reflections in graphite described the crystalline structure of organic materials with a high carbon content, such as CNTs. The occurrence of aromatic rings is confirmed by a symmetrical (002) peak centered at 2θ = 26.1° in the XRD diffractogram of rGO and CNTs-rGO with an interplanar distance of ~3.39 Å. For more information, see the Supplementary Information (Figure S3). The non-appearance of the plane (001) suggests that the reduction process was effective in eliminating oxygen functionalities and restoring the π-π network of the rGO.

3.7. Mechanical Properties

Four to six test specimens used to calculate the mechanical properties of films reinforced with rGO and rGO-CNTs were analyzed, and the parameters reported in this work are directly related to the structural cohesivity of the films and imply their resistance to breaking under tension and fracture strain after the film has been broken [43,44]. Geng et al. developed graphene fibers through wet spinning and applied it to the study of interfacial electrolyte complexation by proposing to in situ construct a hierarchical assembly structure composed of well-ordered rGO sheets and chitosan molecules. It has been reported that graphene fibers exhibit a TS value of 875.5 MPa, which is attributed to the robust ionic bonds and hydrogen interactions between rGO and chitosan. Molecular dynamic simulations revealed that a decrease in rGO sheets and Cs molecules was accompanied by the breaking and reforming of interfacial bonds [45]. Our findings demonstrate that the films produced in this work display interesting TS and elongation-at-break (EB) values influenced by the incorporation of rGO and rGO-CNTs into chitosan. The rGO-CNTs/HCs and rGO-CNTs/LCs films showed a maximum and minimum percentage of EB of 8.37% and 2.71%, respectively. rGO-CNTs/Cs films showed improved mechanical performance with enhanced electrical conductivity, giving the films a suitable prospective application. An inadequate dispersion of CNTs is a critical factor in the improvement of mechanical properties due to the formation of nano cracks. As dispersions undergo ultra-homogenization, polymer molecules experience significant mechanical stress due to the simultaneous presence of shear, turbulence, impacts, and cavitation forces [46]. Table 4 summarizes the mechanical performance of the rGO/Cs and rGO-CNTs/Cs films.

Table 4. Summary of mechanical performance of rGO/Cs and rGO-CNTs/Cs films.

Films	Tensile Strength (MPa)	Elongation at Break (%)	Young Modulus (MPa)	Modulus of Toughness (MJ/m^3)
LCs	72.13 ± 20.03	4.75 ± 1.47	3189.59 ± 524.34	2.06 ± 0.99
rGO/LCs	83.76 ± 20.94	4.39 ± 1.16	4241.22 ± 1344.57	1.92 ± 0.80
rGO-CNTs/LCs	39.88 ± 10.35	2.71 ± 0.68	2225.41 ± 292.00	0.66 ± 0.26
MCs	35.75 ± 2.88	4.75 ± 0.86	1620.60 ± 232.64	1.17 ± 0.32
rGO/MCs	80.43 ± 14.31	5.30 ± 1.81	3897.96 ± 920.32	2.58 ± 1.04
rGO-CNTs/MCs	45.77 ± 5.86	6.33 ± 2.68	2405.75 ± 559.05	1.64 ± 0.59
HCs	49.60 ± 1.55	4.02 ± 0.81	2344.19 ± 466.55	1.21 ± 0.65
rGO/HCs	51.02 ± 9.58	5.22 ± 0.65	2447.75 ± 249.02	1.94 ± 0.39
rGO-CNTs/HCs	74.38 ± 13.65	8.57 ± 2.43	3729.1 ± 487.65	4.93 ± 1.96

Maximum tensile strength of ~83 MPa was achieved by the rGO/LCs films. The mechanical performance of the films shows that the incorporation of rGO and rGO-CNTs into the chitosan significantly influences the tensile strength and Young's modulus. From the data shown in Table 4, it can be observed that by increasing the molecular weight of the chitosan, the TS of the rGO/Cs films decreases by 60% (rGO/HCs < rGO/MCs < rGO/LCs), while the TS of the rGO-CNTs/Cs films increases by 186% (rGO/HCs > rGO/MCs > rGO/LCs). Although it is important to mention that the SEM micrographs show platelet-like structures in the rGO/Cs films, the rGO-CNTs/MCs films tend to have a maximum stress that allows them to withstand the wind without breaking, perhaps due to the interaction between the rGO and the CNTs, which modifies the crystalline structure and results in a more crystalline structure. The scenario is very similar to Young's modulus since as the molecular weight of the chitosan increases, the rGO/Cs films show a decrease of 87.5%, while the rGO-CNTs/Cs films show an increase of 16%. The platelet-like structure of the rGO/Cs film could be responsible for the formation of condensed structures that could influence the TS and EB improvement. The dispersion and strong interaction between the chitosan and rGO may result in interfacial stress transfer in the polymer film containing particles, creating a center of stress concentration [47]. The decreased mechanical properties are linked to a loss of rigidity and, in this work, could be attributed to rGO and rGO-CNTs incorporation. The rGO/LCs and rGO/MCs films displayed the highest TS values, 83 and 80 MPa, respectively, but dramatically dropped for rGO/MCs (51 MPa). These TS values could be attributed to a more densely packed structure that is favored by LCs chitosan. On the other hand, the mechanical performance of rGO-CNTs/Cs films is enhanced by the synergistic effect of CNTs and rGO, leading to an increased TS value from LCs to HCs (39, 45, 75 MPa) in suitable agreement with other reports [16,48,49]. More details of the mechanical performance are depicted in the Supplementary Information (Figure S4).

Positively charged chitosan repels each other and is extended in an acidic solution owing to the protonated -NH_3^+ groups that lead to wrapping along carbon nanotubes. The

$-NH_3^+$ could interact with defected CNT and rGO, inducing a synergistic reinforcement of the physical properties of chitosan. When combined with a suitable TS value, the rGO-CNTs/HCs films displayed the highest EB value, implying a better-quality film. The role of chitosan molecular weight in elongation at break was studied by correlating the $-NH_3^+$ group vibration from the FTIR spectra. We found that if the $-NH_3^+$ group intensity decreases, the elongation at break increases; as the band becomes more intense, the EB value is lower. The highest EB value and the lower intensity of the $-NH_3^+$ group were observed in the rGO-CNTs/HCs film. A synergistic effect of carbon materials into chitosan occurred with an effective interaction between polar groups reacting preferentially in HCs chitosan.

In comparison to pure chitosan films, the toughness value increases with the incorporation of rGO, but not for the films made with rGO-CNTs mixture. This might be due to the wrapping of the chitosan along the surface of the CNTs. In contrast, more cohesive structures are formed when rGO-CNTs are mixed into chitosan. The mechanical properties, such as tensile strength, elongation at break, and elastic modulus, agree with previous reports, thereby enabling films to be utilized in diverse applications. Currently, published studies on chitosan films do not use the synthesis method followed in this research, resulting in variations in the toughness data reported for chitosan films filled with rGO and rGO-CNTs. In addition, there is a scarcity of literature where toughness calculations are performed. Although previously reported toughness values are higher in comparison to chitosan films reported in this work, the methods of synthesis employed by them differ from the one employed in this study [50,51].

3.8. X-ray Photoelectron Spectroscopy (XPS)

XPS measurements allow us to detect O 1s, C 1s, and N 1s in rGO/MCs and rGO-CNTs/MCs films. We settled the photoelectron binding energy (BE) from their corresponding peak positions in the XPS spectra. The binding energy of the C 1s line at 284.5 eV was employed as the reference energy for spectral alignment. XPS spectra films display the changes in the C 1s spectrum; for more information, see Supplementary Information (Figure S5). Figure 5a,b show the core level spectra of N 1s. XPS spectra of the rGO-CNTs/MCs films show the π-π^* shake-up feature, which could be thought to restore the sp^2 network of carbon-based materials. The internal components of the N 1s of Cs at 399.4, 400.5, and 401.7 eV of BE were linked to the amine, amide, and protonated amine groups, respectively, according to [52]. The N 1s core level contains three components for rGO hybrid film (Figure 5a), centered at 398.48, 399.60, and 400.87 eV related to the amine and amide groups and $-NH_3^+$, the peak at higher BE is related to the amine groups (Table 5). According to Figure 5b, the N 1s contains three inner components for rGO-CNTs/MCs at 399.3, 399.8, and 400.5 eV, related to tertiary amine, secondary amine, and protonated amine, respectively. Table 6 depicts that the area ratios of A_{399} and A_{400} gradually increase because of the increase in amide groups. The ratio increases with the addition of CNTs and rGO (Table 5). The XPS analysis could support the FTIR results by confirming the interaction between chitosan amine groups and the oxygen functionalities of the rGO. The positively charged chitosan could contribute to electrostatic interaction with the rGO, which is consistent with previous reports [53,54].

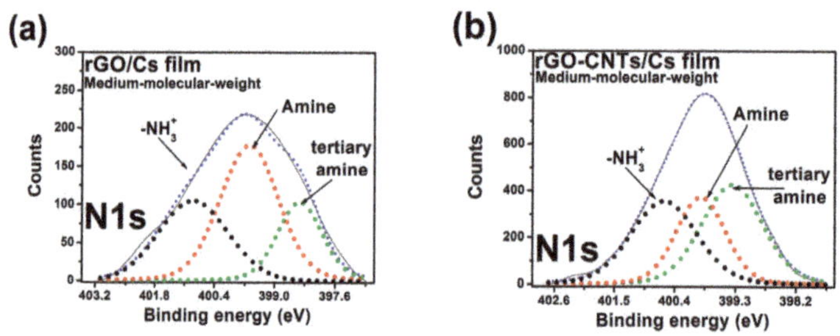

Figure 5. XPS core level N 1s of (**a**) rGO/MCs film and (**b**) rGO-CNTs/MCs film.

Table 5. XPS core level C 1s and N 1s of rGO/MCs and rGO-CNTs/MCs film.

XPS Core Level	Films	
	rGO/MCs	rGO-CNTs/MCs
C 1s	285.05 285.83 286.3 287.01	284.91 286.02 287.05 288.65
N 1s	398.32 399.63 400.93	399.37 399.93 400.60

Table 6. C/O and N/O ratios of rGO/MCs and rGO-CNTs/MCs films.

Films	C (%)	O (%)	N (%)	C/O Ratio	C/N Ratio
rGO/MCs	78.2	14.2	6.2	5.50	12.61
rGO-CNTs/MCs	68.1	25.5	5.0	2.67	13.62

3.9. SEM Analysis

A Scanning Electron Microscope (SEM) was employed to analyze chitosan films with medium molecular weight (rGO/MCs and rGO-CNTs/MCs). The SEM rGO/Cs and rGO-CNTs/Cs images and the zoomed-in section are displayed in Figure 6. The distribution and/or agglomeration level of rGO and rGO-CNTs within the chitosan matrix was confirmed through SEM analysis. According to Lau et al. [55], the micrographs of films exhibit heterogeneous surfaces that consist of protuberances and condensed structures. The aggregation behavior is attributed to the favorable interaction between chitosan and carbon-based materials, as previously mentioned.

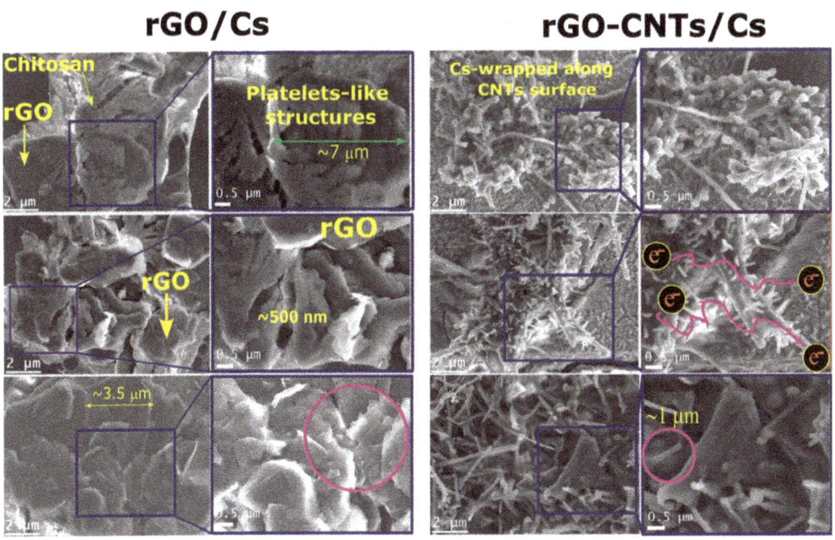

Figure 6. Scanning electron microscopy (SEM) of rGO/Cs and rGO-CNTs/Cs films and zoomed-in sections.

Cs bundles observed in SEM images are because of their high polar nature and perhaps responsible for Cs wrapping capabilities. The active polar groups NH_3^+ in chitosan can interact with CNTs surface through van der Waals interactions, promoting spontaneous wrapping around CNTs, taking, and helical conformation, leading to the formation of a thin layer on the CNTs surface [56]. The interaction between polymers with inflexible backbones and single-walled carbon nanotubes was investigated by Tallury et al. [57] through molecular dynamics simulations. The wrapping of SWCNTs with polymers having stiff backbones is characterized by more distinct conformations than those with flexible backbones. The decrease in stiffness was observed to be responsible for the reduction in intra-chain coiling. According to Liu et al., the CNTs display a strong tendency to form bundles, which is a key factor in achieving an appropriate homogenization before deposition for the solvent cast. Chitosan interacts with carbon nanotubes, causing them to be randomly distributed within the polymer. According to previous reports, surface decoration led to the complete coverage of carbon nanotube surfaces with chitosan bumps [58]. In this research, the pristine structure of the CNTs was unaffected through a functionalization process. Figure 6 demonstrates the differences between the rGO/MCs and rGO-CNTs/MCs films. Several visible rGO sheets were observed, and a successful blend of reduced graphene oxide with chitosan was used to produce films [16]. It seems that the fabrication method is responsible for the folded layered assembly of rGO/Cs films. SEM images of rGO/Cs films demonstrate that Cs covers the rGO sheets, and the rGO is evenly distributed and forms condensed structures throughout the composite film. The TS and EB values may be explained by the protrusions of chitosan that were formed on the CNTs surfaces and eventually wrapped around the carbon nanotubes, as depicted in Figure 6. We attribute the enhanced flexibility of the films to the better dispersion of carbon-based materials with Cs, and the dispersion behavior of rGO into the Cs matrix could be responsible for the improved mechanical properties.

3.10. Resistivity Measurements

A source-measure unit (Keithley Model 4200-SCS system) was employed to measure the resistivity of all films at room temperature. Every measurement was carried out using equidistant tip configurations of 0.040 inches. A gate voltage between −0.001 and +0.001 V was used for a 10 s data collection period. The resistivity (ρ) was calculated from the characteristic I–V curves. Electrical conductivity (σ) was calculated as the inverse of the resistivity (ρ) considering the sample geometry and the thickness, according to Equation (7).

Table 7 displays the electrical conductivity values of the as-prepared films. From previous reports [59], the incorporation of a small quantity of CNTs resulted in a significant enhancement of the film's electrical conductivity from 1.05×10^{-5} to 9.43×10^{-3} S/m. The enhanced electrical conductivity could be attributed to the synergistic interaction between CNTs and chitosan, along with dispersion and overlapped structures. According to the SEM images reported elsewhere in this document, it can be concluded that the enhanced electrical characteristics of rGO-CNTs/Cs films are due to the contacts between the tubes and their non-uniform distribution in chitosan, which is an essential factor for ensuring effective transfer of electrical charge. The rGO/LCs and rGO/MCs films displayed conductivity values below 10^{-8} S/cm that decreased for rGO/HCs films. Surprisingly, rGO/Cs films displayed low electrical values in comparison with the rGO-CNT/Cs films, which could be influenced by a strong interaction between the rGO-CNTs complex and chitosan. From SEM micrographs, we observed that rGO and Cs develop platelet-like structures that could hinder the π-π network of rGO, thus decreasing the conductive properties of rGO by diminishing the charge transport along the film due to the strong interaction of the rGO and the chitosan. Thus, conductivity measurements show that increasing the long chain of chitosan diminishes the electrical conductivity, as was observed in the rGO/Cs films as molecular weight increases (1.55×10^{-9}, 7.45×10^{-8}, 7.6×10^{-10} S/cm). The positively charged chitosan and oxygen functionalities in rGO lead to the formation of a condensed structure with appropriate mechanical performance, but that hinders the charge transport along the film.

Table 7. Electrical conductivity values of the rGO/Cs and rGO-CNTs/Cs films.

Films	σ (S/cm)
rGO/LCs	1.55×10^{-9}
rGO-CNTs/LCs	1.1×10^{-4}
rGO/MCs	7.45×10^{-8}
rGO-CNTs/MCs	1×10^{-9}
rGO/HCs	7.6×10^{-10}
rGO-CNTs/HCs	1.03×10^{-9}

The rGO-CNTs do not contribute to the charge transport due to the condensed structures and the strong interaction between rGO-CNTs and chitosan. Yet, CNTs and rGO are highly conductive materials, and the formation of condensed structures rGO-CNTs/Cs seems to decrease in electrical conductivity values as the molecular weight of chitosan increases (1.1×10^{-4}, 1×10^{-9}, 1.03×10^{-9} S/cm). Interestingly, the rGO-CNTs/LCs film displayed an increase in electrical conductivity (1.1×10^{-4} S/cm) among all the films analyzed in this study. An investigation into the formation of electrospray microcapsules was conducted by Gómez et al. [60], who analyzed the effects of chitosan's molecular weight, ranging from 25 to 300 kDa. The study evaluated the viscosity, surface tension, and electrical conductivity of solutions. It is important to emphasize that the conductivity values of chitosan were collected in liquid solutions, which improved the mobility of charges, and the conductivity values of chitosan were reported to be around 6–7 mS/cm. The conductivity was increased until viscosity delayed charge mobility. A decrease in the viscosity of the solutions at low acetic acid concentrations was associated with a maximum at higher concentrations. Higher concentrations of LCs were associated with a decrease in the viscosity of the solutions at low acetic acid concentrations. The concentration of charges and their mobility, which are critical to the conductivity of the films, are both significantly influenced by the chitosan long chain. As it increases, so does the protonated species. From our findings, we report that *IEC* values increase as Cs molecular weight increases for rGO-CNTs/Cs films, but the rGO/Cs films dramatically dropped the *IEC* values, maybe attributed to the highly condensed structures. The interaction between Cs and rGO may be attributed to the availability of NH_3^+ groups and active defect sites in rGO, such as hydroxyl or epoxide groups. These groups promote the adsorption of NH_3^+ groups through OH–N and O–HN hydrogen bonds. This leads to an enhanced charge transfer toward hydrogen abstraction from absorbed NH_3^+ groups toward the epoxide rings and, consequently, to an enhanced charge transfer to the rGO [61], inducing a more cohesive structure as observed in the SEM analysis, which may be responsible for the *IEC* and electrical conductivity values reported in this paper. The latter may be attributed to the dispersion of rGO, which can be effectively improved by chitosan chains. In the study of Oliva et al. [62] analyzed the adsorption of NH_3^+ on CNTs using QM/MM calculations. Their work aimed to investigate the effect of structural defects, such as pentagon–pentagon pairs or Stone–Wales (SW) defects, on NH_3^+ adsorption. According to the findings, the NH^{3+} molecules were physisorbed onto SW defects and could form long-range bonds approximately 3 Å from nanotube surfaces. In contrast, the same molecules were chemisorbed onto a vacancy defect by covalent bonding at 1.55 Å. This is because the NH_3^+ group can bind to CNTs with either short-range or long-range interactions, according to the defect type on the surface. The features have significant implications for the properties displayed by the rGO/Cs and rGO-CNTs/Cs films discussed in this paper. Cs, with their distinct hydrophilicity due to a high proportion of amino and hydroxyl groups, exhibit a high degree of polymerization that plays a crucial role in the structural arrangement of the modified chitosan [63] can interact with rGO and CNTs through defect sites, which would explain the crystalline arrangement from XRD patterns, which in turn is responsible for the enhanced electrical [64] and PL properties. It is important to consider that non-functionalized CNTs were used in this study. However, vacancies or defects may be intrinsically introduced into

CNTs during the fabrication process, enabling NH_3^+ molecules to adsorb onto the surface, leading to a restructuring of the CNTs framework [65].

Dynamic molecular simulations have shown that extensive wrapping of chitosan can significantly improve the dispersion and solubility behavior of CNTs. This is possible by utilizing the emulsifying capacity of Cs, which is attributed to its peculiar solubility behavior. Oxygen-containing groups serve as acceptors, enhancing conductivity and behaving as a *p*-type semiconductor [66]. The electrical conductivity of Cs is significantly influenced by its molecular weight attributed to the increased steric hindrance resulting from a higher degree of polymerization, which may lead to interchain interactions and hinder the acceptance of electrons by *p*-type carbon-based materials, eventually provoking a diminution in electrical conductivity. This outcome indicates that increased conductivity is counteracted by chemically bonded carbon-based materials and Cs. Considering the behavior of electrical conductivity in this study about previous reports, we conjecture that oxygen molecules act as acceptors in the *p*-type rGO and rGO-CNTs, increasing the conductivity after the interaction with Cs by adsorbing onto the basal plane due to positively charged molecules of the polymer, allowing to the oxygen-containing groups adsorbed at the defect site acting as a *p*-type semiconductor. Due to the intercalation of the polymer molecules within the graphene layers, the platelets would provide sufficient mechanical and functional performance to polymer matrices. According to Shi et al. [67], graphene is the most used filler for polymers due to its high surface area and exceptional mechanical and electrical conductivity. The rGO and Cs interface is crucial in determining film structure and properties. Under-loading a weak interface can lead to stress concentration and platelet pulling out, which is a severe defect. Surface modification can lead to the desired interface strength for high-performance composites. Chitosan helps to build up a strong interface, enhancing interfacial interactions that provide desired physical–chemical properties. The activity of organic molecules and experimental parameters, including solvent, temperature, and polymer molecular weight, may impact the chemisorption of Cs with carbon atoms on the basal plane of graphene layers, as well as the non-covalent attachment of Cs to the CNTs basal plane through a defect-site carbon nanotube.

3.11. Photoluminescence Measurement (PL)

Figure 7a displays the PL emission spectra of rGO/Cs and rGO-CNTs/Cs films, both excited about a wavelength of 375 nm. For comparison, the laser emission is shown. Samples exhibit a broadband luminescence covering almost the whole visible spectral range. The integrated intensity (area under the curve) is presented in Figure 7b, where samples rGO-CNT/MCs and rGO/LCs exhibit both the lowest and the highest emissions, respectively. For a better understanding of their spectral composition, Luminescence spectra were deconvoluted. As an example, the PL spectrum of the rGO/LCs sample (Figure 7c) was deconvoluted into four emission bands centered at 441 nm (2.81 eV), 491 nm (2.52 eV), 538 nm (2.3 eV), and 618 nm (2.0 eV), respectively. Both the number and the center position of the deconvolution peaks were carefully selected, considering the second derivative minima of the measured luminescence (see Supplementary Information, Figure S6). Furthermore, during the deconvolution process, the center position, width, and height of deconvolution peaks varied in such a way that the simulated curve best fits with the experimental data, thus maximizing the R square adjusting coefficient. As an attempt to explain the observed luminescence, an energy level diagram is presented in Figure 7d. The diagram depicts the center positions, beginning, and ending of the observed luminescence bands. After 375 nm excitation, the electrons are pumped to the LE level (3.3 eV). However, the results show that the luminescence starts at 414 nm (2.99 eV). This indicates that electrons lose ~0.31 eV energy through non-radiative transitions before they start to decay radiatively to the ground state. Furthermore, the fact that no emissions were detected in the IR region suggests that there are no radiative transitions between the observed luminescence bands, in which case the emissions would be in the IR region ($\lambda > 707$ nm; $E < 1.75$ eV). This suggests that once in LE, electrons lose energy via non-radiative (phonon) transitions to reach energy levels corresponding to bands centered at L1, L2, L3, and L4. To reach these levels, multiphonon transitions (two or three) with energies Eph~0.4 eV should occur, leading the electrons from LE to the observed emission levels. It is known that multiphonon transitions are less likely, so phonons with higher energy Eph ~0.4 eV (2700 cm^{-1} to 3800 cm^{-1}) are most likely to reduce the energy of the electrons and populate the energy levels from which the electrons radiatively decay toward the ground state. In the case of lower-energy phonons (0.2–0.1 eV), at least 7–8 phonons would be required for the same transition to occur, which is highly unlikely.

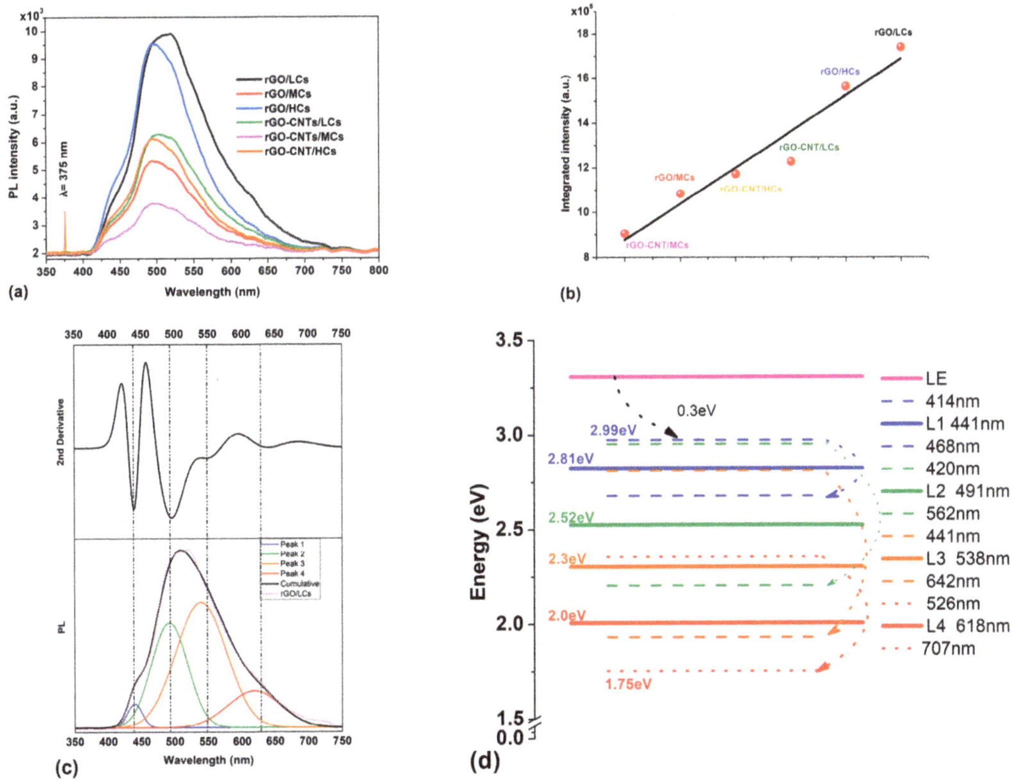

Figure 7. (**a**) PL spectra of rGO/(LCs, MCs, HCs) and rGO-CNTs/(LCs, MCs, HCs) films. For comparison, the laser emission line is shown. (**b**) integrated PL intensity (**c**) PL deconvolution of sample rGO/LCs. (**d**) Energy level diagram depicting the center positions, beginning, and end of the four observed luminescence bands.

The CIE-1931 chromaticity diagram was produced from the PL spectra data. The chromaticity coordinates of the films demonstrated a gradual shift toward the white region, and the purity of color was calculated (see Figure S7 for more details in the Supplementary Materials section). We found that rGO/Cs and rGO-CNTs/Cs chromaticity values are placed near the locus point in the CIE diagram. Incorporating CNTs and rGO into the chitosan matrix provokes a shift of the wavelength for maximum emission toward the white region of the spectrum in almost all the films (excluding rGO/LCs and rGO/HCs).

Furthermore, the fact that no emissions were detected in the IR region suggests that there are no radiative transitions between the observed luminescence bands, in which case the emissions would be in the IR region (λ > 800 nm; E < 1.55 eV). This suggests that once in LE, electrons lose energy via non-radiative (phonon) transitions to reach energy levels corresponding to bands centered on L1, L2, and L3. To reach these levels, multiphonon transitions (two or three) with energies E_{ph} ~0.4 eV should occur, leading the electrons from LE to the observed emission levels. It is known that multiphonon transitions are less likely, so phonons with higher energy E_{ph} ~0.4 eV (2700 cm^{-1} to 3800 cm^{-1}) are most likely to reduce the energy of the electrons and populate the energy levels from which the electrons decay radiatively toward the ground. In the case of lower-energy phonons (0.2–0.1 eV), at least 7–8 phonons would be required for the same transition to occur, which is highly unlikely.

From reference [68], carbon nanotubes can be identified as luminescent centers. From our findings, the films induced the emission of green and white light, which is attributed to an abundance of oxygen atoms and acetamide groups that are responsible for the luminescence of natural polymers such as chitosan [69]. These atoms act as luminescent centers, emitting light due to the high electron density in oxygen-containing groups. The green emission observed between 500 and 525 nm could be due either to donor–acceptor recombination or to a transition from the conduction band to the oxygen antisites, according to Sedky et al. [70]. Several types of defects, including antisite defects and Stone–Wales (SW) defects, can occur within a unit cell. Topological defects, such as the pentagon–heptagon defects generated by the CNTs and rGO processing methods, can create energy levels within the band gap due to the small amount of lattice strain. The defects in graphene and CNTs can be responsible for the observed emission in our research (504–520 nm), causing a shift in the wavelength toward higher values. Amplification of the photoluminescence phenomenon was due to the presence of free electrons in CNTs, which prevented oxygen vacancy defects and promoted the radiative emission mechanism. Kishi et al. reported that the emission of PL is predominantly from the inner walls of carbon nanotubes. The optical and electrical properties of CNTs may be attributed to the interlayer interaction between the inner and outer walls of the material. As a perfect π-conjugated single sheet, graphene lacks electronic band gaps, and it is not photoluminescent; therefore, the creation of energy bandgaps by manipulating the π-electronic network to form quantum-confined sp^2 "islands" in a graphene sheet involves the formation of structural defects, which could be responsible for the creation of bandgaps for emissive electronic transitions, but also directly contribute the bright PL emissions of carbon-based materials. The observed PL emission in graphene materials could be originated from created or induced energy bandgaps in a single graphene sheet and defected multi-layer graphene [71]. More details are depicted in the Supplementary Material (Figure S6). However, based on the above, further work is needed to fully understand the effect of oxygen vacancies on the PL spectra of chitosan-based materials, which are strongly influenced by molecular weight. From the FTIR results, it can be concluded that the oxygen-containing groups belonging to Cs contribute to the high-intensity value observed in the spectrum and that the effect of incorporating rGO and/or CNTs modifies the PL properties of chitosan films. Low-molecular-weight chitosan chains apparently interact more effectively with rGO and CNTs, facilitating the transport of charge carriers and thereby enhancing the electrical conductivity properties reported in this work.

4. Conclusions

All films were manufactured using a simple solvent-casting procedure. Afterward, all films were analyzed and characterized to disclose the effect of carbon nanotubes and reduced graphene oxide in chitosan from low to high molecular weight. The results demonstrated that filler addition had a significant effect on the physical, optical, and electrical properties of chitosan films. We found that HCs is an appropriate matrix for the enhancement of mechanical properties of films, but it has an impact on the intrinsic electrical properties of carbon-based materials when added to chitosan; the films behave as insulators with electrical conductivity values around 10^{-10} S/cm. Agglomerated reduced graphene oxide sheets and carbon nanotubes that interact with chitosan molecules due to hydrogen-bonded interactions may have contributed to this unpredicted performance, limiting the charge transport. We have demonstrated that the incorporation of carbon nanotubes and reduced graphene oxide influence the crystalline structure of chitosan, while the average crystallite diameter is affected. The rGO-CNTs/Cs films analysis suggests that carbon nanotubes are the structures that promote electron transfer along the hexagonal network. This was observed in the electrical conductivity values. SEM micrographs showed that chitosan was homogeneously wrapped along carbon nanotubes, which could be the key to the enhanced electrical and PL properties.

In contrast, the addition of rGO does not significantly enhance the electrical properties of chitosan, and it appears that there is no contribution of the reduced graphene oxide to the electrical conductivity. This might be related to the platelet-like structures observed in the SEM micrograph. The late could be due to the interaction of oxygen-containing groups in the rGO surface with amine or hydroxyl in the chitosan molecule. A homogeneous dispersion of rGO into the chitosan matrix significantly improved tensile strength, Young's module, and elongation at break. Transparent and yellow chitosan films turned opaque, highly influenced by the addition of carbon nanotubes and reduced graphene oxide. rGO-CNTs films significantly reduced the L^* parameter more than rGO/Cs films, possibly due to the efficient process of carbon nanotubes to absorb incident light, directing it deeper into voids and contributing to high absorbance. Contrarily, the rGO addition makes a less browning film, as was demonstrated by the opacity analysis.

However, PL measurements showed relevant changes in the PL plot, and the CIE system offered a more precise color measurement. Our results strongly suggest that carbon nanotubes and reduced graphene oxide in chitosan films induce white and green light emissions attributed to the oxygen-containing groups, which are considered luminescent centers due to lone pairs electrons. The color attributes of chitosan films are influenced by adding reduced graphene oxide and reduced graphene–carbon nanotube oxide. From CIELAB analysis, we found that rGO/HCs films a redshift. Moreover, rGO/MCs show a lighter film, as was observed in the CIELAB analysis. PL studies revealed emissions in the blue, green, and red regions. These variances are attributed to the nature of the photoluminescence due to the antisite oxygen-containing groups, the excitation wavelength, and the film production method. The results suggest that CNT and rGO hybrids might have the potential to develop electrically and photoluminescent materials. The resultant broad emission from the films and their characteristics are suitable for numerous valuable applications, including those in bioengineering.

Supplementary Materials: The following supporting information can be downloaded at https://www.mdpi.com/article/10.3390/polym16131827/s1, Table S1: Labels for films produced by solvent-casting method. Table S2: Color purity percentage calculated from PL curves. Figure S1: FTIR analysis from 3700 to 2700 cm^{-1} region and the second derivative of rGO/Cs and rGO-CNTs/Cs films. Figure S2: Analysis of the XRD pattern of rGO/Cs and rGO-CNTs/Cs films by the deconvolution method using the Voight approximation. Figure S3: Analysis of the (002) plane of rGO/Cs films and rGO-CNTs/Cs films. Figure S4: Mechanical properties of rGO/Cs films and rGO-CNTs/Cs films from LCs to HCs chitosan (a) stress–strain curves, (b) Young's modulus, (c) tension strength, and (d) elongation-at-break percentages. Figure S5: XPS core-level C 1s of (a) rGO/MCs film, and (b) rGO-CNTs/MCs film. Figure S6: PL spectra deconvolution of the rGO/Cs and rGO-CNTs/Cs films using the multiple Gaussian peak fit. The second derivative of the cumulative peak is added as additional information in relation to the deconvolution process. Results reveal that second derivative minima are in well agreement with the center wavelength of the subpeaks maxima. The R^2adjusting coefficient was higher than 0.999 for the deconvolution of all samples. Figure S7: CIE diagram extracted from PL data.

Author Contributions: Conceptualization, J.R.G.-M., A.B.L.-O., D.D.Á.-L. and E.R.-G.; Data curation, J.R.G.-M., A.B.L.-O., D.D.Á.-L., C.G.-G., E.R.-G., E.U.P.-B., F.B.-D., A.G.M.-V. and P.J.M.-C.; Formal analysis, J.R.G.-M., A.B.L.-O., D.D.Á.-L., C.G.-G., E.R.-G., E.U.P.-B., A.G.M.-V. and P.J.M.-C.; Funding acquisition, A.B.L.-O., C.G.-G. and E.R.-G.; Investigation, A.B.L.-O., D.D.Á.-L., E.R.-G. and F.B.-D.; Methodology, J.R.G.-M., A.B.L.-O., D.D.Á.-L., E.U.P.-B. and A.G.M.-V.; Project administration, A.B.L.-O.; Resources, J.R.G.-M., A.B.L.-O., C.G.-G., E.R.-G. and F.B.-D.; Software, C.G.-G.; Supervision, A.B.L.-O. and D.D.Á.-L.; Validation, A.B.L.-O. and D.D.Á.-L.; Writing—original draft, A.B.L.-O., D.D.Á.-L. and P.J.M.-C.; Writing—review and editing, A.B.L.-O., D.D.Á.-L., E.R.-G. and F.B.-D. All authors have read and agreed to the published version of the manuscript.

Funding: This research was funded by the Consejo Nacional de Humanidades Ciencias y Tecnologías (Conahcyt) grant number (315884), and the APC was funded by Tecnologico de Monterrrey.

Institutional Review Board Statement: Not applicable.

Data Availability Statement: The original contributions presented in the study are included in the article/Supplementary Material. Further inquiries can be directed to the corresponding author/s.

Acknowledgments: González-Martínez, Magallanes-Vallejo, and Mogica-Cantú are grateful for their postgraduate fellowship to CONAHCyT. The authors express their gratitude to CONAHCyT for the financial support received through projects 315884, 1905, postdoctoral fellowship, and SNII-CONAHCyT. We would like to thank Sebastian Pacheco for his assistance with the XRD measurements and the CTAOV laboratory for their invaluable assistance.

Conflicts of Interest: The authors declare no conflicts of interest.

References

1. Santiago-Castillo, K.; Del Angel-López, D.; Torres-Huerta, A.M.; Domínguez-Crespo, M.A.; Palma-Ramírez, D.; Willcock, H.; Brachetti-Sibaja, S.B. Effect on the processability, structure and mechanical properties of highly dispersed in situ ZnO:CS nanoparticles into PVA electrospun fibers. *J. Mater. Res. Technol.* **2021**, *11*, 929–945. [CrossRef]
2. Santiago-Castillo, K.; Torres-Huerta, A.M.; del Ángel-López, D.; Domínguez-Crespo, M.A.; Dorantes-Rosales, H.; Palma-Ramírez, D.; Willcock, H. In Situ Growth of Silver Nanoparticles on Chitosan Matrix for the Synthesis of Hybrid Electrospun Fibers: Analysis of Microstructural and Mechanical Properties. *Polymers* **2022**, *14*, 674. [CrossRef] [PubMed]
3. Nisha, U.M.; Meeran, M.N.; Sethupathi, R.; Subha, V.; Rajeswaran, P.; Sivakarthik, P. Studies on Chitosan modified ZrO_2–CeO_2 metal oxide and their Photo catalytic activity under solar light irradiation for water remediation. *J. Indian Chem. Soc.* **2023**, *100*, 101061. [CrossRef]
4. Dinakaran, K.; Senthamilselvi, S.; Gayathri, T.; Kavitha, N.; Roy, D. Enhanced electromagnetic interference shielding in nickel oxide and graphene composite nanostructures loaded polyvinylidene difluoride thin films. *J. Indian Chem. Soc.* **2023**, *100*, 101091. [CrossRef]
5. Buinov, A.S.; Gafarova, E.R.; Grebenik, E.A.; Bardakova, K.N.; Kholkhoev, B.C.; Veryasova, N.N.; Nikitin, P.V.; Kosheleva, N.V.; Shavkuta, B.S.; Kuryanova, A.S.; et al. Fabrication of Conductive Tissue Engineering Nanocomposite Films Based on Chitosan and Surfactant-Stabilized Graphene Dispersions. *Polymers* **2022**, *14*, 3792. [CrossRef] [PubMed]
6. Talebi, A.; Labbaf, S.; Karimzadeh, F. A conductive film of chitosan-polycaprolcatone-polypyrrole with potential in heart patch application. *Polym. Test.* **2019**, *75*, 254–261. [CrossRef]
7. Macias-Almazan, A.; Lois-Correa, J.A.; Dominguez-Crespo, M.A.; Lopez-Oyama, A.B.; Torres-Huerta, A.M.; Brachetti-Sibaja, S.B.; Rodriguez-Salazar, A.E. Influence of operating conditions on proton conductivity of nanocellulose films using two agroindustrial wastes: Sugarcane bagasse and pinewood sawdust. *Carbohydr. Polym.* **2020**, *238*, 116171. [CrossRef]
8. Pulido-Barragán, E.U.; Morales-Cepeda, A.B.; Castro-Guerrero, C.F.; Koschella, A.; Heinze, T. Upgrading Euphorbia Antisyphilitica fiber compost: A waste material turned into nanocrystalline cellulose. *Ind. Crops Prod.* **2020**, *160*, 113111. [CrossRef]
9. Barragán, E.U.P.; Guerrero, C.F.C.; Zamudio, A.M.; Cepeda, A.B.M.; Heinze, T.; Koschella, A. Isolation of Cellulose Nanocrystals from Typha domingensis Named Southern Cattail Using a Batch Reactor. *Fibers Polym.* **2019**, *20*, 1136–1144. [CrossRef]
10. Umnyakova, E.S.; Zharkova, M.S.; Berlov, M.N.; Shamova, O.V.; Kokryakov, V.N. Human antimicrobial peptides in autoimmunity. *Autoimmunity* **2020**, *53*, 137–147. [CrossRef] [PubMed]
11. Shi, S.; Jiang, Y.; Ren, H.; Deng, S.; Sun, J.; Cheng, F.; Jing, J.; Chen, Y. 3D-Printed Carbon-Based Conformal Electromagnetic Interference Shielding Module for Integrated Electronics. *Nano-Micro Lett.* **2024**, *16*, 85. [CrossRef] [PubMed]
12. Dominguez-Crespo, M.A.; Rodriguez, E.; Torres-Huerta, A.M.; Soni-Castro, I.J.; Brachetti-Sibaja, S.B.; Narro-Garcia, R.; Lopez-Oyama, A.B. Production of BN nanostructures by pulsed laser ablation in liquids: Influence of the applied Nd:YAG harmonics on the structural, optical and photoluminescence properties. *Ceram. Int.* **2020**, *46*, 21667–21680. [CrossRef]
13. López-Oyama, A.B.; Silva-Molina, R.A.; Ruíz-García, J.; Gámez-Corrales, R.; Guirado-López, R.A. Structure, electronic properties, and aggregation behavior of hydroxylated carbon nanotubes. *J. Chem. Phys.* **2014**, *141*, 174703. [CrossRef] [PubMed]
14. López-Oyama, A.B.; Domínguez-Crespo, M.A.; Torres-Huerta, A.M.; Onofre-Bustamante, E.; Gámez-Corrales, R.; Cayetano-Castro, N. Electrochemical alternative to obtain reduced graphene oxide by pulse potential: Effect of synthesis parameters and study of corrosion properties. *Diam. Relat. Mater.* **2018**, *88*, 167–188. [CrossRef]
15. López-Oyama, A.B.; Domínguez-Crespo, M.A.; Torres-Huerta, A.M.; Onofre-Bustamante, E.; Gámez-Corrales, R.; Cayetano-Castro, N.; Ferrel-Álvarez, A.C. Dataset on electrochemical reduced graphene oxide production: Effect of synthesis parameters. *Data Brief* **2018**, *21*, 598–603. [CrossRef] [PubMed]

16. Barra, A.; Ferreira, N.M.; Martins, M.A.; Lazar, O.; Pantazi, A.; Jderu, A.A.; Neumayer, S.M.; Rodriguez, B.J.; Enăchescu, M.; Ferreira, P.; et al. Eco-friendly preparation of electrically conductive chitosan-reduced graphene oxide flexible bionanocomposites for food packaging and biological applications. *Compos. Sci. Technol.* **2019**, *173*, 53–60. [CrossRef]
17. Du, X.; Qin, Z.; Li, Z. Free-Standing rGO-CNT Nanocomposites with Excellent Rate Capability and Cycling Stability for Na_2SO_4 Aqueous Electrolyte Supercapacitors. *Nanomaterials* **2021**, *11*, 1420. [CrossRef] [PubMed]
18. Cho, H.; Yun, J.-H.; Back, S.Y.; Lee, J.-S.; Kang, N.; Jang, Y.-I.; Lim, J.; Son, J.-H.; Park, J.-Y.; Kim, J.; et al. Superior thermoelectric cooling performance by suppressing bipolar diffusion effect and enhancing anisotropic texture in p-/n-type Bi_2Te_3 based compounds. *J. Alloys Compd.* **2021**, *888*, 161572. [CrossRef]
19. Sadiq, M.; Khan, M.A.; Sarvar, M.; Raza, M.M.H.; Aalam, S.M.; Zulfequar, M.; Ali, J. High electrochemical performance of rGO-CNTs composites as an electrode material for supercapacitors applications. *Hybrid Adv.* **2023**, *3*, 100051. [CrossRef]
20. Kong, L.; Yin, X.; Xu, H.; Yuan, X.; Wang, T.; Xu, Z.; Huang, J.; Yang, R.; Fan, H. Powerful absorbing and lightweight electromagnetic shielding CNTs/RGO composite. *Carbon* **2019**, *145*, 61–66. [CrossRef]
21. Wang, Y.; Liu, Y.; Yu, Y.; Huang, H. Influence of CNT-rGO composite structures on their permeability and selectivity for membrane water treatment. *J. Membr. Sci.* **2018**, *551*, 326–332. [CrossRef]
22. Pan, Y.; Bao, H.; Li, L. Noncovalently Functionalized Multiwalled Carbon Nanotubes by Chitosan-Grafted Reduced Graphene Oxide and Their Synergistic Reinforcing Effects in Chitosan Films. *ACS Appl. Mater. Interfaces* **2011**, *3*, 4819–4830. [CrossRef] [PubMed]
23. Mergen, Ö.B.; Arda, E.; Evingür, G.A. Electrical, optical and mechanical properties of chitosan biocomposites. *J. Compos. Mater.* **2019**, *54*, 1497–1510. [CrossRef]
24. Vega-Cázarez, C.A.; López-Cervantes, J.; Sánchez-Machado, D.I.; Madera-Santana, T.J.; Soto-Cota, A.; Ramírez-Wong, B. Preparation and Properties of Chitosan–PVA Fibers Produced by Wet Spinning. *J. Polym. Environ.* **2018**, *26*, 946–958. [CrossRef]
25. Han, J.H.; Floros, J.D. Casting Antimicrobial Packaging Films and Measuring Their Physical Properties and Antimicrobial Activity. *J. Plast. Film Sheeting* **1997**, *13*, 287–298. [CrossRef]
26. Escárcega-Galaz, A.A.; Sánchez-Machado, D.I.; López-Cervantes, J.; Sanches-Silva, A.; Madera-Santana, T.J.; Paseiro-Losada, P. Mechanical, structural and physical aspects of chitosan-based films as antimicrobial dressings. *Int. J. Biol. Macromol.* **2018**, *116*, 472–481. [CrossRef] [PubMed]
27. *ASTM D882*; Standard Test Method for Tensile Properties of Thin Plastic Sheeting. ASTM International: Conshohocken, PA, USA, 2018.
28. Shipovskaya, A.B.; Fomina, V.I.; Kazmicheva, O.F.; Rudenko, D.A.; Malinkina, O.N. Optical activity of films based on chitosan of various molecular masses and modifications. *Polym. Sci. Ser. A* **2017**, *59*, 330–341. [CrossRef]
29. Miao, J.; Liu, H.; Li, Y.; Zhang, X. Biodegradable Transparent Substrate Based on Edible Starch–Chitosan Embedded with Nature-Inspired Three-Dimensionally Interconnected Conductive Nanocomposites for Wearable Green Electronics. *ACS Appl. Mater. Interfaces* **2018**, *10*, 23037–23047. [CrossRef] [PubMed]
30. Pan, Y.; Wu, T.; Bao, H.; Li, L. Green fabrication of chitosan films reinforced with parallel aligned graphene oxide. *Carbohydr. Polym.* **2011**, *83*, 1908–1915. [CrossRef]
31. Xia, Y.; Zuo, H.; Lv, J.; Wei, S.; Yao, Y.; Liu, Z.; Lin, Q.; Yu, Y.; Yu, W.; Huang, Y. Preparation of multi-layered microcapsule-shaped activated biomass carbon with ultrahigh surface area from bamboo parenchyma cells for energy storage and cationic dyes removal. *J. Clean. Prod.* **2023**, *396*, 136517. [CrossRef]
32. Aztatzi-Pluma, D.; Castrejón-González, E.O.; Almendarez-Camarillo, A.; Alvarado, J.F.J.; Durán-Morales, Y. Study of the Molecular Interactions between Functionalized Carbon Nanotubes and Chitosan. *J. Phys. Chem. C* **2016**, *120*, 2371–2378. [CrossRef]
33. Carson, L.; Kelly-Brown, C.; Stewart, M.; Oki, A.; Regisford, G.; Luo, Z.; Bakhmutov, V.I. Synthesis and characterization of chitosan–carbon nanotube composites. *Mater. Lett.* **2009**, *63*, 617–620. [CrossRef] [PubMed]
34. Marroquin, J.B.; Rhee, K.Y.; Park, S.J. Chitosan nanocomposite films: Enhanced electrical conductivity, thermal stability, and mechanical properties. *Carbohydr. Polym.* **2013**, *92*, 1783–1791. [CrossRef] [PubMed]
35. Aryaei, A.; Jayatissa, A.H.; Jayasuriya, A.C. Mechanical and biological properties of chitosan/carbon nanotube nanocomposite films. *J. Biomed. Mater. Res. Part A* **2014**, *102*, 2704–2712. [CrossRef] [PubMed]
36. Aslam, M.M.A.; Kuo, H.-W.; Den, W.; Sultan, M.; Rasool, K.; Bilal, M. Chapter 10—Recent trends of carbon nanotubes and chitosan composites for hexavalent chromium removal from aqueous samples. In *Separation Science and Technology*; Ahuja, S., Ed.; Academic Press: Cambridge, MA, USA, 2022; Volume 15, pp. 177–207.
37. Nhung, L.T.T.; Kim, I.Y.; Yoon, Y.S. Quaternized Chitosan-Based Anion Exchange Membrane Composited with Quaternized Poly(vinylbenzyl chloride)/Polysulfone Blend. *Polymers* **2020**, *12*, 2714. [CrossRef]
38. Lewandowska, K.; Szulc, M. Rheological and Film-Forming Properties of Chitosan Composites. *Int. J. Mol. Sci.* **2022**, *23*, 8763. [CrossRef] [PubMed]
39. Azevedo, E.P.; Saldanha, T.D.P.; Navarro, M.V.M.; Medeiros, A.C.; Ginani, M.F.; Raffin, F.N. Mechanical properties and release studies of chitosan films impregnated with silver sulfadiazine. *J. Appl. Polym. Sci.* **2006**, *102*, 3462–3470. [CrossRef]
40. González, J.A.; Mazzobre, M.F.; Villanueva, M.E.; Díaz, L.E.; Copello, G.J. Chitin hybrid materials reinforced with graphene oxide nanosheets: Chemical and mechanical characterisation. *RSC Adv.* **2014**, *4*, 16480–16488. [CrossRef]

41. Podgorbunskikh, E.; Kuskov, T.; Rychkov, D.; Lomovskii, O.; Bychkov, A. Mechanical Amorphization of Chitosan with Different Molecular Weights. *Polymers* **2022**, *14*, 4438. [CrossRef] [PubMed]
42. Barbosa, G.P.; Debone, H.S.; Severino, P.; Souto, E.B.; da Silva, C.F. Design and characterization of chitosan/zeolite composite films—Effect of zeolite type and zeolite dose on the film properties. *Mater. Sci. Eng. C* **2016**, *60*, 246–254. [CrossRef] [PubMed]
43. Suderman, N.; Isa, M.I.N.; Sarbon, N.M. Characterization on the mechanical and physical properties of chicken skin gelatin films in comparison to mammalian gelatin films. *IOP Conf. Ser. Mater. Sci. Eng.* **2018**, *440*, 012033. [CrossRef]
44. Ma, J.; Liu, C.; Li, R.; Wang, J. Properties and structural characterization of chitosan/graphene oxide biocomposites. *Bio-Med. Mater. Eng.* **2012**, *22*, 129–135. [CrossRef] [PubMed]
45. Geng, L.; Lin, Y.; Chen, S.; Shi, S.; Cai, Y.; Li, L.; Peng, X. Superior strength and toughness of graphene/chitosan fibers reinforced by interfacial complexation. *Compos. Sci. Technol.* **2020**, *194*, 108174. [CrossRef]
46. Rodríguez Llanos, J.H.; De Oliveira Vercik, L.C.; Vercik, A. Physical Properties of Chitosan Films Obtained after Neutralization of Polycation by Slow Drip Method. *J. Biomater. Nanobiotechnol.* **2015**, *6*, 15. [CrossRef]
47. Güldiken, Ç.G.; Karaosmanoğlu, O.; Sivas, H.; Gerçel, H.F. ZnO microparticle-loaded chitosan/poly(vinyl alcohol)/acacia gum nanosphere-based nanocomposite thin film wound dressings for accelerated wound healing. *J. Appl. Polym. Sci.* **2020**, *137*, 48445. [CrossRef]
48. Valencia-Sullca, C.; Ben Messaoud, G.; Sánchez-González, L.; Tehrany, E.A.; Vargas, M.; Atarés, L.; Chiralt, A. Chitosan films containing encapsulated eugenol in alginate microspheres. *Food Hydrocoll.* **2024**, *151*, 109791. [CrossRef]
49. Leceta, I.; Guerrero, P.; de la Caba, K. Functional properties of chitosan-based films. *Carbohydr. Polym.* **2013**, *93*, 339–346. [CrossRef] [PubMed]
50. Zhang, Q.; Chen, Y.; Wei, P.; Zhong, Y.; Chen, C.; Cai, J. Extremely strong and tough chitosan films mediated by unique hydrated chitosan crystal structures. *Mater. Today* **2021**, *51*, 27–38. [CrossRef]
51. Miles, K.B.; Ball, R.L.; Matthew, H.W.T. Chitosan films with improved tensile strength and toughness from N-acetyl-cysteine mediated disulfide bonds. *Carbohydr. Polym.* **2016**, *139*, 1–9. [CrossRef] [PubMed]
52. Yang, X.; Tu, Y.; Li, L.; Shang, S.; Tao, X.M. Well-dispersed chitosan/graphene oxide nanocomposites. *ACS Appl. Mater. Interfaces* **2010**, *2*, 1707–1713. [CrossRef] [PubMed]
53. Li, P.-C.; Liao, G.M.; Kumar, S.R.; Shih, C.-M.; Yang, C.-C.; Wang, D.-M.; Lue, S.J. Fabrication and Characterization of Chitosan Nanoparticle-Incorporated Quaternized Poly(Vinyl Alcohol) Composite Membranes as Solid Electrolytes for Direct Methanol Alkaline Fuel Cells. *Electrochim. Acta* **2016**, *187*, 616–628. [CrossRef]
54. Gong, Y.; Yu, Y.; Kang, H.; Chen, X.; Liu, H.; Zhang, Y.; Sun, Y.; Song, H. Synthesis and Characterization of Graphene Oxide/Chitosan Composite Aerogels with High Mechanical Performance. *Polymers* **2019**, *11*, 777. [CrossRef] [PubMed]
55. Lau, Y.-T.R.; Yamaguchi, M.; Li, X.; Bando, Y.; Golberg, D.; Winnik, F.M. Facile and Mild Strategy Toward Biopolymer-Coated Boron Nitride Nanotubes via a Glycine-Assisted Interfacial Process. *J. Phys. Chem. C* **2013**, *117*, 19568–19576. [CrossRef]
56. Molla-Abbasi, P.; Ghaffarian, S.R. Decoration of carbon nanotubes by chitosan in a nanohybrid conductive polymer composite for detection of polar vapours. *RSC Adv.* **2014**, *4*, 30906–30913. [CrossRef]
57. Tallury, S.S.; Pasquinelli, M.A. Molecular Dynamics Simulations of Polymers with Stiff Backbones Interacting with Single-Walled Carbon Nanotubes. *J. Phys. Chem. B* **2010**, *114*, 9349–9355. [CrossRef] [PubMed]
58. Liu, Y.; Tang, J.; Chen, X.; Xin, J.H. Decoration of carbon nanotubes with chitosan. *Carbon* **2005**, *43*, 3178–3180. [CrossRef]
59. Falamarzpour, P.; Ghaffarian Anbaran, S.R. Synergistic effect of carbon nanotubes on chitosan-graphene oxide supramolecular structure. *J. Polym. Res.* **2022**, *29*, 211. [CrossRef]
60. Gómez-Mascaraque, L.G.; Sanchez, G.; López-Rubio, A. Impact of molecular weight on the formation of electrosprayed chitosan microcapsules as delivery vehicles for bioactive compounds. *Carbohydr. Polym.* **2016**, *150*, 121–130. [CrossRef] [PubMed]
61. Peng, Y.; Li, J. Ammonia adsorption on graphene and graphene oxide: A first-principles study. *Front. Environ. Sci. Eng.* **2013**, *7*, 403–411. [CrossRef]
62. Oliva, C.; Strodel, P.; Goldbeck, G.; Maiti, A. Unravelling the Interaction of Ammonia with Carbon Nanotubes. 2008. Available online: https://www.researchgate.net/publication/220002326_Unravelling_the_interaction_of_ammonia_with_carbon_nanotubes (accessed on 14 June 2024).
63. Wang, S.F.; Shen, L.; Zhang, W.D.; Tong, Y.J. Preparation and mechanical properties of chitosan/carbon nanotubes composites. *Biomacromolecules* **2005**, *6*, 3067–3072. [CrossRef] [PubMed]
64. Charlier, J.C. Defects in carbon nanotubes. *Acc. Chem. Res.* **2002**, *35*, 1063–1069. [CrossRef] [PubMed]
65. Omoto, Y.; Morita, H.; Sato, Y.; Nishida, T.; Motomiya, K.; Katsui, H.; Goto, T. Are Non-Six-Membered Ring Defects Formed in Single-Walled Carbon Nanotubes Treated by a Fluorination–Defluorination Process? *Nanomaterials* **2023**, *13*, 1086. [CrossRef] [PubMed]
66. Collins, P.G.; Bradley, K.; Ishigami, M.; Zettl, A. Extreme Oxygen Sensitivity of Electronic Properties of Carbon Nanotubes. *Science* **2000**, *287*, 1801–1804. [CrossRef] [PubMed]
67. Shi, G.; Araby, S.; Gibson, C.T.; Meng, Q.; Zhu, S.; Ma, J. Graphene Platelets and Their Polymer Composites: Fabrication, Structure, Properties, and Applications. *Adv. Funct. Mater.* **2018**, *28*, 1706705. [CrossRef]
68. Shimamoto, D.; Muramatsu, H.; Hayashi, T.; Kim, Y.A.; Endo, M.; Park, J.S.; Saito, R.; Terrones, M.; Dresselhaus, M.S. Strong and stable photoluminescence from the semiconducting inner tubes within double walled carbon nanotubes. *Appl. Phys. Lett.* **2009**, *94*, 083106. [CrossRef]

69. Kumirska, J.; Czerwicka, M.; Kaczyński, Z.; Bychowska, A.; Brzozowski, K.; Thöming, J.; Stepnowski, P. Application of Spectroscopic Methods for Structural Analysis of Chitin and Chitosan. *Mar. Drugs* **2010**, *8*, 1567–1636. [CrossRef] [PubMed]
70. Sedky, A.; Ali, A.M.; Somaily, H.H.; Algarni, H. Electrical, photoluminescence and optical investigation of ZnO nanoparticles sintered at different temperatures. *Opt. Quantum Electron.* **2021**, *53*, 243. [CrossRef]
71. Cao, L.; Meziani, M.J.; Sahu, S.; Sun, Y.-P. Photoluminescence Properties of Graphene versus Other Carbon Nanomaterials. *Acc. Chem. Res.* **2013**, *46*, 171–180. [CrossRef] [PubMed]

polymers

MDPI

Article

Formation of Olive-like TiO_2 Nanospheres in a Polymeric Mesh by Sol-Gel Method

Claudia López Melendez [1], Humberto Alejandro Monreal Romero [2,*], Caleb Carreño-Gallardo [3], Guillermo Martinez Mata [2], Rosaura Pacheco Santiesteban [2], Teresa Pérez Piñon [2], Dagoberto Pérez Piñon [4], Héctor Alfredo López Aguilar [1], Marvin Elco Estrada Macias [2] and José Guadalupe Chacón-Nava [3]

[1] Department of Engineering and Materials, La Salle University, Avenue Lómas de Majalca 1120, Chihuahua CP 31625, Chih., Mexico; clopez@ulsachihuahua.edu.mx (C.L.M.); hlopez@ulsachihuahua.edu.mx (H.A.L.A.)

[2] Department of Biomaterials Science and Nanotechnology, University of Chihuahua (UACH), Avenue University, Chihuahua CP 31000, Chih., Mexico; gmata@uach.mx (G.M.M.); rpacheco@uach.mx (R.P.S.); mperez@uach.mx (T.P.P.); meestrada@uach.mx (M.E.E.M.)

[3] Advanced Materials Research Center, S.C. (CIMAV) and National Nanotechnology Laboratory, Avenue M. Cervantes 120, Industrial Complex Chihuahua, Chihuahua CP 31136, Chih., Mexico; caleb.carreno@cimav.edu.mx (C.C.-G.); jose.chacon@cimav.edu.mx (J.G.C.-N.)

[4] Department of Polymers and Biomaterials, University of Chihuahua (UACH), University Circuit Campus II, Chihuahua CP 31000, Chih., Mexico; dperezp@uach.mx

* Correspondence: hmonreal@uach.mx

Abstract: Olive-like TiO_2 (titanium dioxide), nanospheres compounds were synthesized. Polysaccharide (1–3 linked β-D galactapyranose and 1.4-linked 3.6 anyhdro-α-L-galactopyranose and titanium isopropoxide (IV) was used as a precursor in its formation. The powder sample was evaluated by scanning tunneling microscope, X-ray diffraction pattern, power spectral density, fast Fourier transform, differential thermal analysis, continuous wavelet transform, and isotropy texture analysis. The results demonstrate that these nanospheres can successfully be synthesized in a solution using a polysaccharide network by means of the sol-gel method. The synthesized olive-like TiO_2 nanospheres have diameters ranging from 50 nm to 500 nm. The synthesis parameters, such as temperature, time, and concentration of the polysaccharide, were controlled in solution.

Keywords: titanium dioxide; olive-like TiO_2 nanospheres; polysaccharide

1. Introduction

The synthesis of materials using various organic compounds has increased today to study aspects related to molecular self-assembly to generate structures of diverse morphology, such as nanoparticles, coatings, and organic-based nanomaterials [1]. The formation of olive-type structures, with different properties, such as drug release, hydrogen storage reservoirs, and the use of organic molecules such as vitamins, amino acids, and proteins, has led to the exploration of new routes to its synthesis. The use of metal oxides such as titanium represents a favorable alternative for these activities due to their chemical stability within organic systems. It has been used in various areas, such as clinical medicine in the application of prostheses, alternative energy, and biosensors. Titanium has proven to be very versatile within these areas since its physical, chemical, and electronic properties can increase the kinetic reactions between the surface of the titanium and the molecules that interact in the aqueous medium. This immobilizes the ionic density in the molecular self-assembly processes, reducing fluid resistance when used as transport vehicles within different complex systems such as the cell, muscle fibers, and nanoelectromechanical systems. The elaboration of olive-type morphologies in the presence of polysaccharides can help bind titanium through electrostatic interactions, controlling the growth of particles along the three-dimensional network of the polysaccharide. Due to the presence of pores, nanospheres have good photocatalytic activity and have several applications, such as solar cells, wastewater treatment, pharmacology, and biomedical applications. Additionally, there are different methods in the preparation of nanospheres, such as arc discharge, solvothermal method, laser pyrolysis, directed assembly method, sol-gel method, and tempering method. Likewise, the different strategies of synthesis highlight the importance of the preparation of nanospheres for the formation of olive-like structures using different methods such as the electrodeposition technique, hydrothermal technique, glycolated precursors, and mini-emulsion processes [2–5]. One of the most interesting and versatile processes for this purpose is related to the use of TiO_2 (titanium dioxide) with different polymeric precursors such as Polyvinyl alcohol (PVA), poly(ethyleneimine) (PEI), poly(sodium4-styrene sulfonate), polymethylmethacrylate, polyamide [6–11]. In this manner, the importance of the elaboration of new nanostructures allows the use of different techniques, such as the solvothermal method, green synthesis, and sol-gel [12–14]. Other techniques used to synthesize nanoporous nanostructured materials and nanoparticles is the use of polymeric membranes and cylindrical micelles, which exhibit special characteristics [15,16]. The synthesis of nanoparticles and nanospheres in the presence of polymeric networks plays an important role in the elaboration of various compounds that can be used in areas such as the pharmacology, tissue engineering, and ceramics industries [17–19]. Furthermore, a very interesting area in the development of different structures designed through organic molecules is the design of bioinspired materials using biological organisms, such as biopolymers, bone, and fibers [20]. These structures have interesting mechanical properties that make them suitable for the formation of spatial geometric figures, although they present some limitations due to their natural origin, such as being able to imitate their biomimetic properties [21]. Nevertheless, the goal is to combine the properties of each organic and inorganic material to obtain materials with specific properties, such as the use of repellent compounds in different solutions or materials with high mechanical strength [22]. In this manner, several strategies in the development of nanospheres can be engaged with the generation of bioinspired materials with various biological components, such as agarose, cellulose, and rhodamine [23–25]. Likewise, other types of materials have been used, such as SiO_2, carbon, and $CuCo2O_4$, for the synthesis of nanospheres using techniques such as protein separation, chemical vapor deposition, and heterogeneous contraction [26–28]. Many of these works attempt to explore molecular recognition processes to study chirality mechanisms when molecules of organic origin interact. Thus, studies with chiral polymers are highly promising for catalytic and optoelectronic activities. Furthermore, various nanostructures, such as nanoparticles, nanofibers, and nanospheres, can be explored using diverse technologies, such as scanning tunneling microscopy, scanning electron microscopy, and scanning probe microscopy. It is important to highlight the use of polysaccharides in the synthesis of nanospheres due to the multiple benefits of biocompatibility, non-toxicity, and ease of acquisition. In addition, polysaccharides have several physical, chemical, thermal, and optical characteristics that allow them to be used to generate a wide variety of nanoscale structures. Most of the physical, chemical, thermal, and optical processes are carried out at the nanometric scale, using a great diversity of biomolecules, inorganic molecules, and polypeptides. The resulting nanostructures, in some cases, are assembled to form complexes and materials with different geometric shapes. The forces involved are electrostatic, non-covalent interactions, ionic and hydrogen bonds, the hydrophobic and hydrophilic effect, and van der Waals forces. In this way, an important factor to consider regarding the olive-like TiO_2 nanospheres synthesized with the polysaccharide network is that the interaction of hydrogen bonds, in aqueous means, improves the electrostatic bonding during the synthesis process. In this work, we have prepared olive-like TiO_2 nanospheres, and the resulting nanospheres showed a heterogeneous size using the sol-gel method.

2. Materials and Methods

2.1. Synthesis and Characterization of Olive-like TiO_2 Nanospheres

The TiO_2 nanospheres were synthesized using the following methodology. The compounds were prepared with the addition of 1 g of polysaccharide (1–3-linked β-D galactapyranose and 1.4-linked 3.6 anyhdro-α-L-galactopyranose 99% Sigma-Aldrich St. Louis MO 63178, USA) as a template and 100 mL of bidistilled water in a solution of 300 mL of 99.8% ethanol (Sigma-Aldrich, St. Louis MO 63178 USA). Subsequently, the solution with a pH of 7.0 was subjected to a heating process at 50 °C for 5 min, 25 mL of titanium (IV) isopropoxide (Ti [OCH(CH3)2]4—97% (Sigma-Aldrich, St. Louis MO 63178 USA), 0.5 M, under constant stirring and drying at room temperature for gel formation for 8 min in the mold. In this way, the polymerization of the agarose occurs slowly, forming an intramolecular network controlling the growth of the spheres by minimizing the free energy caused by the surface tension of the polysaccharide fibers. Once the gel was formed, it was placed in ethanol, and the gel obtained was subjected to aging for 24 h. Afterward, the ethanol was exchanged 3 times over the course of 5 days, and the gel was heated to 500 °C for approximately 10 min. The heating purpose is to obtain the powders and establish the beginning of the transition of the crystalline phases of TiO_2. Furthermore, it increases the density of TiO_2 and slows down the degradation of the polysaccharide network without compromising the formation of the geometric shape of the nanospheres. It also preserves the cross-linking of their fibers.

The powders obtained were characterized with the following techniques.

2.2. Characterization of Olive-like TiO_2 Nanospheres by FESEM

The olive-like TiO_2 nanospheres were studied with a field emission scanning electron microscope (FESEM) JEOL-JCSM-7401F Peabody, MA 01960 USA. The specimens were placed in a holder with an acceleration voltage of 2.0 kV and a magnification of 30,000.

2.3. Characterization of Olive-like TiO_2 Nanospheres by X-ray

For the characterization by X-ray diffraction, the samples of the powders began to be prepared and were placed in the sample holder, being careful to place them in the center and without the sample protruding from the walls of the latter so that the diffraction phenomenon occurred without problems, and the results were optimal for later study. The sample was placed in the first space of the diffractor, and it was verified that it was completely well placed so that the container or the sample would not fall. Next, the parameters in which the sample was to be studied were placed, as well as the place where the sample was located, and diffraction was carried out. The crystalline phase of the nanostructures was detected by an X-ray diffraction (XRD) pattern using a CuK (α) at 0.1542 nm on a Phillips X'. PERT Eindhoven AMS. X-ray diffractometer.

2.4. Characterization of Olive-like TiO_2 Nanospheres by TGA-DTA-DTG Analysis

The TGA-DTA-DTG analysis of the nanospheres was processed on DTA-TGA TA, CDMX equipment at a heating rate of 10 °C/min in an air atmosphere.

2.5. Characterization of Olive-like TiO_2 Nanospheres by STM

The characterization of the compounds was carried out through a Nanosurf easyScan 2, Liestal, Switzerland, instrument scanning tunneling microscope equipped with Pt/Ir tips (BT00400), and the images were processed using an easyScan 2 imaging software version 1-6-0-0 and a Mountains software version 9.0 United States Laboratory. The preparation of the samples for STM was carried out as follows: the powders were placed in 5 mL of bidistilled water and stirred in a Thermo scientific 16715 vortex to disperse the nanospheres; subsequently, the nanospheres were dried at room temperature for 24 h. Once dried, they were placed on a microscope glass slide-Esco of 1.2 mm thickness. The parameters used in STM were continuous mode at 20 nN, the tip voltage was 1 V, the sweep speed was 0.1 s, and the operating mode was static force. The image resolution area was 161 × 161 nm.

2.6. Characterization of Olive-like TiO_2 Nanospheres by Mountains Lab Software

The analysis of power spectral density (PSD), fast Fourier transform (FFT), surface geometry analysis, fractal dimension analysis, coefficient of determination R2, continuous wavelet transform analysis, decomposition of the continuous signal, profile of holes analysis, and texture isotropy analysis was executed utilizing the Mountains Lab 9.0 image processing software. The images obtained through STM and FESEM were processed in this software using the following parameters: structural analysis, texture direction, control chart, and critical dimension analysis.

3. Results and Discussion

3.1. Characterization of Olive-like TiO_2 Nanospheres by SEM

Figure 1a shows the SEM micrograph of TiO_2 in the absence of the polysaccharide matrix, it can be seen that there are no nanospheres present, and the surface is irregular. Figure 1b shows the image in SEM of the olive-like TiO_2 nanospheres in the presence of a polysaccharide matrix with diameters of 50–500 nm. This type of morphology offers the advantage of being used as a vehicle to store different substances, such as drugs, proteins, and electronic nanodevices, compared to those existing in the literature. TiO_2 nanospheres have been reported with sizes between 56 and 100 nm using polysaccharides such as agarose and alginate [23–29]. The image presents heterogeneous morphological characteristics, and the hole of the nanospheres is probably due to the polysaccharide matrix. Also, these images suggest that the electrical behavior of the electrostatic junction composed of polysaccharide and TiO_2 is controlled by an electron transfer mechanism characteristic of molecular self-assembly systems in solution. Molecular self-assembly is a process of spontaneous organization with electrostatic interactions under several conditions of equilibrium, such as hydrogen bonds of the polysaccharides, non-covalent bonds, hydrophobic and hydrophilic interactions, and polymerization of processes.

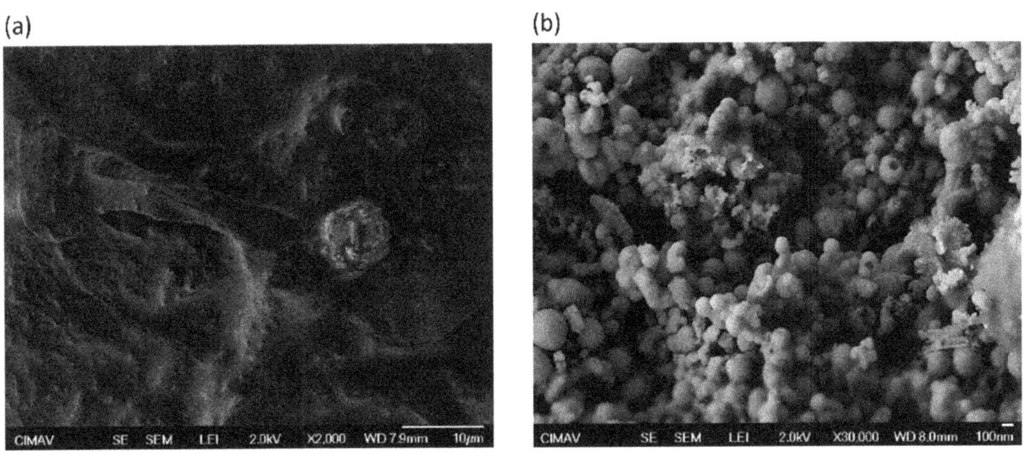

Figure 1. (**a**) SEM image of TiO_2 (titanium dioxide) in absence of polysaccharide; (**b**) SEM micrograph of olive-like TiO_2 nanospheres in presence of polysaccharide network.

3.2. Particle Size Distribution

Figure 2 shows the particle size distribution with ranges of 50–500 nm. Approximately 30% of the nanospheres obtained are in the range of about 100 nm in diameter. In another study, a particle size in the range of 100–200 nm was reported using the biphasic system [30]. Moreover, the average particle size, utilizing the polymeric system, permits the biological activity to be fixed on the surface of the nanospheres or in the aqueous medium. In other words, the control of the properties of the linear polymer and the particle size determine the characteristics of each sphere.

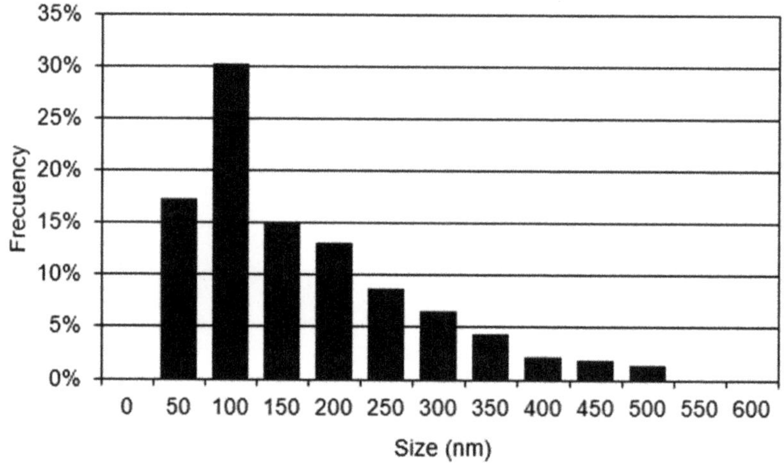

Figure 2. Chart of particle size distribution of olive-like TiO_2 nanospheres.

3.3. *X-ray Diffraction Pattern*

The crystalline phase of the olive-like TiO_2 nanospheres was identified by X-ray diffraction patterns at 500 °C, Figure 3. In these patterns, the reflections at 2θ of 25°, 31°, 38°, 48°, 55°, 57°, 63°, 65°, 70°, and 75° visibly correspond to the anatase phase of TiO_2. The system used allowed the obtaining of crystalline nanospheres, avoiding subsequent sintering methods that could induce TiO_2 to crystallize, affecting the ability to adsorb the polysaccharide network. These results are similar to those found in other works during the synthesis of TiO_2 nanospheres [31–33]. The degree of the crystalline phase is approximately 98%, and the intensity of the reflections of the anatase phase can be observed. A separation of the amorphous phase can be observed before the first reflection, and later, the other reflections appear, which remain during the sintering process.

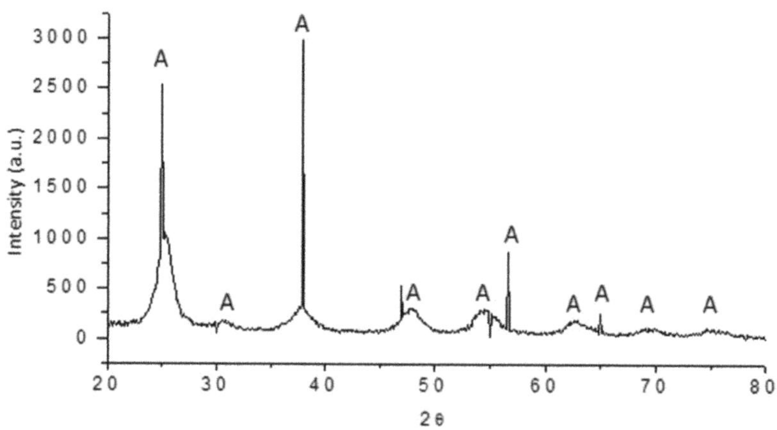

Figure 3. X-ray diffraction pattern of olive-like TiO_2 nanospheres.

3.4. *Thermo-Gravimetric and Differential Thermal Analysis*

In Figures 4 and 5, the results of the TGA-DTA study of the olive-like TiO_2 nanospheres are shown. The temperature region of the TGA curve, Figure 4, shows a weight loss at 100–200 °C. This behavior corresponds to the loss of water and alcohol that causes the appearance of an endothermic peak at 60 °C and an exothermic peak at 300 °C, Figure 5, caused by the combustion of organic matter and the titanium precursor. The polysaccharide fibers have oxidation degrees of the residual galactose and agarobiose groups, which means that there is no loss of weight at 45 °C. This behavior is due to a process of direct coagulation or chemical gelification of the polysaccharide, which allows it to increase the density of TiO_2 and decrease the fast degradation of the polysaccharide.

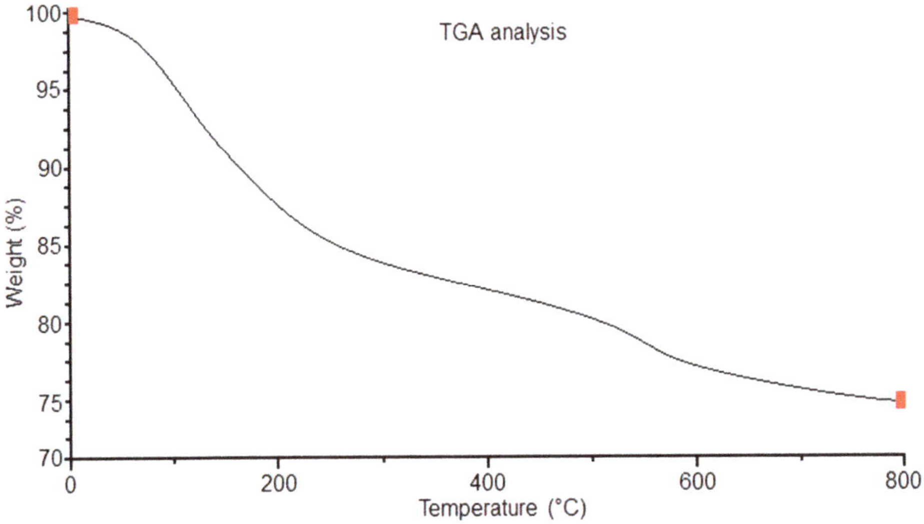

Figure 4. TGA analysis of the olive-like TiO_2 nanospheres.

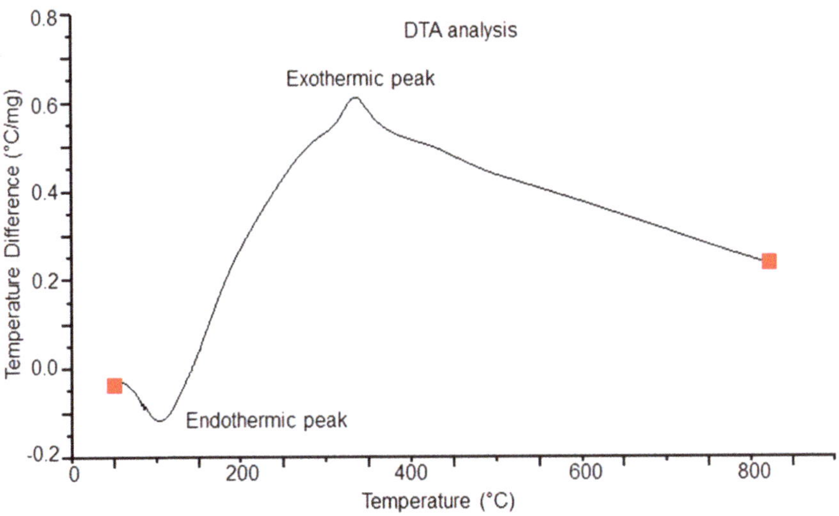

Figure 5. DTA analysis of the olive-like TiO_2 nanospheres.

In addition, there is a slow and continuous degradation of the polysaccharide matrix at temperatures between 65 and 600 °C. The permanence and evolution of the anatase phase is at approximately 500 °C. This prevalence may be due to the kinetic processes of phase transformation, in which the polysaccharide matrix is capable of retaining the anatase phase, inhibiting the nucleation processes for the formation of the rutile phase. Figure 6 shows the DTG analysis of the nanospheres. In the first, second, and third points of the curve, the mass loss occurs in a slow process between 100 and 200 °C. This loss may be due to the desorption of water adsorbed on the surface of the nanospheres. A fourth point is seen at approximately 250 °C; according to this, as the temperature increases, the loss of mass becomes even slower, reaching a curve at 300 °C. We report that the total weight loss is 22.5%. Subsequently, the stabilization of the curve begins at approximately 500 °C. The observed behavior may be due to the breaking of chemical bonds, causing the three-dimensional network of the polysaccharide to decompose and resulting in the thermal degradation of TiO_2. In this way, the presence of these curves may represent the stability of TiO_2 in relation to the three-dimensional network of the polysaccharide carried out by a complex thermal degradation interaction.

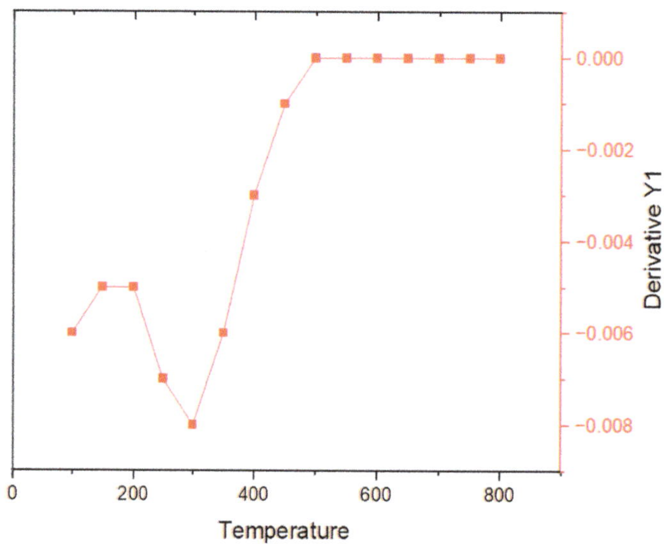

Figure 6. DTG analysis of the olive-like TiO_2 nanospheres.

3.5. Characterization of Olive-like TiO_2 Nanospheres by Scanning Tunneling Microscopy (STM)

Figure 7 shows a complex of nanospheres as part of the integration into a polysaccharide matrix, thereby demonstrating that compatibility exists and electronic activity is prevalent. The observed behavior of nanospheres at low temperatures in STM suggests, however, that the interaction of the adatom would be located in the central region between the rest of the atoms, as shown in Figure 7. In this manner, for the various partially occupied orbitals that are present in the unit cell of the reconstruction of Ti (110), an element more electronegative than the surface (electron acceptor) would have a preference to interact with the polysaccharide. Therefore, if there is an interaction with an electron acceptor component (more electronegative than the surface), there will be a charge transfer from the surface to the element. In both cases, there is a flow of electrons characterized by a variation in the chemical potential of electrons. This will have a model system to transfer (donate or accept) electrons. The study also suggests that the increase in the function of the polysaccharide matrix might be due to the similarity of the fibers and the TiO_2 crystals found in the nanospheres. The binding of Ti-O-H molecules within a solution, such as ethanol, increases the additional energy of the Fermi level, causing the nanospheres to increase electrical conductivity. The difficulty of performing surface topography analysis on monosaccharides derives from the electrostatic charges emitted by STM. These results demonstrate the self-assembly of the nanospheres in the polysaccharide matrix, reducing the generation of unnecessary electrical charges and, thus, allowing the effective union of oxygen with Ti. In this way, the electronegative OR group of the alkoxide renders the metal very susceptible to nucleophilic attack, leaving a partial negative charge. This, during the reaction, allows the polysaccharide to provide a system capable of forming hydrogen bonds and reacting with the hydrolyzed Ti-OH species. During the polymerization process, in the presence of the polysaccharide network, the Ti-OR is converted into a Ti-OH bond. The results also demonstrate that the nanospheres induce a fusion process within the solution by modifying the supramolecular structure of the interpenetrated networks of the polysaccharide. Nanospheres allow the miniaturization and functionality of the networks, allowing the force fields between molecules to be optimal for interaction. This characteristic is what allows, through electrostatic interactions, in an aqueous medium, for the compounds in the solution, mainly the precursors of the titanium and water used, and the polysaccharide fibers, to interact, finally forming the structures and giving rise to the chemical modification of the surface results, favoring adhesion to the TiO_2 surface. Several studies have been developed in which the properties of metal oxides such as TiO_2 and the activity of oxygen vacancies in nanospheres have been analyzed, but this activity has not been reported with polysaccharides such as those performed in this study [34–36].

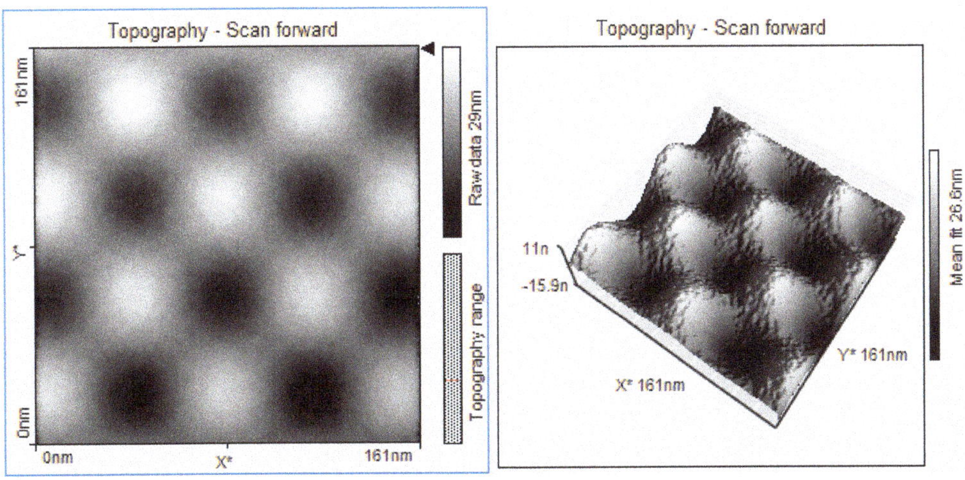

Figure 7. STM image of olive-like TiO_2 nanospheres.

3.6. Characterization of Olive-like TiO_2 Nanospheres by FFT

The frequency spectrum of the fast Fourier transform is observed in Figure 8; the nanospheres produce high frequencies in their intensity levels, which are observed as parts with greater luminosity due to the presence of TiO_2. They can also cause low frequencies that are shown as less intense parts that are caused by the presence of the polysaccharide matrix. The value of power relative to carrier (dBc) analysis of −3522 indicates that the nanospheres have a higher power in the carrier signal than in the power of the input signal. In this way, greater power allows greater interaction in the molecular self-assembly process, leading to a stronger union. In this way, when the aggregates in each pore of the polysaccharide network align to form the nanospheres, there is an unequal number of negatively charged monomers in each pore, resulting in a molecular dipole throughout the matrix. Depending on the orientation of this molecular dipole, an applied voltage can mobilize and destabilize the aggregates, providing voltage-dependent ionic transport. In this manner, the amplitude of the fast Fourier transform is taken by the power spectral density to normalize the frequency width.

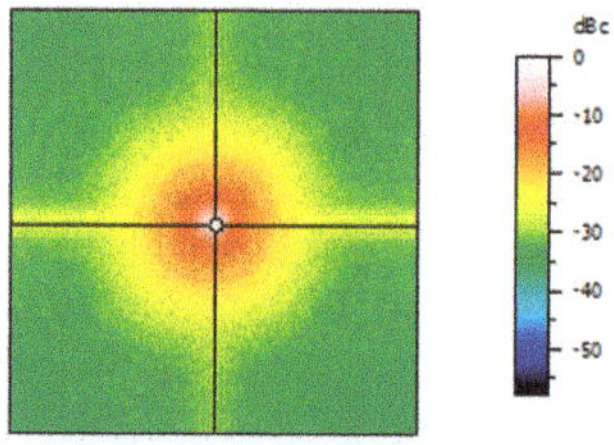

Information			
Window function	None		
Parameters	Value	Unit	
X	20.03	µm	
Y	−33.39	µm	
Wavelength	17.18	µm	
Angle	−30.96	°	
Magnitude	−3.522	dBc	
Phase	19.15	°	

Figure 8. The frequency spectrum of FFT.

3.7. Characterization by Power Spectral Density (PSD) Analysis of Olive-like TiO_2 Nanospheres

The image of Figure 9 shows the characterization of the nanospheres by means of the force spectral density analysis; the area of the square of the curve that corresponds to the energy of the signal, both vertical and lateral, is shown. The results are distinctive in representing the function of the moisture absorption capacity and the property of the nanospheres to scatter light emissions. The dominant energy corresponding to the wavelength presents a maximum peak of 0.4244 µm with an amplitude of 1.581 gL; the graph shows a decrease in its curve with the presence of two peaks, one at 0.6 µm and the other at 0.9 µm. The interaction of TiO_2, ethanol, and polysaccharide facilitates the appearance of peaks that contain force periodicity signals, controlling the existence of electron vibrations and canceling the presence of noise during the temperature application process in the gels. In this context, the level of polarization may depend on the electromagnetic power of the nanospheres and the dipole moment at high and low frequencies, with changes in the dielectric balance.

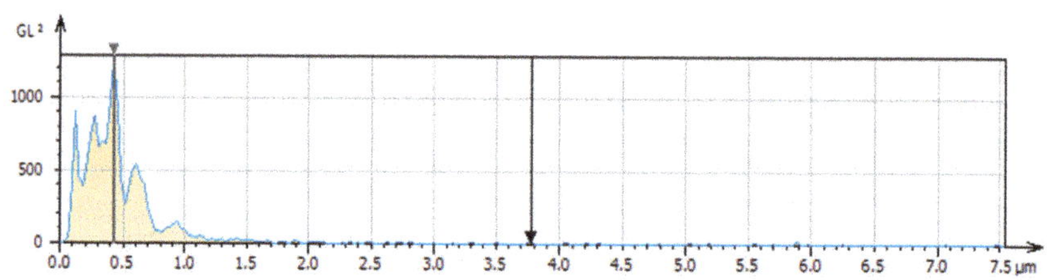

Information		
Zoom factor	x4	
Smoothing	None	
Windows function	Hanning	
Parameters	Value	Unit
Wavelength	3.789	µm
Amplitude	1.581	GL
Dominant wavelength	0.4244	µm
Maximum amplitude	35.78	GL

Figure 9. Power spectral density (PSD) analysis.

3.8. Fractal Dimension Analysis

The values of the analysis of the fractal dimension are shown in Figure 10. The olive-like TiO_2 nanospheres were analyzed using the following equation:

$$\log S = D \log L = D = \log \frac{S}{\log L} \tag{1}$$

where S (8) is the size of the fractal, L is the measurement scale (2.246), and D is the unknown fractal dimension. The nanospheres present a homogeneous morphology, and the values of the fractal dimension of 2.569, as well as the R2 correlation data, establish the existence of a positive relationship with a value of 1, which demonstrates that the nanosphere formation system can be 100% replicated; see Figure 11. In this way, the system evolves and organizes into a state or structure that is consistent with the decrease in the total free energy of the system, including the free energies present in the volume as well as the free energies present at the interface. The energies can be elastic strain energies due to the mismatch between the various phases as well as between the real system and the substrate. The interface energy leads to the selection of those surfaces and interfaces that have the lowest energies. Kinematically controlled nanospheres are governed by variables such as deposition, time, and diffusion fluxes from the metal surface. The structure that is being formed is the result of the interaction of these processes, which act simultaneously during the deposition process of the metal precursor. Finally, depending on the different precursors, the growth rate is minimal, even in the presence of other acidic or basic media. Furthermore, the growth rate also depends on the concentration of the dispersing medium. Likewise, the result of the fractal dimension is related to the morphology of the edges since it presents an irregular surface that is continuous and characteristic of its geometric shape, which is due to the participation of the polysaccharide matrix. The study of particle addition and the formation of fractals in TiO_2 nanostructures have been reported in several works [37–41].

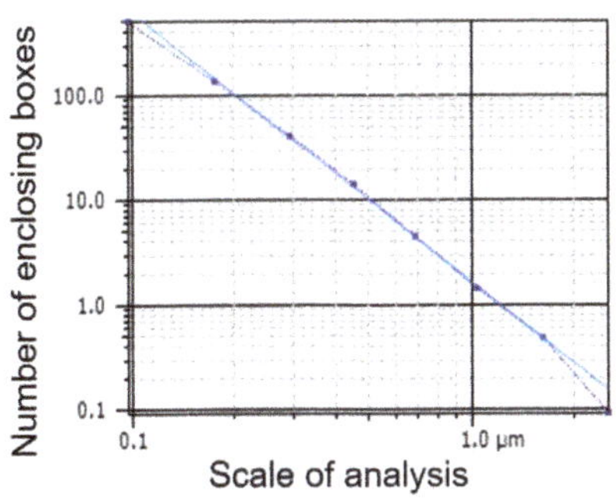

Information	
Method	Enclosing boxes
Parameters	value
Fractal dimension	2.569

Figure 10. Fractal dimension analysis of olive-like TiO_2 nanospheres.

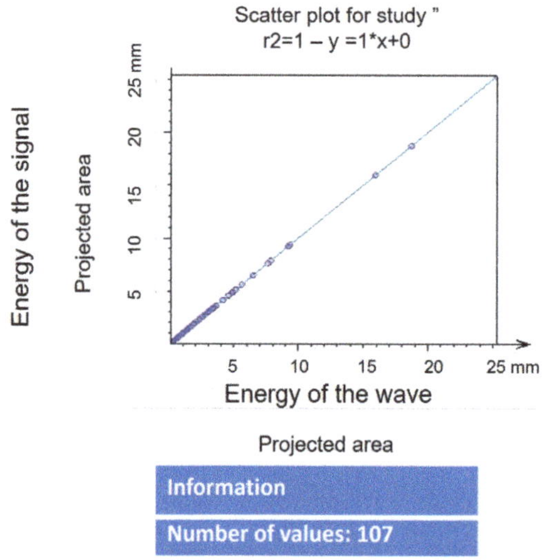

Figure 11. Coefficient of determination R2.

3.9. *Characterization by Continuous Wavelet Transform (CWT)*

The continuous wavelet transform was used for the determination of the origin of the wave propagation of the mother signal in the composite materials of polysaccharide and Ti, and the textural characterization of the nanosphere surface to detect anomalies under the presence of wavelets. Figure 12 shows orthonormal wavelets, which means that they meet the orthonormality conditions within the same signal scale. The x-axis represents the position along the signal (time), the y-axis represents the scale, and the color represents the magnitude of the wave coefficient. The continuous wavelet transform is determined by the following equation:

$$Xwt(a,b) = AA\infty x(t)\Psi a, b(t)dt \quad (2)$$

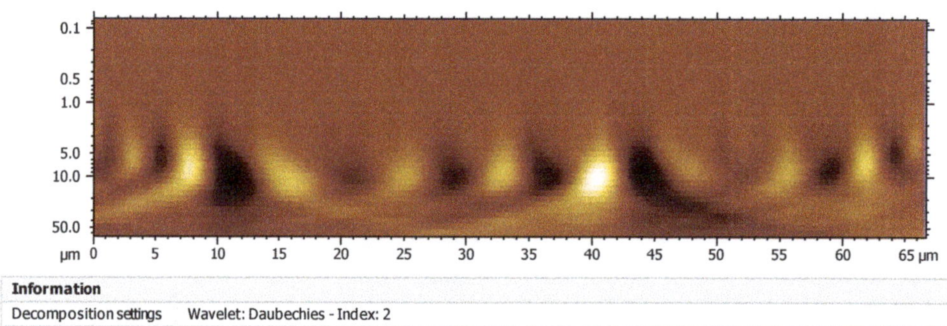

Figure 12. Wavelets of olive-like TiO_2 nanospheres.

The mother wavelet generates several daughter wavelets that are added to the time of the signal $f(t)$, originating sinusoidal components of the original signal. The image shows many of the wavelet coefficients (front and back), which are related to the scale and position of the mother wavelet. The peaks found in the wavelets can be caused by the effect of temperature at the time of sintering, which generates wear in the polymer chains. Figure 13 shows the histogram of the decomposition of the continuous signal. The signal values are greater for a scale factor of 8, 9, and 10 µm, which means that with that scale and at those positions, the similarity in the signals is greater, denoting that the energy of the signal and the energy of the wave are equal to one; this can be interpreted as a correlation coefficient (Figure 11). Likewise, the proposed system originates high- and low-frequency signals that vary in the time–frequency plane, causing high and low temporal resolutions, resulting in no discontinuity on the surface of the nanospheres. The columns of Figure 13 indicate non-periodic signal coefficients (wavelets) with short-duration pulses that make it possible to rapidly detect molecular self-assembly events between the compounds. The results do not show the presence of irregular edges on the surface; these data are conclusive in their determination of great compatibility between the interpenetrated networks of the polysaccharide with titanium. Other studies have analyzed the behavior of dispersive waves and frequency timing by means of cwt in TiO_2 films, circular structures, and other nanostructures [42–44].

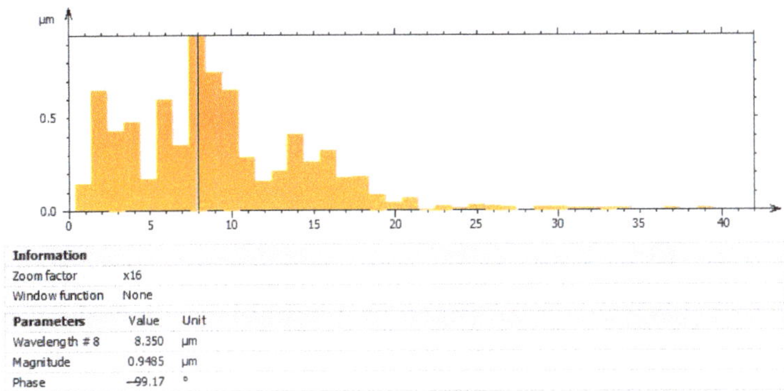

Figure 13. Histogram of signal decomposition of olive-like TiO_2 nanospheres.

3.10. Characterization of the Profile of the Holes of Olive-like TiO_2 Nanospheres

The profile of the holes in the center of the nanospheres is shown in Figure 14. The holes are represented in red, while the green color corresponds to the titanium precursor. The formation of the holes takes place by means of the polysaccharide host molecule, which has cavities that can retain the TiO_2. The nanospheres have a maximum depth of 32.32 gL and an area of 36.59 µm gL. In this way, the proposed system generates supramolecular architectures from metallic TiO_2 ions and organic ligands such as polysaccharide polymer networks. In a study developed by Nora et al., the shell thickness of the surface of the TiO_2 nanospheres was analyzed; however, their depth and area were not measured [45].

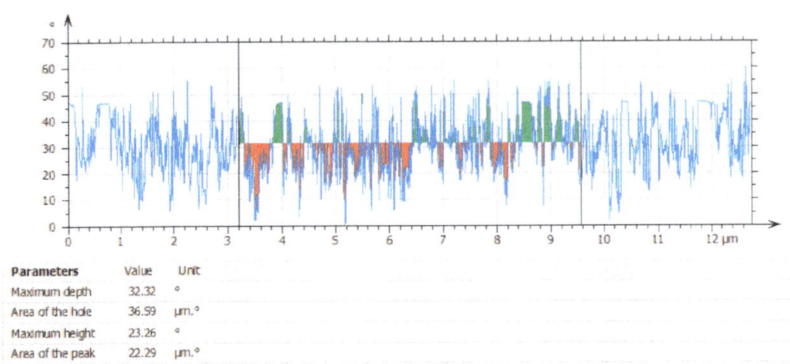

Figure 14. Profile of holes of olive-like TiO_2 nanospheres.

3.11. Texture Isotropy Analysis

To measure the directional uniformity parameters of the surface planes of the olive-like TiO_2 nanospheres, a textural isotropy analysis was performed. The characteristics of the direction of the surface texture can be seen in Figure 15. The results of the values found were isotropy 55.31%, first direction 60.26°, second direction 138.7°, and third direction 148.5°, respectively. These results show the isotropic nature of the system, maintaining the uniformity of the surface features in all directions. Various structures, such as Cu/Ni nanoparticles, collagen fibers, cells, and carbon nanotubes, have been analyzed, but systems such as the one proposed in this work have not been studied [46–50]. Additionally, the location of each point on the surface of the sphere is represented by the angular degrees corresponding to the specific rotation coordinates of the number of grains in the sample in several directions. This means that the diverse molecules that contain the nanospheres are evenly distributed, which corresponds to the isotropic texture of the olive-type TiO_2 nanospheres.

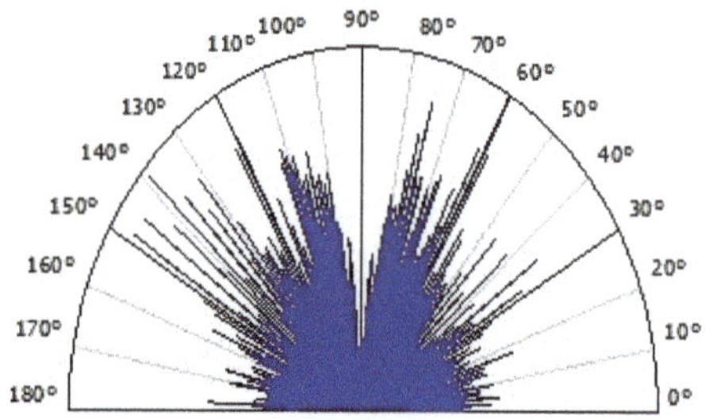

Parameters	Value	Unit
Isotropy	55.31	%
First direction	60.26	°
Second direction	138.7	°
Third direction	148.5	°

Figure 15. Texture isotropy analysis of olive-like TiO_2 nanospheres.

4. Conclusions

In this work, we have reported the formation of olive-like TiO_2 nanospheres in a range of 50–500 nm in diameter. The compounds synthesized can be replicated in other areas, such as biomaterials, nanotechnology, or materials science. The crystallinity of the nanospheres calcined at 500 °C is confirmed by X-ray. In summary, the olive-like TiO_2 nanospheres were examined to determine their effects on the viability and adhesion of the polysaccharide matrix. Additionally, the proposed mechanism and the characteristics mentioned are factors of great relevance to studying the effects of the polysaccharide network and to be able to analyze variables, such as thermodynamic activity, electrostatic interactions, and organic ligands. The fractal dimension values with a value of 2.569 found in the fractal dimension analysis highlight the binding between the titanium and the polysaccharides network system. The functional activity of olive-like TiO_2 nanospheres was confirmed by STM and FESEM in terms of morphological and size characteristics. Other studies of interest include corrosion studies and biocompatibility studies of metallic oxides.

Author Contributions: Methodology, H.A.M.R., C.C.-G., G.M.M., and T.P.P.; software, R.P.S.; formal analysis, D.P.P., H.A.L.A., M.E.E.M., and J.G.C.-N.; investigation, C.L.M.; data curation, M.E.E.M. All authors have read and agreed to the published version of the manuscript.

Funding: This research received no external funding.

Institutional Review Board Statement: Not applicable.

Informed Consent Statement: Not applicable.

Data Availability Statement: Data are contained within the article. For more information can be directed to the corresponding author.

Conflicts of Interest: The authors declare no conflict of interest.

References

1. Jaison, J.; Ahmed, B.; Yen, S.; Alain, D.; Michael, D. Review on nanoparticles and nanostructured materials: History, sources, toxicity and regulations. *Beilstein J. Nanotechnol.* **2018**, *9*, 1050–1074.
2. Artur, M.; Natalia, Ł.; Maciej, S.; Izabela, M.; Wojciech, S. Electrodeposition of Copper and Brass Coatings with Olive-Like Structure. *Materials* **2021**, *14*, 1762. [CrossRef]
3. Reghunath, S.; Pinheiro, D.; KR, S.D. A review of hierarchical nanostructures of TiO_2: Advances and applications. *Appl. Surf. Sci. Adv.* **2021**, *3*, 100063. [CrossRef]
4. Lu, Y.; Gan, Z.; Xia, J.; Du, K.; Peng, Z.; Cao, Y.; Hu, G.; Xiao, J. Hydrothermal Synthesis of Tunable Olive-Like $Ni_{0.8}Co_{0.1}Mn_{0.1}CO_3$ and its Transformation to $LiNi_{0.8}Co_{0.1}Mn_{0.1}O_2$ Cathode Materials for Li-Ion Batteries. *ChemElectroChem* **2019**, *6*, 5661–5670. [CrossRef]
5. Fröschl, T.; Hörmann, U.; Kubiak, P.; Kucerová, G.; Pfanzelt, M.; Weiss, C.; Behm, R.; Hüsing, N.; Kaiser, U.; Landfester, K.; et al. High surface area crystalline titanium dioxide: Potential and limits in electrochemical energy storage and catalysis. *Chem. Soc. Rev.* **2012**, *41*, 5313–5360. [CrossRef]
6. Ye, L.; Zhenping, Q.; Hongxia, G.; Hanxiao, Y.; Guojun, Z.; Shulan, J.; Tingying, Z. Low-temperature synthesis of anatase TiO_2 nanoparticles with tunable surface charges for enhancing photocatalytic activity. *PLoS ONE* **2014**, *9*, 114638.
7. Satheesh, J.; Goud, D.; Narsimlu, N. Synthesis and Characterization of TiO_2 nanoparticles in PVA Polymer Matrix. *J. Appl. Phys.* **2021**, *13*, 50–55.
8. Yufan, Z.; Fan, F.; Yuzhou, L.; Desuo, Z.; Yuyue, C. One-Step Synthesis of Ag@TiO_2 Nanoparticles for Enhanced Photocatalytic Performance. *Nanomaterials* **2018**, *12*, 1032. [CrossRef]
9. Alamgir, M.; Ashis, M.; Nayak, S.; Tiwari, K. Development of PMMA/TiO_2 nanocomposites as excellent dental materials. *J. Mech. Sci. Technol.* **2019**, *33*, 4755–4760. [CrossRef]
10. Shalini, A.; Vibhav, S. Synthesis and Thermal Characterization of PMMA-TiO_2 Nanocomposites. *Mater. Sci. Res.* **2014**, *11*, 168–172.
11. Tian, Y.; Yang, D.; Wang, Y.; Jiang, Z. Template-free synthesis of TiO_2 microcages in agarose gels with improved photocatalytic activity. *J. Nanoparticle Res.* **2013**, *15*, 2141. [CrossRef]
12. Venkatraman, M.; Rama, K.; Muthukumara, S.; Natarajana, A.; Santhanama, V.; Asokan, S.; Dhayalan, V. Size controlled synthesis of TiO_2 nanoparticles by modified solvothermal method towards effective photo catalytic and photovoltaic applications. *Mater. Res. Bull.* **2018**, *97*, 351–360.
13. Aravind, M.; Amalanathan, M.; Sony, M. Synthesis of TiO_2 nanoparticles by chemical and green synthesis methods and their multifaceted properties. *Appl. Sci.* **2021**, *3*, 409. [CrossRef]
14. Mohammad, H.; Riccardo, C. Sol-Gel Synthesis of TiO_2 Nanocrystalline Particles with Enhanced Surface Area through the Reverse Micelle Approach. *Adv. Mater. Sci. Eng.* **2019**, *2019*, 1567824.
15. Takeshi, S.; Takafumi, O.; Hiroki, N.; Koji, N.; Masato, M.; Bartolome, S.; Ken, K.; Masafumi, Y.; Hiroki, T.; Takao, S.; et al. Development of Nanostructured Water Treatment Membranes Based on Thermotropic Liquid Crystals: Molecular Design of Sub-Nanoporous Materials. *Adv. Sci.* **2018**, *5*, 1700405.
16. Meagan, S.; Menachem, E.; Chinedum, O. Nanocomposites of vertically aligned single-walled carbon nanotubes by magnetic alignment and polymerization of a lyotropic precursor. *ACS Nano* **2010**, *4*, 6651–6658.
17. Zeynep, A.; Iva, C.; Gabriella, D.; Sabine, P.; Soccorso, G.; Ibtisam, E. Development of functionalized nanostructured polymeric membranes for water purification. *Chem. Eng. J.* **2016**, *300*, 358–366.
18. Antonio, G.; Mani, D.; Monique, K.; John, A.; Jeroen, J.; Fang, Y. Development of a PCL-silica nanoparticles composite membrane for Guided Bone Regeneration. *Mater. Sci. Eng. C Mater. Biol. Appl.* **2018**, *85*, 154–161.
19. Dina, E.; Norhan, A.; MhdAmmar, H.; Abdelbaki, B.; Alaa, H. Nanoparticles functionalized ceramic membranes: Fabrication, surface modification, and performance. *Environ. Sci. Pollut. Res.* **2021**, *10*, 12256–12281.
20. Meyers, M.; McKittrick, P.Y.; Chen, P. Structural biological materials: Critical mechanics-materials connections. *Science* **2013**, *339*, 773–779. [CrossRef] [PubMed]
21. Wegst, U.; Bai, H.; Saiz, E.; Tomsia, A.; Ritchie, R. Bioinspired structural materials. *Nat. Mater.* **2015**, *14*, 23–36. [CrossRef]

22. Eder, M.; Amini, S.; Fratzl, S. Biological composites-complex structures for functional diversity. *Science* **2018**, *362*, 543–547. [CrossRef]
23. Wang, Y.; Yang, D.; Shi, Y.; Jiang, Z. Bio-inspired synthesis of TiO_2 hollow nanospheres in agarose gels. *J. Alloys Compd.* **2013**, *560*, 42–48. [CrossRef]
24. Saba, H.; Masoud, J. Synthesis of TiO_2 nanoparticles coated on cellulose nanofibers with different morphologies: Effect of the template and sol-gel parameters. *Mater. Sci. Semicond. Process.* **2020**, *109*, 104927.
25. Gao, M.; Li, L.; Liu, Q.; Xue, Y.; Wang, W.; Zhao, Y.; Zou, X. Synthesis of multifunctional silica composite nanospheres and their application in separation of MBP-tagged protein. *Mater. Lett.* **2022**, *318*, 132–222. [CrossRef]
26. Li, B.; Zhang, Y.; Cao, L.; Yonghui, W.; Zhao, Y. Synthesis of silica nanospheres with Ni^{2+}-iminodiacetic acid for protein separation. *J. Sol-Gel Sci. Technol.* **2014**, *71*, 80–384.
27. Gonzalez-Reyna, M.; Rodriguez-Lopez, A.; Pérez-Robles, J.F. One-step synthesis of carbon nanospheres with an encapsulated iron-nickel nanoalloy and its potential use as an electrocatalyst. *Nanotechnology* **2020**, *32*, 095706. [CrossRef]
28. Feng, Y.; Liao, J.; Chen, X.; Wang, H.; Guo, B.; Li, H.; Zhou, L.; Huang, J.; Li, H. Synthesis of rattle-structured $CuCo_2O_4$ nanospheres with tunable sizes based on heterogeneous contraction and their ultrahigh performance toward ammonia borane hydrolysis. *J. Alloys Compd.* **2021**, *863*, 158089. [CrossRef]
29. Paques, J.P.; van der Linden, E.; van Rijn, C.J.; Sagis, L.M. Preparation methods of alginate nanoparticles. *Adv. Colloid Interface Sci.* **2014**, *209*, 163–171. [CrossRef] [PubMed]
30. Yuan, Y.; Xu, H.; Meng, X.; Han, Q. Two-phase synthesis of olive-like NiS particles and chain-like Bi_2S_3 nanowires. *Mater. Sci.-Pol.* **2015**, *33*, 1–5. [CrossRef]
31. Ghosh, O.S.N.; Gayathri, S.; Alagarasan, D.; Kumar, K.V.P.; Viswanath, A.K. One pot transparent sol-gel synthesis of TiO_2 nanospheres with pristine anatase phase and their physicochemical characteristics. *Appl. Sci. Lett.* **2016**, *2*, 23–30. [CrossRef]
32. Urbano, M.A.V.; Vargas, A.; Muñoz, Y.H.O.; Ochoa, H.; Fernandez, Y.O.; Mosquera, P.; Paez, J.E.R.; Rodríguez, E.; Amado, R.J.C.; Camargo, J. Nanoparticles of TiO_2, anatase phase, synthesized by chemical methods. *Ing. Des.* **2011**, *29*, 186–201.
33. Gowthambabu, V.; Deshpande, M.; Govindaraj, R.; Nithesh Krishna, V.K.; Leela Charumathi, M.; Manish Kumar, J.; Dhilip Vignesh, M.S.; Isaac Daniel, R.; Ramasamy, P. Synthesis of anatase TiO_2 microspheres and their efficient performance in dye-sensitized solar cell. *J. Mater. Sci. Mater. Electron.* **2021**, *32*, 26306–26317. [CrossRef]
34. Chen, H.; Wu, T.; Li, X.; Lu, S.; Zhang, F.; Wang, Y.; Zhao, H.; Liu, Q.; Luo, Y.; Asiri, A.M.; et al. Modulating Oxygen Vacancies of TiO2 Nanospheres by Mn-Doping to Boost Electrocatalytic N2 Reduction. *ACS Sustain. Chem. Eng.* **2021**, *9*, 1512–1517. [CrossRef]
35. Scheiber, P.; Fidler, M.; Dulub, O.; Schmid, M.; Diebold, U.; Hou, W.; Aschauer, U.; Selloni, A. (Sub)surface mobility of oxygen vacancies at the TiO_2 anatase (101) surface. *Phys. Rev. Lett.* **2012**, *109*, 136103. [CrossRef]
36. Arenas-Hernandez, A.; Zuñiga Islas, C.; Moreno, M.; Calleja Arriaga, W.; Mendoza-Cervantes, J.C.; Carlos, N.; Ascencio-Hurtado, C.R.; Heredia Jiménez, A. Study of Oxygen Vacancies in TiO_2 Nanostructures and Their Relationship with Photocatalytic Activity. *Appl. Sci.* **2022**, *12*, 3690. [CrossRef]
37. Emanuele, C.; Giulio, B.; Giuseppe, C.; Damiano, A.; Pasqualantonio, P.; Gabriele, F.; Luca, G. Aggregation and fractal formation of Au and TiO_2 nanostructures obtained by fs-pulsed laser deposition: Experiment and simulation. *J. Nanopart. Res.* **2017**, *19*, 311.
38. Cavaliere, E.; Ferrini, G.; Pingue, P.; Gavioli, L. Fractal TiO_2 nanostructures by non-thermal laser ablation at ambient pressure. *J. Phys. Chem. C* **2013**, *117*, 23305–23312. [CrossRef]
39. Celardo, G.; Archetti, D.; Ferrini, G.; Gavioli, L.; Pingue, P.; Cavaliere, E. Evidence of diffusive fractal aggregation of TiO_2 nanoparticles by femtosecond laser ablation at ambient conditions. *Mater. Res. Express* **2017**, *4*, 015013. [CrossRef]
40. Lavisse, L.; Le, G.; Hallo, L.; Jouvard, J.; Carles, S.; Perez, J.; Mitchell, J.; Decloux, J.; Girault, M.; Potin, V.; et al. In-situ small-angle X-ray scattering study of nanoparticles in the plasma plume induced by pulsed laser irradiation of metallic targets. *Appl. Phys. Lett.* **2012**, *16*, 164103. [CrossRef]
41. Dobrescu, G.; Papa, F.; State, R.; Ballint, I. Characterization of bimetallic nanoparticles by fractal analysis. *Powder Technol.* **2018**, *338*, 905–914. [CrossRef]
42. Lin, Y.; Rui, X.; Xue, X.; Li, P.; Wen, X. Wavelet analysis of the surface morphologic of nanocrystalline TiO_2 thin films. *Surf. Sci.* **2005**, *579*, 37–46. [CrossRef]
43. Kim, Y.; Kim, E. Effectiveness of the continuous wavelet transform in the analysis of some dispersive elastic waves. *J. Acoust. Soc. Am.* **2001**, *110*, 86–94. [CrossRef]
44. Tkachenko, I.; Kozhanova, E.; Belyaev, E.; Yazbeck, H. Comparative analysis of application of wavelet analysis for the recognition of element composition nanostructures. *J. Phys. Conf. Ser.* **2019**, *1309*, 012020. [CrossRef]
45. Justh, N.; Bakos, L.P.; Hernádi, K.; Kiss, G.; Réti, B.; Erdélyi, Z.; Parditka, B.; Szilágyi, I.M. Photocatalytic hollow TiO_2 and ZnO nanospheres prepared by atomic layer deposition. *Sci. Rep.* **2017**, *7*, 4337. [CrossRef]
46. Ghaderi, A.; Shafiekhani, A.; Solaymani, S.; Ţălu, Ş.; da Fonseca Filho, H.D.; Ferreira, N.S.; Matos, R.S.; Zahrabi, H.; Dejam, L. Advanced microstructure, morphology and CO gas sensor properties of Cu/Ni bilayers at nanoscale. *Sci. Rep.* **2022**, *12*, 12002. [CrossRef]
47. Josso, B.; Burton, D.R.; Lalor, M.J. Texture orientation and anisotropy calculation by Fourier transform and Principal Component Analysis. *Mech. Syst. Signal Process.* **2005**, *19*, 1152–1161. [CrossRef]

48. Hu, W.; Li, H.; Wang, C.; Gou, S.; Fu, L. Characterization of collagen fibers by means of texture analysis of second harmonic generation images using orientation-dependent gray level co-occurrence matrix method. *J. Biomed. Opt.* **2012**, *17*, 026007. [CrossRef]
49. Efremov, Y.M.; Velay-Lizancos, M.; Weaver, C.J.; Athamneh, A.I.; Zavattieri, P.D.; Suter, D.M.; Raman, A. Anisotropy vs isotropy in living cell indentation with AFM. *Sci. Rep.* **2019**, *9*, 5757. [CrossRef]
50. Jiang, H.; He, X.; Ma, Y.; Fu, B.; Xu, X.; Subramanian, B.; Hu, C. Isotropic Hedgehog-Shaped-TiO_2/Functional-Multiwall-Carbon Nanotube Micromotors with Phototactic Motility in Fuel-Free Environments. *ACS Appl. Mater. Interfaces* **2021**, *13*, 5406–5417. [CrossRef]

polymers

MDPI

Article

Covalent Pectin/Arabinoxylan Hydrogels: Rheological and Microstructural Characterization

Claudia Lara-Espinoza [1], Agustín Rascón-Chu [1,*], Valérie Micard [2], Carole Antoine-Assor [2], Elizabeth Carvajal-Millan [1], Rosalba Troncoso-Rojas [1], Federico Ohlmaier-Delgadillo [1] and Francisco Brown-Bojorquez [3]

1 Research Center for Food and Development, CIAD, A.C., Carretera Gustavo Enrique Astiazaran Rosas No. 46, Col. La Victoria, Hermosillo 83304, Sonora, Mexico; claudialaraes@gmail.com (C.L.-E.); ecarvajal@ciad.mx (E.C.-M.)
2 IATE, INRAE, Institut SupAgro, University Montpellier, 34000 Montpellier, France
3 Departamento de Investigación en Polímeros y Materiales, University of Sonora, Rosales y Blvd. Luis D. Colosio, Hermosillo 83000, Sonora, Mexico; francisco.brown@unison.mx
* Correspondence: arascon@ciad.mx; Tel.: +52-622-289-2400

Abstract: This research aimed to evaluate the gelation process of ferulated pectin (FP) and ferulated arabinoxylan (AXF) in a new mixed hydrogel and determine its microstructural characteristics. FP from sugar beet (*Beta vulgaris*) and arabinoxylan from maize (*Zea mays*) bran were gelled via oxidative coupling using laccase as a crosslinking agent. The dynamic oscillatory rheology of the mixed hydrogel revealed a maximum storage modulus of 768 Pa after 60 min. The scanning electron microscopy images showed that mixed hydrogels possess a microstructure of imperfect honeycomb. The ferulic acid content of the mixed hydrogel was 3.73 mg/g, and ferulic acid dimer 8-5′ was the most abundant. The presence of a trimer was also detected. This study reports the distribution and concentration of ferulic acid dimers, and the rheological and microstructural properties of a mixed hydrogel based on FP and AXF, which has promising features as a new covalent biopolymeric material.

Keywords: mixed hydrogel; ferulic acid; oxidative coupling; microstructural characteristics; rheological properties

1. Introduction

Ferulated pectin (FP) and arabinoxylan (AXF) are polysaccharides found in the diet as fiber from fruits and cereals, respectively [1,2]. Each biopolymer has gained significant relevance due to its wide applications and health benefits. Both polysaccharides share the structural characteristic of having covalently linked ferulic acid (FA) in their carbohydrate chains [3]. In this context, FP is a set of polysaccharides present in the cell wall of terrestrial plants and consists of linear polymers formed mainly by galacturonic acid units linked glycosidically by α-(1,4) bonds (Figure 1a). Commercially, pectin is obtained at a large scale from citrus peels and apple bagasse, although they can also be obtained from minor sources such as sweet potato (*Ipomoea batatas*), sunflower (*Helianthus annuus* L.) residues, mango peel (*Mangifera indica* L.), and sugar beet (*Beta vulgaris*) [1]. Sugar beet has been widely studied as an alternative source of pectin [4,5].

FP present in *B. vulgaris* have FA bound to galactose (O-6, position) and arabinose (O-2 position) residues in the Rhamnogalacturonan I chains (RG-I), which enables a third mechanism of gelation for pectin: oxidative coupling, in addition to the two main gelling mechanisms of pectin, which depend mainly on its degree of esterification. Thus, high-methoxy pectin requires the presence of co-solutes such as sucrose to form a gel and a pH lower than the pKa of their monomeric unit, galacturonic acid. In contrast, low-methoxy pectin require the presence of divalent ions such as calcium and a pH higher than the galacturonic acid pKa [6]. The ability of these polysaccharides to form hydrogels has become relevant in the design of matrices for the controlled release of molecules of therapeutic interest [7,8]. However, it has been observed that FP hydrogels have little gel strength. Theoretically, the latter observation is attributed to the high content of acetyl groups, apparently lower molecular mass, and long side chains [9,10].

On the contrary, ferulated arabinoxylan (AXF) are obtained from cereals, such as corn, wheat, barley, and oats. A linear chain of β forms AXF-(1,4)-D-xylanopyranose units, to which substituents of α-L-arabinofuranoses in position O-2 and O-3 are attached (Figure 1b). These polysaccharides have FA in their structure bound in the O-5 position of arabinose through an ester bond [3], allowing them to form covalent hydrogels resistant to changes in pH, temperature, and ionic strength. This ability of AXF to form hydrogels has been widely studied [2]. However, these hydrogels can have relatively extended gelling times.

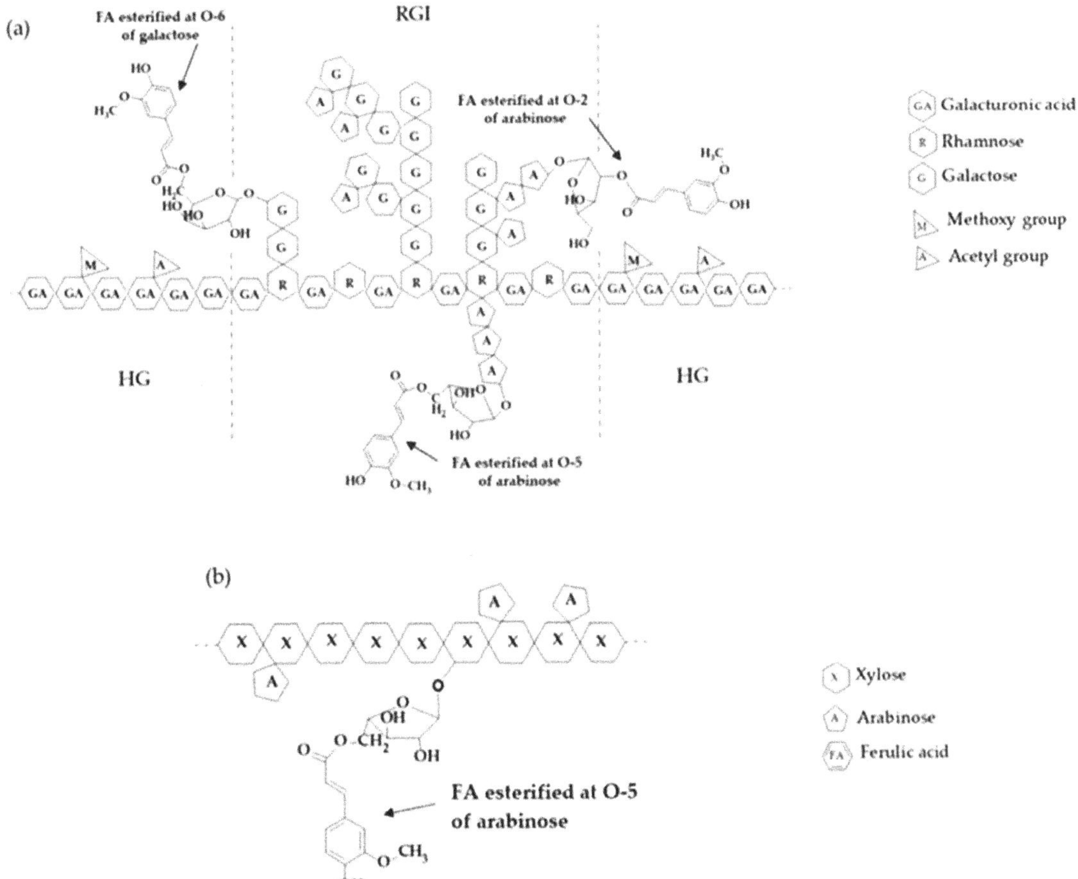

Figure 1. Schematic representation of the chemical structure of FP (**a**) and AXF (**b**). HG = Homogalacturonan region and RGI = Rhamnogalacturonan I region. Adapted from [11].

Due to the feature that both polysaccharides have covalently linked FA in their structures, the ability to form FA dimers (di-FA), where two units of ferulic acid residues are covalently linked, and trimers (tri-FA), where three units of ferulic acid residues are covalently linked, are schematically represented in Figure 2 as theoretical structures between both chains, and their formation is determined in the present work. A mixed hydrogel of such nature has never been reported yet. It is unknown whether the structural differences prevent the coupling of FA from FP and FA from AXF. In this sense, this work proposes a new composite material. The characterization will eventually envision potential applications for the food industry and biomedicine. Thus, the present investigation aims to determine the rheological and microstructural characteristics of FP/AXF mixed hydrogels.

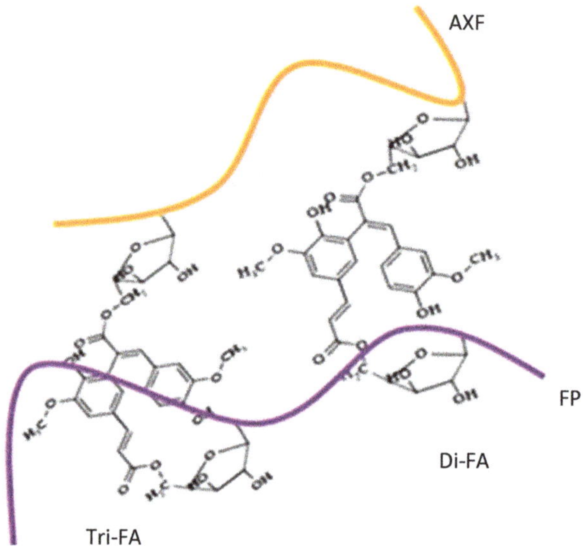

Figure 2. Schematics of theoretical covalent crosslinkings between the ferulated polysaccharides. A theoretical dimer of FA (di-FA) and a theoretical trimer named (tri-FA) are represented between two lines representing ferulated pectin (FP) and another representing ferulated arabinoxylan (AXF) as the expected covalent crosslinking structure in FP/AXF mixed hydrogel. Adapted and modified from [11].

2. Materials and Methods

2.1. Plant Material Preparation and Reagents

Sugar beet roots (*B. vulgaris*) were grown and supplied by the Agronomy Faculty of Sonora University (coordinates: 29_00048″ NL, 111_08007″ WL, and 151 masl) for the obtention of FP. Briefly, samples were washed with tap water and soap to remove dirt and then were sliced in pieces 1 cm thick in a slicing machine (Hobart, Troy, OH, USA) and dried in an oven (Enviro-Pak Inc., Clackamas, OR, USA) at 50 °C with a forced airflow of 2.5 $m^3 s^{-1}$ for 20 h. Finally, the samples were kept at −40 °C until use.

Maize (*Z. mays*) bran was provided by an agricultural company in Sinaloa, Mexico, to extract AXF. The maize bran was separated from the kernel by a mechanical process. Neither chemical nor enzymatic treatments were used.

Commercial laccase from *Trametes versicolor* (0.99 U/mg) was obtained from Sigma-Aldrich Chemical Co. (St. Louis, MO, USA). All the reagents used were of analytical grade.

2.2. FP and AXF Extraction

FP was extracted following the methodology reported by Li et al. [12], with minor variations. Briefly, 150 g of dried sugar beet was suspended in 1.5 L of 0.1 M hydrochloric acid (1:10 *w/v*), and the pH was adjusted to 1.5. The mixture was homogenized, heated on a plate with stirring at 85 °C for 1 h, and allowed to cool at room temperature. Next, the mixture was filtered through a nylon cloth (pore size 60 µm) and centrifuged at 10,000× *g* for 20 min in a centrifuge (Thermo Scientific™, Waltham, MA, USA). The samples were ethanol-precipitated (2:1 *v/v*) and dried by solvent exchange with 80% *v/v* ethanol, absolute ethanol, and acetone. Once dry, the samples were ground in a coffee grinder (Hamilton

Beach®, Glen Allen, VA, USA) to a fine powder and stored in a sealed container, in the dark at room temperature until use.

AXF was extracted as characterized by Morales-Burgos et al. [13]. Briefly, maize bran was milled, suspended in ethanol, and agitated on a rotatory shaker for 12 h at 25 °C. The ethanol-treated bran was filtered and subjected to enzyme inactivation (boiling water for 30 min). Afterward, maize bran was recovered by filtration and then treated with a NaOH solution at 25 °C in darkness under shake conditions. Then, the mixture was filtrated to eliminate residual solids. The filtrate was centrifuged, and the obtained supernatant was acidified to pH 4 with HCl and centrifuged. Finally, the supernatant was precipitated in ethanol for 4 h at 4 °C, then recovered and dried by solvent exchange. It was stored as powder in a sealed container in the dark at room temperature until use.

2.3. FP/AXF Mixed Hydrogel Formation

Mixed hydrogels were formed, keeping FA from each polysaccharide in a 1:1 proportion to a final total concentration of 26 mg/mL or 2.6% (*w*/*v*). Polysaccharide solutions were dissolved in acetate buffer pH 5.5, separated, and then mixed, and laccase enzyme was added as a crosslinking agent (83.33 nkat enzyme/mg of FA). FP and AXF solutions at 3% and 2.3% (*w*/*v*) were prepared as mentioned above and used as controls. The hydrogels were left to form at 25 °C for 2 h in the dark.

2.4. Phenolic Acid Content

The FA, di-FA, and tri-FA content were analyzed by high-performance liquid chromatography. The mixed hydrogels (50 mg) were saponified (2 mL of 2 N NaOH, 100 rpm, 35 °C, and 2 h in darkness). The pH was adjusted to 2.0 using 4 N HCl. Phenolic acid was recovered twice in 5 mL of diethyl ether and evaporated under N_2 flux at 40 °C. The extract was resuspended in 1 mL methanol/water (50:50 *v*/*v*) and filtered (0.45 μm, Millipore, St. Louis, MO, USA). The samples were injected in an HPLC (Waters™, Milford, MA, USA) equipped with a C18 column (Alltech Inc., Lexington, KY, USA; 250 mm × 2.6 mm), previously filtered in a 0.45 μm membrane [13,14]. Detection was performed at 320 nm and 3,4,5-trimethoxycinnamic acid (TMCA) was used as internal standard.

2.5. Rheological Measurements

The rheological properties of the mixed hydrogels were determined by a small-amplitude oscillatory shear using a strain-controlled rheometer (Discovery HR-2 rheometer, TA Instruments, New Castle, DE, USA), with a parallel-plate geometry (40 mm in diameter), following the methodology proposed by Vansteenkiste et al. [14]. The gelation process of the polysaccharide mixture was monitored for 1 h at 25 °C, following the storage (G′) and loss (G″) modulus. During gelation rheological measurements, a silicone oil layer was used to cover the polysaccharide mixture to prevent evaporation. All measurements were performed at a frequency of 0.25 Hz and 5% strain (linearity range of viscoelastic behavior). After the gelation process, a mechanical sweep was conducted in the 0.01–10 Hz range, with 5% strain and a temperature of 25 °C.

2.6. Microstructural Characterization

The mixed hydrogels were frozen to −20 °C and then lyophilized at −80 °C with 0.008 mbar in a Freezone 6 freeze drier (Labconco, Kansas, MO, USA). The obtained hydrogels were then coated with a thin layer of gold. The microstructure of the mixed hydrogels was observed via scanning electron microscopy (SEM) using a microscope (JEOL 5410LV, JEOL, Peabody, MA, USA) with an accelerating voltage of 20 kV.

2.7. Statistical Analysis

All results are expressed as mean values ± standard deviation based on triplicate measurements. A one-way analysis of variance (ANOVA) was performed to determine significant differences between the respective values, and averages were compared by the

Tuckey–Kramer test with a $p \leq 0.05$. The statistical software NCSS 12 (Setup_v12_0_18.exe, 30 August 2021) was used.

3. Results and Discussion

3.1. Chemical Characterization of FP and AXF

The chemical composition of sugar beet FP is presented in Table 1. FP showed a galacturonic acid content of 55 g/100 g of pectin. This compound was the major component present in the structure of FP. The total neutral sugars accounted for 21.6 g/100 g of pectin, mainly rhamnose (6.3 g/100 g), galactose (7.1 g/100 g), and arabinose (3.3 g/100 g), which are the main sugars reported for the RG-I region in the structure of FP [9,10,15]. In addition, other neutral sugars such as xylose, glucose, fucose, and mannose were detected in lower amounts.

Table 1. Chemical characterization of FP from sugar beet.

Compound	Content
Galacturonic acid [1]	55.0 ± 3
Rhamnose [1]	6.3 ± 0.5
Arabinose [1]	3.3 ± 0.4
Xylose [1]	1.4 ± 0.2
Galactose [1]	7.1 ± 0.5
Glucose [1]	1.0 ± 0.01
Mannose [1]	1.3 ± 0.2
Fucose [1]	1.2 ± 0.1
Total neutral sugars [1]	21.6 ± 0.9
Protein [1]	10.3 ± 0.5
Ash [1]	2.13 ± 0.06
DM [1]	57.4 ± 4.1
DA [1]	26.1 ± 2.4
Ferulic acid [2]	5.5 ± 0.1
Diferulic acid [2]	0.26 ± 0.03

DM, degree of methylation; DA, degree of acetylation. [1] Results expressed in g/100 g pectin; [2] Expressed in mg/g pectin.

The FA content was 5.5 mg/g of pectin, and the presence of di-FA was 0.26 mg/g. The presence of tri-FA was not detected. The studied FP also showed a high content of protein (10.3 g/100 g of pectin), which was higher than those generally reported in FP obtained from sugar beet [16]. The latter, due to proteases, were not used in this extraction process. The ash content was 2.13 g/100 g of pectin. Pectin also showed a degree of methylation and acetylation at 57.4% and 26.1%, respectively (Table 1).

The AXF used in this investigation was previously characterized by Morales-Burgos et al. [13] and presented an arabinose-to-xylose ratio of 0.72 and an FA and di-FA content of 7.18 and 0.44 mg/g AXF, respectively.

3.2. Ferulic Acid, Dimer, and Trimer Content in Mixed Hydrogels

To cast hydrogels where FA was in equal amounts from each polysaccharide, a mix of FP at 1.5% and arabinoxylan at 1.15% were mixed to a final total polysaccharide percentage of 2.6%. As FA content is the common residue for all hydrogels, polymers' concentration and laccase were fixed as a function of FA (83.33 nkat enzyme/ mg of FA). To some extent, the latter seemed to be the logical way to control laccase, as FA was the main substrate present. In Table 2, the resulting phenolic acid content in the mixed hydrogels are shown. Initially, the FP/AXF mixture (before gelation) exhibited an FA concentration of 6.2 ± 0.3 mg/g, while the mixed hydrogel (after gelation) showed a total FA content of 3.7 ± 0.2 mg/g. The FA content diminished consistently with the crosslinking formation process where di-FA and tri-FA were generated. It can be noted how the proportions of total FA dimer and trimer content increased after the gelling process and correlated with the oxidation of FA by the action of the enzyme laccase. Only 40% of the initial FA was oxidized (2.48 mg/g), and 60% remained free FA, corresponding to the 3.7 mg FA/g dw obtained in the mixed hydrogel.

Table 2. FA, FA dimer, and FA trimer content in FP/AXF mixed before and after 2 h of gelation.

	FA	Di-FA	Tri-FA	FA Oxidized [1]
		(mg/g)		(%)
Before gelation	6.2 ± 0.3	0.34 ± 0.03	0.010 ± 0.002	-
After gelation	3.7 ± 0.2	3.6 ± 0.3	0.66 ± 0.07	40

[1] FA oxidized, calculated from the initial total FA oxidized during the gelling process.

Similarly, Mendez-Encinas et al. [17] studied AXF hydrogels and reported that laccase action oxidized only 35% of FA original content. This percentage was slightly lower than that found in this investigation for the mixed hydrogels. On the contrary, Martinez-Lopez et al. [18] reported that 70% of the initial FA was oxidized to form dimers and trimers of FA in AXF microspheres. These discrepancies may be due to further structural differences such as molecular weight, neutral-sugar ratio, viscosity, enzyme concentration, and laccase properties.

Nevertheless, remaining free FA can represent an advantage in mixed hydrogels since it has been widely reported that FA has antioxidant capacities. Figure 3 shows a schematic representation of the structure of the most-reported dimers and a trimer. Their names derive from the carbon position of the ferulic acid units that participate in the covalent linkage. In Figure 4, the results for the distribution and concentration of the di-FA and a tri-FA in the mixed hydrogels after FA oxidation are shown. From the total number of the newly formed dimers, the 8-5′ structure was the predominant form after gelation, representing 69% of the total content of dimers, while the 8-O-4′ and 5-5′ structures represented 19% and 12%, respectively. Several authors report the 8-5′ dimer as the main isomer in FP [19,20] and AXF hydrogels [13]. Moreover, a similar pattern was recently observed in the data obtained by Ohlmaier-Delgadillo et al. [11] for a mixed hydrogel. This hydrogel was composed of a low-methoxy FP obtained from a sugar beet solid waste and AXF from maize dried distiller's grains, where the dimers 8-5′, 8-O-4′, and 5-5′ were found, with the 8-5′ isomer being the main one. The former di-FA values confirmed the results presented in this work. In this regard, several studies have evaluated the antioxidant activity of AXF [21,22] and AXF hydrogels [17]. Mendez-Encinas et al. [17] studied the antioxidant capacity of AXF hydrogels and observed that FA exhibited antioxidant capacity even after gelling. These authors reported a concentration of 3.52 mg/g in AXF hydrogel.

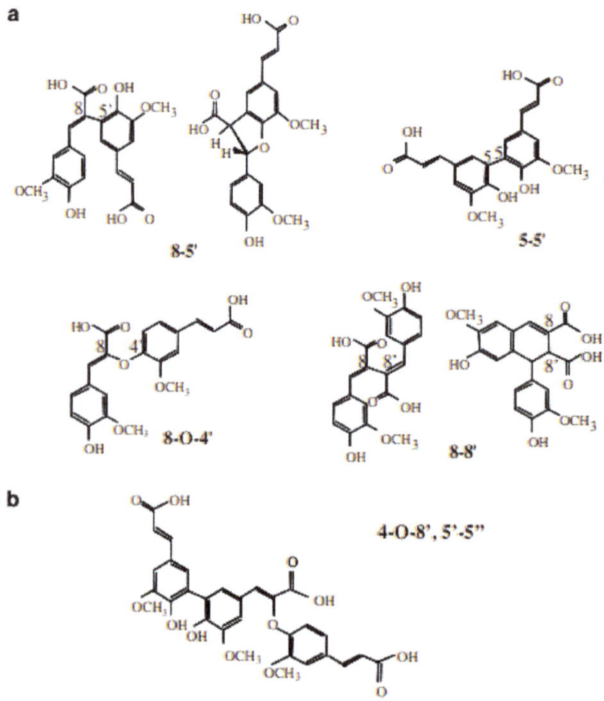

Figure 3. Schematic chemical structures of dehydrodimers of ferulic acid (di-FA) (**a**) and a dehydrotrimer of ferulic acid (tri-FA) (**b**). Adapted from [23].

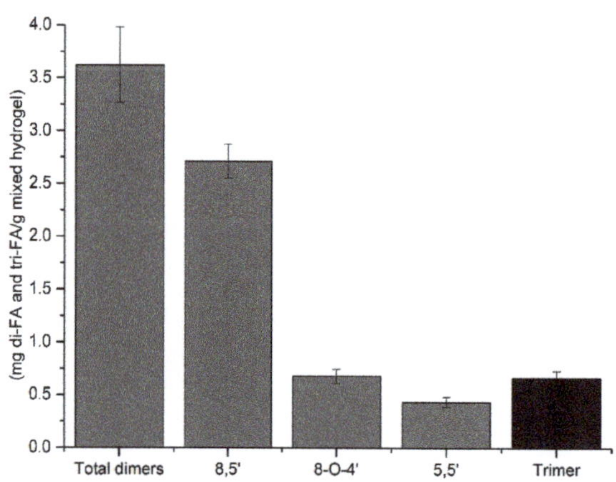

Figure 4. Dimers (▒) and trimer (■) of ferulic acid after gelation in the mixed hydrogel of FP and AXF. Results are presented as mean ± SD based on triplicate measurements.

Furthermore, the antioxidant activity of the FA oligomers formed after the dimerization of FA has also been reported by several authors [24–26]. These authors infer that the antioxidant activity is superior in FA dimers than in its substrate (FA). In this context, Garcia-Conesa et al. [24] studied the antioxidant capacity of ferulic acid dimers, chemically synthesized, concluding that their antioxidant capacity is dependent on the free phenolic hydroxyl groups present in their structure as well as the nature of the dimer formed. These same authors also studied the antioxidant activity of ferulic acid dimers isolated from wheat bran, inferring that ferulic acid dimers, specifically 8-O-4′, showed better antioxidant activity than free ferulic acid [25]. The FA dimers reported in this study for a mixed hydrogel could exert additional antioxidant activity on the abovementioned FA content. However, further research is required to support this hypothesis conclusively.

The major presence of the 8-5′ dimer is a desirable feature for our high-methoxy FP, given its spatial conformation contributes to a more accessible way for the crosslinking between polysaccharide chains and the elasticity of the hydrogel. In addition, this isoform type is reported to be thermodynamically more stable [2]. Furthermore, a tri-FA was also detected in the mixed hydrogel (Figure 1). This isomer was also detected and identified for AXF from maize (*Z. mays*) and sugar beet (*B. vulgaris*) pectin. Indeed, 8-O-4′, 5-5′ was the first tri-FA reported in natural maize bran [27,28] and AXF hydrogels [29]. In addition, more regioisomers of tri-FA have been identified as naturally occurring in maize and sugar beet [30,31]. The latter is an opportunity for new research in mixed hydrogels and for new FA trimers, tetramers, and pentamers proposed in cereal cell walls, may also be present in this new proposed material for the spatial distribution of pectin RGI; and RGII "hairy" regions may affect the FA residues' encounter with the final gel scaffolding.

3.3. Rheological Characterization of Mixed Hydrogels

The G′ (storage modulus) and G″ (loss modulus) of the mixed solution (FP 1.5%; AXF 1.15%) (*w*/*v*) were monitored via oxidative gelation by laccase over time. The rheological results are shown in Figure 5a. At the beginning of the gelling process, G″ was higher than G′, but an increment in G′ occurred rapidly and prevailed over G″ until the end. The gelation time of the mixed hydrogel was ~4 min, which was extrapolated from the crossover of G′ and G″ (tan δ = 1), indicating the sol/gel transition point. The mixed hydrogel showed maximum G′ and G″ values of 768 and 3.30 Pa, respectively, with a final tan δ value (G″/G′) of 0.0043 at 60 min. The increment in G′ over time corresponds to di-FA and tri-FA formation; evidently, the modulus of elasticity or storage is directly proportional to the number of crosslinks, which in turn influences the strength of the hydrogel. Also, as mentioned above, the predominant content of 8-5′ di-FA contributed to the elasticity observed in the hydrogel.

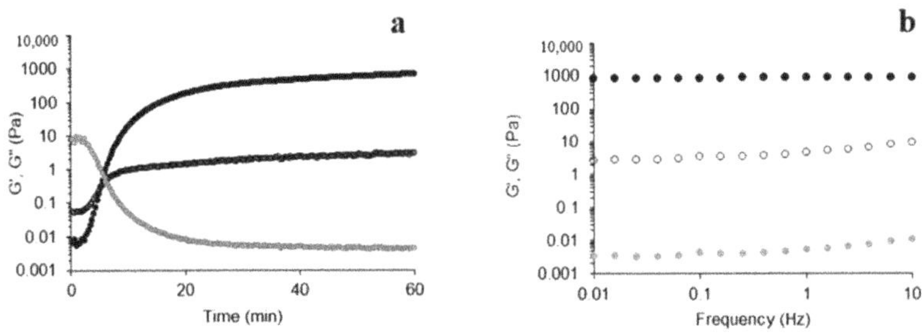

Figure 5. Rheological kinetics of 1.5 and 1.15% (*w*/*v*) FP and AXF mix dispersion during gelation at 0.25 Hz and 5% strain (**a**) and mechanical spectra of mixed gel after 1 h gelation at 5% strain (**b**). (G′●, G″○, tan δ●). Experiments were performed at 25 °C.

The mechanical spectrum of the mixed hydrogel is shown in Figure 5b, showing a typical behavior of a solid-like material with a prevalence of G′ over G″ and a lineal G′ independent of frequency and a much smaller G″ dependent on frequency [32]. Also, the hydrogels showed stable behavior under the tested conditions.

The general results of small oscillatory shear rheology and density of crosslinks values [33], including AXF and FP hydrogels, are summarized in Table 3. The G′ value was statistically higher for the arabinoxylan hydrogel control, and it can be observed that the mixed hydrogels presented a higher storage modulus concerning the pectin hydrogel.

Regarding G″, its value was lower for arabinoxylan hydrogel, indicating less viscous behavior at the end of the 60 min run. Interestingly, the mixed-hydrogel G′ and density of crosslink values are midway, resembling more of a proportionate behavior although not linear. The rheology parameters reflect the 1:1 FA from each polymer and the multiple interactions of their chains. This may explain why the G′ and the density of crosslinks values of the mixed gels are close to the middle value corresponding to AXF and FP hydrogels per the logarithmic additivity rule.

Table 3. Storage modulus (G′), loss modulus (G″), and tan delta (tan δ) of AXF, FP, and FP/AXF mixed gels.

	G′ (Pa)	G″ (Pa)	Tan δ (G″/G′)	Density of Crosslinks (mol/cm^3) × 10^{-7}
AXF hydrogel (2.3%)	1540 ± 280 [b]	2.53 ± 0.26 [a]	0.0021 ± 0.0007 [a]	6.21
FP hydrogel (3%)	425 ± 11 [a]	4.64 ± 0.16 [c]	0.012 ± 0.001 [b]	1.71
FP/AXF (2.6%) mixed hydrogel	768 ± 2 [a]	3.30 ± 0.04 [b]	0.00430 ± 0.00005 [a]	3.10
(log additivity rule) [33]	809	3.4	0.0050	3.3

Values are presented as mean ± SD of triplicate analyses. Different letters in the same column express differences ($p < 0.05$) between the hydrogels.

Complementarily, the tan δ data showed no significant difference between arabinoxylan and the mixed hydrogels (0.0021 and 0.0043, respectively). Conversely, the FP hydrogel showed tenfold a value for tan δ (0.012), which was related to a higher G″ observed for this hydrogel. The latter suggests that some FP chains were only dispersed in the polymeric network and did not participate in the crosslinking scaffold, contributing mainly to the viscous module. In sum, the values of tan δ for the AXF hydrogels and mixed FP/AXF hydrogels denote higher participation of the polysaccharide chains in the polymer network than the FP hydrogels. The "hairy" regions play a significant role in this rheological observation and the resulting repulsion due to a pH value of 5.5 favoring charged residues.

Structural properties such as the xylose/arabinose ratio in AXF [2], the number of lateral chains in pectin [34], and physicochemical properties such as FA content, molecular weight, and the viscosity of polysaccharides have a notable effect on the gelling capacity of hydrogel matrices [2,11].

In this context, in a study conducted by Schooneveld-Bergmans et al. [35], the oxidative crosslinking of AXF from both wheat bran and flour was evaluated, showing that both polysaccharides participated in the crosslinking and had a synergistic effect on viscosity; nevertheless, in our study, the results observed do not present this behavior. So, an apparent interaction of both the polysaccharides (FP and AXF) on their rheology can be noted as logarithmic additivity. Also, FA has an important effect on a polysaccharide's physical and functional characteristics.

3.4. Microstructure of Hydrogels

The microstructure of the hydrogels, evaluated at a scale of 100 µm by SEM, is shown in Figure 6. The AX hydrogel presented a regular, highly porous microstructure resembling an imperfect honeycomb. This microstructure has been observed widely in other investigations in arabinoxylan from maize [14]. In the FP hydrogel, a highly porous microstructure was also observed, but its cavities (alveolate) were considerably irregular and resembled an imperfect honeycomb. The observed microstructure agrees with the reported elsewhere [36,37].

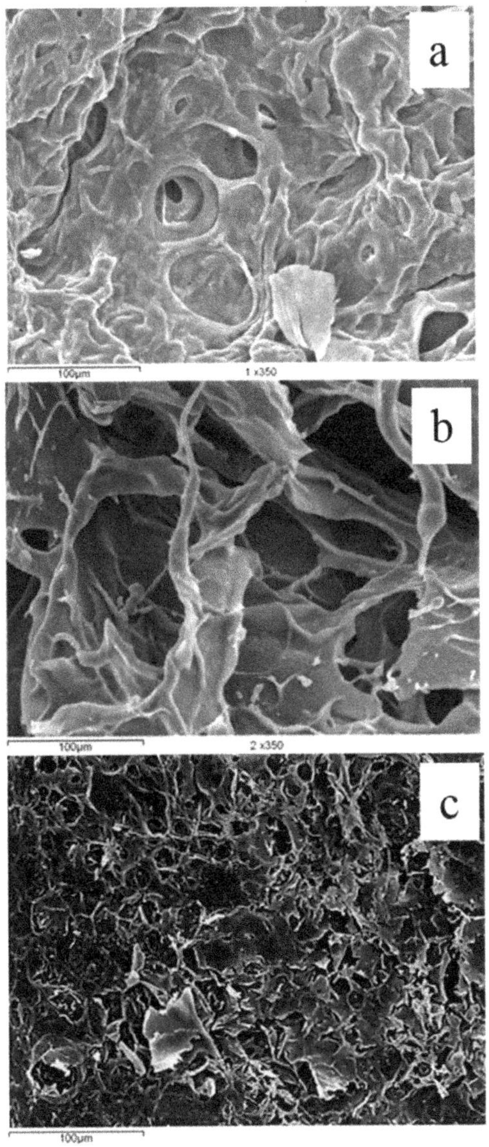

Figure 6. SEM images of Arabinoxylan gel (2.3%) (**a**), pectin gel (3%) (**b**), and mixed gel (pectin 1.5%/AX 1.15%) (**c**). Magnification 350×.

The SEM image of the mixed hydrogel conserved a porous microstructure with somewhat comparable cavities. Compared to pure pectin or arabinoxylan hydrogels, a more compact microstructure can be observed. Similarly, previous studies of low-methoxy pectin hydrogels showed a comparable microstructure [37]. The evidence in our study suggests that ferulated pectin covalent crosslinking in the microstructure observed in the hydrogels resembles an imperfect honeycomb more instead. Further research is needed to conclude this fully. The current data suggest that different polysaccharide ratios and FA content might be critical to designing materials with different pore sizes and alveoli. In this regard, research has been undertaken to fully assess the potential for bioactive-compound encapsulation in combination with probiotics. Hypothetically, pore size and alveolate pattern regulation with polysaccharide ratios and FA content would be feasible. Nevertheless, we are not at that point yet.

4. Conclusions

The FP/AXF mixed hydrogel was successfully created via oxidative coupling, showing a higher storage modulus than FP hydrogel. Through tan delta, it could be inferred that a close interchain interaction between both polysaccharides affects rheological properties. The microstructure of the mixed hydrogels showed microstructural characteristics for both polysaccharides resembling honeycombs. A more compact structure gives this material an advantage over pectin or arabinoxylan hydrogels by separating the design of bio-composites with suitable properties for encapsulating bioactive compounds. This investigation is one of the first to report the FA dimers, trimer distribution, and the rheological and microstructural characteristics of an FP/AXF mixed hydrogel with promising features as a new covalent biopolymeric material.

Author Contributions: Investigation and writing—original draft preparation, C.L.-E.; writing—review and editing, F.O.-D.; experimental review and editing, F.B.-B.; writing—review and editing, R.T.-R.; formal analysis, supervision, and writing—review and editing E.C.-M.; investigation and writing—review and editing, V.M. and C.A.-A.; formal analysis, supervision, writing—review and editing, funding acquisition, and project administration, A.R.-C. All authors have read and agreed to the published version of the manuscript.

Funding: This research was funded by CONACYT, grant number CB-2015-01-254297, to A.R.-C.

Institutional Review Board Statement: Ethical review and approval were waived for this study due to experimental protocols did not involve humans or animals in any manner.

Data Availability Statement: Publicly available datasets were analyzed in this study. These data can be found here: https://ciad.repositorioinstitucional.mx/jspui/. (accessed on 5 September 2024).

Acknowledgments: The authors wish to thank Jorge Nemesio Mercado Ruiz for technical assistance, Alma Consuelo Campa Mada, and Karla Guadalupe Martinez Robinson for technical assistance.

Conflicts of Interest: The authors declare no conflicts of interest.

References

1. Srivastava, P.; Rishabha, M. Sources of pectin, extraction and its applications in pharmaceutical industry—An overview. *Indian J. Nat. Prod. Resour.* **2011**, *2*, 10–18. Available online: https://www.researchgate.net/profile/Pranati-Srivastava/publication/280016869_Sources_of_pectin_extraction_and_its_applications_in_pharmaceutical_industry_-_An_overview/links/5923dee7aca27295a8aa779e/Sources-of-pectin-extraction-and-its-applications-in-p (accessed on 5 September 2024).
2. Niño-Medina, G.; Carvajal-Millan, E.; Rascon-Chu, A.; Marquez-Escalante, J.; Guerrero, V.; Salas-Muñoz, E. Feruloylated arabinoxylans and arabinoxylan gels: Structure, sources and applications. *Phytoch. Rev.* **2010**, *9*, 111–121. [CrossRef]
3. Saulnier, L.; Thibault, J. Ferulic acid and diferulic acids as components of sugar beet pectins and maize bran heteroxylans. *J. Sci. Food Agric.* **1999**, *79*, 396–402. [CrossRef]
4. Niu, H.; Hou, K.; Chen, H.; Fu, X. A review of sugar beet pectin-stabilized emulsion: Extraction, structure, interfacial self-assembly and emulsion stability. *Crit. Rev. Food Sci. Nutr.* **2022**, *64*, 852–872. [CrossRef] [PubMed]
5. Fishman, M.; Chau, H.; Cooke, P.; Hotchkiss, A. Global Structure of Microwave-assisted Flash-extracted Sugar Beet Pectin. *J. Agric. Food Chem.* **2008**, *56*, 1471–1478. [CrossRef]
6. Ström, A.; Schuster, E.; Goh, S.M. Rheological characterization of acid pectin samples in the absence and presence of monovalent ions. *Carbohydr. Polym.* **2014**, *113*, 336–343. [CrossRef]
7. Li, J.; Mooney, D. Designing hydrogels for controlled drug delivery. *Nat. Rev. Mater.* **2016**, *1*, 16071. [CrossRef]
8. Thang, N.H.; Chien, T.B.; Cuong, D.X. Polymer-Based Hydrogels Applied in Drug Delivery: An Overview. *Gels* **2023**, *9*, 523. [CrossRef]
9. Levigne, S.; Ralet, M.C.; Thibault, J.F. Characterisation of pectins extracted from fresh sugar beet under different conditions using an experimental design. *Carbohydr. Polym.* **2002**, *49*, 145–153. [CrossRef]
10. Yapo, B.; Wathelet, B.; Paquot, M. Comparison of alcohol precipitation and membrane filtration effects on sugar beet pulp pectin chemical features and surface properties. *Food Hydrocoll.* **2007**, *21*, 245–255. [CrossRef]

11. Ohlmaier-Delgadillo, F.; Carvajal-Millan, E.; López-Franco, Y.L.; Islas-Osuna, M.A.; Micard, V.; Antoine-Assor, C.; Rascón-Chu, A. Ferulated Pectins and Ferulated Arabinoxylans Mixed Gel for *Saccharomyces boulardii* Entrapment in Electrosprayed Microbeads. *Molecules* **2021**, *26*, 2478. [CrossRef] [PubMed]
12. Li, D.; Du, G.; Jing, W.; Li, J.; Yan, J.; Liu, Z. Combined effects of independent variables on yield and protein content of pectin extracted from sugar beet pulp by citric acid. *Carbohydr. Polym.* **2015**, *129*, 108–114. [CrossRef] [PubMed]
13. Morales-Burgos, A.M.; Carvajal-Millan, E.; Lopez-Franco, Y.; Rascon-Chu, A.; Lizardi-Mendoza, J.; Sotelo-Cruz, N.; Brown-Bojorquez, F.; Burgara-Estrella, A.; Pedroza-Montero, M. Syneresis in Gels of Highly Ferulated Arabinoxylans: Characterization of Covalent Cross-Linking, Rheology, and Microstructure. *Polymers* **2017**, *9*, 164. [CrossRef] [PubMed]
14. Vansteenkiste, E.; Babot, C.; Rouau, X.; Micard, V. Oxidative gelation of feruloylated arabinoxylan as affected by protein. Influence on protein enzymatic hydrolysis. *Food Hydrocoll.* **2004**, *18*, 557–564. [CrossRef]
15. Peng, X.; Mu, T.; Zhang, M.; Sun, H.; Chen, J.; Yu, M. Effects of pH and high hydrostatic pressure on the structural and rheological properties of sugar beet pectin. *Food Hydrocoll.* **2016**, *60*, 161–169. [CrossRef]
16. Chen, H.M.; Fu, X.; Luo, Z.G. Effect of molecular structure on emulsifying properties of sugar beet pulp pectin. *Food Hidrocoll.* **2016**, *54*, 99–106. [CrossRef]
17. Mendez-Encinas, M.; Carvajal-Millan, E.; Rascon-Chu, A.; Astiazaran-Garcia, H.; Valencia-Rivera, D.E.; Brown-Bojorquez, F.; Alday, E.; Velazquez, C. Arabinoxylan-Based Particles: In Vitro Antioxidant Capacity and Cytotoxicity on a Human Colon Cell Line. *Medicina* **2019**, *55*, 349. [CrossRef]
18. Martinez-Lopez, A.L.; Carvajal-Millan, E.; Sotelo-Cruz, N.; Micard, V.; Rascon-Chu, A.; Lopez-Franco, Y.L.; Lizardi-Mendoza, J.; Canett-Romero, R. Enzymatically cross-linked arabinoxylan microspheres as oral insulin delivery system. *Int. J. Biol. Macromol.* **2019**, *126*, 952–959. [CrossRef]
19. Ralet, M.-C.; André-Leroux, G.; Quéméner, B.; Thibault, J.-F. Sugar beet (*Beta vulgaris*) pectins are covalently cross-linked through diferulic bridges in the cell wall. *Phytochemistry* **2005**, *66*, 2800–2814. [CrossRef]
20. Lara-Espinoza, C.L.; Sanchez-Villegas, J.A.; Lopez-Franco, Y.; Carvajal-Millan, E.; Troncoso-Rojas, R.; Carvallo-Ruiz, T.; Rascon-Chu, A. Composition, Physicochemical Features, and Covalent Gelling Properties of Ferulated Pectin Extracted from Three Sugar Beet (*Beta vulgaris* L.) Cultivars Grown under Desertic Conditions. *Agronomy* **2020**, *11*, 40. [CrossRef]
21. Ayala-Soto, F.E.; Serna-Saldivar, S.O.; Garcia-Lara, S.; Perez-Carrillo, E. Hydroxycinnamic Acids, Sugar Composition and Antioxidant Capacity of Arabinoxylans Extracted from Different Maize Fiber Sources. *Food Hydrocoll.* **2014**, *35*, 471–475. [CrossRef]
22. Malunga, L.N.; Beta, T. Antioxidant Capacity of Water-Extractable Arabinoxylan from Commercial Barley, Wheat, and Wheat Fractions. *Cereal Chem.* **2015**, *92*, 29–36. [CrossRef]
23. Niño-Medina, G.; Carvajal-Millan, E.; Lizardi, J.; Rascon-Chu, A.; Marquez-Escalante, J.A.; Gardea, A.; Martinez-Lopez, A.L.; Guerrero, V. Maize processing waste water arabinoxylans: Gelling capability and cross-linking content. *Food Chem.* **2009**, *115*, 1286–1290. [CrossRef]
24. Garcia-Conesa, M.T.; Plumb, G.W.; Kroon, P.A.; Wallace, G.; Williamson, G. Antioxidant properties of ferulic acid dimers. *Redox Rep.* **1997**, *3*, 239–244. [CrossRef] [PubMed]
25. Garcia-Conesa, M.T.; Plumb, G.W.; Waldron, K.W.; Ralph, J.; Williamson, G. Ferulic Acid Dehydrodimers from Wheat Bran: Isolation, Purification and Antioxidant Properties of 8-O-4-diferulic Acid. *Redox Rep.* **1997**, *3*, 319–323. [CrossRef]
26. Adelakun, O.E.; Kudanga, T.; Parker, A.; Green, I.R.; Roes-Hill, M.; Burton, S.G. Laccase-catalyzed dimerization of ferulic acid amplifies antioxidant activity. *J. Mol. Catal. B Enzym.* **2012**, *74*, 29–35. [CrossRef]
27. Rouau, X.; Cheynier, V.; Surget, A.; Gloux, D.; Barron, C.; Meudec, E.; Louis-Montero, J.; Criton, M. A dehydrotrimer of ferulic acid from maize bran. *Phitochemistry* **2003**, *63*, 899–903. [CrossRef]
28. Bunzel, M.; Ralph, J.; Funk, C. Isolation and identification of a ferulic acid dehydrotrimer from saponified maize bran insoluble fiber. *Eur. Food Res. Technol.* **2003**, *217*, 128–133. [CrossRef]
29. Carvajal-Millan, E.; Guigliarelli, B.; Belle, V.; Rouau, X.; Micard, V. Storage stability of laccase induced arabinoxylan gels. *Carbohydr. Polym.* **2005**, *59*, 181–188. [CrossRef]
30. Waterstraat, M.; Bunzel, D.; Bunzel, M. Identification of 8-O-4/8-5(Cyclic)- and 8-8(Cyclic)/5-5-Coupled Dehydrotriferulic Acids, Naturally Occurring in Cell Wall of Mono and Dicotyledonous Plants. *J. Agric. Food Chem.* **2016**, *64*, 7244–7250. [CrossRef]
31. Waterstraat, M.; Bunzel, M. A Multi-Step Chromatographic Approach to Purify Radically Generated Ferulate Oligomers Reveals Naturally Ocurring 5-5/8-8(Cyclic)-,8-8(Noncyclic)/8-*O*-4-, and 5-5/8-8(Noncyclic)-Coupled Dehydrotriferulic Acids. *Front. Chem.* **2018**, *6*, 190. [CrossRef] [PubMed]
32. Ilyin, S.O. Structural Rheology in the Development and Study of Complex Polymer Materials. *Polymers* **2024**, *16*, 2458. [CrossRef] [PubMed]
33. Chen, H.; Gan, J.; Ji, A.; Song, S.; Yin, L. Development of double network gels based on soy protein isolate and sugar beet pectin induced by thermal treatment and laccase catalysis. *Food Chem.* **2019**, *292*, 188–196. [CrossRef] [PubMed]
34. Funami, T.; Nakauma, M.; Ishihara, S.; Tanaka, R.; Inoue, T.; Phillips, G.O. Structural modifications of sugar beet pectin and the relationship of structure to functionality. *Food Hidrocoll.* **2011**, *25*, 221–229. [CrossRef]
35. Schooneveld-Bergmans, M.E.F.; Dignum, M.J.W.; Grabber, J.H.; Beldman, G.; Voragen, A.G.J. Studies on the oxidative cross-linking of feruloylated arabinoxylans from wheat flour and wheat bran. *Carbohydr. Polym.* **1999**, *38*, 309–317. [CrossRef]

36. Hou, J.J.; Guo, J.; Wang, J.M.; He, X.T.; Yuan, Y.; Yin, S.W.; Yang, X.Q. Edible double-network gels based on soy protein and sugar beet pectin with hierarchical microstructure. *Food Hydrocoll.* **2015**, *50*, 94–101. [CrossRef]
37. Kyomugasho, C.; Christiaens, S.; Van de Walle, D.; Van Loey, A.M.; Dewettinck, K.; Hendrickx, M.E. Evaluation of cation-facilitated pectin-gel properties: Cryo-SEM visualization and rheological properties. *Food Hydrocoll.* **2016**, *61*, 172–182. [CrossRef]

polymers

MDPI

Article

Surface Modification of Gold Nanorods (GNRDs) Using Double Thermo-Responsive Block Copolymers: Evaluation of Self-Assembly and Stability of Nanohybrids

Jesús E. Márquez-Castro [1], Angel Licea-Claverie [1,*], Carlos Guerrero-Sánchez [2,3,*] and Eugenio R. Méndez [4]

[1] Centro de Graduados e Investigación en Química, Tecnológico Nacional de México/Instituto Tecnológico de Tijuana, Tijuana 22454, Mexico; eduardo.mc@tectijuana.edu.mx

[2] Laboratory of Organic and Macromolecular Chemistry (IOMC), Friedrich Schiller University Jena, 07743 Jena, Germany

[3] Jena Center for Soft Matter (JCSM), Friedrich Schiller University Jena, 07743 Jena, Germany

[4] División de Física Aplicada, Centro de Investigación Científica y Educación Superior de Ensenada, Ensenada 22860, Mexico; emendez@cicese.edu.mx

* Correspondence: aliceac@tectijuana.mx (A.L.-C.); carlos.guerrero.sanchez@uni-jena.de (C.G.-S.); Tel.: +52-664-6234043 (A.L.-C.); +49-364-1948598 (C.G.-S.)

Abstract: A series of copolymers containing a thermo-responsive biocompatible first block of poly[di(ethylene glycol) methyl ether methacrylate)-*co*-(oligo(ethylene glycol) methyl ether methacrylate], P(DEGMA-*co*-OEGMA) were chain-extended to incorporate either poly(*N*-isopropylacrylamide), PNIPAAm or poly(*N*-isopropylacrylamide-*co*-butyl acrylate), P(NIPAAm-co-BA) as second thermo-responsive block using reversible addition–fragmentation chain transfer (RAFT) polymerization. P(DEGMA-*co*-OEGMA)-*b*-PNIPAAm copolymers showed two response temperatures at 33 and 43 °C in an aqueous solution forming stable aggregates at 37 °C. In contrast, P(DEGMA-*co*-OEGMA)-*b*-P(NIPAAm-*co*-BA) copolymers showed aggregation below room temperature due to the shift in response temperature provoked by the presence of hydrophobic butyl acrylate (BA) units, and shrinkage upon heating up to body temperature, while maintaining the second response temperature above 40 °C. The terminal trithiocarbonate group of the block copolymers was modified to a thiol functionality and used to stabilize gold nanorods (GNRDs) via the "grafting to" approach. The Localized Surface Plasmon Resonance (LSPR) absorption band of GNRDs with an aspect ratio of 3.9 (length/diameter) was located at 820 nm after surface grafting with block copolymers showing a hydrodynamic diameter of 160 nm at 37 °C. On the other hand, the stability of the P(DEGMA-*co*-OEGMA)-*b*-PNIPAAm@GNRDs and P(DEGMA-*co*-OEGMA)-*b*-P(NIPAAm-*co*-BA)@GNRDs nanohybrids was monitored for 8 days; where the LSPR absorption band did not shift or show any broadening. Aqueous dispersed nanohybrids were irradiated with a near-infrared laser (300 mW), where the temperature of the surroundings increased 16 °C after 16 min, where conditions for no precipitation were determined. These tailored temperature-responsive nanohybrids represent interesting candidates to develop drug nanocarriers for photo-thermal therapies.

Keywords: gold nanorods; RAFT polymerization; thermo-responsive polymers; photo-thermal therapy; nanohybrids

1. Introduction

Gold nanoparticles (AuNPs) display optical and electronic properties complementary to various applications due to their many advantages, such as compatibility, enhanced long thermo-stability, simple method of preparation, and multifunctionality [1,2], as well as their wavelength selective optical absorption. The development of AuNPs with different morphologies has received much attention in recent years, especially with the advent of plasmonic photo-thermal therapy (PPTT), also known as photo-thermal ablation, which employs these AuNPs as photo-thermic agents [3,4].

PPTT is currently one of the most promising research approaches in treating cancer cells and infectious diseases. The selective absorption properties of the photo-thermal agent are due to the excitation of localized surface plasmon resonance (LSPR), whose frequency is highly influenced by the size of the nanoparticles, their morphology, and the surrounding environment [5]. Among the diverse morphologies of AuNPs, special interest has been focused on gold nanorods (GNRDs) due to their outstanding physicochemical properties, such as light scattering, possibility to modulate their surface plasmon resonance and to couple to biomolecules or polymers via Au-S bonds, and their high surface-to-volume-ratio [6,7]. GNRDs exhibit the unique optical properties of strong light absorption and heat emission in the near-infrared (NIR) region (750–1000 nm) depending on their size and aspect ratio. However, GNRDs are cytotoxic due to the presence of hexadecyltrimethyl ammonium (CTAB) moieties on their surface, which are used to maintain dispersion and prevent aggregation [8,9].

To circumvent their cytotoxicity, researchers have used polymers, bioactive molecules, and phospholipid molecules to graft them onto the surface of GNRDs. In addition to the photo-therapeutic applications, NIR light-induced heating of GNRDs could also be utilized for drug and gene delivery applications, which are usually immobilized on the surface of GNRDs and loaded into the external concavities of functional polymer coatings [10,11]. On the other hand, due to the rather simple surface functionalization of GNRDs, a thermo-responsive copolymer coating can be introduced via a "grafting to" approach; commonly, GNRDs are grafted with polymers containing thiol (-SH)-reactive functional groups [12]. Polymers synthesized via reversible addition–fragmentation chain transfer (RAFT) polymerization are distinguished by the presence of functional end groups, such as trithiocarbonate, dithioester, or xanthate groups, which are derived from the chain transfer agent (CTA) used for their synthesis and can be transformed to thiol groups [13,14]. In turn, thiol-functionalized polymers can be conveniently used to decorate the surface of AuNPs via Au-S conjugation to form polymer@GNRDs nanohybrids [15], while some reports directly use trithiocarbonate terminated polymers for the same purpose [16].

Recently, thermo-responsive double hydrophilic block copolymers (TDHBCs) have gained a great deal of attention as a subclass of stimuli-responsive materials due to promising applications in the biomedical field, including drug delivery and stabilization of gold nanoparticles [17]. TDHBCs are composed of two blocks with different lower critical solution temperatures (LCST), such as poly(*N*-isopropylacrylamide)-*block*-poly(oligo ethylene glycol methyl ether acrylate) (PNIPAAm-*b*-POEGA) [18] and poly(diethylene glycol) methyl ether methacrylate)-*block*-poly(poly(ethylene glycol) methyl ether methacrylate) (PO_2-*b*-PO_{300}) [19]. In this context, PNIPAAm is probably the most extensively investigated thermo-responsive polymer in biomedical applications, whose LCST value in aqueous media is ca. 32 °C [20,21]. For example, for aqueous dispersions of PNIPAAm-*b*-POEGA, as the temperature rises above the LCST value of the PNIPAAm segment, the copolymer self-assembles to form micelles consisting of an insoluble core of PNIPAAm surrounded by a corona of a solvated POEGA block. As the temperature of the system further increases above the LCST of the POEGA segment, it becomes insoluble and collapses onto the surface of the PNIPAAm core [18]. The formation and dissociation of thermo-responsive diblock copolymers nano-assemblies can be controlled by temperature and, thus, without the need to use organic solvents [22,23]. Meanwhile, copolymers containing short oligo(ethylene glycol) methacrylates (OEGMAS) have also been identified for their multi-functionality as they exhibit both stimuli-responsive and biocompatible properties [24]. For instance, investigations on the poly[(di(ethylene glycol) methyl ether methacrylate)-*co*-(oligo(ethylene glycol)] methyl ether methacrylate)] (P(DEGMA-*co*-OEGMA)) copolymer system showed that the hydrophilic/hydrophobic thermal transition could be adjusted by varying the comonomer composition [25].

Herein, we report the use of copolymers poly(DEGMA-*co*-OEGMA) with an adjusted cloud point temperature (T_{cp}) prepared via RAFT polymerization using a trithiocarbonate RAFT agent (CTA) and chain-extended with NIPAAm and with NIPAAm/butyl acrylate (BA) comonomer mixtures to yield doubly thermo-responsive block copolymers. In addition, the self-assembly of these materials in aqueous media was evaluated against temperature. The T_{cp} of the first block was designed to be above 40 °C by varying the comonomer content of P(DEGMA-*co*-OEGMA); similarly, the second T_{cp} was adjusted to values below 33 °C using NIPAAm and BA as comonomers. Furthermore, the trithiocarbonate end-group in these copolymers was reduced using hexylamine, and the obtained thiol-terminated block copolymers were immobilized on the surface of GNRDs following a "grafting to" approach to yield block copolymer@GNRDs nanohybrids with a double thermo-responsive coating (see Scheme 1). Moreover, NIR-induced heating experiments on the nanohybrids in aqueous dispersions were carried out with the aid of a NIR laser diode (785 nm wavelength, output power up to 300 mW) where the behavior of self-assemblies as a function of temperature was also evaluated under physiological temperature conditions. The tailored temperature-dependent nanohybrids proposed herein may represent interesting candidates to develop drug loading and delivery strategies for medical therapies.

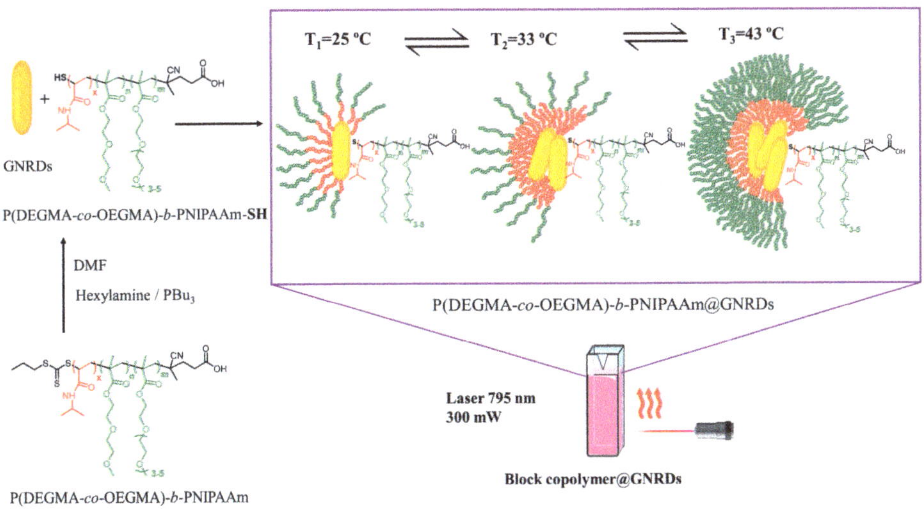

Scheme 1. Schematic illustration of the preparation of P(DEGMA-*co*-OEGMA)-*b*-PNIPAAm@GNRDs nanohybrids.

2. Materials and Methods

2.1. Materials

Oligo(ethylene glycol) methyl ether methacrylate (OEGMA, 99%, M_w = 300 g/mol), di(ethylene glycol) methyl ether methacrylate (DEGMA, 99%, M_w = 188.22 g/mol), and butyl acrylate (BA, 99%) were purified through a column packed with inhibitor removers for hydroquinone, while *N*-isopropylacrylamide (NIPAAm, 99%) was recrystallized in petroleum ether. 4,4-Azobis(4-cyano-valeric acid) (ACVA, 98%), tetrachloroauric acid trihydrate ($HAuCl_4.3H_2O$, 99%), sodium borohydride ($NaBH_4$, 99%), ascorbic acid ($C_6H_8O_6$, 99%), hexadecyltrimethylammonium bromide (CTAB, 99%), silver nitrate ($AgNO_3$, 99%), sulfuric acid (H_2SO_4, 98%), n-hexylamine ($C_6H_{15}N$, 96%) and tributylphosphine [$CH_3(CH_2)3P$, 97%)] were used as received. Unless specified otherwise, all the chemicals used were acquired from Sigma-Aldrich (Toluca, Mexico). Petroleum ether, diethyl ether, dichloromethane (DCM), and dimethyl formamide (DMF) were purchased from FERMONT (Monterrey, Mexico). Purification of synthetic products was carried out via chromatography using silica gel (70–230 mesh, Across Organics, Geel, Belgium).

2.2. Synthesis of P(DEGMA-co-OEGMA) Copolymers

A series of copolymers (first block) were prepared using different ratios of monomer: CTA with a feed of DEGMA:OEGMA 70:30 mol% using 4-cyano-4-(propylthiocarbothioylthio) pentanoic acid as a trithiocarbonate-based CTA, which in turn was synthesized as described elsewhere [25]. The synthesis of the first block is briefly described: DEGMA (1.97 g, 10.5 mmol), OEGMA (1.35 g, 4.5 mmol), CTA (31 mg, 0.119 mmol), and ACVA (3.13 mg, 0.0199 mmol) were dissolved in dry DMF at 25% v/v, the mixture was placed into 20 mL glass ampoules containing a magnetic stir bar, argon was sparged into the solution for 25 min to displace oxygen and each ampoule was sealed. All reactions were carried out under magnetic stirring at 70 °C for 4 h. The polymerization was stopped by cooling to room temperature. Afterward, the copolymer was purified by precipitation into an excess of diethyl ether/petroleum ether (1:1 v/v) under constant agitation. Then, the copolymer was re-dissolved in the minimum amount of DCM and precipitated into an excess of diethyl ether (X3); the precipitate was recovered by discarding the supernatant, followed by drying under reduced pressure in a vacuum oven at room temperature.

2.3. Synthesis of P(DEGMA-co-OEGMA)-b-PNIPAAm Block Copolymers

The P(DEGMA-*co*-OEGMA) first blocks were used as macro-RAFT agents to incorporate the second thermo-responsive PNIPAAm blocks. The synthesis of one block copolymer is described in detail: P(DEGMA-*co*-OEGMA) ($M_{n\,GPC}$ = 17,500 g mol^{-1}) (0.200 g, 0.016 mmol), NIPAAm (115 mg, 1.025 mmol), and ACVA (0.286 mg, 0.00102 mmol) were dissolved in 2 mL of dry DMF. The mixture was placed into 10 mL glass ampoules containing a magnetic stir bar, argon was sparged into the solution for 25 min to displace oxygen and each ampoule was sealed; all reactions were carried out under magnetic stirring at 70 °C for 2 h. The polymerization was stopped by cooling to room temperature. Afterward, the diblock copolymer was purified by precipitation into an excess of diethyl ether/petroleum ether (1:1 v/v) under constant agitation. Then, the copolymer was re-dissolved in the minimum amount of DCM and precipitated into an excess of diethyl ether (X3), the precipitate was recovered by discarding the supernatant followed by drying under reduced pressure in a vacuum oven at room temperature.

2.4. Synthesis of P(DEGMA-co-OEGMA)-b-P(NIPAAm-co-BA)

A similar procedure was followed for the case of P(DEGMA-*co*-OEGMA)-*b*-P(NIPAAm-*co*-BA) block copolymers, i.e., utilizing P(DEGMA-*co*-OEGMA) as a macro-RAFT agent, with the difference that BA was added in amounts of 10 mol% with respect to NIPAAm. An example is described: P(DEGMA-*co*-OEGMA) ($M_{n\,SEC}$ = 17,500 g mol^{-1}) (0.200 g, 0.016 mmol), NIPAAm (102 mg, 0.923 mmol), BA (13 mg, 0.102 mmol), ACVA (0.286 mg, 0.00102 mmol) were dissolved in 2 mL of dry DMF. The mixture was placed into 10 mL glass ampoules containing a magnetic stir bar, argon was sparged into the solution for 25 min to displace oxygen and each ampoule was sealed. All reactions were carried out under magnetic stirring at 70 °C for 2 h. The polymerization was stopped by cooling to room temperature. Afterward, the diblock copolymer was purified by precipitation into an excess of diethyl ether/petroleum ether (1:1 v/v) under constant agitation. Then, the copolymer was re-dissolved in the minimum amount of DCM and precipitated into an excess of diethyl ether (X3), the precipitate was recovered by discarding the supernatant followed by drying under reduced pressure in a vacuum oven at room temperature.

2.5. Transformation of Trithiocarbonate End-Groups of the Block Copolymers into Thiol Functionality

The trithiocarbonate end-groups of the obtained copolymers were modified to thiols via aminolysis utilizing hexylamine adapting a reported methodology [26]. One example is described in detail: In a 20 mL Schlenk flask, (0.200 g, 0.0081 mmol) of P(DEGMA-*co*-OEGMA)-*b*-PNIPAAm ($M_{n\,SEC}$ = 24,500 g mol^{-1}) was dissolved in 2 mL of dry DMF, and the mixture was stirred for 20 min until homogenization, nitrogen was sparged into the solution for 25 min to displace oxygen. Then, hexylamine and tributyl phosphine were added into the reaction mixture in a molar ratio [hexylamine]:[copolymer-CTA]:[tributyl phosphine] of [50]:[1]:[10]. The mixture was stirred for 4 h and the color of the solution changed from pale yellow to transparent. Then, the copolymer-SH was purified by precipitation into an excess of diethyl ether/petroleum ether (1:1 v/v) under constant agitation. Then, the copolymer was re-dissolved in the minimum amount of DCM and precipitated into an excess of diethyl ether (X3), the precipitate was recovered by discarding the supernatant followed by drying under reduced pressure in a vacuum oven at room temperature. The product was obtained with a yield of approximately 85 wt%.

2.6. Synthesis and Stabilization of Gold Nanorods (GNRDs)

The GNRDs synthesis was performed using the seed-mediated growth method described elsewhere [26,27]. First, CTAB-coated gold seeds were obtained by chemical reduction of $HAuCl_4$ utilizing $NaBH_4$. Briefly, (0.25 mL, 10 mM) of $HAuCl_4$ was slowly added into a CTAB solution (7.5 mL, 100 mM) under stirring. Then, (0.6 mL, 10 mM) $NaBH_4$ was incorporated and the solution quickly changed from yellow to brown indicating the formation of gold seeds. Second, the GNRDs were prepared from gold seeds within a growth solution. The growth solution was prepared by adding (5 mL, 10 mM) of $HAuCl_4$, (100 mL, 10 mM) of CTAB, (1 mL, 10 mM) of $AgNO_3$, and (1 mL, 1 M) of H_2SO_4 into a 250 mL round-bottom flask at 30 °C under stirring. Next, (0.8 mL, 100 mM) of ascorbic acid was added under stirring until the dispersion became colorless. Then, 0.250 mL of gold seed was added under vigorous stirring and left undisturbed for 6 h. The GNRDs were purified via centrifugation (10,000 rpm for 30 min) discarding the supernatant; this purification step was repeated twice. The purified GNRDs were dispersed in distilled water and stored in a refrigerated amber vial until use.

To graft P(DEGMA-*co*-OEGMA)-*b*-PNIPAAm-SH onto GNRDs, 5 mL of GNRDs in dispersion and 5 mg of copolymers were mixed under magnetic stirring for 3 h; the resulting dispersion was let to settle for 24 h, and precipitated GNRDs were removed by decantation as the product can be found in the supernatant. A similar procedure was followed in the case of P(DEGMA-*co*-OEGMA)-*b*-P(NIPAAm-*co*-BA)-SH; 5 mg of the copolymer was dissolved in DMF and slowly added into 5 mL of GNRDs dispersion. This mixture was dialyzed for 24 h, the resulting dispersion was let to settle for 24 h, and precipitated GNRDs were removed by decantation as the product can be found in the supernatant.

2.7. Characterization of Polymers

Proton nuclear magnetic resonance (^{1}H-NMR) spectra were collected at 400 MHz on a Bruker AMX-400 (Bruker Corporation, Billerica, MA, USA) spectrometer at 298 K. Deuterated chloroform ($CDCl_3$) was used as an analysis solvent for all samples. Chemical shifts are reported in parts per million (ppm) using tetramethylsilane (TMS) as the internal standard. Ultraviolet-visible (UV-Vis) spectra were recorded using a UV-Vis Varian Cary 100 spectrophotometer (Agilent Technologies, Santa Clara, CA, USA) in the wavelength range from 200 to 900 nm.

Size exclusion chromatography (SEC) was performed on a Waters 515 HPLC chromatograph (Waters Corp., Milford, MA, USA) equipped with a series of columns (Shodex: GF-1G, 7B, GF-510 HQ, GF-310 HQ, Tokyo, Japan), a Wyatt Technology refractive index detector (Optilab DSP, Santa Barbara, CA, USA) and a Wyatt Technology multiangle light scattering (LS) detector (Dawn DSP, Santa Barbara, CA, USA) equipped with a red laser (λ = 633 nm). Poly(ethylene oxide) standards of narrow dispersity ($Đ$) were used to check the accuracy of the LS detector in DMF as a mobile phase. DMF was used as the mobile phase at a 0.3 mL min^{-1} flow rate at 40 °C. Sample solutions were prepared using 20 mg mL^{-1} concentration and filtered through a 0.2 μm Nylon membrane.

2.8. Characterization of Aggregates and Nanohybrids

Dynamic light scattering (DLS) measurements were carried out using Malvern Instruments Nano-ZS Nanosizer (ZEN 3690), (Malvern, Worcestershire, UK). The instrument is equipped with a helium-neon laser (λ = 633 nm) with a size detection range of 0.6 nm to 5 μm. DLS experiments were performed at the scattering angle of 90°, and the distribution of sizes was calculated using Malvern instruments dispersion technology Zetasizer software (version 7.1) based on the CONTIN analysis and the Stokes–Einstein equation for spheres. The D_h reported is the average of three measurements.

The T_{cp} was estimated via DLS using 1 mg mL^{-1} copolymer solutions and a temperature program that increased in two-degree steps with an equilibration time of 1 min between each step.

Morphological studies of GNRDs and nanohybrids block copolymer@GNRDs were performed using a H7500 transmission electron microscope (TEM) (Hitachi Co. Ltd., Tokyo, Japan), operating at an accelerating voltage of 80 kV. A drop of dispersed GNRDs or nanohybrids at a concentration value of 0.1 mg mL^{-1} was poured over a 75-mesh copper grid coated with a thin layer of carbon, followed by the removal of the excess liquid at room temperature.

Thermogravimetric analysis (TGA) was performed using TA-instruments Discovery-TGA equipment (TA-instrument, New Castle, DE, USA). Measurements were performed for block copolymers and nanohybrids block copolymer@GNRDs under a nitrogen atmosphere and by heating from room temperature up to 600 °C using a heating rate of 10 °C min^{-1}.

2.9. NIR-Induced Heating Investigations on Nanohybrid Dispersions

Light-induced heating studies of nanohybrid dispersions were performed using a fiber-coupled diode laser (CNI, MDL-III-785, Changchun, China) with variable output power (a maximum of 300 mW) and wavelength of 785 nm. Two mL of nanohybrid dispersion was placed inside a quartz cell at room temperature without any kind of thermal isolation. A thermocouple was introduced into the sample to record temperature changes in intervals of 1 s for a total of 16 min.

3. Results and Discussion

3.1. Synthesis and Characterization of P(DEGMA-co-OEGMA) Macro-RAFT Agent

The synthesis of the P(DEGMA-*co*-OEGMA)-*b*-PNIPAAm and P(DEGMA-*co*-OEGMA)-*b*-P(NIPAAm-*co*-BA) block copolymers was accomplished by a two steps sequential RAFT polymerization procedure. A trithiocarbonate type CTA was chosen as it is known to produce well-defined polymers when used for the polymerization of methacrylates and acrylamides [28,29].

A series of P(DEGMA-*co*-OEGMA) copolymers of different molar masses were prepared keeping the DEGMA:OEGMA molar ratio constant at 70:30 in the feed (see Scheme 2). This particular comonomer composition was chosen to attain a LCST close to 40 °C [30]. The chemical structure and actual composition of the synthesized copolymers were determined by ^{1}H-NMR spectroscopy. Among the most distinctive signals in the ^{1}H-NMR spectrum of the P(DEGMA-*co*-OEGMA) macro-RAFT agent (P2 in Figure 1) is the signal at 4.1 ppm corresponding to the methylene protons "c" attached to the methacrylate in the OEGMA units; the methylenes "d" of the DEG units in the range from 3.75 to 3.55 ppm and the double signal (e) at 3.35 ppm corresponding to the terminal methyl proton "e" of both repetitive units. The ^{1}H-NMR spectra of the CTA and of the other macro-CTA's can be found in the Supplementary Materials file (Figures S1 and S2).

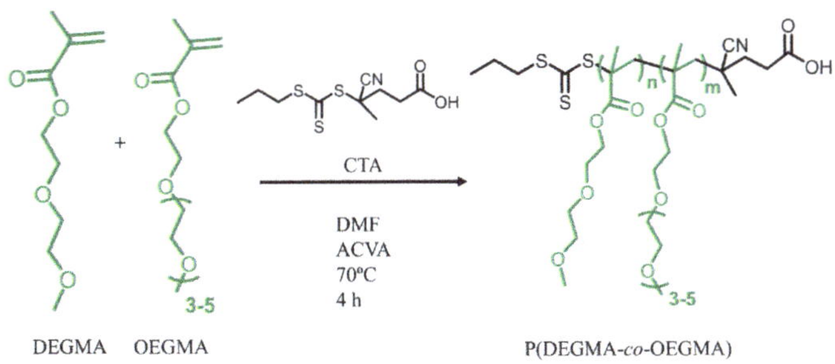

Scheme 2. Synthesis route for P(DEGMA-*co*-OEGMA) copolymers vía RAFT polymerization.

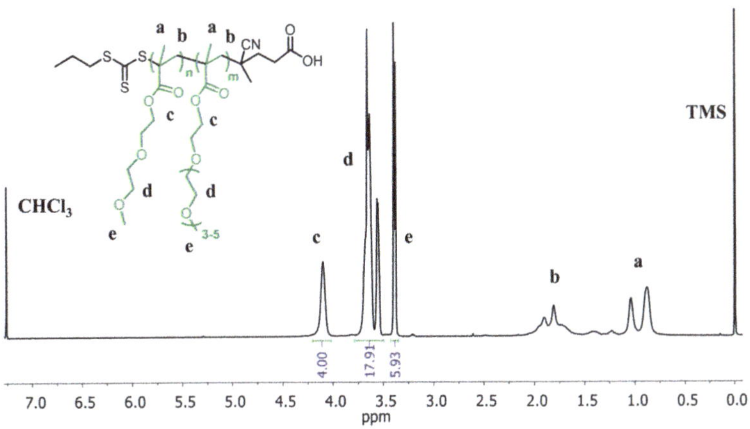

Figure 1. ^{1}H-NMR spectrum (400 MHz) in $CDCl_3$ of the P(DEGMA-*co*-OEGMA) copolymer P2.

Four P(DEGMA-*co*-OEGMA) macro-RAFT agents (first block) of different molar mass were prepared varying the monomer:CTA ratio, where the theoretical molar masswas estimated by using the ideal RAFT equation (Table 1), the molar composition of the copolymers was very similar to that initially utilized for the reaction due to the use of co-monomers with very similar reactivity towards radicals [31,32]. SEC analysis of these first blocks delivered experimental molar mass values close to the theoretical ones as well as low *Đ* values (1.05–1.10) (see Table 1), and monomodal and narrow elution time traces (see Figure S3 in Supplementary Materials file) suggesting a good control of the RAFT polymerization. The actual copolymer composition of the obtained materials, as determined by ^{1}H-NMR using the integration of signals "d" in the spectra and the equations reported in [30] for PDEGMA and POEGMA homopolymers, are also summarized in Table 1, whose composition values were close to the comonomer ratio (70:30) used in the feed.

Table 1. Conditions for RAFT polymerization of P(DEGMA-*co*-OEGMA) in DMF at 70 °C for 4 h in the presence of a trithiocarbonate CTA.

Entry	Sample [1]	Feed Molar Ratio [M]:[CTA]:[I]	P(D:O) [2] (mol%)	Conv. [3] (%)	M_n Theo [4] (g/mol)	M_n SEC [5] (g/mol)	Đ [5] M_w/M_n
P1	P(DEGMA$_{72\%}$-*co*-OEGMA$_{28\%}$)	[66]:[1]:[0.1]	72:28	76	11,700	11,100	1.10
P2	P(DEGMA$_{73\%}$-*co*-OEGMA$_{27\%}$)	[134]:[1]:[0.1]	73:27	75	22,800	17,500	1.05
P3	P(DEGMA$_{70\%}$-*co*-OEGMA$_{30\%}$)	[200]:[1]:[0.1]	70:30	63	27,900	25,800	1.07
P4	P(DEGMA$_{71\%}$-*co*-OEGMA$_{29\%}$)	[270]:[1]:[0.1]	71:29	58	34,700	30,100	1.06

[1] The percentage subscripts numbers represent the composition of the copolymers by ^{1}H-NMR (400 MHz). [2] The composition was determined by ^{1}H-NMR. [3] The conversion was determined by ^{1}H-NMR. [4] The theoretical molecular weight was calculated by using the ideal RAFT equation: $M_{n,theo} = \left[\left(\frac{[Mon]}{[CTA]}\right) \cdot Mon_w \cdot Conv\right] + M_{n\ CTA}$, where [Mon] is the sum of [D] + [O], Mon_w is the molar mass of the monomer, in this case $Mon_w = [(0.7\ M_w DEGMA) + (0.3\ M_w OEGMA)]$ and M_{wCTA} is the molar mass of CTA. [5] Determined by SEC in DMF at 40 °C with RI and LS detectors using poly(ethylene oxide) linear standards for verification of the LS detector. P(DEGMA-*co*-OEGMA) dn/dc = 0.091 mL g^{-1}.

3.2. *Temperature-Responsive Behavior of P(DEGMA-co-OEGMA) Copolymers in Aqueous Media*

For the thermo-responsive studies, solutions were prepared by directly dissolving P(DEGMA-*co*-OEGMA) copolymers in deionized water (1 mg mL^{-1}) at room temperature and subjected to DLS analysis. The cloud point (T_{cp}) is a well-known parameter for the evaluation of thermo-responsiveness in polymer solutions. In our investigations, T_{cp} was estimated by determining the temperature at which the D_h value experienced a sharp increase at a specific concentration [29]. The hydrophilic oligo(ethylene glycol) units may form hydrogen bonds with water molecules, while the backbones of the principal chains might undergo hydrophobic interactions. At the T_{cp}, the hydrogen bonds break, and the hydrophobic interactions start predominating, leading to the formation of aggregates and an increase in turbidity. On the other hand, it is well known that the T_{cp} of aqueous dispersions containing DEGMA/OEGMA copolymers can be modulated in a wide temperature range [30,33]. Nonetheless, this behavior is not always straightforward to elucidate since combinations of OEGMA with a very similar monomer 2-hydroxyethyl methacrylate (HEMA) result in copolymers with either LCST or upper critical solution temperature (UCST) [34]. When the P(DEGMA-*co*-OEGMA) copolymer has a composition of around 70:30, the dispersion has a T_{cp} value of 43 °C (as shown in Figure 2). However, the presence of salts influences the behavior of P(DEGMA-*co*-OEGMA) copolymers in aqueous media. Therefore, the T_{cp} can also be displaced depending on the composition of the aqueous medium. For instance, T_{cp} values of P(DEGMA-*co*-OEGMA) copolymers PBS buffer, are lower than in pure water [30,35].

Note that the T_{cp} value of sample P1 is 45 °C (see Figure 2) and higher than that of the other three samples. This sample has the lowest molar mass, 11,700 g mol^{-1}; half of that of sample P2. The effect of higher LCST for lower molar mass samples of the same polymer is reported, for instance, for the cases of poly(*N*-vinylcaprolactam) [36] and poly(2-dimethyl(aminoethyl) methacrylate) aqueous solutions [37].

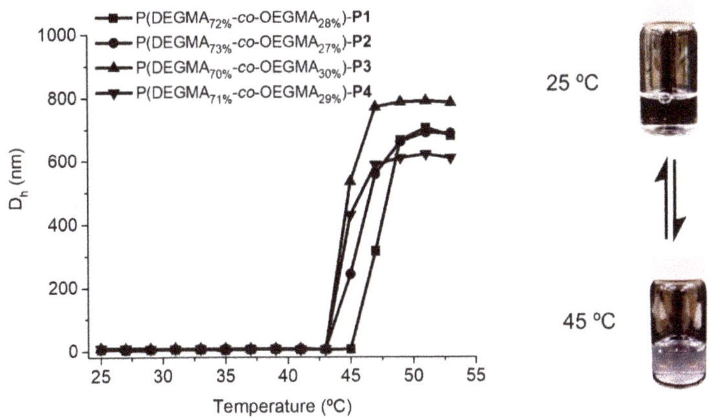

Figure 2. Analysis of T_{cp} of aqueous solutions of P(DEGMA-*co*-OEGMA) copolymers (inserted pictures correspond to sample P(DEGMA$_{70\%}$-*co*-OEGMA$_{30\%}$), P3.

3.3. Synthesis and Characterization of P(DEGMA-co-OEGMA)-b-PNIPAAm and P(DEGMA-co-OEGMA)-b-P(NIPAAm-co-BA)

Double thermo-responsive P(DEGMA-*co*-OEGMA)-*b*-PNIPAAm block copolymers were prepared using P(DEGMA-*co*-OEGMA) as a macro-RAFT agent (see Section 2 for details). To lower the T_{cp} value of the PNIPAAm segment in the corresponding aqueous solution, 10 mol% of BA was incorporated as a co-monomer during the RAFT synthesis of the second block (see Scheme 3). The aim was to obtain self-assembled core–shell structures at 25 °C in aqueous media featuring a hydrophilic P(DEGMA-*co*-OEGMA) shell and a hydrophobic P(NIPAAm-*co*-BA) core.

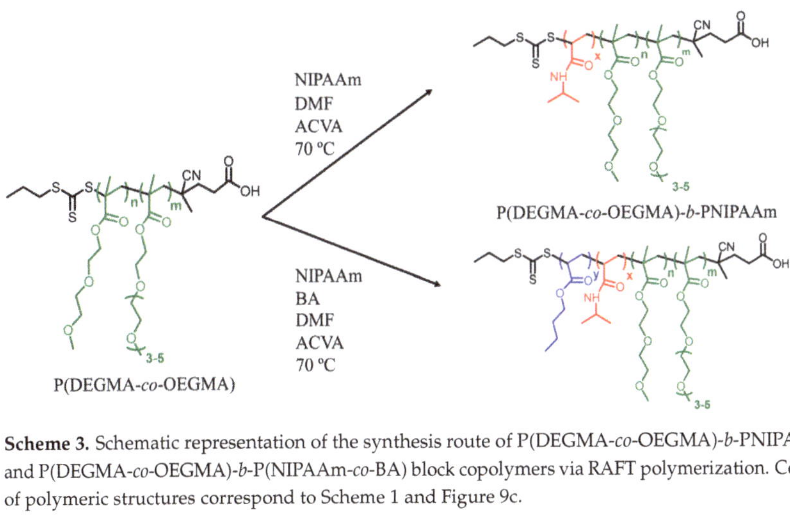

Scheme 3. Schematic representation of the synthesis route of P(DEGMA-*co*-OEGMA)-*b*-PNIPAAm and P(DEGMA-*co*-OEGMA)-*b*-P(NIPAAm-*co*-BA) block copolymers via RAFT polymerization. Colors of polymeric structures correspond to Scheme 1 and Figure 9c.

Figure 3 displays the ^{1}H-NMR spectra of the P(DEGMA-*co*-OEGMA)-*b*-PNIPAAm sample P2-2 and of the P(DEGMA-*co*-OEGMA)-*b*-P(NIPAAm-*co*-BA) sample P2-3. A part from the same P(DEGMA-*co*-OEGMA) signals observed at the same chemical shift values as those found in Figure 2, Figure 3a,b also show new signals ascribed to the PNIPAAm and the P(NIPAAm-*co*-BA) segments, respectively. The most prominent new signals are: the methine proton of the isopropyl groups of PNIPAAm observed at 3.97 ppm, which overlaps with the methylene protons (2) of PBA; a broad signal between 5.7 and 7.1 ppm attributed to the amide proton (h) of PNIPAAm; and, at 1.13 ppm, the methyl protons (i) assigned to the isopropyl groups of PNIPAAm. The methyl protons of the PBA units overlap with the methyl groups of the methacrylate moieties of the P(DEGMA-*co*-OEGMA) block [38]. The molar ratio of block copolymers was determined by comparing the integral value of signal (e) for the P(DEGMA-*co*-OEGMA) block to that of signal (h) of PNIPAAm. In the case of the segments containing PBA, the molar ratio of PBA to PNIPAAm was determined by comparing the integral value of signal (h) to the integral value between 3.8 and 4.25 ppm (signals c, j, and 3). The ^{1}H-NMR spectra of the other prepared block copolymers can be found in the Supplementary Materials file (Figure S4).

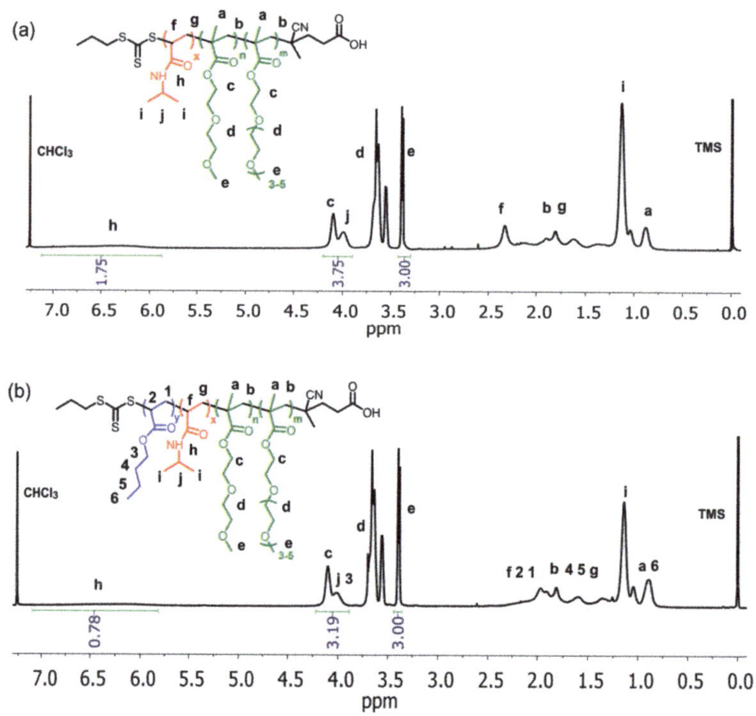

Figure 3. ^{1}H-NMR spectra (400 MHz) in $CDCl_3$ of block copolymers: (**a**) P(DEGMA-*co*-OEGMA)-*b*-PNIPAAm (P2-2), (**b**) P(DEGMA-*co*-OEGMA)-*b*-P(NIPAAm-*co*-BA) (P2-3). Colors of polymeric structures correspond to Scheme 1 and Figure 9c.

A series of block copolymers featuring different molar mass was prepared by varying NIPAAm:BA:macro-CTA ratio; the corresponding theoretical molar masses were estimated using an adaptation of the ideal RAFT equation, which considers the M_n of the respective P(DEGMA-*co*-OEGMA) macro-RAFT agent. Table 2 summarizes the results of these copolymerization reactions, where it can be observed that the experimental molar masses obtained were in proper agreement with the theoretical values due to the similar reactivity ratios of NIPAAm and BA; thus, a statistical copolymer block would be expected [39]. In general, the measured $Đ$ values were below 1.13, suggesting proper control during the RAFT polymerizations. The obtained block copolymers featured a monomodal, narrow, and well-defined molar mass distribution (see Table 2 and Figure S5 in Supplementary Materials file).

Table 2. Experimental conditions and results of the RAFT polymerizations for the synthesis of P(DEGMA-*co*-OEGMA)-*b*-PNIPAAm and P(DEGMA-*co*-OEGMA)-*b*-P(NIPAAm-*co*-BA) in DMF at 70 °C for 2 h using P(DEGMA-*co*-OEGMA)s as macro-RAFT agents (mCTAs).

Entry	Sample [1]	Feed Molar Ratio [Mon]:[mCTA]:[I] [2]	Composition PDO:PNIPAAm:PBA [3] (mol %)	Conv. [4] (%)	M_n Theo [5] (g mol^{-1})	M_n SEC [6] (g mol^{-1})	Đ [6]
P1	P(D$_{72\%}$-*co*-O$_{28\%}$)	------	100:0:0	76	11,700	11,100	1.10
P1-1	P(D-*co*-O)$_{68\%}$-*b*-PNIPAAm$_{32\%}$	[50]:[1]:[0.1]	68:32:0	58	14,500	17,500	1.05
P1-2	P(D-*co*-O)$_{47\%}$-*b*-PNIPAAm$_{53\%}$	[100]:[1]:[0.1]	47:53:0	71	20,100	18,900	1.04
P1-3	P(D-*co*-O)$_{49\%}$-*b*-P(NIPAAm$_{40\%}$-*co*-BA$_{11\%}$)	[100]:[1]:[0.1]	49:40:11	65	18,600	19,300	1.05
P2	P(D$_{73\%}$-*co*-O$_{27\%}$)	------	100:0:0	75	22,500	17,500	1.05
P2-1	P(D-*co*-O)$_{63\%}$-*b*-PNIPAAm$_{37\%}$	[100]:[1]:[0.1]	63:37:0	72	30,400	24,500	1.05
P2-2	P(D-*co*-O)$_{37\%}$-*b*-PNIPAAm$_{63\%}$	[200]:[1]:[0.1]	37:63:0	71	35,500	32,600	1.09
P2-3	P(D-*co*-O)$_{48\%}$-*b*-P(NIPAAm$_{43\%}$-*co*-BA$_{9\%}$)	[200]:[1]:[0.1]	48:43:9	62	31,700	27,900	1.10
P3	P(D$_{70\%}$-*co*-O$_{30\%}$)	------	100:0:0	63	27,900	25,800	1.07
P3-1	P(D-*co*-O)$_{68\%}$-*b*-PNIPAAm$_{32\%}$	[100]:[1]:[0.1]	68:32:0	75	34,300	31,100	1.05
P3-2	P(D-*co*-O)$_{45\%}$-*b*-PNIPAAm$_{55\%}$	[200]:[1]:[0.1]	45:55:0	75	42,700	36,200	1.13
P3-3	P(D-*co*-O)$_{50\%}$-*b*-P(NIPAAm$_{44\%}$-*co*-BA$_{6\%}$)	[200]:[1]:[0.1]	50:44:6	59	39,300	34,200	1.09

[1] The percentage subscripts numbers represent the composition of the copolymers as determined via ^{1}H-NMR (400 MHz). [2] The block copolymers P1-3, P2-3 and P3-3, 10 mol% of BA was added according to NIPAAm. [3] The composition was determined by ^{1}H-NMR. [4] The conversion was determined by ^{1}H-NMR for the P(DEGMA-*co*-OEGMA)-*b*-PNIPAAm and determined gravimetrically for the P(DEGMA-*co*-OEGMA)-*b*-P(NIPAAm-*co*-BA). [5] The theoretical molar mass was calculated by the equation: $M_{n,theo} = \left[\left(\frac{[Mon]}{[macroCTA]}\right) \cdot Mon_w \cdot Conv\right] + M_{n\ macroCTA}$, where [Mon] = [NIPAAm] + [BA], Mon_w is the molar mass of the corresponding monomers, in this case $Mon_w = [(X\ M_wNIPAAm) + (Y\ M_wBA)]$ and $X + Y = 100\%$ and M_n macroCTA is the molar mass of macroCTA. [6] Determined by SEC in DMF at 40 °C with RI and LS detectors using poly(ethylene oxide) linear standards of narrow dispersity for verification of LS detector; dn/dc = 0.091X + 0.070Y (mL g^{-1}), being X and Y de molar ratios of P(DEGMA-*co*-OEGMA) = X and PNIPAM + PBA = Y in the copolymer.

The thermal properties of the synthesized copolymers were examined via DSC in the temperature range from −80 to 200 °C. The glass transition temperature (T_g) of linear PNIPAAm can be found between 85 and 130 °C depending on its molar mass [40], whereas it has been reported that the T_g of P(DEGMA-*co*-OEGMA) copolymers can be adjusted between −53 and −59 °C with the co-monomer composition [35]. The DSC investigations of the P(DEGMA-*co*-OEGMA)$_{37\%}$-*b*-PNIPAAm$_{63\%}$ and P(DEGMA-*co*-OEGMA)$_{48\%}$-*b*-P(NIPAAm$_{43\%}$-*co*-BA$_{9\%}$) block copolymers yielded two glass transitions due to phase separation between the different segments of the block copolymers. The T_g's of PNIPAAm and P(DEGMA-*co*-OEGMA) segments were observed at 73 and −45 °C, respectively, and that of PNIPAAm shifted to lower values when BA units were incorporated as comonomer (see Figure S6 in the Supplementary Materials file). The experimentally determined two T_g's are, to a large extent, comparable to those of the respective (homo-) co-polymer segments, which confirms the microphase-separation of the block segments further indicating the successful formation of block copolymer materials. The recorded thermograms of the other investigated block copolymers can also be found in the Supplementary Materials file (Figures S7 and S8).

3.4. *Thermo-Responsive Behavior of P(DEGMA-co-OEGMA)-b-PNIPAAm and P(DEGMA-co-OEGMA)-b-P(NIPAAm-co-BA) Block Copolymers in Aqueous Media*

The self-assembly behavior of the block copolymers against temperature in aqueous media was studied by monitoring their aggregation via DLS. P(DEGMA-*co*-OEGMA)-*b*-PNIPAAm block copolymers were directly dissolved in deionized water (1 mg mL^{-1}), while the P(DEGMA-*co*-OEGMA)-*b*-P(NIPAAm-*co*-BA) block copolymers were first dissolved in DMF and slowly dripped into deionized water followed by dialysis for 24 h to remove DMF; the final concentration was adjusted to a value of 1 mg mL^{-1}. Figure 4 shows the temperature-induced self-assembly behavior of samples P(DEGMA-*co*-OEGMA)$_{37\%}$-*b*-PNIPAAm$_{63\%}$ and P(DEGMA-*co*-OEGMA)$_{48\%}$-*b*-P(NIPAAm$_{43\%}$-*co*-BA$_{9\%}$) (subscript numbers indicate molar fractions in the block copolymers) in terms of DLS measurements.

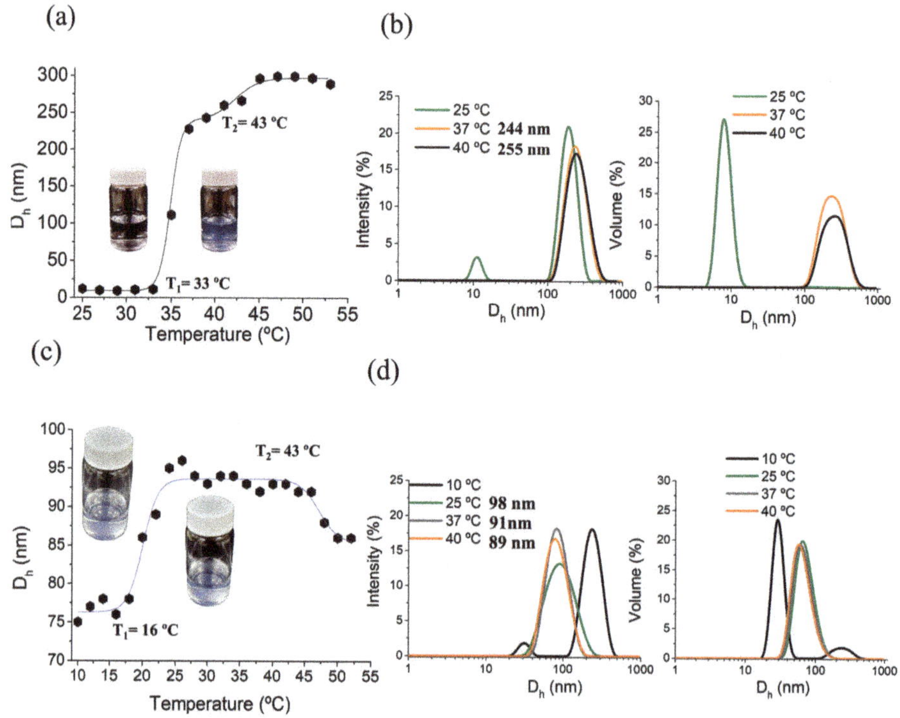

Figure 4. Analysis of size change in aqueous media and size distribution at different temperatures for block copolymers (1 mg mL^{-1}): (**a**,**b**) P(DEGMA-*co*-OEGMA)$_{37\%}$-*b*-PNIPAAm$_{63\%}$; (**c**,**d**) P(DEGMA-*co*-OEGMA)$_{48\%}$-*b*-P(NIPAAm$_{43\%}$-*co*-BA$_{9\%}$).

Figure 4a shows a two-step increase in particle size with a first increase in D_h at a temperature close to the LCST of PNIPAAm (33 °C) suggesting the formation of aggregates. When temperature further increases above this latter value, the PNIPAAm segments are dehydrated, and the dispersion undergoes a self-assembly process to form aggregates consisting presumably of an insoluble PNIPAAm core surrounded by a corona of the solvated segments of the P(DEGMA-*co*-OEGMA) [19]. When temperature further increases above the T_{cp} of the P(DEGMA-*co*-OEGMA) block (whose chains turn insoluble and accumulate on the surface of the PNIPAAm core), it leads to the formation of larger aggregates as evidenced by a further increase in D_h starting at 43 °C; these experimental values were also fit to double-Boltzmann model (Figure 4a). Both blocks solubilize again when the temperature gradually decreases to room temperature. The particle size distribution at three different temperatures is also displayed in Figure 4b. The size distributions, as obtained via DLS, are plotted by intensity and by volume. When the block copolymers form uniform aggregates, both distributions, by intensity and by volume, are similar. However, when aggregates disassemble at room temperature (25 °C), two distributions of sizes can be observed presumably corresponding to single polymer chains and remaining aggregates. The size distribution by volume might be a better representation of the relative number of single chains as compared to that of aggregates (Figure 4b, right plot) because the size distribution by intensity overestimates the number of aggregates due to the strong dependence of scattering intensity on particle size [41] (Figure 4b, middle plot).

In the case of the double thermo-responsive block copolymers containing hydrophobic BA units (P(DEGMA-*co*-OEGMA)$_{48\%}$-*b*-P(NIPAAm$_{43\%}$-*co*-BA$_{9\%}$)), Figure 4c shows that relatively stable aggregates are formed below 16 °C. Presumably, these are also core–shell structures; however, they aggregate at room temperature, and their stability above 27 °C results in a slight shrinkage of aggregates upon heating (without further aggregation) until the T_{cp} of the P(DEGMA-*co*-OEGMA) blocks is reached (43 °C). Further heating to temperatures higher than 50 °C leads to stronger shrinkage and finally to precipitation; this same behavior has been reported for double thermo-responsive PNIPAM-*b*-PDEMAEMA-*b*-PS triblock copolymers [42]. Adding hydrophobic units into thermo-responsive polymers shifts their corresponding LCST to lower values; for example, the T_{cp} of PNIPAAm can be shifted between 10–16 °C depending on the copolymer composition. This phenomenon has already been reported for hydrophobic BA units contributing to the formation of polymer aggregates that self-assemble at 25 °C featuring monomodal size distributions (with a particle size of ca. 89 nm) [38]. However, when the temperature decreases to 10 °C, the polymer aggregates dissociate leading to the rise in two size distributions featuring small and larger aggregates (Figure 4d, middle plot). The relative amount of small and large aggregates can be assessed with the aid of the corresponding size distribution by volume, which shows that small aggregates are preferably formed (Figure 4d, right plot). It can be summarized that heating from room temperature up to the physiological relevant temperature of 37 °C, the investigated P(DEGMA-*co*-OEGMA)-*b*-PNIPAAm block copolymers self-assemble into nanostructures of ca. ~200 nm, while the investigated P(DEGMA-*co*-OEGMA)-*b*-P(NIPAAm-*co*-BA) block copolymers shrink to form stable nanostructures in the range of ca. ~95 nm. Upon cooling from room temperature, the investigated P(DEGMA-*co*-OEGMA)-*b*-PNIPAAm block copolymers dissociate to single block copolymer chains of ca. ~8 nm, while the investigated P(DEGMA-*co*-OEGMA)-*b*-P(NIPAAm-*co*-BA) block copolymers form more persistent aggregates of relatively small size (ca. ~30 nm). The Supplementary Materials file shows the thermo-responsive behavior of all the investigated block copolymers (Figures S9 and S10).

3.5. Aminolysis of Trithiocarbonate End-Groups of Block Copolymers to Form Block Copolymers with Thiol End-Groups

The preparation of reactive thiol-terminated polymers using aminolysis reaction as a post-functionalization strategy is of relevance in polymer synthesis as these materials can be used in different applications including GNRDs stabilizers [12]. However, it is important to avoid the formation of undesired disulfide bonds between two polymer chain ends. Among the contributing factors are the chemical structure of the corresponding (co)polymer, the reaction time and solvent, the type of amine used, and/or the presence of oxygen [15,43,44]. It has been reported that the presence of phosphine during the modification decreases the coupling between polymer chains as a consequence of the formation of disulfide bonds [8] (see Scheme 4).

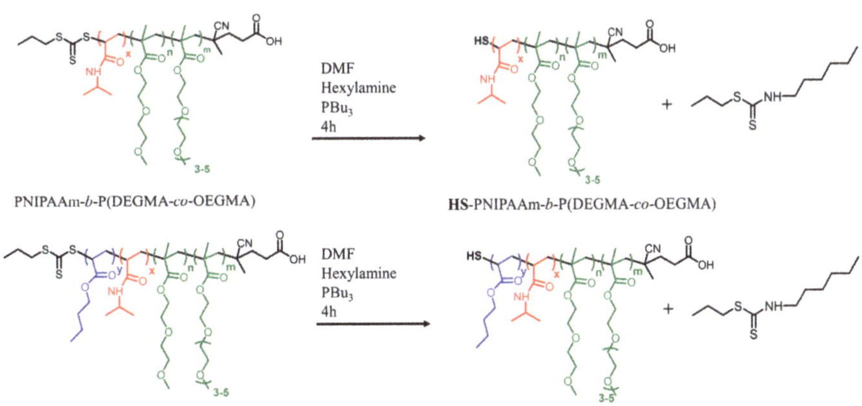

Scheme 4. Schematic representation of the preparation of thiol-terminated block copolymers via aminolysis of trithiocarbonates. Colors of polymeric structures correspond to Scheme 1 and Figure 9c.

Copolymers prepared via RAFT polymerization exhibit a strong absorption band at 310 nm due to the presence of a terminal trithiocarbonate group featuring a n-π^* transition [44]. Table 3 shows a list of the block copolymers selected for post-functionalization based on their molar mass, self-assembly behavior, and size distribution.

Table 3. Block copolymers selected for aminolysis and some of their properties.

Sample [1]	M_n [2] SEC [1] (g mol^{-1})	$Đ$ [2]	T_{cp} 1 [3] (°C)	T_{cp} 2 [3] (°C)	T_g 1 [4] (°C)	T_g 2 [4] (°C)
P(D-*co*-O)$_{47\%}$-*b*-PNIPAAm$_{53\%}$	19,800	1.04	33	43	−40	79
P(D-*co*-O)$_{37\%}$-*b*-PNIPAAm$_{63\%}$	32,600	1.09	33	43	−45	74
P(D-*co*-O)$_{45\%}$-*b*-PNIPAAm$_{55\%}$	36,200	1.13	35	43	−43	73
P(D-*co*-O)$_{48\%}$-*b*-P(NIPAAm$_{43\%}$-*co*-BA$_{9\%}$)	27,900	1.10	16	43	−35	64
P(D-*co*-O)$_{50\%}$-*b*-P(NIPAAm$_{44\%}$-*co*-BA$_{6\%}$)	34,200	1.09	16	43	−30	58

[1] Subscript numbers represent the composition of the copolymers in mol%. [2] Determined via SEC in DMF at 40 °C with RI and LS detector using poly(ethylene oxide) linear standard for verification of LS detector. [3] Determined via DLS in aqueous media (1 mg mL^{-1}). [4] Determined via DSC.

Representative results obtained from the aminolysis reactions of the investigated copolymers are displayed in Figure 5a (i.e., UV-vis spectra). The first product of the performed RAFT polymerizations, namely P(DEGMA$_{73\%}$-*co*-OEGMA$_{27\%}$) macro-CTA, exhibited a strong absorption band, whose intensity decreased as the chain-extension polymerization was performed (i.e., synthesis of the PNIPAAm block). After aminolysis of the trithiocarbonate end-groups, the absorption band vanished indicating the formation of thiol-terminated polymers.

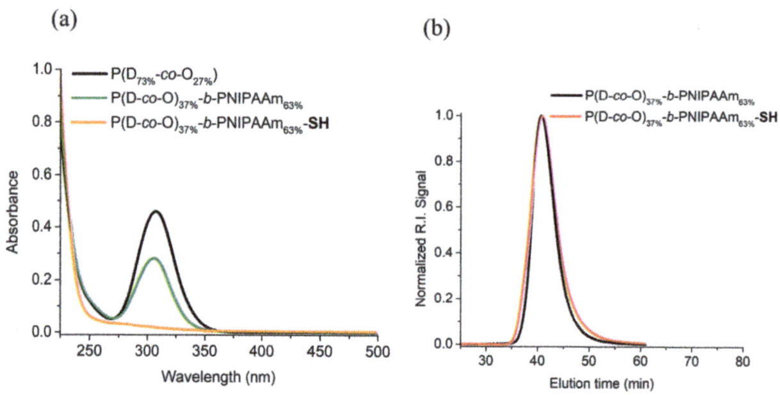

Figure 5. (**a**) UV-vis spectra (c = 1 mg mL^{-1} in ethanol), and (**b**) SEC traces in DMF for P(DEGMA-*co*-OEGMA)$_{37\%}$-*b*-PNIPAAm$_{63\%}$ (M_n = 32,600 g mol^{-1}, $Đ$ = 1.09) and P(DEGMA-*co*-OEGMA)$_{37\%}$-*b*-PNIPAAm$_{63\%}$-SH (M_n = 35,800 g mol^{-1}, $Đ$ = 1.11).

Figure 5b displays the corresponding SEC traces of P(DEGMA-*co*-OEGMA)$_{37\%}$-*b*-PNIPAAm$_{63\%}$ with trithiocarbonate end-group and with thiol end-group. The elution time trace ascribed to the thiol-terminated copolymer practically remains unaltered as compared to the trace of its precursor material. Nevertheless, a slight broadening can be observed, indicating a slight change in the molar mass and dispersity of the copolymer. The UV-Vis spectra of all the modified block copolymers can be found in the Supplementary Materials file (Figure S11).

3.6. *Preparation of Double Thermo-Responsive Copolymer@GNRDs Nanohybrids: Self-Assembly and Stability Investigations*

GNRDs were synthesized via a seed-mediated growth method [26,27]. Figure 6a displays the UV-vis spectra of the GNRDs, where the extinction spectra of the GNRDs in water exhibit two absorption bands at λ_{max} = 510 nm corresponding to the transverse surface plasmon resonance in the visible range, and the longitudinal surface plasmon resonance (LSPR) in the near-infrared (NIR) range at ca. λ_{max} = 820 nm. It can also be observed that the LSPR is slightly modified as the CTAB stabilizer is removed from the system. The LSPR in the NIR region is particularly relevant for this current investigation as well as the relationship of its intensity and position with the surface modification. Figure 6b shows a TEM micrograph featuring well-defined rods, whose average dimensions were 56.18 ± 5.38 nm in length and 14.08 ± 1.40 nm in width yielding an aspect ratio (AR) of 3.98 (length/diameter).

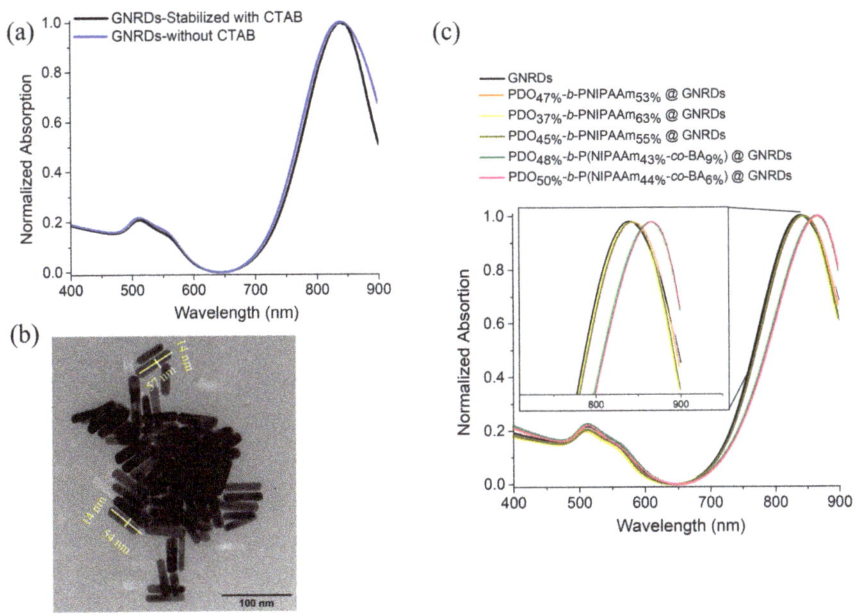

Figure 6. Characterization of the synthesized GNRDs featuring an aspect ratio of 3.98: (**a**) UV-Vis spectra with and without CTAB; (**b**) TEM micrograph of GNRDs dispersed after removing CTAB; (**c**) UV-Vis spectra of the block copolymer@GNRDs nanohybrids in an aqueous medium.

After synthesis and purification, the GNRDs were further modified with the previously prepared thiol end-functionalized block copolymers. A series of P(DEGMA-*co*-OEGMA)-*b*-PNIPAAm-SH and P(DEGMA-*co*-OEGMA)-*b*-P(NIPAAm-*co*-BA)-SH block copolymers were immobilized onto the surface of GNRDs via a "grafting to" approach. The aim of this was to protect GNRDs from self-aggregation and provide them with a thermo-responsive aggregation/disaggregation property by employing the proposed block copolymers. The grafting reaction was carried out in an aqueous medium at room temperature using a simple mixing. Figure 6c shows that the LSPR band presented a slight shift upon polymer coating. This behavior is more evident in the case of P(DEGMA-*co*-OEGMA)-*b*-P(NIPAAm-*co*-BA)@GNRDs nanohybrids presumably because they exhibit more hydrophobic characteristics. This shift in the LSPR band might be due to changes in the local dielectric coefficient and surface refractive index [45]. It is also evident that there is no broadening of the LSPR band, which suggests the absence of GNRDs aggregation after modification with the block copolymers.

On the other hand, it is also relevant for biomedical applications to investigate the stability of the nanohybrids in dispersion over time, which can be monitored by assessing possible changes in the UV-Vis absorption spectra of the nanohybrids. Thus, this optical stability was examined for 8 days via the analysis of the position and broadness of the LSPR band at

25 °C. Figure 7a shows that the LSPR band did not present any significant displacement to higher wavelengths or change in broadness during the investigated period of time. Although the intensity of the absorbance band decreases over time, the dispersions did not show any precipitation at room temperature (see Figure S12 in the Supplementary Materials file). Meanwhile, TEM micrographs of P(DEGMA-*co*-OEGMA)$_{37\%}$-*b*-PNIPAAm$_{63\%}$@GNRDs dried at room temperature are displayed in Figure 7b, in which the negatively stained uranyl acetate polymer shell can be observed as a halo featuring the copolymer coating (See Figure S13 in Supplementary Materials file for additional micrographs).

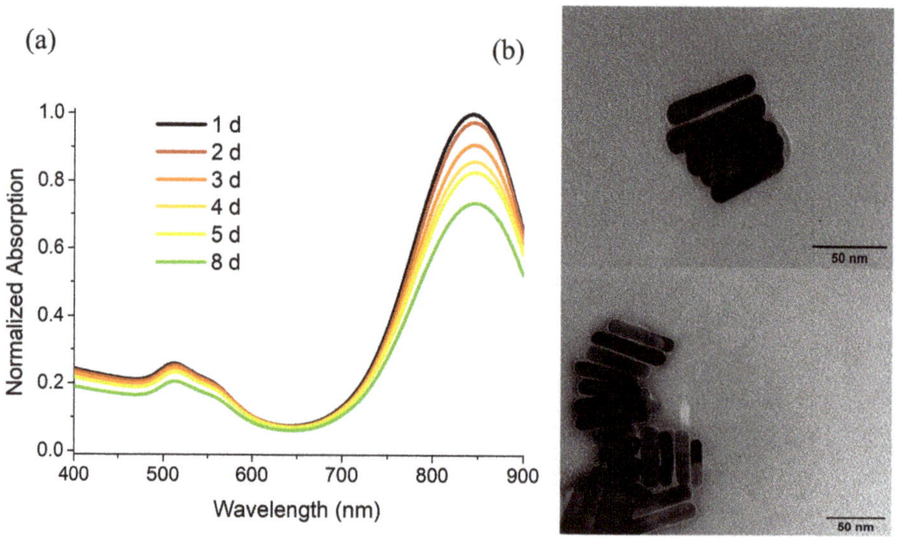

Figure 7. Characterization of P(EGMA-*co*-OEGMA)$_{37\%}$-*b*-PNIPAAm$_{63\%}$@GNRDs nanohybrid: (**a**) UV-Vis spectra of dispersions monitored for 8 days at room temperature, (**b**) Representative TEM micrographs of block copolymer@GNRDs in dry state (as stained with uranyl acetate).

3.7. Behavior of Copolymer@GNRDs Nanohybrids upon Heating and by NIR-Irradiation

The block copolymer@GNRDs nanohybrids are expected to retain the thermo-responsiveness of P(DEGMA-*co*-OEGMA)-*b*-PNIPAAm and P(DEGMA-*co*-OEGMA)-*b*-P(NIPAAm-*co*-BA) block copolymers [15,26,46].

Changes in temperature of the block copolymer@GNRDs nanohybrids in aqueous suspension were probed using DLS. It is known that for rod-like particles, the estimation of sizes in terms of the diffusion coefficients produces bimodal distributions (Figure 8a), whose peaks do not correspond to either the longitudinal or transverse dimensions of the rods [45]. The small size peak is attributed to the rotational diffusion coefficient of the rods and the large size peak simply means that the nanorods have the same diffusion coefficient as a spherical gold nanoparticle of that size. Nevertheless, the bimodal distribution can be taken as an indication of the presence of rod-like particles in a sample. Such is the case observed for our block copolymer@GNRDs nanohybrids at 25 °C (Figure 8a). In this situation, the P(DEGMA-*co*-OEGMA)-*b*-PNIPAAm polymeric chains are hydrated, forming a solvated polymer brush-type coating on the GNRDs.

At higher temperatures, on the other hand, the suspensions are expected to form aggregates with shapes that are closer to that of a sphere, leading to monomodal distributions in the DLS estimations (Figure 8b,c); in such cases, the estimated sizes are expected to be more reliable. Thus, changes from bimodal to a monomodal distribution in the DLS estimations were taken as a signature of aggregation, and the peaks of the monomodal distributions were used to estimate the aggregate sizes.

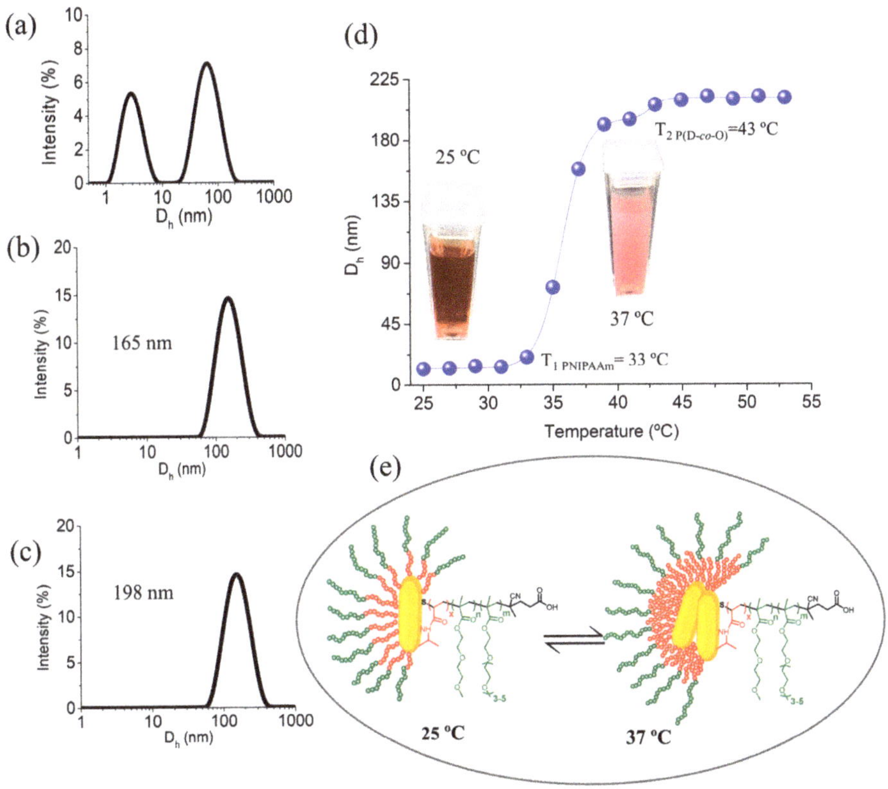

Figure 8. Size distributions of P(EGMA-*co*-OEGMA)$_{37\%}$-*b*-PNIPAAm$_{63\%}$@GNRDs nanohybrids in an aqueous medium at different temperatures: (**a**) 25 °C; (**b**) 27 °C; (**c**) 40 °C; (**d**) self-assembly behavior of the nanohybrid in an aqueous medium upon increasing temperature; (**e**) schematic representation of the self-assembly behavior. Colors of polymeric structures correspond to Scheme 1.

We observe that at 37 °C the DLS estimation exhibits a monomodal size distribution with a maximum at D_h =165 nm. The increase in particle size can be explained by the dehydration of the PNIPAAm block above its T_{cp}, which forms a semispherical aggregate with a hydrophobic core of PNIPAAm@GNRDs and a P(DEGMA-*co*-OEGMA) hydrophilic shell (Figure 8b). Increasing the temperature further to 40 °C, which is close to the T_{cp} of the P(DEGMA-*co*-OEGMA) thermo-responsive block (Figure 8c), increases the estimated D_h of the aggregates to 198 nm. Figure 8d,e illustrate the self-assembly behavior of the P(EGMA-*co*-OEGMA)$_{37\%}$-*b*-PNIPAAm$_{63\%}$@GNRDs in aqueous suspension. An increase in the turbidity of the suspension due to the aggregation of the copolymer coating can be clearly observed in the photographs, but it is important to mention that these dispersions did not show any precipitation until 50 °C. More information on this is presented in Table S1 of the Supplementary Materials file.

The molar mass of the P(DEGMA-*co*-OEGMA) segment has an important effect on the stabilization of the nanohybrids at temperatures above the T_{cp} of the PNIPAAm segment as it contributes to preserving the stability of the hybrid aggregates featuring D_h values <200 nm at 37 °C (see Figure 9b for the cases P(EGMA-*co*-OEGMA)$_{37\%}$-*b*-PNIPAAm$_{63\%}$@GNRDs and P(EGMA-*co*-OEGMA)$_{45\%}$-*b*-PNIPAAm$_{55\%}$@GNRDs).

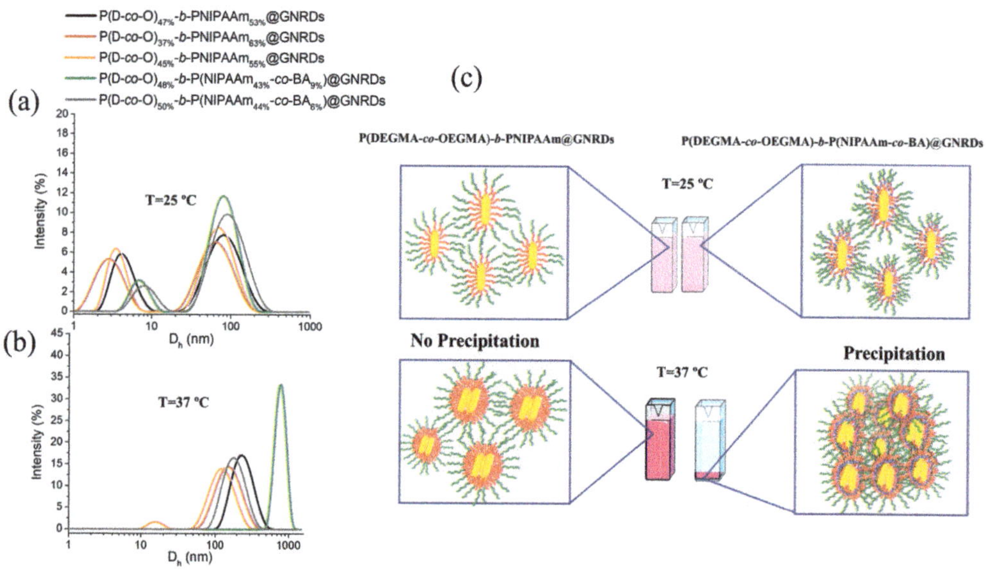

Figure 9. Size distributions of block copolymers@GNRDs nanohybrids: (**a**) at 25 °C; (**b**) at 37 °C; (**c**) schematic representation of the behavior of the block copolymer@GNRDs nanohybrids in the temperature range between 25 and 37 °C.

However, for the P(DEGMA-*co*-OEGMA)-*b*-P(NIPAAm-*co*-BA)@GNRDs suspensions (containing hydrophobic BA units), the thermo-responsiveness derived from the PNIPAAm copolymer segment is already active at 16 °C leading to the formation of stable aggregates at room temperature, and heating up to 37 °C increases hydrophobic polymer-polymer-GNRDS interactions leading to the destabilization of nanohybrids aggregates and to partial precipitation when the BA content is relatively high (9 mol%), but not when is lower (6 mol%). The fact that there are two apparent size distributions of nanohybrids at room temperature (Figure 9a) may be ascribed to the anisotropy of the block copolymer-coated GNRDs as already discussed above [47]. A schematic representation describing this hypothesized temperature-driven self-assembly behavior at 25 and 37 °C is displayed in Figure 9c. Additional DLS results for measurements performed at 37 °C can be found in Table S2 in the Supplementary Materials file. Considering that after activating the thermo-responsive behavior the self-aggregation phenomena upon heating lead ultimately to a phase separation, it is not common to find in the literature, a report on the self-aggregation behavior of nanohybrids above their aggregation temperature. This current investigation uses block copolymers featuring a double thermo-responsiveness hypothesizing that between both transition temperatures, there may be a sweet spot where such nanohybrids remain stable.

In our case, the D_h of the nanohybrids can be maintained below 200 nm at 37 °C via the manipulation of various contributing factors such as the molar mass of the block copolymers and/or their copolymer composition (Figure 10a). In the case of P(DEGMA-*co*-OEGMA)-*b*-PNIPAAm, the second transition temperature ascribed to the P(DEGMA-*co*-OEGMA) block does not lead to a change in the size of the nanohybrids when the copolymer chains are attached to the GNRDs. However, in the case of the more hydrophobic copolymers containing BA units, this second transition temperature shifts to lower values and provokes further aggregation and even precipitation of the entire nanohybrids (i.e., when the copolymer chains are attached to GNRDs) (Figure 10a). On the other hand, it is also relevant to investigate how the LSPR band changes as the block copolymer self-assembles on the GNRDs surface upon increasing temperature. Upon heating the block copolymer@GNRDs dispersion above the T_{cp} of the PNIPAAm segment, the latter undergoes a phase transition and precipitates over the GNRDs surface. This change in aqueous medium leads to an increase in the refractive index of the dispersion, which in turn causes a shift in the LSPR band. A representative behavior of the extinction spectra of P(DEGMA-*co*-OEGMA)$_{37\%}$-*b*-PNIPAAm$_{63\%}$@GNRDs in the temperature range from 25 up to 40 °C is shown in Figure 10b.

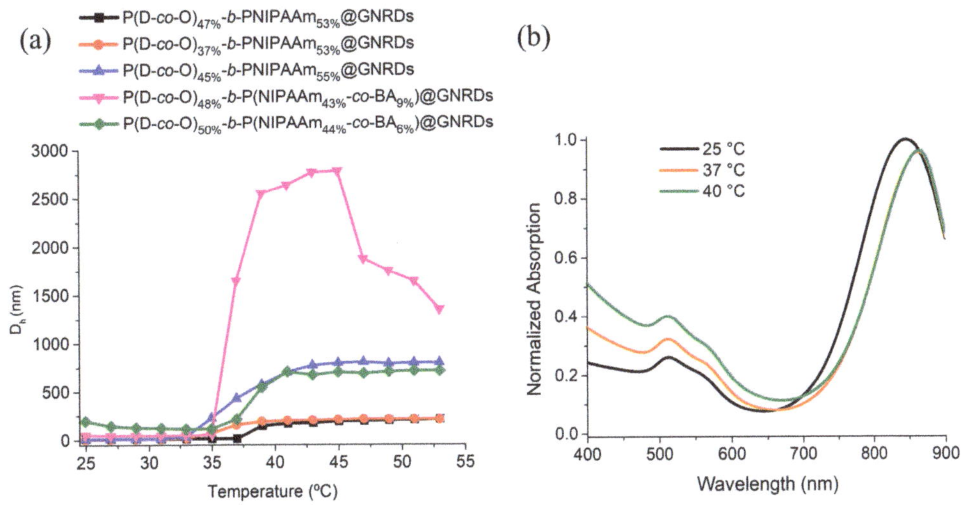

Figure 10. (**a**) Evolution of size vs. temperature for different block copolymer@GNRDs nanohybrids in an aqueous medium; (**b**) UV-Vis absorption spectra of P(EGMA-*co*-OEGMA)$_{37\%}$-*b*-PNIPAAm$_{63\%}$@GNRDs in an aqueous medium at 25, 37 and 40 °C.

A shift in the LSPR was observed with increasing temperature, this effect is not different at 40 °C as the T_{cp} of the first block P(DEGMA-*co*-OEGMA) is still not reached and the aggregates remain stable. The situation changes when the polymer interactions increase causing precipitation of the samples that contain BA-units. UV-Vis spectra of the other nanohybrids can be found in the Supplementary Materials file (Figure S14).

The photo-thermal behavior of the P(DEGMA-*co*-OEGMA)-*b*-PNIPAAm@GNRDs and P(DEGMA-*co*-OEGMA)-*b*-P(NIPAAm-*co*-BA)@GNRDs nanohybrids in an aqueous medium was investigated with the aid of a 785 nm NIR laser. The measurements consisted of irradiating the sample (2 mL) for 16 min with a laser output power of 300 mW while recording the temperature of the system every 1 s. Pristine distilled water was irradiated at 300 mW as a control experiment to ensure that the temperature evolution recorded in the case of nanohybrid dispersion was not due to energy absorption of water or of the measuring instrument (see Figure 11a). The GNRDs stabilized without CTAB dispersed in an aqueous medium showed a temperature increment of 16 °C after the experiment. For the block copolymer@GNRDs nanohybrid dispersions, the temperature rise was slightly lower than that recorded for GNRDs presumably due to the shift in the LSPR band and the change in the refractive index of the dispersion as the transition temperature of the copolymer is reached. For example, a temperature increment of 12 °C was observed for the P(EGMA-*co*-OEGMA)$_{47\%}$-*b*-PNIPAAm$_{53\%}$@GNRDs system as well as an increase in turbidity of the suspension during the NIR-irradiation. However, this temperature increase is still considerable and may be suitable for photo-thermal therapies. Only in the case of the already detected and discussed P(DEGMA-*co*-OEGMA)$_{45\%}$-*b*-P(NIPAAm$_{55\%}$-*co*-BA$_{9\%}$)@GNRDs system featuring a significant precipitation of the nanohybrid, the heating of the surrounding is not sufficient for the requirements of the intended application. Figure 11b reveals the content of GNRDs in the nanohybrid systems as estimated via TGA. For this purpose, the wt.% residue of each block copolymer@GNRDs system was determined at 600 °C and the obtained wt.% residue values ranged between 8 and 26 wt.% depending on the preparation method and copolymer material utilized. The content of the GNRDs was estimated by the difference in the residue values of the block copolymer with and without GNRDs measured at the same temperature (600 °C). However, the GNRD content does not seem to influence the measured temperature increase provoked by NIR-irradiation. The onset of the decomposition step of the block copolymers can be observed at ca. 350 °C for all the investigated cases. TGA thermograms of block copolymers without GNRDs can be found in Figure S15 (Supplementary Materials file).

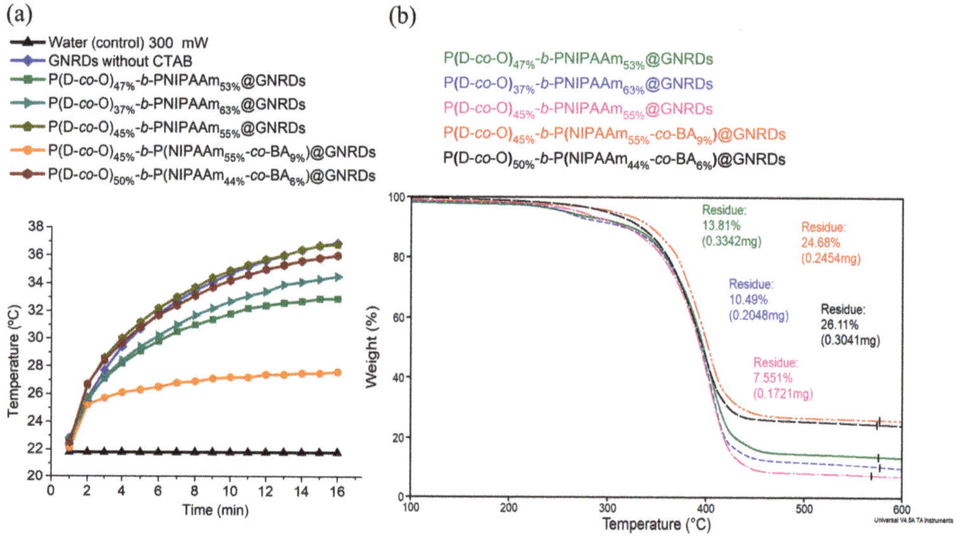

Figure 11. (**a**) Surrounding temperature of block copolymer@GNRDs and GNRDs aqueous dispersions vs. NIR-irradiation time (at 795 nm, 300 mW); (**b**) TGA thermograms of the investigated block copolymer@GNRDs.

4. Conclusions

In this contribution, double thermo-responsive block copolymers of different compositions and molar mass were synthesized via RAFT copolymerization using a trithiocarbonate CTA. The composition of the first block (composed of DEGMA and OEGMA units) was adjusted accordingly to feature a response temperature above 40 °C, while the second block (composed of NIPAAm) was varied to show a response temperature above room temperature, but below the physiologically relevant temperature of 37 °C. In a similar case to the latter, a transition temperature below room temperature was also achieved via the incorporation of BA units into the NIPAAm block.

Thiol-terminated block copolymers were prepared via aminolysis of the trithiocarbonate end-groups. Thereafter, the thiol-reactive groups were used for grafting the double thermo-responsive block copolymers onto GNRDs. The GNRDs grafted with block copolymers, where the second block is composed of NIPAAm units, remained stable in an aqueous dispersion for at least 8 days at room temperature; while at temperatures above their first response temperature, e.g., at 37 °C, the nanohybrids self-assemble to form aggregates with a D_h < 200 nm without precipitating. In contrast, GNRDs grafted with block copolymers, where the second block is composed of NIPAM and BA units, already self-assembled at room temperature as the first response temperature is at 16 °C. At 37 °C, these nanohybrids show agglomeration between polymer aggregates leading to a partial precipitation, which strongly depends on the BA content (e.g., 9 mol% led to precipitation while 6 mol% did not) suggesting that the second transition temperature shifts to lower values. The photo-thermal effect remained active when the investigated block copolymers@GNRDs nanohybrids were irradiated in the NIR region. After 16 min of NIR-irradiation (795 nm, 300 mW), the nanohybrids showed a temperature increment of the systems in the range from 12 to 16 °C at a given concentration (except for the system containing 9 mol % of BA, which precipitated). All in all, the proposed block copolymer@GNRDs nanohybrids with two-temperature responses may be interesting candidates for a temperature-specific drug loading at their first response temperature and a specific release at their second response temperature under NIR-irradiation, which may open new possibilities for photo-thermal therapies.

Supplementary Materials: The following supporting information can be downloaded at: https://www.mdpi.com/article/10.3390/polym16233293/s1, Figure S1: ^{1}H-NMR spectrum (400 MHz) in $CDCl_3$ of chain transfer agent (CTA); Figure S2: ^{1}H-NMR spectra (400 MHz) in $CDCl_3$ of the P(DEGMA-*co*-OEGMA) copolymers: (a) P1; (b) P3; (c) P4; Figure S3: Normalized SEC traces (RI detector) for P(DEGMA-*co*-OEGMA) macro-CTA´s; Figure S4: ^{1}H-NMR spectra (400 MHz) in $CDCl_3$ of the block copolymers: (a) P1-2; (b) P1-3; (c) P3-2; (d) P3-3; Figure S5: Normalized SEC traces (RI detector) in DMF for P(DEGMA-*co*-OEGMA)-*b*-PNIPAAm and P(DEGMA-*co*-OEGMA)-*b*-P(NIPAAm-*co*-BA) block copolymers; Figure S6: DSC thermograms of the P(DEGMA-*co*-OEGMA)-*b*-PNIPAAm (P2-2) and P(DEGMA-*co*-OEGMA)-*b*-P(NIPAAm-*co*-BA) (P2-3) block copolymers; Figure S7: DSC thermograms of the block copolymers; (a) P1-2, (b) P3-2; Figure S8: DSC thermograms of the block copolymer P3-3; Figure S9: Determination of T_{cp} of block copolymers (1 mg mL^{-1}); a) P(DEGMA-*co*-OEGMA)$_{47\%}$-*b*-PNIPAAm$_{53\%}$, (b) P(DEGMA-*co*-OEGMA)$_{45\%}$-*b*-PNIPAAm$_{55\%}$; Figure S10: Determination of T_{cp} of block copolymers (1 mg mL-1); (c) P(DEGMA-*co*-OEGMA)$_{50\%}$-*b*-P(NIPAAm$_{44\%}$-*co*-BA$_{6\%}$), (d) P(DEGMA-*co*-OEGMA)$_{49\%}$-*b*-P(NIPAAm$_{40\%}$-*co*-BA$_{11\%}$); Figure S11: UV-Vis spectra (c = 1 mg mL^{-1} in ethanol) of the block copolymers and thiol-terminated block copolymers; Figure S12: Optical stability with time: (a) P(DEGMA-*co*-OEGMA)$_{47\%}$-*b*-PNIPAAm$_{53\%}$@GNRD; (b) P(DEGMA-*co*-OEGMA)$_{45\%}$-*b*-PNIPAAm$_{55\%}$@GNRDs; (c) P(DEGMA-*co*-OEGMA)$_{48\%}$-*b*-P(NIPAAm$_{43\%}$-*co*-BA$_{9\%}$)@GNRDs; (d) P(DEGMA-*co*-OEGMA)$_{50\%}$-*b*-P(NIPAAm$_{44\%}$-*co*-BA$_{6\%}$)@GNRDs; Figure S13: TEM micrographs of the nanohybrids: (a) P(DEGMA-*co*-OEGMA)$_{45\%}$-*b*-PNIPAAm$_{55\%}$@GNRDs; (b) P(DEGMA-*co*-OEGMA)$_{50\%}$-*b*-P(NIPAAm$_{44\%}$-*co*-BA$_{6\%}$)@GNRDs; Figure S14: UV-Vis absorption spectra at 25, 37 and 40 °C for nanohybrids: (a) P(DEGMA-*co*-OEGMA)$_{47\%}$-*b*-PNIPAAm$_{53\%}$@GNRDs, (b) P(DEGMA-*co*-OEGMA)$_{45\%}$-*b*-PNIPAAm$_{55\%}$@GNRDs; Figure S15: TGA-thermograms of the block copolymers; Table S1: Aggregation of sample P(DEGMA-*co*-OEGMA)$_{37\%}$-*b*-PNIPAAm$_{63\%}$@GNRDs at different temperatures; Table S2: Aggregation data of block copolymers@GNRDs at 37 °C.

Author Contributions: Investigation, Methodology, Writing—original draft, J.E.M.-C. Conceptualization, Formal analysis, Funding acquisition, Project Administration, Supervision, Writing—review and editing, A.L.-C. Validation, Formal Analysis, Supervision Writing—review and editing, C.G.-S. Supervision, Writing—review and editing, E.R.M. All authors have read and agreed to the published version of the manuscript.

Funding: This investigation was funded by the Tecnológico Nacional de Mexico under grant TNM-19652.24-P. J.E.M.-C. thanks CONAHCYT Mexico for a scholarship. C.G.-S. thanks to the financial support of the Deutsche Forschungsgemeinschaft (DFG, Germany) Collaborative Research Center 1278 "PolyTarget" (project number 316213987; projects B02 and Z01).

Institutional Review Board Statement: Not applicable.

Data Availability Statement: Data are available upon reasonable request.

Acknowledgments: The author thanks the following people for their contribution to the project: Valentin Miranda-Salazar (IT-Tijuana) for NMR measurements, Gabriela Guzmán (LNMA-CICESE) for TEM micrographs, Miguel Angel González-Mandujano (CICESE) and Dannareli Barrón-Ortiz (CICESE) for NIR-irradiation analysis.

Conflicts of Interest: The authors declare no conflicts of interest.

References

1. Nurakhmetova, Z.A.; Azhkeyeva, A.N.; Klassen, I.A.; Tatykhanova, G.S. Synthesis and Stabilization of Gold Nanoparticles Using Water-Soluble Synthetic and Natural Polymers. *Polymers* **2020**, *12*, 2625. [CrossRef] [PubMed]
2. Kesharwani, P.; Ma, R.; Sang, L.; Fatima, M.; Sheikh, A.; Abourehab, M.A.S.; Gupta, N.; Chen, Z.S.; Zhou, Y. Gold Nanoparticles and Gold Nanorods in the Landscape of Cancer Therapy. *Mol. Cancer* **2023**, *22*, 98. [CrossRef] [PubMed]

3. Mackey, M.A.; Ali, M.R.K.; Austin, L.A.; Near, R.D.; El-Sayed, M.A. The Most Effective Gold Nanorod Size for Plasmonic Photothermal Therapy: Theory and in Vitro Experiments. *J. Phys. Chem. B* **2014**, *118*, 1319–1326. [CrossRef] [PubMed]
4. Liao, S.; Yue, W.; Cai, S.; Tang, Q.; Lu, W.; Huang, L.; Qi, T.; Liao, J. Improvement of Gold Nanorods in Photothermal Therapy: Recent Progress and Perspective. *Front. Pharmacol.* **2021**, *12*, 664123. [CrossRef] [PubMed]
5. Kermanshahian, K.; Yadegar, A.; Ghourchian, H. Gold Nanorods Etching as a Powerful Signaling Process for Plasmonic Multicolorimetric Chemo-/Biosensors: Strategies and Applications. *Coord. Chem. Rev.* **2021**, *442*, 213934. [CrossRef]
6. Abbasian, M.; Mahmoodzadeh, F.; Salehi, R.; Amirshaghaghi, A. Chemo-Photothermal Therapy of Cancer Cells Using Gold Nanorod-Cored Stimuli-Responsive Triblock Copolymer. *New J. Chem.* **2017**, *41*, 12777–12788. [CrossRef]
7. Liu, L.Z.; Sun, X.Y.; Yan, Z.Y.; Ye, B.F. NIR Responsive AuNR/PNIPAM/PEGDA Inverse Opal Hydrogel Microcarriers for Controllable Drug Delivery. *New J. Chem.* **2021**, *45*, 7893–7899. [CrossRef]
8. Kwon, Y.; Choi, Y.; Jang, J.; Yoon, S.; Choi, J. Nir Laser-Responsive Pnipam and Gold Nanorod Composites for the Engineering of Thermally Reactive Drug Delivery Nanomedicine. *Pharmaceutics* **2020**, *12*, 204. [CrossRef]
9. Bao, Y.; Oluwafemi, A. Recent Advances in Surface Modified Gold Nanorods and Their Improved Sensing Performance. *Chem. Commun.* **2023**, *60*, 469–481. [CrossRef]
10. DiazDuarte-Rodriguez, M.; Cortez-Lemus, N.A.; Licea-Claverie, A.; Licea-Rodriguez, J.; Méndez, E.R. Dual Responsive Polymersomes for Gold Nanorod and Doxorubicin Encapsulation: Nanomaterials with Potential Use as Smart Drug Delivery Systems. *Polymers* **2019**, *11*, 939. [CrossRef]
11. Ghamkhari, A.; Mahmoodzadeh, F.; Ghorbani, M.; Hamishehkar, H. A Novel Gold Nanorods Coated by Stimuli-Responsive ABC Triblock Copolymer for Chemotherapy of Solid Tumors. *Eur. Polym. J.* **2019**, *115*, 313–324. [CrossRef]
12. Dhara, M. Nanohybrid Materials Using Gold Nanoparticles and RAFT-Synthesized Polymers for Biomedical Applications. *J. Macromol. Sci. Part A Pure Appl. Chem.* **2023**, *60*, 841–855. [CrossRef]
13. Perrier, S. 50th Anniversary Perspective: RAFT Polymerization—A User Guide. *Macromolecules* **2017**, *50*, 7433–7447. [CrossRef]
14. Moad, G.; Rizzardo, E.; Thang, S.H. End-Functional Polymers, Thiocarbonylthio Group Removal/Transformation and Reversible Addition-Fragmentation-Chain Transfer (RAFT) Polymerization. *Polym. Int.* **2011**, *60*, 9–25. [CrossRef]
15. Li, C.; Wang, C.; Ji, Z.; Jiang, N.; Lin, W.; Li, D. Synthesis of Thiol-Terminated Thermoresponsive Polymers and Their Enhancement Effect on Optical Limiting Property of Gold Nanoparticles. *Eur. Polym. J.* **2019**, *113*, 404–410. [CrossRef]
16. Han, F.; Armstrong, T.; Andres-Arroyo, A.; Bennett, D.; Soeriyadi, A.; Chamazketi, A.A.; Bakthavathsalam, P.; Tilley, R.D.; Gooding, J.J.; Reece, P.J. Optical tweezers-based characterisation of gold core–satellite plasmonic nano-assemblies incorporating thermo-responsive polymers. *Nanoscale* **2020**, *12*, 1680–1687. [CrossRef]
17. Pietsch, C.; Mansfeld, U.; Guerrero-Sanchez, C.; Hoeppener, S.; Vollrath, A.; Wagner, M.; Hoogenboom, R.; Saubern, S.; Thang, S.H.; Becer, C.R.; et al. Thermo-Induced Self-Assembly of Responsive Poly(DMAEMA-*b*-DEGMA) Block Copolymers into Multi- and Unilamellar Vesicles. *Macromolecules* **2012**, *45*, 9292–9302. [CrossRef]
18. Vagias, A.; Papagiannopoulos, A.; Kreuzer, L.P.; Giaouzi, D.; Busch, S.; Pispas, S.; Müller-Buschbaum, P. Effects of Polymer Block Length Asymmetry and Temperature on the Nanoscale Morphology of Thermoresponsive Double Hydrophilic Block Copolymers in Aqueous Solutions. *Macromolecules* **2021**, *54*, 7298–7313. [CrossRef]
19. Hu, N.; Mi, L.; Metwalli, E.; Bießmann, L.; Herold, C.; Cubitt, R.; Zhong, Q.; Müller-Buschbaum, P. Effect of Thermal Stimulus on Kinetic Rehydration of Thermoresponsive Poly(Diethylene Glycol Monomethyl Ether Methacrylate)-*Block*-Poly(Poly(Ethylene Glycol) Methyl Ether Methacrylate) Thin Films Probed by In Situ Neutron Reflectivity. *Langmuir* **2022**, *38*, 8094–8103. [CrossRef]
20. Shao, G.; Liu, Y.; Cao, R.; Han, G.; Yuan, B.; Zhang, W. Thermo-Responsive Block Copolymers: Assembly and Application. *Polym. Chem.* **2023**, *14*, 1863–1880. [CrossRef]
21. Philipp, M.; Kyriakos, K.; Silvi, L.; Lohstroh, W.; Petry, W.; Krüger, J.K.; Papadakis, C.M.; Müller-Buschbaum, P. From Molecular Dehydration to Excess Volumes of Phase-Separating PNIPAM Solutions. *J. Phys. Chem. B* **2014**, *118*, 4253–4260. [CrossRef] [PubMed]
22. Chen, S.; Wang, K.; Zhang, W.A. New Thermoresponsive Polymer of Poly(*N*-Acryloylsarcosine Methyl Ester) with a Tunable LCST. *Polym. Chem.* **2017**, *8*, 3090–3101. [CrossRef]
23. Suzuki, N.; Koyama, S.; Koike, R.; Ebara, N.; Arai, R.; Takeoka, Y.; Rikukawa, M.; Tsai, F.Y. Palladium-Catalyzed Mizoroki–Heck and Copper-Free Sonogashira Coupling Reactions in Water Using Thermoresponsive Polymer Micelles. *Polymers* **2021**, *13*, 2717. [CrossRef] [PubMed]
24. Hinton, T.M.; Guerrero-Sanchez, C.; Graham, J.E.; Le, T.; Muir, B.W.; Shi, S.; Tizard, M.L.V.; Gunatillake, P.A.; McLean, K.M.; Thang, S.H. The effect of RAFT-derived cationic block copolymer structure on gene silencing efficiency. *Biomaterials* **2012**, *33*, 7631–7642. [CrossRef]
25. Montoya-Villegas, K.A.; Licea-Claveríe, Á.; Zapata-González, I.; Gómez, E.; Ramírez-Jiménez, A. The Effect in the RAFT Polymerization of Two Oligo(Ethylene Glycol) Methacrylates When the CTA 4-Cyano-4-(Propylthiocarbonothioylthio) Pentanoic Acid Is Auto-Hydrolyzed to Its Corresponding Amide. *J. Polym. Res.* **2019**, *26*, 71. [CrossRef]
26. Márquez-Castro, J.E.; Licea-Claverie, A.; Licea-Rodriguez, J.; Quiroga-Sánchez, L.P.; Méndez, E.R. Surface Grafted Gold Nanorods (GNRDs) Using Thermosensitive Copolymers with Various Transition Temperatures: Nanomaterials with Potential Application for Photothermal Therapy. *Eur. Polym. J.* **2023**, *197*, 112341. [CrossRef]

27. Liu, J.; Detrembleur, C.; De Pauw-Gillet, M.C.; Mornet, S.; Duguet, E.; Jérôme, C. Gold Nanorods Coated with a Thermo-Responsive Poly(Ethylene Glycol)-b-Poly(N-Vinylcaprolactam) Corona as Drug Delivery Systems for Remotely near Infrared-Triggered Release. *Polym. Chem.* **2014**, *5*, 799–813. [CrossRef]
28. Selianitis, D.; Pispas, S. PDEGMA-b-PDIPAEMA Copolymers via RAFT Polymerization and Their PH and Thermoresponsive Schizophrenic Self-Assembly in Aqueous Media. *J. Polym. Sci.* **2020**, *58*, 1867–1880. [CrossRef]
29. Wang, H.; Ullah, A. Synthesis and Evaluation of Thermoresponsive Renewable Lipid-Based Block Copolymers for Drug Delivery. *Polymers* **2022**, *14*, 3436. [CrossRef]
30. Ramírez-Jiménez, A.; Montoya-Villegas, K.A.; Licea-Claverie, A.; Gonzalez-Ayon, M.A. Tunable Thermo-Responsive Copolymers from DEGMA and OEGMA Synthesized by RAFT Polymerization and the Effect of the Concentration and Saline Phosphate Buffer on Its Phase Transition. *Polymers* **2019**, *11*, 1657. [CrossRef]
31. Weaver, L.G.; Stockmann, R.; Postma, A.; Thang, S.H. Multi-responsive (diethylene glycol) methyl ether methacrylate (DEGMA)-based copolymer systems. *RSC Adv.* **2016**, *6*, 90923–90933. [CrossRef]
32. Kasprów, M.; Machnik, J.; Otulakowski, L.; Dworak, A.; Trzebicka, B. Thermoresponsive P(HEMA-co-OEGMA) copolymers: Synthesis, characteristics and solution behavior. *RSC Adv.* **2019**, *9*, 40966–40974. [CrossRef] [PubMed]
33. Szweda, D.; Szweda, R.; Dworak, A.; Trzebicka, B. Thermoresponsive Poly[Oligo(Ethylene Glycol) Methacrylate]s and Their Bioconjugates—Synthesis and Solution Behavior. *Polimery* **2017**, *62*, 298–310. [CrossRef]
34. Tymetska, S.; Shymborska, Y.; Stetsyshyn, Y.; Budkowski, A.; Bernasik, A.; Awsiuk, K.; Donchak, V.; Raczkowska, J. Thermoresponsive Smart Copolymer Coatings Based on P(NIPAM-*co*-HEMA) and P(OEGMA-*co*-HEMA) Brushes for Regenerative Medicine. *ACS Biomater. Sci. Eng.* **2023**, *9*, 6256–6272. [CrossRef] [PubMed]
35. López, Á.M.; Tirado-Guizar, A.; Licea-Claverie, A.; Ramírez-Jiménez, A. Thermo and PH-Responsive Poly(DEGMA-Co-OEGMA)-b-Poly(DEAEM) Synthesized by RAFT Polymerization and Its Self-Assembly Study. *Macromol. Res.* **2022**, *30*, 917–929. [CrossRef]
36. Cortez-Lemus, N.; Licea-Claverie, A. Poly(N-vinylcaprolactam), a comprehensive review on a thermoresponsive polymer becoming popular. *Prog. Polym. Sci.* **2016**, *53*, 1–51. [CrossRef]
37. Yañez-Macias, R.; Alvarez-Moises, I.; Perevyazko, I.; Lezov, A.; Guerrero-Santos, R.; Schubert, U.S.; Guerrero-Sanchez, C. Effect of the Degree of Quaternization and Molar Mass on the Cloud Point of Poly[2-(dimethylamino)ethyl methacrylate] Aqueous Solutions: A Systematic Investigation. *Macromol. Chem. Phys.* **2017**, *218*, 1700065. [CrossRef]
38. Škvarla, J.; Zedník, J.; Šlouf, M.; Pispas, S.; Štěpánek, M. Poly(N-Isopropyl Acrylamide)-Block-Poly(n-Butyl Acrylate) Thermoresponsive Amphiphilic Copolymers: Synthesis, Characterization and Self-Assembly Behavior in Aqueous Solutions. *Eur. Polym. J.* **2014**, *61*, 124–132. [CrossRef]
39. Yin, F.; Behra, J.S.; Beija, M.; Brûlet, A.; Fitremann, J.; Payré, B.; Gineste, S.; Destarac, M.; Lauth-de Viguerie, N.; Marty, J.D. Effect of the microstructure of n-butyl acrylate/N-isopropylacrylamide copolymers on their thermo-responsiveness, self-organization and gel properties in water. *J. Colloid Interface Sci.* **2020**, *578*, 685–697. [CrossRef]
40. Yin, F.; Laborie, P.; Lonetti, B.; Gineste, S.; Coppel, Y.; Lauth-De Viguerie, N.; Marty, J.D. Dual Thermo- and PH-Responsive Block Copolymer of Poly(N-Isopropylacrylamide)-Block-Poly(N,N-Diethylamino Ethyl Acrylamide): Synthesis, Characterization, Phase Transition, and Self-Assembly Behavior in Aqueous Solution. *Macromolecules* **2023**, *56*, 3703–3720. [CrossRef]
41. Stetefeld, J.; McKenna, S.A.; Patel, T.R. Dynamic light scattering: A practical guide and applications in biomedical sciences. *Biophys. Rev.* **2016**, *8*, 409–427. [CrossRef] [PubMed]
42. Li, Q.; Gao, C.; Li, S.; Huo, F.; Zhang, W. Doubly Thermo-Responsive ABC Triblock Copolymer Nanoparticles Prepared through Dispersion RAFT Polymerization. *Polym. Chem.* **2014**, *5*, 2961–2972. [CrossRef]
43. Kim, J.; Jung, H.Y.; Park, M.J. End-Group Chemistry and Junction Chemistry in Polymer Science: Past, Present, and Future. *Macromolecules* **2020**, *53*, 746–763. [CrossRef]
44. Scherger, M.; Räder, H.J.; Nuhn, L. Self-Immolative RAFT-Polymer End Group Modification. *Macromol. Rapid Commun.* **2021**, *42*, 2000752. [CrossRef]
45. Zygadlo, K.; Liu, C.H.; Bernardo, E.R.; Ai, H.; Nieh, M.P.; Hanson, L.A. Correlating Structural Changes in Thermoresponsive Hydrogels to the Optical Response of Embedded Plasmonic Nanoparticles. *Nanoscale Adv.* **2023**, *6*, 146–154. [CrossRef]
46. Rohleder, D.; Vana, P. Near-Infrared-Triggered Photothermal Aggregation of Polymer-Grafted Gold Nanorods in a Simulated Blood Fluid. *Biomacromolecules* **2021**, *22*, 1614–1624. [CrossRef]
47. Liu, H.; Pierre-Pierre, N.; Huo, Q. Dynamic light scattering for gold nanorod size characterization and study of nanorod–protein interactions. *Gold Bull.* **2012**, *45*, 187–195. [CrossRef]

polymers

MDPI

Article

Impact of Combined Thermo- and Photo-Oxidation on the Physicochemical Properties of Oxo-Biodegradable Low-Density Polyethylene Films

M. F. Rojas-Trejo [1], A. Valadez-Gonzalez [1,*], L. Veleva [2,*], R. Benavides [3], M. T. Rodriguez-Hernandez [3] and M. V. Moreno-Chulim [1]

1 Unidad de Materiales, Centro de Investigación Científica de Yucatán, Calle 43 No. 130 Col. Chuburná de Hidalgo, Merida 97205, Mexico; mafer.rojas.trejo@gmail.com (M.F.R.-T.); veromor@cicy.mx (M.V.M.-C.)
2 Applied Physics Department, Center for Research and Advanced Studies (Cinvestav-Mérida), Merida 97310, Mexico
3 Centro de Investigación en Química Aplicada (CIQA), Blvd. Enrique Reyna Hermosillo 140, Saltillo 25294, Mexico; roberto.benavides@ciqa.edu.mx (R.B.); teresa.rodriguez@ciqa.mx (M.T.R.-H.)
* Correspondence: avaladez@cicy.mx (A.V.-G.); veleva@cinvestav.mx (L.V.)

Abstract: This research addresses the study of the combined effect of two abiotic treatments, a thermo-oxidative treatment followed by a photo-oxidative treatment with ultraviolet light, on the physicochemical properties of commercially available low-density polyethylene films with an oxo-degradant additive (OXOLDPE) and without (LDPE). The change in the oxidized film properties was characterized using FTIR, XRD, TGA, GPC, and SEM analytical techniques. The results indicated that the increment in carbonyl index (CI) and crystallinity percentage (X_{XRD}) was higher for those films that received the combined oxidative treatments compared with those that received only one of them, thermo- or photo-oxidative treatment. Moreover, the combined oxidative treatments produced more ester and carboxylic groups on the degradation products than the other single treatments. An analysis of variance (ANOVA) was carried out, and a synergistic effect was observed between the thermo- and photo-oxidative treatments for both ester and carboxylic degradation products. TGA results revealed that the loss of thermal stability in the films was more significant after their exposure to the combined thermo- and photo-oxidative treatments compared with those which received only one. The GPC results showed that the combined oxidative treatment is necessary to decrease the Mz and M_{z+1} average molecular weight of degraded films containing an oxo-degradant additive to the same extent as MW and Mn. The SEM surface appearance of the films changed more drastically after their exposure to the combined thermo- and photo-oxidative treatments, and they seemed to erode with the presence of inorganic fillers ($CaCO_3$). These results suggest that the combined oxidative treatments produced degradation products with lower molecular weight and greater content of ester and carboxylic groups that should enhance its environmental biodegradability.

Keywords: oxo-degradable LDPE; thermo- and photo-oxidative treatments; abiotic degradation

1. Introduction

The accumulation of plastics in the environment continues to increase at ever higher rates, mainly because they are materials with very low biodegradation rates. In contrast, the manufacture of plastic materials increases year after year. Polyolefins represent a significant fraction of the synthetic plastics manufactured annually, and low-density polyethylene (LDPE), in particular, stands out among them because most of products composed of it are single-use [1,2]. To increment the biodegradability of plastics once their useful life has ended, additives based on stearates of transition metals called oxo-degradants are added during their manufacture. Incorporating oxo-degradant additives can accelerate the degradation process of LDPE since transition metals can act as catalysts for photo- and thermal-degradation processes. The catalytic degradation of polyethylene in the presence of transition metals has been attributed to its ability to generate free radicals on its surface. These radicals later react with oxygen to generate low-molecular-weight oxidation products as oxygen is introduced into the carbon chain as hydroxyls, carbonyls, and peroxides [3–5]. The stearates of Fe^{3+}, Mn^{2+}, or Co^{2+} are the most commonly used oxo-degradants, while Ca^{2+} and Cu^{2+} stearates are used with less frequency [3–8]. These materials are called oxo-biodegradable, but there is currently controversy over whether oxo-biodegradable plastics are actually biodegradable and usable as a carbon source by microorganisms. Many studies confirm that oxo-biodegradable plastics are biodegradable [9–14], while many others consider that they are not [15–20]. This discrepancy has even led the EU to prohibit the use of such materials [21]. In addition, several studies indicate that oxo-biodegradable plastics tend to generate a more significant amount of microplastics when they degrade in the environment compared to conventional plastics because the formation of these microparticles is closely related to the cleavage of the polymeric chains, the chemical structure of the degradation products formed and the increase in the percentage of crystallinity [22–24]. It is known that biodegradation of a synthetic polymer is a process that consists of two stages, biotic and abiotic degradation, that lead to the formation of low molecular weight fragments bearing ester and carboxylic groups that can be used as a carbon source by microorganisms [9,11,12,25–30]. Likewise, it is known that the products of thermo-oxidative degradation or photo-oxidative degradation of polyethylene give rise to shorter polymer chains (decrease in molecular weight) bearing ester and carboxylic groups. The vast majority of research on the abiotic degradation of oxo-biodegradable polyethylene has separately studied the influence of thermo-oxidative degradation [31–39] or photo-oxidative degradation [40–46] on its physicochemical properties and/or its biodegradability. The results of these studies have shown that although the decrease in molecular weight and the presence of ester and carboxylic groups does indeed increase the rate and quantity of LDPE assimilated by microorganisms, these are still low. Moreover, what has been regularly observed is that microorganisms can only metabolize short oxidized molecular chains. These findings highlight how crucial the abiotic degradation stage is in the biodegradation of plastics, such as LDPE, with and without oxo-degradant additives.

This work aims to study the combined effect of thermo-oxidative and photo-oxidative degradation on the physicochemical properties of commercial LDPE films with and without an oxo-degradant additive.

For this purpose, commercial low-density polyethylene films with an oxo-degradant additive (OXOLDPE) and without one (LDPE) were subjected to different oxidative treatments: 90 days at 70 °C, 1000 h of UV irradiation, and subsequently a combination of both treatments. The FTIR, XRD, TGA, GPC, and SEM-EDS analytical techniques were used to assess the changes in the films' physical and chemical properties.

2. Materials and Methods

2.1. Materials

Commercial low-density polyethylene film (LDPE, 0.035 mm thickness) and an oxo-biodegradable LDPE film (OXOLDPE, 0.035 mm thickness) were obtained from a local producer (POLIMERIDA, YUC, MX). Specimens (100 mm × 150 mm × 0.035 mm) were cut, conditioned at 25 °C and 25% RH, and kept in sealed bags until exposure to experimental tests.

2.2. Experimental Tests

2.2.1. Thermo-Oxidative Test

The samples in triplicate were exposed to a thermo-oxidative environment in a forced air convection oven (EQUATHERM, Seattle, WA, USA) at 70 °C for 90 days. Then, they were withdrawn from the oven and conditioned at 25 °C and 25% RH and kept in sealed bags until exposure to experimental tests.

2.2.2. Photo-Oxidative Test

The samples in triplicate were exposed to an accelerated UV aging camera (QUV-SPRAY, Westlake, OH, USA) and continuously irradiated with UVB-313 lamps (0.71 W/m^2) at 60 °C for 1000 h. Subsequently, they were conditioned at 25 °C and 25% RH, and stored in sealed bags in a desiccator for later analysis.

2.2.3. Combined Thermo- and Photo-Oxidative Test

In order to assess the combined effect of both oxidative tests, the tested samples were previously thermo-oxidized (at 70 °C for 90 days) and posteriorly irradiated (1000 h, at 0.71 W/m^2 and 60 °C) with a UV aging camera. After that, the specimens were conditioned at 25 °C and 25% RH in a desiccator and stored in sealed bags for later analysis. The tested samples were coded as indicated in Table 1.

Table 1. Sample codes.

Sample Code	Description
LDPE	As-received LDPE film (without treatment)
LDPE-THERMO	LDPE film thermo-oxidized
LDPE-PHOTO	LDPE film photo-oxidized
LDPE-THERMO+PHOTO	LDPE thermo-oxidized and then photo-oxidized
OXOLDPE	As-received Oxo-biodegradable LDPE film (without treatment)
OXOLDPE-THERMO	Oxo-biodegradable LDPE film thermo-oxidized
OXOLDPE-PHOTO	Oxo-biodegradable LDPE film photo-oxidized
OXOLDPE-THERMO+PHOTO	Oxobiodegradable LDPE thermo-oxidized and then photo-oxidized

2.3. Physicochemical Characterization

2.3.1. Fourier Transformed Infrared Spectroscopy (FTIR)

The FTIR characterization was performed using a Nicolet 8700 Thermo Scientific spectrometer (Thermo Electron Scientific® Instruments LLCD, Madison, WI, USA). The spectra were recorded with a resolution of 2 cm^{-1} and 100 scans in a wavenumber range from 4000 cm^{-1} to 500 cm^{-1}. The spectra obtained were normalized using the band at 2020 cm^{-1} as internal standard before they were analyzed [47,48]. The overall carbonyl index (CI) was calculated (Equation (1)), according to the method proposed by Almond et al. [49]. The interval of the FTIR spectra of carbonyl peaks from 1850 to 1650 cm^{-1} was integrated and normalized with the methylene (CH_2) absorption peak integrated area in the range of 1500–1420 cm^{-1}.

$$Carbonyl\ Index\ (IC) = \frac{Area\ of\ the\ arbsorption\ peaks\ of\ carbonyls\,(1850 - 1650\ \mathrm{cm}^{-1})}{Area\ of\ the\ methylene\ absorption\ peak\ \ (1500 - 1420\ \mathrm{cm}^{-1})} \tag{1}$$

The carbonyl region was deconvoluted using the Lorentzian non-linear least squares fitting procedure (Origin Pro 9, OriginLab Corporation, Northampton, MA, USA) to calculate the esters (COOR) and carboxylic acids (COOH) contribution. The ester and carboxylic acid indexes were calculated using the peaks at 1740 cm^{-1} and 1710 cm^{-1}, normalized with the methylene absorption peak, respectively, like CI. The unsaturation (C=C) index was calculated using the peak at 990 cm^{-1} and also normalized with the methylene absorption peak, like CI.

2.3.2. X-Ray Diffraction (XRD)

A Bruker D-8 Advance diffractometer (Billerica, MA, USA) operating at 40 kV and an intensity of 30 mA (Kα of Cu at λ = 1.5406 Å, 2θ = 15–25° (0.02° step size)) and a step time of 0.5 s was used to obtain the XRD spectra of the studied samples. The percentage of crystallinity was calculated using the following relation (2):

$$X_{XRD} = \frac{A_c^{110} + A_c^{200}}{A_a + A_c^{110} + A_c^{200}} \times 100 \quad (2)$$

where A*a*, Ac^{100}, and Ac^{200} are the areas under the amorphous halo, the 100 and the 200 reflections, respectively. The mean crystallite size (L_{110}), corresponding to 110 cell planes, was obtained by XRD data and the Scherrer Equation (3) [44].

$$L_{110} = \frac{K\lambda}{\beta\cos\theta} \quad (3)$$

where L_{110} represents the mean dimension of the crystal normal to the corresponding 110 plane, K is a first-order constant, λ represents the wavelength, 2θ is the scattering angle, and β is the half-height of the scattering peak. In this study, the parameters λ = 1.5418 Å, 2θ = 21.5° were used.

2.3.3. Thermogravimetric Analysis (TGA)

The thermal stability of the studied sample films was assessed using a TGA-7 Perkin-Elmer Inc. (Waltham, MA, USA). The sample (10 mg) was heated at a constant rate of 5 °C/min from 30 °C to 600 °C, under a nitrogen atmosphere.

2.3.4. Gel Permeation Chromatography (GPC)

The molecular weight distributions (MWD) of LDPE and OXOLDPE films were evaluated before and after the thermo- and photo-oxidative treatments using an Agilent PL-GPC 220 High Temperature Chromatograph (Santa Clara, CA, USA) at 140 °C. For this, 20 mg of the sample was weighed and dissolved in 15 mL of 1,2,4-Trichlorobenzene for 1 h at 170 °C. The solution was then filtered through stainless steel filters (0.5 microns).

2.3.5. Scanning Electron Microscopy (SEM-EDS)

The surface morphology changes and elemental analysis of samples before and after being tested were evaluated with SEM-EDS (JEOL Model JSM- 7600F (Akishima, Japan), at 15 kV). The micrographs were obtained at a magnification of x500. The specimens were previously coated with a thin layer of gold/palladium.

3. Results and Discussions

3.1. Fourier Transformed Infrared Spectroscopy (FTIR)

The effect of thermo-oxidative and photo-oxidative treatments, and the combination of both, on the chemical properties of the studied films can be observed in the FTIR spectra (Figure 1). According to the absorbance intensity of the peaks, the degradation of the LDPE and OXOLDPE films due to the thermo-oxidative treatment is lower than that induced by the photo-oxidative treatment and much lower than the combined treatment. This is evident from the carbonyl group region (C=O) peaks between 1600–1800 cm^{-1} and in the hydroxyl group region (–OH) between 3200–3600 cm^{-1}. The formation of these groups has been reported as a consequence of the oxidative degradation of polyolefins [50–54]. Likewise, the absorbance bands of the OXOLDPE films seem to be higher than those of the LDPE. These facts suggest that films containing the oxo-degradant additive have experienced more significant oxidative degradation, which is consistent with what has been reported in the literature [3,8,15,18,31,38–40].

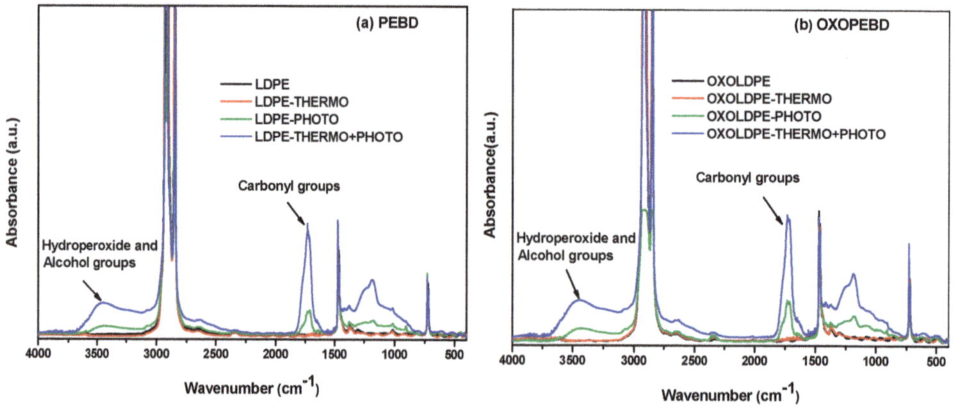

Figure 1. FTIR spectra of LDPE (**a**) and OXOLDPE (**b**) samples after different oxidative tests.

The CI values, calculated using Equation (1) for LDPE and OXOLDPE films due to the oxidative treatments, are presented in Figure 2. As can be observed, the presence of degradation products containing carbonyl groups increases in both types of films in the following order: (without treatment) < thermo < photo << thermo+photo. However, it should be noted that the content of these carbonyl groups is higher in the films containing the oxo-degrading additive (OXOLDPE). It is also evident that the increase in carbonyl group content due to the combined oxidative treatment is higher than expected from an additive behavior. These facts suggest that there is a synergistic effect between the thermo- and photo-oxidative treatments for the formation of degradation products containing carbonyl groups for both LDPE and OXOLDPE.

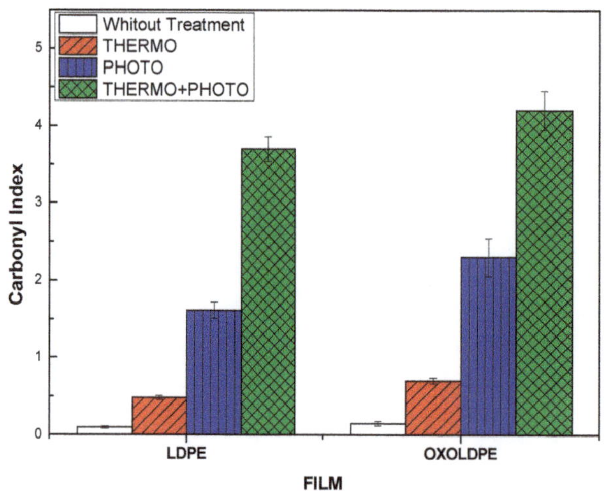

Figure 2. Carbonyl index (CI) of LDPE and OXOLDPE samples after different oxidative tests.

This behavior is interesting since it tells us that the combination of thermo- and photo-oxidative treatments seems to be more relevant than the presence of the oxo-degradant additive.

The Analysis of Variance (ANOVA) was conducted using the MINITAB™ 18 statistical software (Minitab, LLC., State College, PA, USA) to clarify this. In Figure 3a, the Pareto Chart shows that the single oxidative treatments, their combination, and the kind of film (LDPE or OXOLDPE) are all statistically significant ($p < 0.05$) in the formation of carbonyl groups, though the photo-oxidative treatment contributes the most to it. It should be noted that the contribution of the combined treatments is higher than the presence of the oxo-degradant additive.

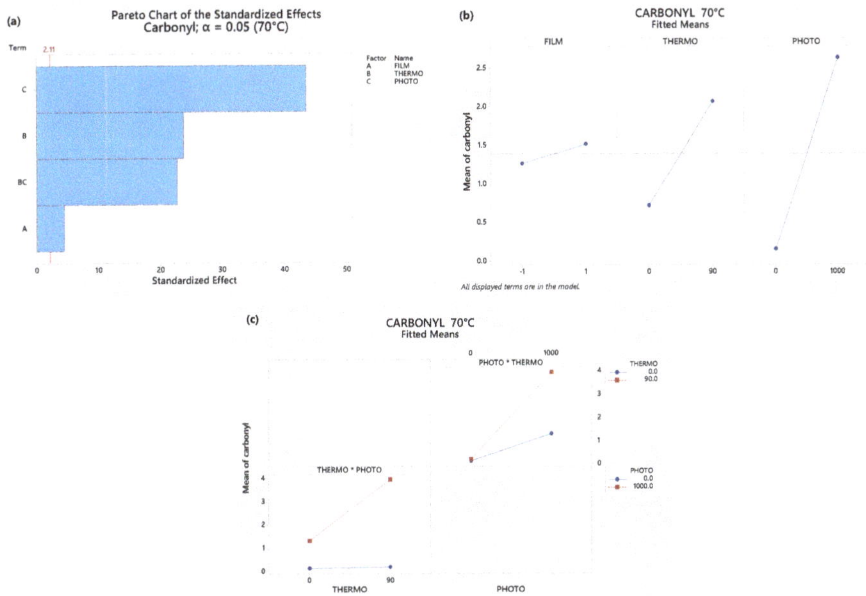

Figure 3. ANOVA results: (**a**) Pareto charts, (**b**) principal effects graph, and (**c**) interaction plot.

In Figure 3b, it can be observed that the carbonyl group formation is more sensitive to both single treatments than the kind of films, i.e., the treatment type contributes to a greater extent than the presence of the oxo-degradant additive. Figure 3c shows the interaction graphs of the thermo- and photo-oxidative treatments. It can be observed that the thermo-oxidative treatment drastically increases the effect of the photo-oxidative treatment on the formation of carbonyl groups. This behavior suggests that there is a synergistic effect between both oxidative treatments [55].

The carbonyl region was deconvoluted using the Lorentzian non-linear least squares fitting procedure (Origin Pro 9) to calculate the esters (COOR) and carboxylic acids (COOH) contribution. Their corresponding specific carbonyl group indexes was calculated as described in Section 2.3.1.

Figure 4 shows the corresponding carbonyl indexes for ester and carboxylic groups that bear the film's degradation products. It can be seen in Figure 4a that the ester group formation predominates in the degradation products for thermo-, photo- and the combined oxidative treatments for LDPE. On the other hand, for OXOLDPE, the formation of carboxylic groups is enhanced by the combined oxidative treatments, as can be seen in Figure 4b. The formation of ester groups and carboxylic groups during thermo-oxidative and photo-oxidative degradation in polyethylene with and without the presence of oxo-degrading additives has been reported by several researchers [7,11,33,39,52,56–60]. The results found in this work suggest that the combination of oxidative treatments in the presence of an oxo-degrading additive promotes the generation of carboxylic groups over that of ester groups.

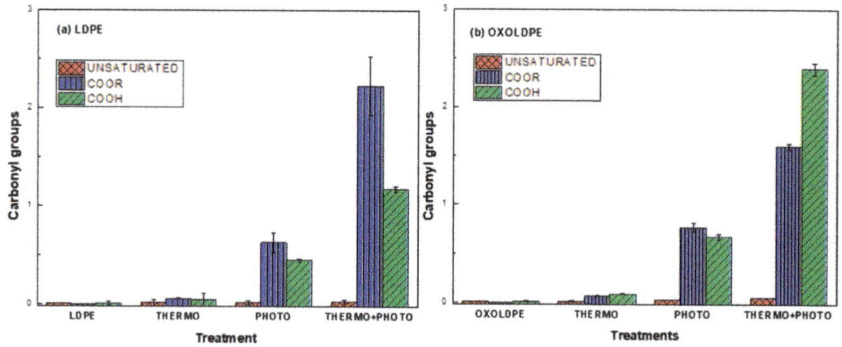

Figure 4. ANOVA analysis of ester and carboxylic group index of LDPE (**a**) and OXOLDPE (**b**) samples after different oxidative treatments.

The results of the ANOVA analysis, using the Minitab 18 software, for the carboxylic and ester carbonyl group indexes are shown in Figure Figure 5 and Figure 6, respectively.

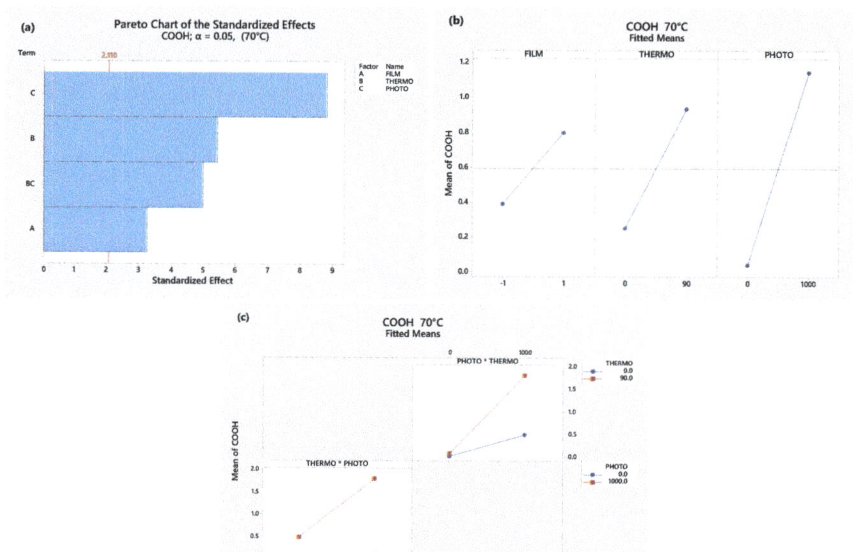

Figure 5. ANOVA analysis of the carboxylic acids (COOH) formed in OXOLDPE samples films after different oxidative treatments: (**a**) Pareto chart, (**b**) principal effects graph, and (**c**) interaction plot.

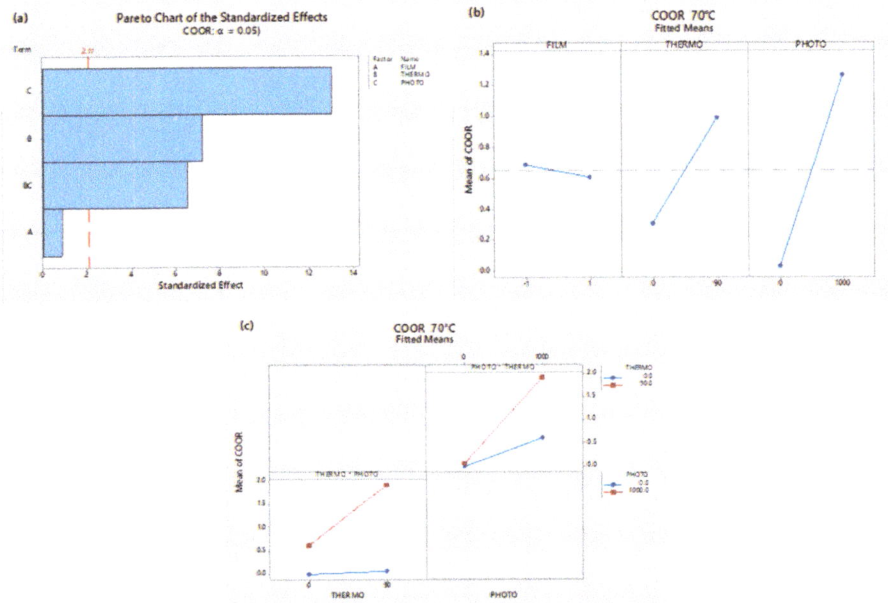

Figure 6. ANOVA analysis of esters (COOR) formed in OXOLDPE sample films after different oxidative treatments: (**a**) Pareto chart, (**b**) principal effects graph, and (**c**) interaction plot.

In Figure 5, it can be seen that thermo-, photo-, the combined oxidative treatments and the oxo-degradant additive (OXOLDPE) are statistically significant. In Figure 5b, it can be seen that the photo-oxidative treatment is more important than the thermo- one, and than the oxo-degradant additive. The synergy between both oxidative single treatments is evident in Figure 5c.

In Figure 6, it can be seen that the single oxidative treatments and their combination, but not the oxo-degradant additive, are statistically significant in the ester group formation. In Figure 6a, it can be seen that the photo-oxidative treatment contributes to a great extent compared to thermo-oxidative treatment in ester formation. In Figure 6c, it is evident that there is a synergistic effect between both single oxidative treatments.

3.2. X-Ray Diffraction (XRD)

Table 2 shows the crystallinity (X_{XRD}) percentage and crystal size (L_{110}) for LDPE and OXOLDPE films corresponding to the oxidative treatments. The results indicate that the thermo-oxidative treatment increased the crystallinity fraction by about 20% for both LDPE and OXOLDPE films compared with the films without treatment. On the other hand, the photo-oxidative treatment increased the percentage of LDPE crystallinity by 20%, and 60% for the OXOLDPE films. Likewise, the combined oxidative treatment increased the crystallinity fraction of the LDPE and OXOLDPE films by 40% and 75%, respectively. These facts suggest that the formation of ester and carboxylic acid groups, shown with the FTIR results, led to the scission of the polymeric chains [3,35,39,46]. Gedde et al. [61] considered that chain scission occurs essentially within the amorphous phase, favoring the release of short-chain segments. These segments, having greater mobility than the initial chains, would have the ability to migrate to the surface of the crystalline phase and participate in secondary crystallization (chemical crystallization).

Table 2. Crystallinity and crystal size of LDPE and OXOLDPE films based on XRD spectra analysis.

SamplesTreatment	X_{XRD} (%)		L_{110} (nm)	
	LDPE	OXOLDPE	LDPE	OXOLDPE
WITHOUT TREATMENT	54	45	19.5	18.7
THERMO-	65	54	22.1	21.7
PHOTO-	66	73	22.6	19.2
THERMO+PHOTO	75	79	21.3	20.5

As for the average size of the crystals (L110), there is not much difference between the results obtained for LDPE and OXOLDPE films. It seems that incorporating the oxo-degradant additive in LDPE does not affect the average size of the formed crystallites. As regards the effect of the oxidative treatments, it is evident that the sizes of the crystallites are very similar regardless of the type of oxidative treatment to which the films were subjected (Table 2). These facts suggested that a homogeneous crystallization process prevailed over time, in addition to forming mobile small chain fragments from the chain scission reaction, subjected to reorganization and re-crystallization, while the existing crystallites did not break apart with increasing time [62].

3.3. *Thermogravimetric Analysis (TGA)*

The thermograms for LDPE and OXOLDPE before and after the oxidative treatments are shown in Figure 7, where it can be observed how the thermal stability of the films decreases with the different oxidative treatments.

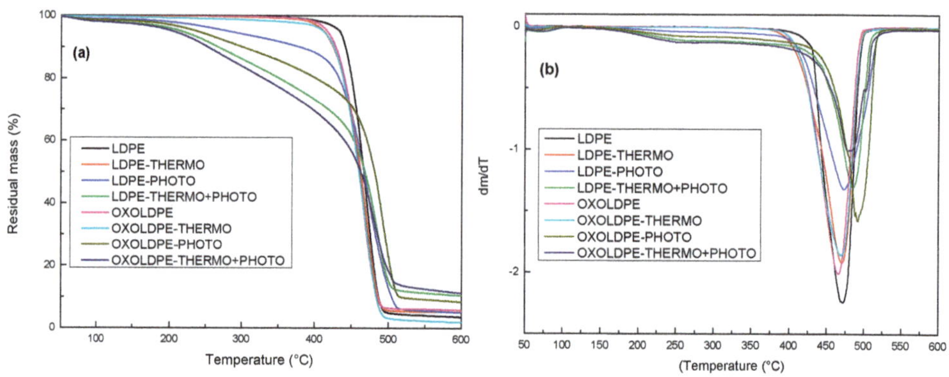

Figure 7. Change in thermal stability of LDPE and OXOLDPE film samples after different treatments: (**a**) residual mass (%) as a function of temperature and (**b**) mass loss rate dm/dT.

Figure 7a reveals that the mass loss of both LDPE and OXOLDPE films started at low temperatures, and after different oxidative treatments their thermal stability decreased because of the abiotic degradation process. For both films, the loss of thermal stability followed the order: thermo+photo>>photo>thermo. This behavior agrees with the FTIR results (Figure 1), which reveal a more significant formation of oxidized degradation products after the combined oxidative treatment for both films. Meanwhile, the mass loss rate as a function of temperature (Figure 7b) shows that the mass loss started around 160 °C for the films subjected to photo-oxidative treatment and those subjected to combined treatment (thermo+photo). This behavior suggests that the products of oxidative degradation have a much lower molecular weight than those formed after the thermo-oxidative treatment of the studied films [19,62–65]. Table 3 compares the temperatures at which 10% and 25% mass losses occurred ($T_{10\%}$, $T_{25\%}$) and the temperature (T_{max}) at which the mass loss rate (dm/dT) was at its maximum.

Table 3. Temperature for selected weight loss of studied LDPE and OXOLDPE film samples after oxidative treatments.

Sample and Treatment	$T_{10\%}$ (°C)	$T_{25\%}$ (°C)	T_{max} (°C) dm/dT (%)
LDPE	442	453	471/40
LDPE+THERMO	425	444	470/35
LDPE+PHOTO	379	441	472/43
LDPE+THERMO+PHOTO	268	387	485/31
OXOLDPE	429	445	465/40
OXOLDPE+THERMO	426	443	468/35
OXOLDPE+PHOTO	302	435	492/35
OXOLDPE+THERMO+PHOTO	254	366	482/33

The results show that the mass loss temperatures of 10% and 25% decrease for both LDPE and OXOLDPE films with the oxidative treatments in the following order: thermo+photo>photo>thermo. This is an indication that there is a decrease in the molecular weight of the films due to the oxidative treatments [4,64,65]. On the other hand, the behavior of T_{max} is different. For LDPE films, T_{max} remained constant after the thermo- and photo-treatments, and increases by about 14 °C after the combined treatment, probably due to the increase in the percentage of crystallinity of the films [62,66,67]. For OXOLDPE films, T_{max} increased by about twice (27 °C) after the photo-treatment with respect to the LDPE films. However, for the films that received the combined oxidative treatment, the increase in T_{max} was 17 °C, even though the percentage of crystallinity of these samples was much higher than those that received only the photo-treatment. These facts suggested that the degradation products formed are less thermally stable than those formed after the combined oxidative degradation process, compared to the photo-oxidative one.

3.4. Gel Permeation Chromatography (GPC)

Figure 8 shows the molecular weight distributions obtained by GPC of LDPE and OXOLDPE after being subjected to the different oxidative treatments.

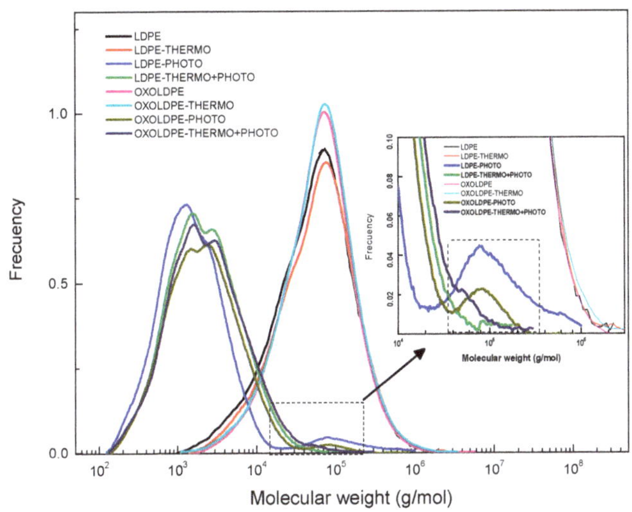

Figure 8. Comparison of the molecular weight distributions obtained by the analysis of the GPC of LDPE and OXOLDPE films subjected to different oxidative treatments.

This figure shows that the thermo-oxidative treatment did not modify significantly the molecular weight distribution (MWD) of any films. On the other hand, the shift towards the region of lower molecular weight is evident after the photo-oxidative and combined oxidative treatments of the films. Several researchers have published similar results, mainly after photo-oxidative treatments of LDPE films with and without oxo-degradant additives [4,12,19,31–33,53,54]. The inserted box in Figure 8 shows that the photo-oxidative treated films present a shoulder around 105 (g/mol). This shoulder practically disappears in those films that received the combined oxidative treatment. This finding is very interesting, since it indicates that the combined treatment decreases the average molecular weight of the LDPE and OXOLDPE films and drops the fraction of higher molecular weights. Table 4 summarizes the average molecular weights in number (M_n), weight (M_W), and viscosity (M_V) of the films, together with their polydispersity index (M_W/M_n), which is frequently used to characterize the MWD of polymeric matrices, as an indicator of its amplitude. Table 4 also incorporates the average molecular weights M_z and M_{z+1}, the values of which are more influenced by the existence of polymer chains of higher molecular weight, compared to the M_n and M_w average weights, calculated from the molecular weight distributions presented in Figure 8. From the point of view of processing and performance of the material during its useful life, the importance of the molecular averages M_z and M_{z+1} have been pointed out [67–70], but not from the point of view of the material degradation, once its useful life has expired. What is widely documented is that during its biodegradation, microorganisms are only able to use as substrate the fractions that contain short molecular chains, less than 5000 g/mol.

Table 4. Molecular weight averages of abiotic-treated LDPE and OXOLDPE films.

Molecular Weight Averages	M_n	M_W	M_Z	M_{Z+1}	M_V	M_W/M_n
LDPE	24,323	99,048	363,924	1,156,245	83,376	4.1
LDPE-THERMO	23,429	103,476	348,152	971,778	87,195	4.4
LDPE-PHOTO	775	6543	30,668	278,466	3860	8.4
LDPE-THERMO+PHOTO	1000	3945	16,574	46,797	3222	3.9
OXOLDPE	31,256	98,630	298,559	1,046,386	85,564	3.16
OXOLDPE-THERMO	33,462	99,721	252,369	553,030	87,419	3
OXOLDPE-PHOTO	1043	9989	170,790	408,802	6146	9.6
OXOLDPE-THERMO+PHOTO	1042	5559	46,025	139,820	4149	5.3

The results presented in Table 4 suggest that the thermo-oxidative treatment does not significantly modify the values of M_n and M_W and, therefore, the M_W/M_n ratio of either LDPE or OXOLDPE films. Regarding the photo-oxidative treatment, the changes are more drastic: the values of M_n, M_W and M_V fall abruptly for both films, while the polydispersity doubles for LDPE and triples for OXOLDPE, i.e., the fraction of high molecular weight chains remained significant, according to the values of M_z and M_{z+1}. On the other hand, the effect of the combined oxidative treatment does not significantly affect the values of M_n and M_w, compared to those after the photo-oxidative film treatment. However, the polydispersity decreased significantly from 8.4 to 3.9 for LDPE and from 9.6 to 5.3 for OXOLDPE. To appreciate these changes in more detail, the percentages of variation of these indicators were calculated by taking the values of the LDPE films as a reference.

Figure 9 compares the change in molecular weights of the tested LDPE and OXOLDPE films. The comparison of the molecular weights confirms that the photo- and combined oxidative treatments significantly reduced the values of M_z and M_{z+1} for the tested films, because of the reduction of the fraction of chains having a higher molecular weight. For LDPE films, the most significant change is in Mz+1, which increased from 75% after the photo-oxidative treatment to 95% during the combined treatment.

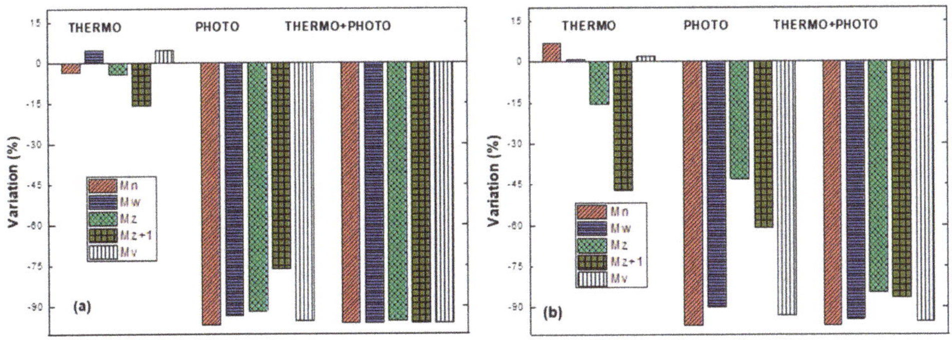

Figure 9. The change in different molecular weights of LDPE (**a**) and OXOLDPE (**b**) films after different abiotic oxidative treatments.

It seems that this effect is less pronounced in the presence of oxo-degradant additives in OXOLDPE (Figure 9b). The M_z and M_{z+1} of the OXOLDPE films decreased by 41% and 61%, respectively, after the photo-oxidative treatment, while with the combined treatment, the values of M_z and M_{z+1} decreased by 85% and 86%, respectively.

3.5. Scanning Electron Microscopy (SEM-EDS)

SEM images of the change in the surface morphology of the OXOLDPE films subjected to different degradation treatments are presented in Figure 10.

The reference OXOLDPE surface (Figure 10a) was relatively smooth and white spots (zones A) were visible, which EDS analysis suggested to be calcium (Ca). The surface of the thermo-oxidized specimen (at 70 °C for 90 days, Figure 10 b) revealed the white Ca-spots, as maintained on the surface. After the photo-oxidative treatment (for 1000 h, Figure 10c), cracks appeared on the film surface because of the crystallinity increase and reduction in the molecular mass [19,28,37,38,65]. The morphology of the films treated with the combined thermo- and photo-oxidative treatments (Figure 10d) changed drastically, i.e., the surface was more eroded and exhibited white spots in different shapes (zones A); EDS analysis confirmed that the films of OXOLDPE had been filled with inorganic filler grains, mainly of Ca and Mg at a very low content. The Ca filler (up to 40%) has been considered as the main oxo-degradant additive (as $CaCO_3$) for the LDPE films. The calcium carbonate ($CaCO_3$) filler is of low cost and may improve the films' structural stability during processing. It has been reported to act as a weathering barrier protective agent against UV light incidence [51].

The LDPE and OXOLDPE films, subjected to thermo- and photo-oxidative treatments as well as to the combination of both treatments, were characterized by means of FTIR, XRD, TGA and GPC analysis. The results revealed that the combined oxidative treatment resulted in the formation of oxidized degradation products with a higher abundance of ester and carboxylic groups, as well as in a greater drop in the average molecular weights M_z and M_{z+1}, compared to those obtained after individual thermo- and photo-oxidative treatments. The effect of the combined oxidative treatment was evident for the LDPE and OXOLDPE films tested, and it was suggested that there was a synergistic effect between both oxidative treatments when the combined test was carried out. It is important to highlight that because of the presence of the oxo-degrading additive, the formation of carboxylic groups was greater than that of ester groups.

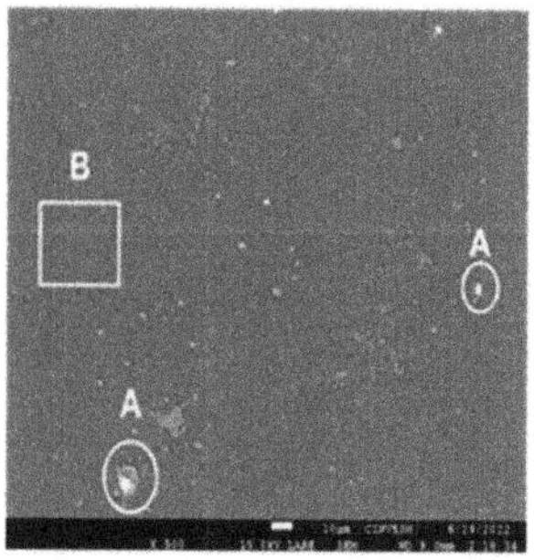

(a)

Zone	C (%W)	O (%W)	Ca (%W)
A	17.93	41.90	40.17
B	88.43	10.50	1.07

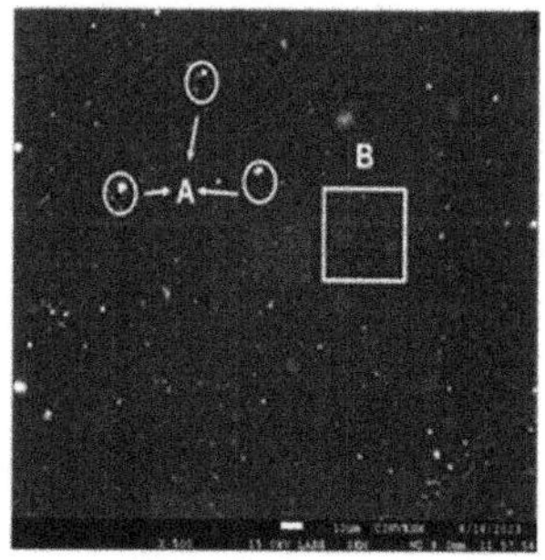

(b)

Zone	C (%W)	O (%W)	Ca (%W)
A	72.47	11.20	5.71
B	98.09	1.74	0.17

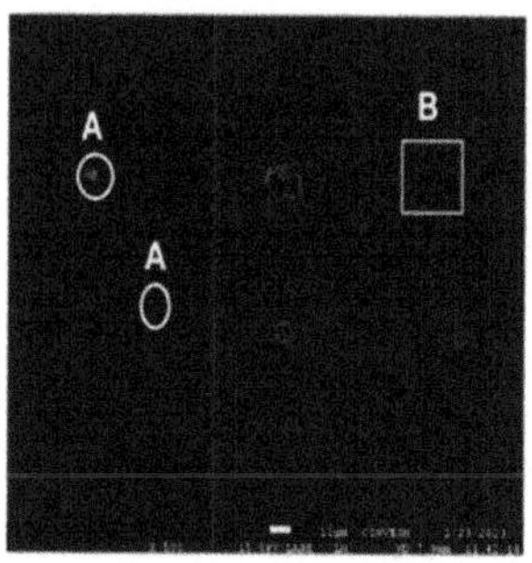

(c)

Zone	C (%W)	O (%W)	Ca (%W)	Mg (%W)
A	53.68	27.21	11.83	7.27
B	93.10	6.90	---	---

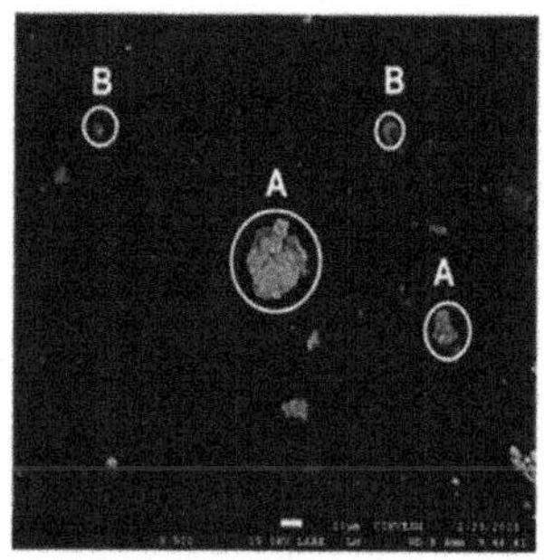

(d)

Zone	C (%W)	O (%W)	Ca (%W)	Mg (%W)
A	14.70	45.12	40.23	0.21
B	86.23	13.39	0.37	---

Figure 10. SEM images (x500) of the surface of (**a**) non-treated OXOLDPE film and after different abiotic oxidative treatments (**b**) thermo-oxidative, (**c**) photo-oxidative and (**d**) thermo+photo-oxidative.

These results are very interesting, and merit further study, because they provide the guidelines for achieving higher percentages of mineralization of LDPE films and even very probably of other types of polyolefins.

4. Conclusions

The study of the combined effect of two abiotic treatments (a thermo-oxidative treatment followed by a photo-oxidative treatment with ultraviolet light) on the physicochemical properties of commercially available low-density polyethylene (LDPE) films and with oxo-degradant additives (OXOLDPE) was carried out. The results indicate that the increment in carbonyl index (CI) and crystallinity percentage (X_{XRD}) was higher for those films that received the combined oxidative treatments, compared with those that received only one treatment, either thermo- or photo-oxidative. Moreover, the combined oxidative treatment produced more ester and carboxylic groups as degradation products than single treatments of the tested films. Deconvolution in the carbonyl band indicated a high concentration of ester groups and carboxylic acids, suggesting that the Norrish I reaction mechanism predominates during the cleavage of the polymer chains. An analysis of variance (ANOVA) was carried out and showed that there was a synergistic effect between the thermo- and photo-oxidative treatments for the ester and carboxylic degradation products. TGA results showed that the loss of thermal stability in the tested films was more significant after their combined thermo- and photo-oxidative treatments, compared with that observed after single treatments. The GPC analysis showed that the combined oxidative treatment led to decrease of the M_z and M_{z+1} average molecular weights of the degraded films containing an oxo-degradant additive, likewise to the extent of the M_w and M_n weights. The SEM images revealed that the morphology of the surface films' appearance changed more drastically after their exposure to the combined thermo- and photo-oxidative treatments. The surface seemed eroded, showing the presence of the inorganic filler of $CaCO_3$ as main filler and, at a low content, Mg.

These results suggest that the combined oxidative treatment (thermo- and photo-oxidative) produced degradation of products with a lower molecular weight and greater content of ester and carboxylic groups, which should enhance the LDPE film's environmental biodegradability.

Author Contributions: Conceptualization, A.V.-G. and L.V.; methodology, M.F.R.-T., A.V.-G., R.B., M.T.R.-H. and M.V.M.-C.; formal analysis, A.V.-G.; investigation, M.F.R.-T., R.B., M.T.R.-H. and M.V.M.-C.; resources, R.B., M.T.R.-H. and M.V.M.-C.; data curation, M.F.R.-T. and A.V.-G.; writing—original draft preparation, A.V.-G. and M.F.R.-T.; writing—review and editing, L.V.; project administration, A.V.-G. All authors have read and agreed to the published version of the manuscript.

Funding: This research received no external funding.

Data Availability Statement: The datasets presented in this article are not readily available because the data are part of an ongoing study.

Acknowledgments: M.F. Rojas-Trejo acknowledges the Mexican Council for Science and Technology for the scholarship for her Ph.D. study. The authors gratefully thank the National Laboratory of Nano and Biomaterials (LANNBIO-Cinvestav) for allowing the use of XRD and SEM-EDS facilities, funded by the projects FOMIX-Yucatán 2008-108160, CONACYT LAB-2009-01-123913, 292692, 294643, 188345 and 204822. The authors thank Daniel Aguilar for obtaining the diffractograms, and Mario Herrera Salvador for corrective maintenance of the diffractometer, Víctor Rejón and Leonardo Hernández for obtaining the SEM micrographs.

Conflicts of Interest: The authors declare no conflicts of interest.

References

1. Dey, S.; Veerendra, G.T.N.; Babu, P.S.S.A.; Manoj, A.V.P.; Nagarjuna, K. Degradation of Plastics Waste and Its Effects on Biological Ecosystems: A Scientific Analysis and Comprehensive Review. *Biomed. Mater. Devices* **2023**, *2*, 70–112. [CrossRef]
2. Tiago, G.A.; Mariquito, A.; Martins-Dias, S.; Marques, A.C. The problem of polyethylene waste—recent attempts for its mitigation. *Sci. Total. Environ.* **2023**, *892*, 164629. [CrossRef] [PubMed]

3. Roy, P.; Surekha, P.; Raman, R.; Rajagopal, C. Investigating the role of metal oxidation state on the degradation behaviour of LDPE. *Polym. Degrad. Stab.* **2009**, *94*, 1033–1039. [CrossRef]
4. Pablos, J.L.; Abrusci, C.; Marín, I.; López-Marin, J.; Catalina, F.; Espí, E.; Corrales, T. Photodegradation of polyethylenes: Comparative effect of Fe and Ca-stearates as pro-oxidant additives. *Polym. Degrad. Stab.* **2010**, *95*, 2057–2064. [CrossRef]
5. Roy, P.K.; Singh, P.; Kumar, D.; Rajagopal, C. Manganese stearate initiated photo-oxidative and thermo-oxidative degradation of LDPE, LLDPE and their blends. *J. Appl. Polym. Sci.* **2010**, *117*, 524–533. [CrossRef]
6. Suresh, B.; Maruthamuthu, S.; Kannan, M.; Chandramohan, A. Mechanical and surface properties of low-density polyethylene film modified by photo-oxidation. *Polym. J.* **2011**, *43*, 398–406. [CrossRef]
7. Antunes, M.C.; Agnelli, J.A.; Babetto, A.S.; Bonse, B.C.; Bettini, S.H. Abiotic thermo-oxidative degradation of high density polyethylene: Effect of manganese stearate concentration. *Polym. Degrad. Stab.* **2017**, *143*, 95–103. [CrossRef]
8. De Beer, J.A.; Focke, W.W. Oxidative Degradation of Polyolefins in the Presence of Cupric and Ferric Stearate Additives. *Macromol. Symp.* **2019**, *384*, 1800149. [CrossRef]
9. Albertsson, A.-C.; Andersson, S.O.; Karlsson, S. The mechanism of biodegradation of polyethylene. *Polym. Degrad. Stab.* **1987**, *18*, 73–87. [CrossRef]
10. Karlsson, S.; Ljungquist, O.; Albertsson, A.-C. Biodegradation of polyethylene and the influence of surfactants. *Polym. Degrad. Stab.* **1988**, *21*, 237–250. [CrossRef]
11. Arnaud, R.; Dabin, P.; Lemaire, J.; Al-Malaika, S.; Chohan, S.; Coker, M.; Scott, G.; Fauve, A.; Maaroufi, A. Photooxidation and biodegradation of commercial photodegradable polyethylenes. *Polym. Degrad. Stab* **1994**, *46*, 211–224. [CrossRef]
12. Chiellini, E.; Corti, A.; Swift, G. Biodegradation of thermally-oxidized, fragmented low-density polyethylenes. *Polym. Degrad. Stab.* **2003**, *81*, 341–351. [CrossRef]
13. Selke, S.; Auras, R.; Nguyen, T.A.; Aguirre, E.C.; Cheruvathur, R.; Liu, Y. Evaluation of Biodegradation-Promoting Additives for Plastics. *Environ. Sci. Technol.* **2015**, *49*, 3769–3777. [CrossRef] [PubMed]
14. Rose, R.-S.; Richardson, K.H.; Latvanen, E.J.; Hanson, C.A.; Resmini, M.; Sanders, I.A. Microbial Degradation of Plastic in Aqueous Solutions Demonstrated by CO_2 Evolution and Quantification. *Int. J. Mol. Sci.* **2020**, *21*, 1176. [CrossRef] [PubMed]
15. Yashchuk, O.; Portillo, F.; Hermida, E. Degradation of Polyethylene Film Samples Containing Oxo-Degradable Additives. *Procedia Mater. Sci.* **2012**, *1*, 439–445. [CrossRef]
16. Sen, S.K.; Raut, S. Microbial degradation of low density polyethylene (LDPE): A review. *J. Environ. Chem. Eng.* **2015**, *3*, 462–473. [CrossRef]
17. Finzi-Quintão, C.M.; Novack, K.M.; Bernardes-Silva, A.C. Identification of Biodegradable and Oxo-Biodegradable Plastic Bags Samples Composition. *Macromol. Symp.* **2016**, *367*, 9–17. [CrossRef]
18. Portillo, F.; Yashchuk, O.; Hermida, É. Evaluation of the rate of abiotic and biotic degradation of oxo-degradable polyethylene. *Polym. Test.* **2016**, *53*, 58–69. [CrossRef]
19. Musioł, M.; Rydz, J.; Janeczek, H.; Radecka, I.; Jiang, G.; Kowalczuk, M. Forensic engineering of advanced polymeric materials Part IV: Case study of oxo-biodegradable polyethylene commercial bag—Aging in biotic and abiotic environment. *Waste Manag.* **2017**, *64*, 20–27. [CrossRef] [PubMed]
20. Al-Salem, S.M.; Sultan, H.H.; Karam, H.J.; Al-Dhafeeri, A.T. Determination of biodegradation rate of commercial oxo-biodegradable polyethylene film products using ASTM D 5988. *J. Polym. Res.* **2019**, *26*, 157. [CrossRef]
21. Directive 2019/904/EU of the European Parliament and of the Council of 5 June 2019 on the Reduction of the Impact of Certain Plastic Products on the Environment (Text with EEA Relevance). *Off. J. Eur. Union* **2019**, *155*, 1–19. Available online: https://eur-lex.europa.eu/eli/dir/2019/904/oj/eng (accessed on 13 November 2024).
22. Markowicz, F.; Szymańska-Pulikowska, A. Analysis of the possibility of environmental pollution by composted biodegradable and oxobiodegradable plastics. *Geosciences* **2019**, *9*, 460. [CrossRef]
23. Napper, I.E.; Thompson, R.C. Environmental Deterioration of Biodegradable, Oxo-biodegradable, Compostable, and Conventional Plastic Carrier Bags in the Sea, Soil, and Open-Air Over a 3-Year Period. *Environ. Sci. Technol.* **2019**, *53*, 4775–4783. [CrossRef]
24. Yang, Y.; Li, Z.; Yan, C.; Chadwick, D.; Jones, D.L.; Liu, E.; Liu, Q.; Bai, R.; He, W. Kinetics of microplastic generation from different types of mulch films in agricultural soil. *Sci. Total. Environ.* **2022**, *814*, 152572. [CrossRef] [PubMed]
25. Potts, J.E.; Clendinning, R.A.; Ackart, W.B.; Niegisch, W.D. The Biodegradability of Synthetic Polymers. In *Polymers and Ecological Problems*; Guillet, J., Ed.; Springer: Boston, MA, USA, 1973; Volume 3, pp. 75–85. [CrossRef]
26. Weiland, M.; Daro, A.; David, C. Biodegradation of thermally oxidized polyethylene. *Polym. Degrad. Stab.* **1995**, *48*, 275–289. [CrossRef]
27. Karlsson, S.; Hakkarainen, M.; Albertsson, A.-C. Dicarboxylic Acids and Ketoacids Formed in Degradable Polyethylenes by Zip Depolymerization through a Cyclic Transition State. *Macromolecules* **1997**, *30*, 7721–7728. [CrossRef]
28. Albertsson, A.-C.; Barenstedt, C.; Karlsson, S. Abiotic degradation products from enhanced environmentally degradable polyethylene. *Acta Polym.* **1994**, *45*, 97–103. [CrossRef]

29. Chiellini, E.; Corti, A.; D'Antone, S. Oxo-biodegradable full carbon backbone polymers—biodegradation behaviour of thermally oxidized polyethylene in an aqueous medium. *Polym. Degrad. Stab.* **2007**, *92*, 1378–1383. [CrossRef]
30. Restrepo-Flórez, J.-M.; Bassi, A.; Thompson, M.R. Microbial degradation and deterioration of polyethylene—A review. *Int. Biodeterior. Biodegradation* **2014**, *88*, 83–90. [CrossRef]
31. Albertsson, A.-C.; Barenstedt, C.; Karlsson, S. Susceptibility of enhanced environmentally degradable polyethylene to thermal and photo-oxidation. *Polym. Degrad. Stab.* **1992**, *37*, 163–171. [CrossRef]
32. Andrady, A.L.; Pegram, J.E.; Tropsha, Y. Changes in carbonyl index and average molecular weight on embrittlement of enhanced-photodegradable polyethylenes. *J. Polym. Environ.* **1993**, *1*, 171–179. [CrossRef]
33. Khabbaz, F.; Albertsson, A.-C. Rapid test methods for analyzing degradable polyolefins with a pro-oxidant system. *J. Appl. Polym. Sci.* **2001**, *79*, 2309–2316. [CrossRef]
34. Weiland, M.; David, C. Thermal oxidation of polyethylene in compost environment. *Polym. Degrad. Stab.* **1994**, *45*, 371–377. [CrossRef]
35. Khabbaz, F.; Albertsson, A.-C.; Karlsson, S. Chemical and morphological changes of environmentally degradable polyethylene films exposed to thermo-oxidation. *Polym. Degrad. Stab.* **1999**, *63*, 127–138. [CrossRef]
36. Bonhomme, S.; Cuer, A.; Delort, A.-M.; Lemaire, J.; Sancelme, M.; Scott, G. Environmental biodegradation of polyethylene. *Polym. Degrad. Stab.* **2003**, *81*, 441–452. [CrossRef]
37. Husarova, L.; Machovsky, M.; Gerych, P.; Houser, J.; Koutny, M. Aerobic biodegradation of calcium carbonate filled polyethylene film containing pro-oxidant additives. *Polym. Degrad. Stab.* **2010**, *95*, 1794–1799. [CrossRef]
38. De la Orden, M.; Montes, J.; Urreaga, J.M.; Bento, A.; Ribeiro, M.; Pérez, E.; Cerrada, M. Thermo and photo-oxidation of functionalized metallocene high density polyethylene: Effect of hydrophilic groups. *Polym. Degrad. Stab.* **2014**, *111*, 78–88. [CrossRef]
39. Moo-Tun, N.M.; Valadez-González, A.; Uribe-Calderon, J.A. Thermo-oxidative aging of low density polyethylene blown films in presence of cellulose nanocrystals and a pro-oxidant additive. *Polym. Bull.* **2017**, *75*, 3149–3169. [CrossRef]
40. Roy, P.; Surekha, P.; Rajagopal, C.; Chatterjee, S.; Choudhary, V. Accelerated aging of LDPE films containing cobalt complexes as prooxidants. *Polym. Degrad. Stab.* **2006**, *91*, 1791–1799. [CrossRef]
41. Roy, P.; Surekha, P.; Rajagopal, C.; Chatterjee, S.; Choudhary, V. Studies on the photo-oxidative degradation of LDPE films in the presence of oxidised polyethylene. *Polym. Degrad. Stab.* **2007**, *92*, 1151–1160. [CrossRef]
42. Magagula, B.; Nhlapo, N.; Focke, W.W. Mn2Al-LDH- and Co2Al-LDH-stearate as photodegradants for LDPE film. *Polym. Degrad. Stab.* **2009**, *94*, 947–954. [CrossRef]
43. Fontanella, S.; Bonhomme, S.; Koutny, M.; Husarova, L.; Brusson, J.-M.; Courdavault, J.-P.; Pitteri, S.; Samuel, G.; Pichon, G.; Lemaire, J.; et al. Comparison of the biodegradability of various polyethylene films containing pro-oxidant additives. *Polym. Degrad. Stab.* **2010**, *95*, 1011–1021. [CrossRef]
44. Hsu, Y.-C.; Weir, M.P.; Truss, R.W.; Garvey, C.J.; Nicholson, T.M.; Halley, P.J. A fundamental study on photo-oxidative degradation of linear low density polyethylene films at embrittlement. *Polymer* **2012**, *53*, 2385–2393. [CrossRef]
45. Corti, A.; Sudhakar, M.; Chiellini, E. Assessment of the Whole Environmental Degradation of Oxo-Biodegradable Linear Low Density Polyethylene (LLDPE) Films Designed for Mulching Applications. *J. Polym. Environ.* **2012**, *20*, 1007–1018. [CrossRef]
46. Gauthier, E.; Laycock, B.; Cuoq, F.-M.; Halley, P.; George, K. Correlation between chain microstructural changes and embrittlement of LLDPE-based films during photo- and thermo-oxidative degradation. *Polym. Degrad. Stab.* **2012**, *98*, 425–435. [CrossRef]
47. Hoekstra, H.; Spoormaker, J.; Breen, J.; Audouin, L.; Verdu, J. UV-exposure of stabilized and non-stabilized HDPE films: Physico-chemical characterization. *Polym. Degrad. Stab.* **1995**, *49*, 251–262. [CrossRef]
48. Graciano-Verdugo, A.; Peralta, E.; González-Ríos, H.; Soto-Valdez, H. Determination of α-tocopherol in low-density polyethylene (LDPE) films by diffuse reflectance Fourier transform infrared (DRIFT-IR) spectroscopy compared to high performance liquid chromatography (HPLC). *Anal. Chim. Acta* **2006**, *557*, 367–372. [CrossRef]
49. Almond, J.; Sugumaar, P.; Wenzel, M.N.; Hill, G.; Wallis, C. Determination of the carbonyl index of polyethylene and polypropylene using specified area under band methodology with ATR-FTIR spectroscopy. *e-Polymers* **2020**, *20*, 369–381. [CrossRef]
50. Arnaud, R.; Moisan, J.Y.; Lemaire, J. Primary hydroperoxidation in low-density polyethylene. *Macromolecules* **1984**, *17*, 332–336. [CrossRef]
51. Valadez-Gonzalez, A.; Cervantes-Uc, J.; Veleva, L. Mineral filler influence on the photo-oxidation of high density polyethylene: I. Accelerated UV chamber exposure test. *Polym. Degrad. Stab.* **1999**, *63*, 253–260. [CrossRef]
52. Tidjani, A. Comparison of formation of oxidation products during photo-oxidation of linear low density polyethylene under different natural and accelerated weathering conditions. *Polym. Degrad. Stab.* **2000**, *68*, 465–469. [CrossRef]
53. Da Cruz, M.; Van Schoors, L.; Benzarti, K.; Colin, X. Thermo-oxidative degradation of additive free polyethylene. Part I. Analysis of chemical modifications at molecular and macromolecular scales. *J. Appl. Polym. Sci.* **2016**, *133*, 43287. [CrossRef]
54. Ranjan, V.P.; Goel, S. Degradation of Low-Density Polyethylene Film Exposed to UV Radiation in Four Environments. *J. Hazard. Toxic Radioact. Waste* **2019**, *23*, 04019015. [CrossRef]

55. Dzul-Cervantes, M.; Herrera-Franco, P.J.; Tábi, T.; Valadez-Gonzalez, A. Using Factorial Design Methodology to Assess PLA-g-Ma and Henequen Microfibrillated Cellulose Content on the Mechanical Properties of Poly(lactic acid) Composites. *Int. J. Polym. Sci.* **2017**, *2017*, 4046862. [CrossRef]
56. Lacoste, J.; Carlsson, D.J. Gamma-, photo-, and thermally-initiated oxidation of linear low density polyethylene: A quantitative comparison of oxidation products. *J. Polym. Sci. Part A Polym. Chem.* **1992**, *30*, 493–500. [CrossRef]
57. Benítez, A.; Sánchez, J.J.; Arnal, M.L.; Müller, A.J.; Rodríguez, O.; Morales, G. Abiotic degradation of LDPE and LLDPE formulated with a pro-oxidant additive. *Polym. Degrad. Stab.* **2013**, *98*, 490–501. [CrossRef]
58. Gardette, M.; Perthue, A.; Gardette, J.; Janecska, T.; Földes, E.; Pukánszky, B.; Therias, S. Photo- and thermal-oxidation of polyethylene: Comparison of mechanisms and influence of unsaturation content. *Polym. Degrad. Stab.* **2013**, *98*, 2383–2390. [CrossRef]
59. Liu, X.; Gao, C.; Sangwan, P.; Yu, L.; Tong, Z. Accelerating the degradation of polyolefins through additives and blending. *J. Appl. Polym. Sci.* **2014**, *131*, 40750. [CrossRef]
60. Eyheraguibel, B.; Traikia, M.; Fontanella, S.; Sancelme, M.; Bonhomme, S.; Fromageot, D.; Lemaire, J.; Lauranson, G.; Lacoste, J.; Delort, A. Characterization of oxidized oligomers from polyethylene films by mass spectrometry and NMR spectroscopy before and after biodegradation by a Rhodococcus rhodochrous strain. *Chemosphere* **2017**, *184*, 366–374. [CrossRef]
61. Gedde, U.W.; Ifwarson, M. Molecular structure and morphology of crosslinked polyethylene in an aged hot-water pipe. *Polym. Eng. Sci.* **1990**, *30*, 202–210. [CrossRef]
62. Mucha, M. Thermogravimetric studies of polymethylenes and polyethylenes. *J. Polym. Sci. Polym. Symp.* **1976**, *57*, 25–31. [CrossRef]
63. Contat-Rodrigo, L.; Ribes-Greus, A.; Imrie, C.T. Thermal analysis of high-density polyethylene and low-density polyethylene with enhanced biodegradability. *J. Appl. Polym. Sci.* **2002**, *86*, 764–772. [CrossRef]
64. Al-Salem, S.; Al-Hazza'A, A.; Karam, H.; Al-Wadi, M.; Al-Dhafeeri, A.; Al-Rowaih, A. Insights into the evaluation of the abiotic and biotic degradation rate of commercial pro-oxidant filled polyethylene (PE) thin films. *J. Environ. Manag.* **2019**, *250*, 109475. [CrossRef] [PubMed]
65. Roy, P.K.; Surekha, P.; Rajagopal, C.; Choudhary, V. Thermal degradation studies of LDPE containing cobalt stearate as pro-oxidant. *Express Polym. Lett.* **2007**, *1*, 208–216. [CrossRef]
66. Antelava, A.; Constantinou, A.; Bumajdad, A.; Manos, G.; Dewil, R.; Al-Salem, S.M. Identification of Commercial Oxo-Biodegradable Plastics: Study of UV Induced Degradation in an Effort to Combat Plastic Waste Accumulation. *J. Polym. Environ.* **2020**, *28*, 2364–2376. [CrossRef]
67. Zeng, S.; Lu, D.; Yang, R. Effects of Crystallinity and Branched Chain on Thermal Degradation of Polyethylene: A SCC-DFTB Molecular Dynamics Study. *Polymers* **2024**, *16*, 3038. [CrossRef] [PubMed]
68. Musil, J.; Zatloukal, M. Experimental investigation of flow induced molecular weight fractionation phenomenon for two linear HDPE polymer melts having same Mn and Mw but different Mz and Mz+1 average molecular weights. *Chem. Eng. Sci.* **2012**, *81*, 146–156. [CrossRef]
69. Roé-Sosa, A.; Estrada, M.R.; Calderas, F.; Sánchez-Arévalo, F.; Manero, O.; de Velasquez, M.T.O.L. Degradation and biodegradation of polyethylene with pro-oxidant aditives under compost conditions establishing relationships between physicochemical and rheological parameters. *J. Appl. Polym. Sci.* **2015**, *132*, 42721. [CrossRef]
70. Meng, K.; Lwanga, E.H.; van der Zee, M.; Munhoz, D.R.; Geissen, V. Fragmentation and depolymerization of microplastics in the earthworm gut: A potential for microplastic bioremediation? *J. Hazard. Mater.* **2023**, *447*, 130765. [CrossRef] [PubMed]

polymers

MDPI

Article

Comparative Study of the Method of Decorticating on Chemical Composition and Physicochemical Properties of Xyloglucan Extracted from *Tamarindus indica* L. Seeds at a Semi-Industrial Scale

Valeria Espíndola-Sotres [1,2], Elsa Gutiérrez-Cortez [3,*], Abraham Mendez-Albores [4], Arturo Aguirre-Gómez [5], Alfredo Maciel-Cerda [6], Moustapha Bah [2], Araceli Ulloa Saavedra [7], Francisco Luna-Vázquez [2], María Eugenia Ramirez-Ortíz [8], Alejandra Rojas-Molina [2] and Isela Rojas-Molina [2,*]

1 Doctorado en Ciencias Químico-Biológicas, Facultad de Química, Universidad Autónoma de Querétaro, Cerro de las Campanas S/N, Las Campanas, Centro Universitario, Santiago de Querétaro C.P. 76010, Querétaro, Mexico; valee.so10@gmail.com
2 Laboratorio de Investigación Química y Farmacológica de Productos Naturales, Facultad de Química, Universidad Autónoma de Querétaro, Cerro de las Campanas S/N, Centro Universitario, Santiago de Querétaro C.P. 76010, Querétaro, Mexico; moubah@uaq.mx (M.B.); fjlunavz@yahoo.com.mx (F.L.-V.); rojasa@uaq.mx (A.R.-M.)
3 Unidad de Investigación Multidisciplinaria (UIM), Laboratorio de Procesos en Ingeniería Agroalimentaria, FES-Cuautitlán, Universidad Nacional Autónoma de México, Cuautitlán Izcalli C.P. 54714, Estado de México, Mexico
4 Unidad de Investigación Multidisciplinaria (UIM), Laboratorio 14-A1, Ciencia y Tecnología de Materiales, FES-Cuautitlán, Universidad Nacional Autónoma de México, Cuautitlán Izcalli C.P. 54714, Estado de México, Mexico; albores@unam.mx
5 Unidad de Investigación Multidisciplinaria (UIM), Laboratorio de Análisis de Investigación en Química Agrícola y Ambiental, FES-Cuautitlán, Universidad Nacional Autónoma de México, Cuautitlán Izcalli C.P. 54714, Estado de México, Mexico; aag@unam.mx
6 Instituto de Investigaciones en Materiales, Universidad Nacional Autónoma de México, Ciudad de México C.P. 04510, Mexico; macielal@unam.mx
7 Unidad de Investigación Multidisciplinaria (UIM), Laboratorio 16, Procesos de Transformación y Tecnologías Emergentes en Alimentos, FES-Cuautitlán, Universidad Nacional Autónoma de México, Cuautitlán Izcalli C.P. 54714, Estado de México, Mexico; araceli_ulloa@yahoo.com.mx
8 Laboratorio Experimental Multidisciplinario (LEM 1), FES-Cuautitlán, Universidad Nacional Autónoma de México, Cuautitlán Izcalli C.P. 54740, Estado de México, Mexico; eugenia.ramirez@cuautitlan.unam.mx
* Correspondence: elsaneqpm123@gmail.com (E.G.-C.); jirojasmolina@gmail.com (I.R.-M.); Tel.: +52-56231919 (ext. 39602) (E.G.-C.); +52-442-192-1200 (ext. 75030) (I.R.-M.); Fax: +52-442-192-1302 (I.R.-M.)

Abstract: Xyloglucan from *Tamarindus indica* seeds (TISs) is a polysaccharide widely used in the food, biomedical, and pharmaceutical sectors. Nevertheless, the challenge in future research for the food processing industry is to provide adequate knowledge regarding natural product extraction, chemical modifications, interactions, and potential applications according to sustainability issues. The goal of this work was to implement a sustainable method for xyloglucan extraction from TISs at a semi-industrial scale and carry out the characterization of this hydrocolloid, to compare the effect of the technique of decorticating of seeds on the chemical composition and physicochemical properties of xyloglucan. The TISs were decorticated using soaking (DS) and roasting (DR) methods, and, then, the xyloglucan was extracted applying a semi-industrial mechanical separation process. Subsequently, the extraction yield, chemical content, Fourier transform infrared analysis, color, morphology, molecular weight (MW), viscosity, texture, Z potential, particle size, and thermal properties were evaluated. Xyloglucan extraction from TISs at a semi-industrial scale was demonstrated for the first time. The xyloglucan yield by DR (44.04%) was significantly higher ($p < 0.05$) compared with DS (41.42%), while separation efficiency was similar in both methods (~97%). Significant differences ($p < 0.05$) in fat, ashes, crude fiber, calcium, total phenolic content, and antioxidant capacity in xyloglucan samples were observed by applying DS and DR. The method of decorticating promoted changes in the MW and particle size of xyloglucan samples, which were reflected in the viscosity, particle size, texture attributes, Z potential, and thermal properties of xyloglucan. These results show that the decorticating method is an important issue to be considered in the resultant chemical and physicochemical properties of xyloglucan extracted from tamarind seeds, for suitable applications of the xyloglucan in the food processing and pharmaceutical industries.

Keywords: *Tamarindus indica* L.; tamarind seeds; xyloglucan; mechanical separation; separation efficiency; physicochemical properties

1. Introduction

Globally, it has been estimated that around 1.3 billion metric tons of annually generated food is wasted, representing approximately one-third of the total food produced [1,2]. For this reason, the generation of food waste and its transformation into value-added products (bioactive compounds, biobased plastics, and biofuels) constitute a global concern to develop sustainable processes for food production [3].

In 2020, the production of tamarind (*Tamarindus indica* L.) in Mexico was 55,752 metric tons [4], mainly destined for pulp extraction. This means that, in this agro-industrial activity, many by-products (shells and seeds) are generated without valorization [5].

The *T. indica* L. seeds represent ~30% of the total composition of this fruit [6], and they are used to obtain xyloglucan (galactoxyloglucan), which is a polysaccharide whose structure is made up of a β-(1,4)-D-glucan backbone substituted with side chains of α-(1,4) D-xylopyranose and (1,6) linked [β-D-galactopyranosyl-(1,2)-α-D-xylopyranosyl] to glucose residues [7].

The physicochemical properties of xyloglucan allow this hydrocolloid to be used in various technological applications such as stabilizer, thickener, viscosity enhancer, emulsifier, gelling, drug carrier, release retardant, and binder in the food and pharmaceutical industries [8].

Several studies have been reported for the xyloglucan extraction, where a sample pretreatment is carried out to eliminate the seed coat [9] and undesirable components (fats and proteins), which contaminate the final product [10–12]. Nevertheless, some of these procedures are very long and require no conventional heat treatments or the use of solvents that are not environmentally friendly. In addition, the yield of xyloglucan obtained is highly variable or not reported [13–17]. Consequently, all these factors have a significant impact on the processing cost [2]. On the other hand, the extraction processes have been carried out at the laboratory level (small scale). Thus, it is necessary to determine whether, on a larger scale, the operating conditions are adequate so that the production of xyloglucan can be profitable to obtain a value-added product [18].

Nowadays, innovations in recovering diverse molecules (compounds and natural products) from various sources are focused on the development of efficient, economical, and versatile techniques [19]. Regarding this, introducing magnetic ionic liquids (MILs) as an advanced pretreatment sample technique in microextractions allows high extraction efficiencies, good selectivities, and easy phase separations avoiding unnecessary steps such as centrifugation and filtration [20]. Aqueous two-phase systems composed of smart polymers (ATPSs) and thermo-separating magnetic ionic liquid aqueous two-phase systems (MILATPS) are other efficient, economical, and highly selective tools for the separation of target bioactive compounds and elimination of contaminants [21,22]. Despite the advantages that these emerging technologies offer, optimal operational parameters should be further explored to ensure the viability and functionality of isolated compounds.

Little attention has been given to the decortication of the *T. indica* seeds for the xyloglucan extraction. Therefore, investigation regarding the comparison of the decorticating method in yield and separation efficiency for xyloglucan extraction is still meager. In this context, it can be assumed that there is no universal protocol available for the xyloglucan extraction, which complies with all requirements for a sustainable process. This means that environmentally friendly solvents are used, easy solvent recovery methods are practiced, and minimal losses in the natural characteristics of xyloglucan are observed. For this reason, another aspect to consider is determining the possible applications of the products obtained from the sustainable extraction process, according to the characteristics that these products show [23,24].

We hypothesized that the method of decorticating *T. indica* seeds (soaking and roasting), before xyloglucan extraction, governs the properties of this biopolymer. Thus, the goal of this work was to implement a sustainable method for xyloglucan extraction from *T. indica* L. seeds at a semi-industrial scale and further characterize this hydrocolloid, to determine the effect of the procedure of decorticating seeds on the chemical composition and physicochemical properties of the extracted xyloglucan.

2. Materials and Methods

2.1. Xyloglucan Extraction

Xyloglucan extraction from *T. indica* seeds was carried out according to the block diagram shown in Figure 1. The decorticating method by the seeds soaking was carried out following the technique described by Khounvilay and Sittikijyothin [25] with minimal modifications. This method consisted of soaking 1 kg of seeds in 5 L of water at 4 °C for 8 days. Subsequently, the seed coat (testa) was removed, and the wet seeds were dried in a solar dryer at 38 °C for two days. Seed decortication by roasting was carried out by placing 1 kg of seeds on a perforated tray in an oven (Duke, model DUK-E102-G, Orldando, FL, USA) at 110 °C for 10 min and stirring the material every 5 min for uniform heating; then, the testa was removed from the seed.

The decorticated seeds were ground in a mill (Pulvex, mod. 200, Mexico City, Mexico) with a feeding rate of 75 g/min using a 2 mm sieve. Subsequently, the pulverized seed was passed through a sieve (No., 8 USA series, Retsch, Newtown, PA, USA). The material obtained was used to prepare 5% (*w*/*w*) suspensions. The suspensions were heated at 70 °C for 40 min, and the pH was determined. Next, the suspensions were centrifuged at 7000 rpm in a disk stack centrifuge (DIDACTA, Mod. TAG1/d, Torino, TO, Italy) to obtain soluble solids. Then, xyloglucan was precipitated from soluble solids by the addition of ethyl alcohol (96%) and water with a 1:2 (*v*/*v*) ratio. The separation of xyloglucan from liquid was carried out by filtration using a stainless-steel sieve with a pore opening of 2 mm. The xyloglucan retained in the sieve was dehydrated in an oven at 30 °C for 60 min. The material was pulverized with a blade mill (Pulvex, mod. 200, Mexico City, Mexico) at a speed of 200 g/min with a 1 mm outlet sieve. The solid granules of xyloglucan were passed through a sieve (No. 100 USA series, Retsch, Newtown, PA, USA) and stored until analysis. The ethyl alcohol used was recovered by distillation. The yield and separation efficiency were calculated using the equations reported by Quintero-García et al. [23].

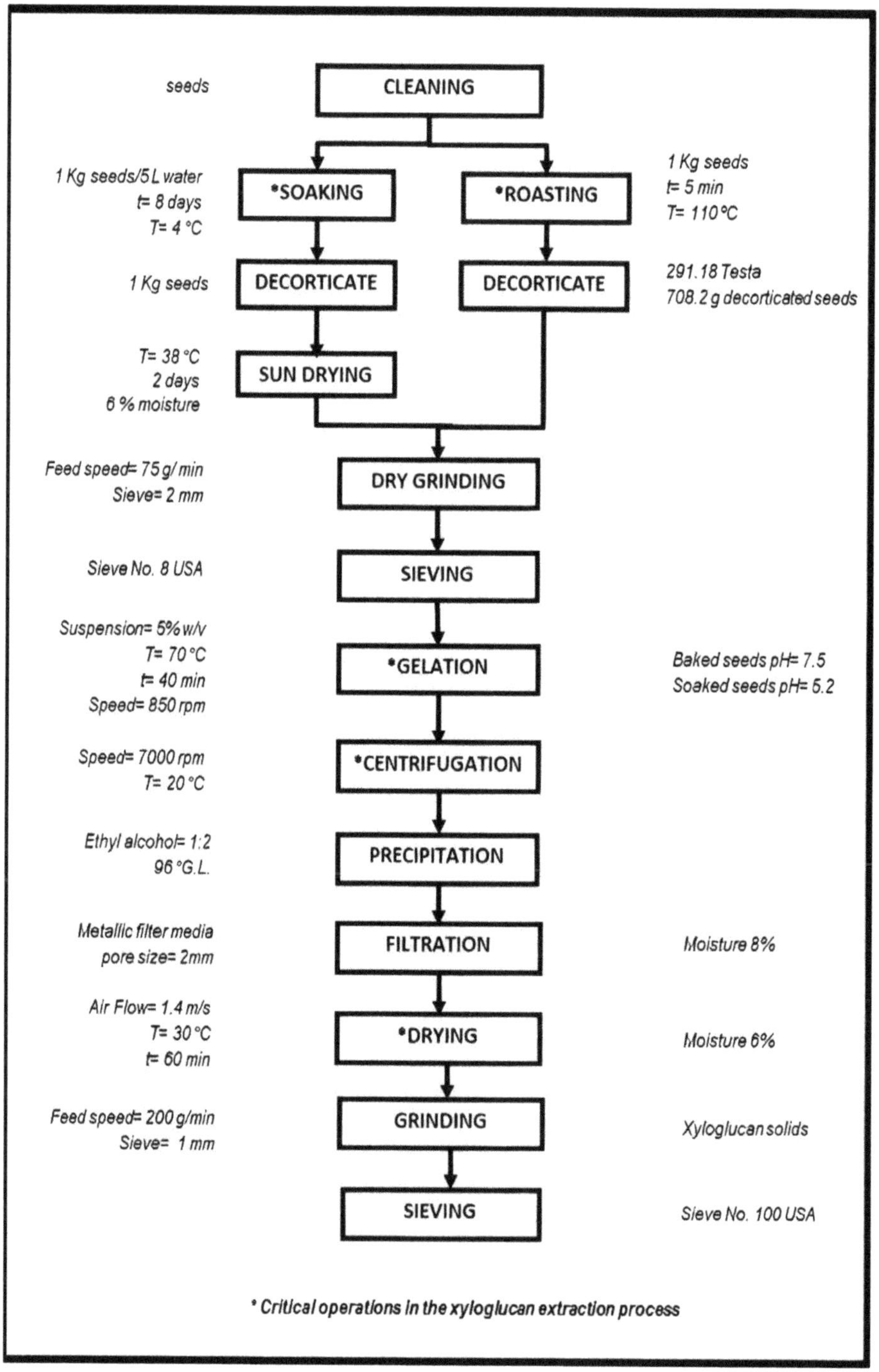

Figure 1. Block diagram for the xyloglucan extraction from *T. indica* seeds.

2.2. Determination of Energy Consumption

The energy (E) required for xyloglucan grinding to obtain a powder was determined using the equation reported by Gutiérrez-Cortez et al. [26].

2.3. Compositional Analysis and Determination of Calcium Content

Dehydrated xyloglucan samples were analyzed following the methods established by the Association of Official Analytical Chemists (AOAC, 2000) [27]. The total carbohydrate content was calculated by difference (% Carbohydrates = 100 − [% moisture + % crude fiber + % crude protein + % ether extract + % ash]). In addition, the calcium content in xyloglucan samples was determined by the method (968.08) [27] with an atomic absorption spectrophotometer (Perkin Elmer, Mod. Analyst 300, Boston, MA, USA) and a flame detector.

2.4. Total Phenolic and Antioxidant Capability

The extraction of soluble phenolic compounds was carried out using 20 mL of a solution of water, methanol, and formic acid (18:80:02 v/v). The quantification of total phenolic content was conducted according to Li et al. [28] with the Folin–Ciocalteu reagent. The results were expressed as micrograms of gallic acid equivalents (μg GAE/g dry weight of sample). The antioxidant activity of xyloglucan was determined according to the method reported by Thivya et al. [29], and the results were expressed as Trolox equivalents (μg Trolox/g dry weight of sample).

2.5. Quantification of Galactose, Xylose, and Glucose by HPLC

The physicochemical properties of xyloglucans are highly dependent on their monosaccharide composition (i.e., galactose/xylose/glucose ratios); therefore, the quantification of galactose, xylose, and glucose was carried out. Initially, 1 g of each sample was hydrolyzed with 33 mL of HCl 4 M at a temperature of 100 °C for 8 h in a block heater (Thermo Scientific™, Waltham, MA, USA) [17,30]; then, the samples were centrifuged at 8000 rpm for 15 min, and the supernatant was filtered with a membrane and used to perform the determinations. The identification and quantification of monosaccharides was performed by a high-performance liquid chromatography (HPLC) system (Waters Co, Milford, CT, USA. Mod. e2695) coupled with a refractive index detector (Waters Co, Milford, CT, USA. Mod. 2414) and a Waters NH_2 column (4.6 mm × 250 mm, 5 μm, Waters Co, Milford, CT, USA) at 80 °C. The mobile phase contained acetonitrile/water (80:20) in the isocratic mode and a total analysis time of 25 min. The sample injection volume was 20 μL, and the mobile phase flow rate was set at 1.5 mL/min. The retention times of samples were compared with monosaccharide standards (Sigma Chemical, Inc., St. Louis, MO, USA), and calibration curves were prepared for quantification. The analyses were conducted in triplicate.

2.6. Characterization by Fourier Transform Infrared (FT-IR) Spectroscopy

A Perkin Elmer Frontier NIR/MIR SP8000 spectrophotometer equipped with an ATR accessory (DuraSamplIR II, Smiths Detection, Warrington, UK) was employed for the analysis. The spectral range of 4000 to 400 cm^{-1} was chosen to cover the MIR region. To ensure robust outcomes, 32 scans were performed, resulting in improved spectrum quality and a favorable signal-to-noise ratio. A resolution of 4 cm^{-1} was selected to capture the relevant FTIR absorption bands. The bands were identified and analyzed using the Spectrum 10.4.2 software.

2.7. Color Measurement

The color values were recorded as L, darkness/lightness (0, black; 100, white); a (−a, greenness; +a, redness); and b (−b, blueness; +b, yellowness) using a colorimeter

(Minolta, CR300, Tokyo, Japan). The color measurements were taken from five different positions [23].

2.8. Morphological Studies

Scanning electron microscopy (SEM) images of the xyloglucan samples were examined using a high-vacuum scanning electron microscope (JEOL, JSM-6060LV, Tokyo, Japan) with resolution in HV mode (5 nm). Each sample was carefully attached to a sample holder with carbon tape and sputtered with gold. The analysis conditions were 20 kV electron acceleration voltage and a 12–20 Pa pressure [23].

2.9. Estimation of Molecular Weight

The molecular weight (MW) was estimated by capillary viscometry following the methodology reported by Sixto-Berrocal et al. [31]. Briefly, 0.2 g of dry xyloglucan was dissolved in 100 mL of a buffer solution (0.3 M of HAc/0.2 M of NaAc) with pH ~4.6 and constant stirring for 24 h, and, then, the solution was filtered. Subsequently, xyloglucan dilutions were prepared using deionized water (0.025, 0.05, 0.1, and 0.2 gdL^{-1}). The viscosity of each dilution was measured using an Ostwald-type capillary viscometer immersed in a water bath at 25 ± 1 °C. The fall times of the buffer solution (t_0) and the biopolymeric solutions (t) were measured, and five measurements were performed for each sample. The relative, specific, and reduced viscosities were calculated to obtain the intrinsic viscosity (η). The average MW was estimated according to the Mark–Houwink Sakurada (MHS) equation, which relates the inherent viscosity of a polymer solution with the molecular weight of the polymer, using a value of $a = 0.76$ [32].

2.10. Viscosity Determination

The viscosity of samples was achieved in xyloglucan solutions prepared at concentrations of 1.5, 2.0, 2.5, 3.0, and 3.5% (*w*/*v*) at 25 °C with magnetic stirring until the hydration (~40 min) and the solutions were cooled at room temperature before the analysis. Measurements were performed at different shear rates in a range from 0 to 8 s^{-1} in a viscosimeter with a concentric cylinder geometry (Mettler-Toledo, Mod. RM180 Rheomat, Columbus, OH, USA) and a rotating cylinder sensor (DIN 1) [23].

2.11. Texture Analysis

Xyloglucan solutions were prepared at 5% (*w*/*v*) in distilled water at 35 °C using a thermo-shaker (Thermo Scientific, Mod. 4625Q, Waltham, MA, USA) at a rate of 250 rpm. The samples were stored until they reached a temperature of 25 °C. The force–time curves for two-cycle compression were obtained using a Universal Texturometer (Shimadzu, Mod. EZ-S, Kyoto, Japan). Then, the samples were placed in a Petri dish (diameter of 6.0 cm, height of 1.5 cm) and compressed with a cylinder (diameter 2.54 cm) at a rate of 1.3 mm/s to a compression depth of 8 mm (20% compression). When the compression stroke was completed, the plunger reversed and started its upward stroke. Subsequently, a second down and up-cycle was run on the same sample after 5 s. Five replicates were made for each sample.

2.12. Z Potential and Particle Size Measurements

Z potential (ζ-potential) and particle size measurements were performed with a particle size analyzer (ZetaSizer Pro, Malvern Instruments, Worcestershire, UK). Before the analysis, all samples were appropriately diluted using Milli-Q water (Millipore, Milli-Q Direct 8, Bay City, MI, USA) to yield an appropriate scattering intensity. Measurements were carried out at 25 °C using a DTS1070 disposable capillary cell (Malvern Panalytical, Westborough, MA, USA) and the particle size was obtained by calculating the average of five replicates. The Zeta potential was determined by electrophoretic light scattering. The conversion of the electrophoretic mobility to Zeta potential was performed using the Helmholtz–Smoluchowski equation [33]. Data were analyzed using the ZS Xplorer software v2.01 (Malvern Panalytical, Westborough, MA, USA).

2.13. Differential Scanning Calorimetry (DSC)

The thermal properties were analyzed utilizing a differential scanning calorimeter (Q100 TA Instruments, New Castle, DE, USA). Briefly, 5 mg of xyloglucan was placed into an aluminum standard pan (40 μL) and hermetically sealed. The plan was submitted to a heating ramp from −50 to 300 °C, with a heating rate of 10 °C/min, and nitrogen at a flow rate of 10 mL/min was employed. An empty pan served as a reference [34].

2.14. Thermogravimetric Analysis (TGA)

The mass loss of xyloglucan samples was evaluated through a TGA Q5000 (TA Instruments, New Castle, DE, USA) under the following conditions: a temperature range of 0–700 °C, a heating rate of 10 °C/min, and a nitrogen atmosphere with a flow of 10 mL/min. TA Universal Analysis software v4.5A was used for data processing [34].

2.15. Statistical Analysis

The results obtained from the xyloglucan extraction, chemical composition, and physicochemical characterization of experimental samples were analyzed by a *t* student's test with $\alpha = 0.05$, and, in each case, the statistical package SPSS version 2.2 was used.

3. Results and Discussion

3.1. Xyloglucan Extraction

Table 1 shows the yield, separation efficiency, and energy consumed by roasting and soaking *T. indica* seeds as decortication methods for xyloglucan extraction. The yield obtained in decorticated seeds with the roasting method (44.04 ± 0.37%) was significantly higher ($p < 0.05$) than the yield observed in the soaking method (41.42 ± 0.10%). Soaking is a culinary process to remove soluble components in legume seeds, i.e., antinutritional factors (α-galacto-oligosaccharides), which are eliminated in the discarded soaking solution [35]. In addition, soaking causes a decrease in carbohydrate contents (starch, total sugars, reducing sugars, total carbohydrates) as a result of the simple diffusion of sugars after being solubilized [36]; consequently, these losses have an impact on the yield. The yields of this current study are higher than those reported by Carvalho et al. [37], who obtained a yield of 7.7 and 32.66% decorticating seeds by soaking in water at 100 °C for 4 h using consecutive extractions with hot and cold water, respectively. Likewise, our yields are higher compared to those observed by Crispin et al. [11] and Alpizar-Reyes et al. [38], who reported values of 27.93 and 29.83%, respectively. All these authors carried out the xyloglucan extraction from tamarind seeds without coat removal. Regarding the yield obtained by the roasting method, Chawananorasest et al. [10] reported yields from 11.15 to 54.65% decorticating the seeds in an oven at 100 °C for 30 min, applying accelerated solvent extraction (ASE) and defatting tamarind seed powder using organic solvents (methanol and hexane) before the extraction of xyloglucan. The xyloglucan yield observed in our extraction process (44.04%) is in the range reported by these authors with the advantage that, in our case, the use of solvents is avoided, and the roasting time is also reduced according to the criteria for an eco-friendly process. Two decorticating methods involving 1) roasting seeds at 40 °C for 24 h and 2) the microwave heat pretreatment of seeds followed by soaking and hot air drying showed maximum yields of 22.12 and 25.8%, respectively [11,17]. Both procedures included defatting and protein soluble removal from the seeds to obtain a xyloglucan free of impurities. As more steps are involved in the xyloglucan extraction, the yield of xyloglucan decreases, as the processing cost increases. Thivya et al. [29] applied different extraction methods to obtain xyloglucan from roasted tamarind seeds, i.e., hot water extraction, cold water extraction, ultrasound-assisted extraction, microwave-assisted extraction, and alkali extraction, and, in all cases, the xyloglucan yield was less than 25%. In contrast, Limsangouan et al. [16] reported xyloglucan yields of 46.44, 53 and 62.28% from decorticated seeds, applying hot water extraction (conventional method before removal of fat and proteins), high-pressure assisted extraction, and super-critical water extraction, respectively. Definitively, higher yields are observed applying novel extraction methods; nevertheless, it is important to evaluate the feasibility of implementing a process easily adapted by small, medium, and large-scale producers, since greater yields for xyloglucan extraction imply increasing costs for manufacturers and issues from an ecological point of view [29].

Table 1. Xyloglucan yield, separation efficiency of the xyloglucan obtained from *T. indica* seeds by using two decortication methods. Total energy consumption and energy cost of the processes.

Decorticate Method	Xyloglucan Yield (%)	Xyloglucan Separation Efficiency (%)	Total Energy Consumption * (kW/kg Dry Raw Material)	Energy Cost USD/kW·h
Soaking	41.42 ± 0.10^{a}	97.36 ± 0.70^{a}	15.82 ± 1.90^{a}	2.27
Roasting	44.04 ± 0.37^{b}	97.11 ± 0.50^{a}	16.18 ± 0.30^{a}	2.31

* Energy consumption includes the first grinding, centrifugation, drying, and second grinding, including oven roasting for the roasting decortication method. Values represent the mean ± standard deviation (SD), $n = 5$. Means in columns with different letters differ significantly ($p < 0.05$).

No significant differences ($p < 0.05$) were detected in separation efficiency between soaking and roasting decorticate (see Table 1); therefore, the decorticating method does not influence this parameter. These values are similar to those reported by Quintero-García et al. [23], who also used a disk stack centrifuge, reporting a separation efficiency of 97% for *O. ficus indica* mucilage extraction. In this regard, the utilization of a disk stack centrifuge, commonly used in the food industry as a result of its mechanical simplicity, attains high G-force, allowing the mechanical separation between two phases with different densities efficiently [39].

The energy expenses to process 1 kg of tamarind seeds (from the first grinding, centrifugation, drying, and second grinding, including oven roasting for the roasting decortication method) were not significantly different ($p < 0.05$), using both decorticating methods since the unit operations for the seed processing after the pre-treatment step do not differ, as is shown in Figure 1. Nonetheless, it is important to mention that, in the pretreatment of seeds, the soaking decorticate method requires a large amount of water (5 m^3/Tn of seed) and additional energy, since the seeds are soaked for 8 days at refrigeration temperature (4 °C). In contrast, with the roasting decorticate method, only the energy consumption for the raw material roasting would be considered (Figure 1). The data shown in Table 1 constitute an initial attempt to scale up the xyloglucan extraction using a semi-industrial disk stack centrifuge, maintaining geometric similarity between the equipment at a target scale. At present, most of the procedures reported for the xyloglucan extraction have been developed at scale [10,11,16,37,38,40]; this fact increases extra uncertainty during scale-up, avoiding identifying potential hazards and associated risks (important solvent requirements, high energy demands, improper control over process conditions). In addition, some still apply the pretreatment of samples using organic solvents and proteases. All of these issues are critical to improving production capacity and resource efficiency and reducing the environmental impact [41].

3.2. Chemical Analysis

The chemical composition of xyloglucan extracted by soaking and roasting is presented in Table 2.

Table 2. Chemical proximate analysis, total phenolic content, antioxidant capacity, and monosaccharide content (galactose, xylose, and glucose) of xyloglucan obtained by using two decorticate methods (dry mass).

Component \ Decorticate Method	Soaking	Roasting
Protein (%)	3.94 ± 0.10 [a]	3.86 ± 0.15 [a]
Fat (%)	0.69± 0.01 [a]	0.26 ± 0.10 [b]
Ashes (%)	0.47± 0.02 [a]	0.70 ± 0.02 [b]
Crude fiber (%)	1.26± 0.08 [a]	1.17 ± 0.12 [b]
Carbohydrates (%)	93.6± 0.90 [a]	93.95 ± 0.70 [a]
Calcium (mg/g)	1.58 ± 0.10 [a]	1.72 ± 0.12 [b]
Total phenolic content (μg GAE/g)	132.13 ± 11.95 [a]	77.27 ± 9.79 [b]
Antioxidant capacity (μg Trolox/g)	105.48 ± 8.72 [a]	52.88 ± 5.24 [b]
Galactose (%)	6.95 ± 0.14 [a]	9.02 ± 0.10 [b]
Xylose (%)	22.50 ± 0.87 [a]	28.10 ± 1.05 [b]
Glucose (%)	32.03 ± 1.2 [a]	43.47 ± 0.65 [b]
Galactose/xylose/glucose ratio by mass	1.0:3.2:4.6	1.0:3.1:4.8

The values represent mean ± standard deviation (SD), $n = 5$. Means in rows with different letters differ significantly ($p \leq 0.05$).

No significant differences were observed ($p < 0.05$) in the protein content of xyloglucan extracted by both decorticate methods. However, these values are below the data reported by other authors [11,38,40], applying or not purification methods to tamarind seeds (defatting and alkaline protein extraction) before xyloglucan extraction. The minimum value reported for this component is 5.04% [16]. In contrast, the fat content in the xyloglucan extracted from decorticated seeds by soaking was significantly higher ($p < 0.05$), in comparison to the fat content in xyloglucan extracted from decorticated seeds by roasting. This difference can be attributed to the loss of fatty acids contained in the tamarind seed during roasting since it has been shown that these compounds are sensitive to heating [42]. In this regard, Ghafoor et al. [43] pointed out that concentrations of linolenic, linoleic, and oleic acids in unroasted *Salvia hispanica* L. seeds are higher than those from roasted ones, so they recommend that chia seeds should be heated at temperatures below or equal to 90 °C to preserve their fatty acid profile. Interestingly, the fat content reported by other authors [9,11,29,30,40] is higher than in Table 2. However, as expected, lower fat values are observed when applying seed defatting as a purifying method (0.10 and 0.38%) [11,16].

Ash content in xyloglucan obtained from soaked seeds decreased significantly ($p < 0.05$) compared to the ash content in the xyloglucan extracted from roasted seeds. This difference has been attributed to leaching solid matter in soaking water, which includes minerals [36]. These results are associated with the calcium content, which was significantly lower ($p < 0.05$) in xyloglucan obtained from soaked seeds than that detected in xyloglucan from roasted seeds. Luo et al. [44] also found an important reduction in certain minerals (Ca, Mg, K, and Na) in faba beans and azuki beans after soaking in distilled water for 6 and 12 h at room temperature.

The soaking of *T. indica* seeds promoted a significant increase ($p < 0.05$) in crude fiber content concerning the roasting of seeds (Table 2). This can be explained as follows: cereals and legume seeds contain phytates, which are associated with bivalent minerals (Ca, Mg, Fe, and Zn) forming a phytate–mineral complex [45]. During soaking, these plant products undergo an important reduction in phytate content at 30 °C for 24 h as a result of hydrolysis from endogenous phytase [46]; then, phytate degradation triggers the release of the minerals aforementioned from the phytate–mineral complex, migrating towards the cell wall, where the soluble pectins are linked to monovalent minerals [47]; nevertheless, the release of divalent minerals, specifically Ca, replaces the monovalent minerals, and the pectin becomes insoluble, forming a strong cell wall [48], therefore being part of the crude fiber, which is an approximation of the cell wall material content in the vegetables [49].

No significant differences ($p < 0.05$) were observed in the carbohydrate content of xyloglucan obtained from seeds decorticated by soaking and roasting (Table 2). These results are higher than those reported previously, with ranges of 62.3–89.56% [11,38,40] and consistent with the data reported by Limsagouan et al. [16]. These authors mentioned that xyloglucan is the main carbohydrate stored in the tamarind seeds; thus, low protein, fat, and ash contents in xyloglucan imply a high purity of the compound and, consequently, a high separation efficiency, as shown in Table 1.

Phenolic content and antioxidant capacity in xyloglucan obtained from seeds decorticated by soaking and roasting methods are shown in Table 2. Total phenolic compounds were higher ($p < 0.05$) in xyloglucan extracted from soaked seeds than roasted seeds. This can be attributed to degradation, oxidation, isomerization, decomposition, or polymerization reactions that are a result of the roasting process, which affects the phenolic content. Although moderate roasting conditions were used in this study, total phenolic content and antioxidant capacity were reduced. Nevertheless, lower temperatures and times during roasting were not enough to eliminate the seed coat.

Several studies have demonstrated that quercetin, rutin, catechin, or chlorogenic acid are thermolabile due to their hydroxylated structure [50]. In contrast, although the soaking treatment of the seeds was in aqueous media, this process was carried out at 4 °C and in a closed container, which decreases the probability of oxidation reactions or an increase in the solubility of phenolic compounds [51,52]. Interestingly, total phenolic compounds in xyloglucan obtained from soaked seeds are higher than those reported in xyloglucan obtained with various methods, involving hot water extraction [53], ultrasound-assisted extraction, microwave-assisted extraction, alkali extraction [9], extractions with subcritical water at different temperatures [16], and methods that include the defatted and deproteinization of the tamarind seeds [11]. Even, in some cases, phenolic compounds have not been detected in the xyloglucan extracted with the aforementioned methods [29,53]. About the antioxidant capacity, it was observed that this parameter was lower in xyloglucan extracted from roasted seed in comparison to xyloglucan obtained from soaked seeds, suggesting that phenolic compounds are mostly responsible for the antioxidant activity samples, which depend on media and heat treatment [16,54,55]. These results are in line with those of Sorourian et al. [56] and Maghsoudlou et al. [50], who pointed out that the antioxidant activity of polysaccharides depends on covalently linked phenolic compounds and other factors (i.e., polysaccharide solubility, sugar ring structure, molecular weight, the occurrence of positive or negatively charged groups and protein moieties).

Significant differences ($p < 0.05$) were detected in galactose, glucose, and xylose content between samples (Table 2); additionally, the galactose/xylose/glucose ratios were in the range reported by Nguyen et al. [17], for xyloglucan extracted form tamarind seeds using microwave treatment for seed decortication analyzed by HPLC. The differences in the glucose/xylose/galactose between experimental samples can be related with the decortication method, which modified physicochemical properties of xyloglucan as is explained in the subsequent sections.

3.3. Fourier Transform Infrared Spectroscopy (FTIR)

The FTIR spectra of xyloglucan extracted from decorticated seeds by soaking and roasting are shown in Figure 2.

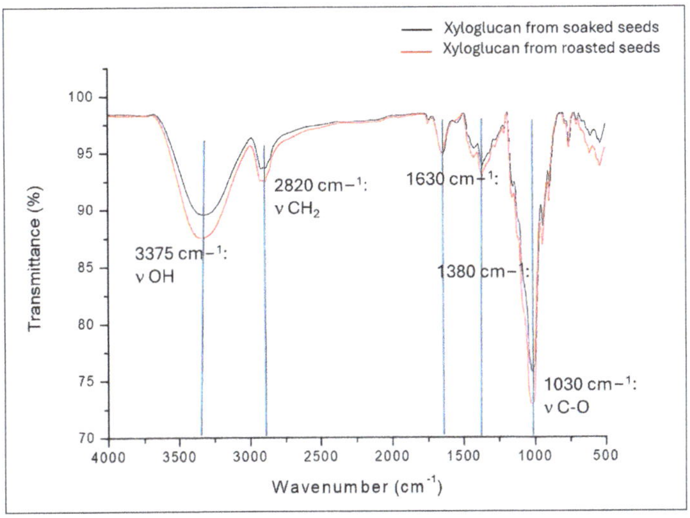

Figure 2. FTIR spectra of xyloglucan extracted from decorticated seeds by soaking and roasting.

Similar bands were observed in the xyloglucan FTIR spectra of xyloglucan obtained by soaking and roasting, implying that the chemical structure of xyloglucan was not altered due to the decorticating methods (Figure 2). Significant bands in the current spectra are those due to the broad stretching band of OH (3375 cm^{-1}), aliphatic C-H (2820 cm^{-1}), and C-O (1030 cm^{-1}), which have been reported previously for xyloglucan [10,11,30,36,42]. The peak at 1380 cm^{-1} has been considered characteristic of xyloglucan too [57], while the band at 1630 cm^{-1} can be attributed to the amide I (-C=O stretching) [58]. No stretching characteristic bands were identified for phenolic compounds; nonetheless, only a very weak band at 630 cm^{-1}, ascribable to the C=C group of these phenolics was detected. Another spectral fact is that the most important bands increased their intensities from the xyloglucan of decorticated seeds by soaking to the xyloglucan of decorticated seeds by roasting. These increases can be attributed to a better xyloglucan yield obtained from roasted tamarind seeds because of xyloglucan losses from seeds during the soaking process (Table 1).

3.4. Color Analysis

Color is a quality attribute that determines the industrial application of raw materials. The color parameters of xyloglucan extracted from decorticated seeds by soaking and roasting are shown in Table 3.

The color parameters differed significantly ($p < 0.05$) between samples. The L^* value in xyloglucan obtained from seeds decorticated by roasting decreased compared with that of xyloglucan extracted from seeds decorticated by soaking. In contrast, a^* and b^* values in xyloglucan from roasted seeds were higher than those observed in xyloglucan from soaked seeds. These results explain the darker color observed in xyloglucan obtained from roasted seeds. The color change in this sample can be explained by browning products derived from Maillard reactions, as a consequence of sugar interactions with amino acids from residual proteins during roasting [59]. These findings are in accordance with Limsangouan et al. [16], who detected an intense color in xyloglucan extracted using subcritical water. Color differences in xyloglucan determine its potential applications; for example, xyloglucan with a dark color can be used as an additive, in which the color of the final product is not a critical attribute for consumer acceptance. On the other hand, xyloglucan with a light color and that is translucent is ideal for biodegradable film development for food and pharmaceutical industries to obtain food packaging and pharmaceutical dosage forms.

Table 3. Color analysis and molecular weight of xyloglucan obtained by using two decorticate methods.

Decorticate Method	L^*	a^*	b^*	Molecular Weight (kDa)
Soaking	$+96.66 \pm 0.10$ [a]	$+0.20 \pm 0.01$ [a]	$+1.88 \pm 0.15$ [a]	816.50 ± 0.35 [a]
Roasting	$+92.31 \pm 0.12$ [b]	$+2.28 \pm 0.16$ [b]	$+10.96 \pm 0.12$ [b]	1926.97 ± 0.13 [b]

The values represent mean ± standard deviation (SD), $n = 5$. Means in columns with different letters differ significantly ($p \leq 0.05$).

3.5. Morphological Analysis

Figure 3a,b show xyloglucan obtained from soaked and roasted seeds, respectively. In both cases, fibrillar and elongated structures are evident; though, in Figure 3a, these fibrillar structures are wider and less rough than in Figure 3b. In the inner part of the fibrillar structures (Figure 3a,b), these structures are not visible (see arrows), while a continuous matrix with a relatively smooth surface is detected as shown in Figure 3c (see arrow). Similar results were observed by Choi et al. [60] and Limsangouan et al. [30]. These authors showed long interconnected fibril structures, as well as a continuous structure slightly porous in xyloglucan extracted from tamarind seeds with boiling water (conventional method). On the other hand, Alpizar-Reyes et al. [38] observed spherical particles with a polymeric appearance in isolated xyloglucan from tamarind seeds and dried with a spay dryer. Differences in xyloglucan morphology are related to degradation processes of this polysaccharide, as a result of extraction and drying methods [19,30,38,60].

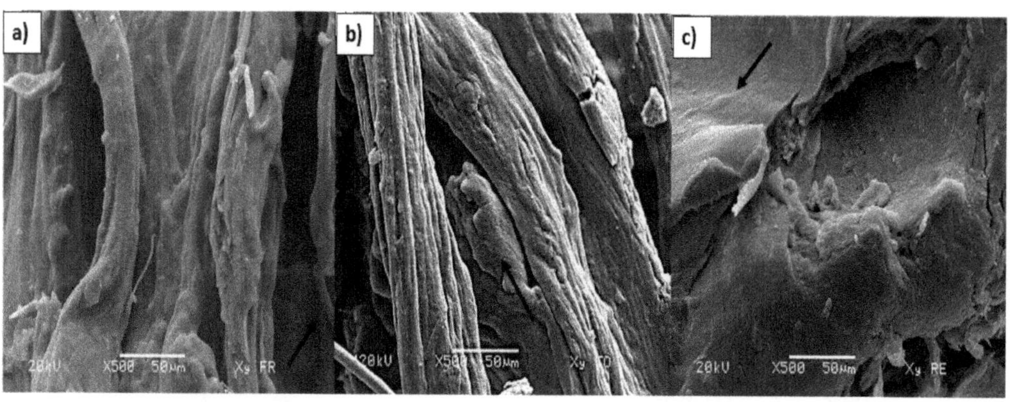

Figure 3. Micrographs of xyloglucan extracted from decorticated seeds by soaking (**a**,**b**) and roasting. (**c**) Detail of xyloglucan in the inner part of the fibrillar structures. Arrows in (**a**,**b**) denote the absence of fibers under the elongated structures and arrow in (**c**) denotes a smooth surface under the fibrillar structures.

3.6. Estimation of Molecular Weight

The average molecular weight of xyloglucan obtained from tamarind seeds decorticated by soaking and roasting methods is shown in Table 3. The values are in the range of 115–2500 kDa reported by Nishinari et al. [61]. Nevertheless, the molecular weight of xyloglucan extracted from soaked and roasted seeds was higher than the maximum values reported previously (770.9, 770.89, and 699 kDa) [16,30,60]. The authors attribute low molecular weights to critical extraction conditions (subcritical water, high pressure, gamma ray, and electron beam irradiation), which caused major deterioration of the molecular chains in the xyloglucan components compared to xyloglucan extracted by conventional methods.

The molecular weight of xyloglucan from roasted seeds was significantly higher ($p < 0.05$) than that observed in xyloglucan from soaked seeds. The method of decorticating affected the molecular weight of xyloglucan, which agrees with Kazemi et al. [62]. These authors mentioned that the molecular weight of hydrocolloids depends on the extraction methods, solvents, and time. The low molecular weight in xyloglucan from soaked seeds can be explained by the action of β-glucanases, which are enzymes involved in the synthesis, remodeling, and turnover of cell wall components during physiological processes [63]. These enzymes cleave cellulosic and non-cellulosic polysaccharides containing contiguous (1,4)-β-glucosyl residues in their backbone, including xyloglucan [64]. The β-1,4-glucanases are involved in plant responses toward abiotic and biotic stresses as a defense mechanism [63]. Soaked seeds were exposed to abiotic stress, such as low temperature and excess moisture (see Figure 1), which probably triggered the activation of (1,4)-β-glucanases favoring xyloglucan depolymerization. Interestingly, the glucose, xylose and galactose content in xyloglucan from soaked seeds was significantly lower than those detected in xyloglucan from roasted seeds (Table 2). These data support that the soaking method promotes the loss of monosaccharides, decreasing the molecular weight of the xyloglucan from soaked seeds. Probably, a decrease in xylose and galactose contents also reduce the branched chains in the xyloglucan molecule extracted with the soaking decorticate method.

3.7. Viscosity Analysis

The viscosity values versus shear rates of xyloglucan extracted from seeds decorticated by soaking and roasting methods at different concentrations are shown in Figures 4a and 4b, respectively.

In both cases, the xyloglucan at concentrations of 1.5 and 2% w/w, viscosity was independent of the shear rate obeying Newton's law. On the contrary, at concentrations of 2.5, 3.0, and 3.5% w/w, the viscosity decreases as shear force increases, showing the behavior of a non-Newtonian fluid (pseudoplastic fluid). The effect of shear rate on viscosity is attributed to the progressive intermolecular network degradation; therefore, greater shear decreases the opposition of the molecules dispersed in the sample [65]. Likewise, the viscosity of dispersions increased as a function of the concentration and similar trends have been reported previously by other authors [11,30,37,66]. The viscosity of xyloglucan obtained from roasted seeds is higher than that observed in xyloglucan from soaked seeds, mainly at the highest concentrations (3 and 3.5% w/w). Frequently, the viscosity of hydrocolloid dispersions increases with molecular weight [67], which correlates with the molecular weight of xyloglucan obtained from roasted seeds, which is higher than its counterpart (Table 3). In branched polymers, the movement of the polymer molecules is prevented or at least drastically suppressed, since the branched chains reduce the translational diffusion coefficient of the polymer, and, consequently, the viscosity is increased, which is more shear rate-dependent than that of linear polymers [68]. Possibly, higher xylose and galactose contents (responsible of the branching pattern of xyloglucan) in xyloglucan from roasted seeds than in xyloglucan from soaked seeds explain higher viscosity values in xyloglucan extracted with the roasting decorticate method.

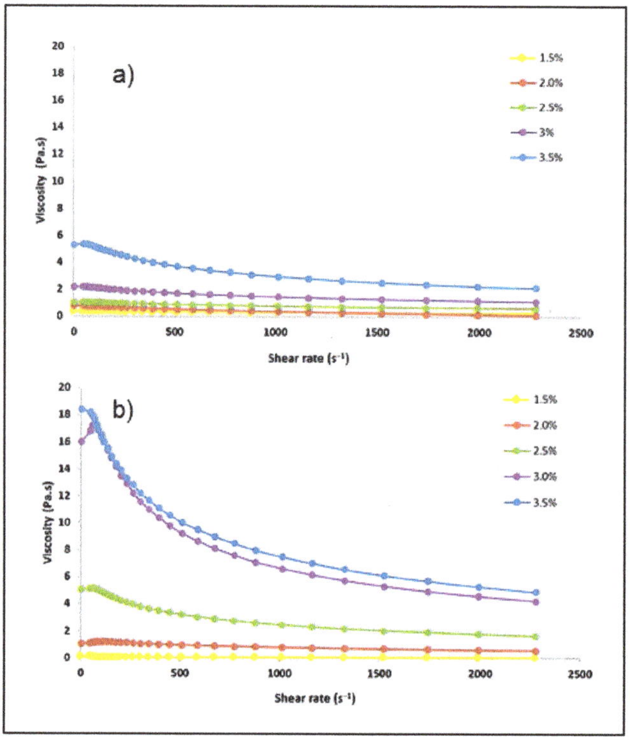

Figure 4. Viscosity values versus the shear rate of (**a**) xyloglucan extracted from seeds decorticated by soaking and (**b**) xyloglucan extracted from seeds decorticated by roasting at different concentrations (1.5, 2.0, 2.5, 3.0, and 3.5% *w*/*v*).

The viscosity values of xyloglucan at low concentrations (1.5 and 2% *w*/*w*) are comparable with the viscosity of xyloglucan extracted by the conventional method and applying spray drying in xyloglucan dispersions tested at the same concentrations [11,66]. However, the viscosity values observed in the present work are higher than the values reported for xyloglucan extracted by chemical, high-pressure, and subcritical water extraction methods [16,30,37]. Critical conditions applied during xyloglucan extraction with the aforementioned methods weaken the hydrogen bonds between molecules and leads to the xyloglucan chain reduction and, consequently, a decrease in viscosity [30].

Differences in the viscosity values of xyloglucan dispersions shown in Figure 4 determine their technological applications. For example, xyloglucan obtained by soaking seeds as a decorticate method can be used in pharmaceutical products with low viscosity (ophthalmological and nasal formulations) [8,69], as well as a thickening agent in juices and nectars [70]. On the other hand, xyloglucan from roasted seeds is adequate in film and coating development and as an excipient in gel formulation [71,72].

3.8. Texture Analysis

The texture attributes of xyloglucan gels obtained at 5% (*w*/*v*) are shown in Table 4.

Table 4. Instrumental texture attributes in xyloglucan obtained by soaked and roasted seeds of *Tamarindus indica*.

Sample	Hardness (N)	Springiness (%)	Adhesiveness (N)
Soaked seeds	1.39 ± 0.04 [a]	51.96 ±1.84 [a]	0.036 ± 0.007 [a]
Roasted seeds	19.66 ± 1.78 [b]	144.45 ± 3.87 [b]	−0.852 ± 0.001 [b]

Values represent the mean ± standard deviation (SD), $n = 5$. Means in columns with different letters differ significantly ($p \leq 0.05$).

In both cases, the xyloglucan samples developed soft gels, and, consequently, it was only possible to measure the parameters of hardness, springiness, and elasticity. Differences between xyloglucan samples in the aforementioned parameters were found to be significant ($p < 0.05$). It has been mentioned that xyloglucan forms a thermoreversible gel in an aqueous solution with the addition of small molecules and large amounts of saccharose and ethylic alcohol, due to these compounds together with xyloglucan developing a network structure in contrast with the native xyloglucan, which is presented as a viscoelastic solution [30,72,73]. Textural properties such as adhesion and hardness of composite gels are attributable to Van der Waals' interactions, electrostatic forces, and molecular interactions via hydrogen bonds [74,75]. Most likely, the low values of textural parameters in xyloglucan from soaked seeds are a consequence of calcium losses (Table 2) and a low molecular weight as a result of possible xyloglucan depolymerization (Table 3), which reduce the molecular interactions between polysaccharide–polysaccharide chains and calcium–polysaccharide chains. It has been reported that the removal of galactose from tamarind xyloglucan promotes the gelation due to this molecular remotion increasing the ratio of hepta- and octa-saccharides and decreasing the ratio of nonasaccharides. The first saccharides do aggregate, and the last ones do not. This gelation is caused by the association of main chains by hydrophobic interaction; though, in aqueous solutions, xyloglucan molecules behave as almost flexible random-coil polymers and do not show any conformational change such as coil–helix transition, as seen in some other polysaccharides. This explains the soft gels developed by this polysaccharide [76].

Differences in the texture attributes of xyloglucans are useful in predicting their possible applications; for example, xyloglucan from roasted seeds is adequate for the development of food packaging and edible coatings, spreadable foods, and the formulation of mucoadhesive gels and films, while xyloglucan from soaked seeds is suitable for the manufacture of cosmetic and pharmaceutical products such as creams, suspensions, ophthalmic medications, and personal hygiene products.

3.9. Zeta Potential and Particle Size Determination

Zeta potential is a physical property, a magnitude of charge attraction or repulsion between particles, exhibited by any particle in suspension; i.e., it is the potential difference across phase boundaries between solids and liquids and is considered to be a key indicator of the stability of colloidal dispersions [77].

Statistical differences ($p \leq 0.05$) were detected in the Z potential of the xyloglucan samples obtained by soaked and roasted seeds in their respective dispersions (Figure 5). However, the Z potential in the first case is in the range of −5 to −11 mV observed in xyloglucan extracted by conventional methods [11].

On the other hand, the Z potential in both samples is higher than the Z potential of mucilage extracted from O. ficus indica [19]; this can be explained as follows: (1) during extraction methods, the impurities, such as proteins that contribute to negative total charge, are removed (see Table 1) [11], and (2) the lack of carboxyl groups in the xyloglucan structure, which, in a pH range from 2.0 to 9.0, increase the negative charge of the Z potential for several biopolymers [78]. A relationship between viscosity and Zeta potential has been studied, and clear conclusions have not been reported; nevertheless, it has been established that there is an inverse relation between them. That is, the viscosity reaches its maximum value when the Zeta potential is zero (or trends to zero) and vice versa in mineral colloidal suspensions [79] and in colloidal apple juices added with hydrocolloids (carboxymethylcellulose and xanthan gum) [80]. Similar trends were observed in xyloglucan from roasted seeds, in which zeta potential is nearest to zero than xyloglucan from soaked seeds (Figure 5), and, in the first one, viscosities reached the highest values (Figure 4). These results suggest that xyloglucan extracted from soaked and roasted seeds can be used as a stabilizing, thickening, and gelling agent in the pharmaceutical and food industries. Furthermore, the negative Z potential values observed in both xyloglucan samples indicate possible interactions between this hydrocolloid and cationic biopolymers, i.e., proteins to obtain micro and nanoparticles by the complex coacervation method, with potential applications in controlled drug delivery systems [81].

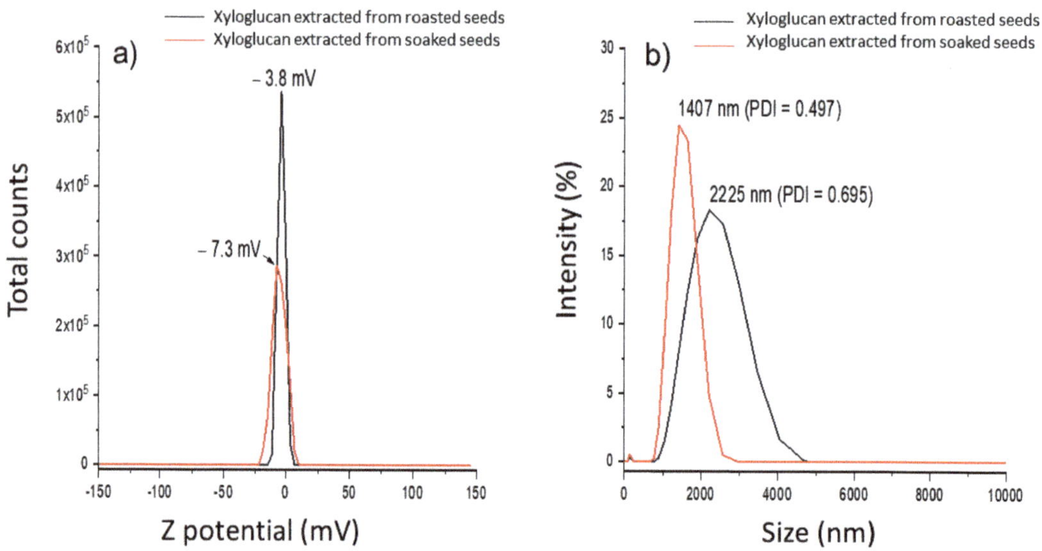

Figure 5. Representative graph of (**a**) the Zeta potential and (**b**) particle size distribution of the xyloglucan extracted from *T. indica* seeds. PDI = polydispersity index.

The particle size distribution of the xyloglucan obtained from roasted seeds was significantly higher ($p \leq 0.05$) than that observed for the xyloglucan from soaked seeds (2225 ± 42 and 1407 ± 151 nm, respectively); nonetheless, these values are above those reported previously (983.2 nm) [11]. High values of the polydispersity index (PDI > 0.5) imply less uniformity of the size distribution of particles [82]. Moreover, it has been pointed out that wide-molecular-weight distribution is due to a high proportion of long-chain xyloglucan molecules [30], and this agrees with molecular weights observed in the experimental samples (Table 3).

3.10. Thermal Analysis

The thermal properties of the xyloglucan samples were determined by thermogravimetric analysis (TGA) and differential scanning calorimetry (DSC). Results are shown in Table 5.

Table 5. Derivative thermogravimetric, residue content, and glass transition (Tg) of xyloglucan obtained by soaked and roasted seeds of *Tamarindus indica*.

Sample	Stage 1				Stage 2				
	T Range °C	Weight Loss %	T Peak °C	ΔH (J/g)	T Range °C	Weight Loss %	T Peak °C	Residues at 700 °C %	Tg °C
Soaked seeds	44–171	7.19 ± 0.36 [a]	93.63	208.8	392–431	76.88 ± 2.1 [a]	330.10	10.8 ± 0.20	254
Roasted seeds	42–138	9.16 ± 0.86 [a]	112.21	236.9	302–398	66.32 ± 1.3 [b]	326.96	13.5 ± 0.50	258

Means in columns with different letters differ significantly ($p \leq 0.05$).

The TGA evidenced two stages of weight loss for both samples. The weight loss at the first stage has been associated with the loss of adsorbed and bound water [42,83], while the second one is related to xyloglucan decomposition [42,83,84]. Likewise, it has been pointed out that a weight loss between 100 and 200 °C relates to the loss of coordinated mineral compounds containing calcium [85]. Interestingly, the highest calcium content in experimental samples was observed in xyloglucan from roasted seeds (Table 2), which means that, during the soaking of tamarind seeds, a fraction of coordinated calcium compounds was removed, and, consequently, the weight loss was minor at this stage compared to xyloglucan from roasted seeds.

At the second stage of decomposition, the xyloglucan from soaked seeds showed a weight loss 10% greater than that observed for xyloglucan from roasted seeds (Table 5). In general, longer polymer chains tend to have greater intermolecular forces and crosslinking (hydrogen bonds) than shorter chains, which, in the first case, corresponds to a greater molecular order (greater crystallinity) [86]. This fact agrees with a higher molecular weight observed in xyloglucan from roasted seeds compared to that detected in xyloglucan from soaked seeds (Table 3). Xyloglucan from roasted seeds showed a lower mass loss up to 700 °C, residues of which were higher compared to those obtained for xyloglucan from soaked seeds (13.5 and 10.8%, respectively); this is according to the higher calcium content in the xyloglucan from roasted seeds (Table 2). Therefore, these data support that xyloglucan from roasted seeds is more resistant to thermal degradation than xyloglucan from soaked seeds. Probably, a higher glucose content in xyloglucan from roasted seeds than in xyloglucan from soaked seeds (Table 2) denotes the prevalence of long glucose linear chains, which show a higher thermal resistance than do short glucose chains in xyloglucan from soaked seeds.

The xyloglucan from roasted seeds shows a temperature of water evaporation starting at 50 °C and a maximum value at 112 °C, while the glass transition temperature was detected with a soft signal at 258 °C. On the other hand, xyloglucan from soaked seeds exhibits a temperature of water evaporation starting at 40 °C and a glass transition temperature at 254 °C with a maximum value of 97 °C. Interestingly, the glass transition temperature observed for experimental samples is similar to data reported by Marais et al. [87] and Del Real et al. [88] (252 and 242 °C, respectively). The enthalpy of vaporization of xyloglucan from roasted seeds was higher (236.9 J/g) than that detected for xyloglucan from soaked seeds (208.8 J/g). Likewise, the maximum temperature of water vaporization for xyloglucan from roasted seeds shows a difference of 11 °C to its counterpart. These results suggest that, in the xyloglucan from roasted seeds, there is a greater interaction between the polysaccharide linear chains, compared to the xyloglucan from soaked seeds, which provides greater rigidity to its structure.

The most important aspects to consider from the results derived from this study are as follows: (1) The yield observed for the xyloglucan extracted from roasted and soaked seeds was similar, (2) the total phenolic content and antioxidant capacity of xyloglucan from soaked seeds was higher compared with xyloglucan from roasted seeds, and (3) the

MW, viscosity, Z potential, particle size distribution, textural properties (adhesiveness and springiness), and thermal resistance in xyloglucan from roasted seeds was higher than xyloglucan from soaked seeds. On the other hand, the xyloglucan yield observed in this current study, compared with those methods most recently reported in the scientific literature, was higher than that reported for xyloglucan extracted from tamarind seeds using microwave treatment for the seed's decortication. The yield detected in our xyloglucan extraction method was similar to that observed by other extraction methods such as the high-pressure assisted process, ultrasound-assisted process combined with γ-irradiation, and electron beam irradiation. The super-critical water extraction method is one of the most efficient methods for the extraction of xyloglucan since the yields are greater than 50%, and the antioxidant capacity, as well as the total phenolic content, is the highest recently reported. The viscosity values and thermal stability of xyloglucan detected in the xyloglucan samples obtained with the two decorticating methods in the current study are higher than those reported with the aforementioned extraction processes. Our results are encouraging, since no methods reported at present have performed extractions on an industrial scale, and most of them do not mention the decorticating method used as a pre-treatment procedure of raw material.

4. Conclusions

In this current study, differences in the chemical composition and physicochemical properties of xyloglucan extracted from tamarind seeds decorticated by soaking and roasting methods were investigated. The yields and separation efficiencies obtained in this work were comparable with those observed previously by using different extraction methods, with the advantage that, in our case, the use of solvents and enzymes is avoided. In addition, the roasting time and, consequently, energy consumption are also reduced according to the criteria for an eco-friendly process. However, higher yields are indeed observed through innovative extraction technologies; in this work, the extraction of xyloglucan from tamarind seeds using a mechanical separation process on a semi-industrial scale was demonstrated for the first time. This fact is important to evaluate the feasibility of implementing a process easily adapted by small, medium, and large-scale producers since greater yields for xyloglucan extraction imply increasing costs for manufacturers and issues from an ecological point of view. The low contents of protein and fat in xyloglucan samples obtained from roasted and soaked seeds indicate that their purities are comparable to xyloglucan extracted by different extraction methods. The viscosity, molecular weight, textural properties (Hardness and springiness), Zeta potential, polydispersity index, and glass transition temperature of the xyloglucan extracted from soaked seeds were significantly lower than those of xyloglucan extracted from roasted seeds. These results support that the decortication method of *T. indica* seeds, as a pretreatment stage of raw material, influences the chemical composition and physicochemical properties of xyloglucan. This fact has not been studied in previous reports in this area and must be considered for suitable applications of xyloglucan in the food processing and pharmaceutical industries. Further studies are required, from an engineering point of view, for scale-up of the xyloglucan extraction at an industrial level, applying the soaking and roasting methods for the seeds decorticating.

Author Contributions: Conceptualization—E.G.-C.; supervision—A.M.-A.; funding acquisition and writing—original draft preparation—I.R.-M.; investigation—A.A.-G.; methodology—A.M.-C., A.U.S., F.L.-V. and M.E.R.-O.; investigation—E.G.-C., V.E.-S. and M.B.; formal analysis—V.E.-S.; writing—review and editing—A.R.-M. All authors have read and agreed to the published version of the manuscript.

Funding: This research was funded by the "Chemistry We are All" Fund 2023 (FQST2023), through the Research Promotion Funding in the Consolidated Modality-FOMIC 2023, Project FQU 202305, and Student Projects Modality-FOPES 2023 of the Autonomous University of Queretaro, Chemistry Department and the Support Program for Research and Technological Innovation Projects (PAPIIT-2022, Project IA207121) of the National Autonomous University of Mexico (DGAPA-UNAM), PAPIME-2022, PE 103922 (DGAPA-UNAM). Cathedra CI2452.

Institutional Review Board Statement: Not applicable.

Data Availability Statement: The original contributions presented in this study are included in the article. Further inquiries can be directed to the corresponding authors.

Acknowledgments: Valeria Espíndola Sotres thanks the CONAHCYT for their Ph.D. scholarship and "Chemistry we are all" Fund 2023 (Research Promotion in de Student Projects Modality-FOPES 2023). The authors also thank the National Laboratory for the Characterization of Materials (CFATA-UNAM) for the facilities provided for the assessment of the morphological properties of experimental samples. The authors also thank Flor González Cataño (Laboratorio Especializado en Análisis de Alimentos. Posgrado de Alimentos-UAQ) and Leticia García Aguilar (Laboratorio de Farmacología. Posgrado en Ciencias Químico Biológicas-UAQ) for the technical support to analyze the monosaccharides content in samples.

Conflicts of Interest: The authors declare no conflicts of interest.

References

1. Paritosh, K.; Kushwaha, S.K.; Yadav, M.; Pareek, N.; Chawade, A.; Vivekanand, V. Food Waste to Energy: An Overview of Sustainable Approaches for Food Waste Management and Nutrient Recycling. *Biomed. Res. Int.* **2017**, *2017*, 2370927. [CrossRef] [PubMed]
2. Sharma, P.; Gaur, V.K.; Sirohi, R.; Varjani, S.; Kim, S.H.; Wong, J.W. Sustainable processing of food waste for production of bio-based products for circular bioeconomy. *Bioresour. Technol.* **2021**, *325*, 124684. [CrossRef] [PubMed]
3. Maurizzi, E.; Bigi, F.; Quartieri, A.; De Leo, R.; Volpelli, L.A.; Pulvirenti, A. The Green Era of Food Packaging: General Considerations and New Trends. *Polymers* **2022**, *14*, 4257. [CrossRef] [PubMed]
4. Bastida, C.O. Estadísticas de Producción de Tamarindo en México. Available online: https://blogagricultura.com/estadisticas-tamarindo-mexico/ (accessed on 25 February 2024).
5. Sánchez-Silva, J.M.; González-Estrada, R.R.; Blancas-Benítez, F.J.; Fonseca-Cantabrana, A. Utilización de subproductos agroindustriales para la bioadsorción de metales pesados. *Rev. Espec. En Cienc. Químico-Biológicas* **2020**, *23*, 1–18. [CrossRef]
6. Kumar, C.S.; Bhattacharya, S. Tamarind seed: Properties, Processing and Utilization. *Crit. Rev. Food Sci. Nutr.* **2008**, *48*, 1–20. [CrossRef]
7. Sharma, M.; Mondal, D.; Mukesh, C.; Prasad, K. Preparation of tamarind gum based soft ion gels having thixotropic properties. *Carbohydr. Polym.* **2014**, *102*, 467–471. [CrossRef]
8. Kulkarni, A.D.; Joshi, A.A.; Patil, C.L.; Amale, P.D.; Patel, H.M.; Surana, S.J.; Belgamwar, V.S.; Chaudhari, K.S.; Pardeshi, C.V. Xyloglucan: A functional biomacromolecule for drug delivery applications. *Int. J. Biol. Macromol.* **2017**, *104 Pt A*, 799–812. [CrossRef]
9. Thivya, P.; Prakash, R.N.; Ramachandran, N.S. Extraction of xyloglucan from tamarind industrial waste by different methods and their potential application in the food sector. *Int. J. Food Sci. Technol.* **2022**, *58*, 2014–2020. [CrossRef]
10. Chawananorasest, K.; Saengtongdee, P.; Kaemchantuek, P. Extraction and Characterization of Tamarind (*Tamarind indica* L.) seed polysaccharides (TSP) from three difference sources. *Molecules* **2016**, *21*, 775. [CrossRef]
11. Crispin-Isidro, G.; Hernández-Rodríguez, L.; Ramírez-Santiago, C.; Sandoval-Castilla, O.; Lobato-Calleros, C.; Vernon-Carter, E. Influence of purification on physicochemical and emulsifying properties of tamarind (*Tamarindus indica* L.) seed gum. *J. Food Hydrocoll.* **2019**, *93*, 402–412. [CrossRef]
12. Al-Jobouri, A.H. Studying Some the functional properties of Tamarind *Tamarindus indica* L. mucilage. *Al-Qadisiyah J. Agr. Sci.* **2020**, *3*, 304–307. [CrossRef]
13. De Souza, C.F.; Lucyszyn, N.; Woehl, M.A.; Riegel-Vidotti, I.C.; Borsali, R.; Sierakowski, M.R. Property evaluations of dry-cast reconstituted bacterial cellulose/tamarind xyloglucan biocomposites. *Carbohydr. Polym.* **2013**, *93*, 144–153. [CrossRef] [PubMed]
14. Nagajothi, M.S.; Balasubramanian, A.; Renganayaki, P.R.; Sudha, P. Optimizing Roasting Standards for Quality Gum Production in Tamarind Seed Gum Production. *Int. J. Curr. Microbiol. App. Sci.* **2017**, *6*, 1964–1968. [CrossRef]

15. Mohan, C.C.; Harini, K.; Aafrin, B.V.; Priya, U.L.; Jenita, P.M.; Babuskin, S.; Sukumar, M. Extraction and characterization of polysaccharides from tamarind seeds, rice mill residue, okra waste, and sugarcane bagasse for its bio-thermoplastic properties. *Carbohydr. Polym.* **2018**, *186*, 394–401. [CrossRef] [PubMed]
16. Limsangouan, N.; Milasing, N.; Thongngam, M.; Khuwijitjaru, P.; Jittanit, W. Physical and chemical properties, antioxidant capacity, and total phenolic content of xyloglucan component in tamarind (*Tamarindus indica*) seed extracted using subcritical water. *J. Food Process. Preserv.* **2019**, *43*, e14146. [CrossRef]
17. Nguyễn, T.T.; Jittanit, W.; Srichamnong, W. Production of xyloglucan component extracted from tamarind (*Tamarindus indica*) seeds using microwave treatment for seed decortication. *J. Food Process. Preserv.* **2019**, *43*, e14055. [CrossRef]
18. Ong, K.L.; Kaur, G.; Pensupa, N.; Uisan, K.; Lin, C.S.K. Trends in food waste valorization for the production of chemicals, materials, and fuels: Case study South and Southeast Asia. *Bioresour. Technol.* **2018**, *248*, 100–112. [CrossRef]
19. Wang, Y.; Wang, S.; Liu, L. Recovery of natural active molecules using aqueous two-phase systems comprising of ionic liquids/deep eutectic solvents. *Green Chem. Eng.* **2021**, *3*, 5–14. Available online: http://www.sciengine.com/doi/10.1016/j.gce.2021.07.007 (accessed on 12 December 2024). [CrossRef]
20. Yao, T.; Song, J.; Zhou, C.; Shi, X. Recent progress of the applications of functionalized magnetic ionic liquids in sample pretreatment. *Sep. Purif. Tech.* **2024**, *341*, 126979. [CrossRef]
21. Buarque, F.S.; Gautério, G.V.; Coelho, M.A.Z.; Lemes, A.C.; Ribeiro, B.D. Aqueous Two-Phase Systems Based on Ionic Liquids and Deep Eutectic Solvents as a Tool for the Recovery of Non-Protein Bioactive Compounds—A Review. *Processes* **2023**, *11*, 31. [CrossRef]
22. Yao, T.; Li, H.; Ren, Y.; Feng, M.; Hu, Y.; Yan, H.; Peng, L. Extraction and recovery of phenolic compounds from aqueous solution by thermo-separating magnetic ionic liquid aqueous two-phase system. *Sep. Purif. Tech.* **2022**, *282 Pt B*, 120034. [CrossRef]
23. Quintero-García, M.; Gutiérrez-Cortez, E.; Bah, M.; Rojas-Molina, A.; Cornejo-Villegas, M.L.A.; Del Real, A.; Rojas-Molina, I. Comparative Analysis of the Chemical Composition and Physicochemical Properties of the Mucilage Extracted from Fresh and Dehydrated *Opuntia ficus indica* Cladodes. *Foods* **2021**, *10*, 2137. [CrossRef] [PubMed]
24. Pirsa, S.; Hafezi, K. Hydrocolloids: Structure, preparation method, and application in the food industry. *Food Chem.* **2023**, *399*, 1333967. [CrossRef] [PubMed]
25. Khounvilay, K.; Sittikijyothin, W. Rheological behavior of tamarind seed gum in aqueous solutions. *Food Hydrocoll.* **2012**, *26*, 334–338. [CrossRef]
26. Gutiérrez-Cortez, E.; Rojas-Molina, J.I.; Zambrano-Zaragoza, M.L.; Quintanar-Guerrero, D.; González-Reza, R.M.; Rojas-Molina, A.; Espinosa-Arbeláez, D.G. Effect of processing conditions on the production of nixtamalized corn flours by the traditional method. *CyTA J. Food.* **2013**, *11*, 46–53. [CrossRef]
27. The Association of Official Analytical Chemists. *Official Methods of Analysis*, 17th ed.; AOAC International: Gaithersburg, MD, USA, 2000; Methods 925.45, 991.20, 945.16, 923.03 and 962.09.
28. Li, H.B.; Cheng, K.W.; Wong, C.; Fan, K.W.; Chen, F.; Jiang, Y. Evaluation of antioxidant capacity and total phenolic content of different fractions of selected microalgae. *Food Chem.* **2007**, *102*, 771–776. [CrossRef]
29. Thivya, P.; Bhosale, Y.K.; Anandakumar, S.; Hema, V.; Sinija, V.R. Exploring the effective utilization of Shallot stalk waste and Tamarind seed for packaging film preparation. *Waste Biomass Valorization* **2021**, *12*, 5779–5794. [CrossRef]
30. Limsangouan, N.; Charunuch, C.; Sastry, S.K.; Srichamnong, W.; Jittanit, W. High-pressure processing of tamarind (*Tamarindus indica*) seed for xyloglucan extraction. *LWT Food Sci. Technol.* **2020**, *134*, 110112. [CrossRef]
31. Sixto-Berrocal, A.M.; Vázquez-Aldana, M.; Miranda-Castro, S.P.; Martínez-Trujillo, M.A.; Cruz-Díaz, M.R. Chitin/chitosan extraction from shrimp shell waste by a completely biotechnological process. *Int. J. Biol. Macromol.* **2023**, *230*, 123204. [CrossRef]
32. Kasaai, M.R. Calculation of Mark-Houwink-Sakurada (MHS) equation viscometric constants for chitosan in any solvent-temperature system using experimentally reported viscometric constants data. *Carbohydr. Polym.* **2007**, *68*, 477–488. [CrossRef]
33. Sze, A.; Erickson, D.; Ren, L.; Li, D. Zeta-potential measurement using the Smoluchowski equation and the slope of the current-time relationship in electroosmotic flow. *J. Colloid Interface Sci.* **2003**, *261*, 402–410. [CrossRef] [PubMed]
34. Daoub, R.M.; Elmubarak, A.H.; Misran, M.; Hassan, E.A.; Osman, M.E. Characterization and functional properties of some natural *Acacia* gums. *J. Saudi Soc. Agric. Sci.* **2008**, *17*, 241–249. [CrossRef]
35. Njoumi, S.; Josephe-Amiot, M.; Rochette, I.; Bellagha, S.; Mouquet-Rivier, C. Soaking and cooking modify the alpha-galacto-oligosaccharide and dietary fiber content in five Mediterranean legumes. *Int. J. Food Sci. Nutr.* **2019**, *70*, 551–561. [CrossRef] [PubMed]
36. El-Safy, S.; Mukhtar, E.; Salem, R. The impact of soaking and germination on chemical composition, antinutritional factors of some legumes and cereals grain seeds. *Alex. Sci. Exch.* **2013**, *34*, 499–513. Available online: https://journals.ekb.eg/article_3112_2748c579b24fb6d6e7b9881d061a0d1c.pdf (accessed on 23 October 2024).
37. Carvalho, R.T.R.; Pedrosa, C.D.; Leal, A.C.P.; Palermo, L.C.M.; Mansur, C.R.E. Extraction, characterization and rheological behavior of Tamarind gum under high salinity. *Braz. J. Anal. Chem.* **2022**, *9*, 62–75. [CrossRef]

38. Alpizar-Reyes, E.; Carrillo-Navas, H.; Gallardo-Rivera, R.; Varela-Guerrero, V.; Álvarez-Ramírez, J.; Pérez-Alonso, C. Functional properties and physicochemical characteristics of tamarind (*Tamarindus indica* L.) seed mucilage powder as a novel hydrocolloid. *J. of Food Eng.* **2017**, *209*, 68–75. [CrossRef]
39. Sutherland, K. Other types of separation equipment. In *Filters and Filtration Handbook*; Elsevier Science: Oxford, UK, 2008; pp. 451–473. Available online: https://fastcdn.pro/filegallery/golcode.com/Downloads/Filters%20and%20Filtration%20Handbook%205th%20Edition.pdf (accessed on 8 July 2024).
40. Hettiarachchi, S.S.; Kumari, W.; Amarakoon, R. Extraction of Xyloglucan polymer from Tamarind (*Tamarindus indica*) seeds. *European J. Adv. Chem. Res.* **2023**, *4*, 15–21. [CrossRef]
41. Zilagvi, S.; Mocan, M. Scaling up resource efficiency and cleaner production for a sustainable industrial development. *Procedia Soc. Behav. Sci.* **2018**, *238*, 466–474. [CrossRef]
42. Jing, B.; Guo, R.; Wang, M.; Zhang, L.; Yu, X. Influence of seed roasting on the quality of glucosinolate content and flavor in virgin rapeseed oil. *LWT-Food Sci Technol.* **2020**, *126*, 109301. [CrossRef]
43. Ghafoor, K.; Ahmed, I.M.; Özcan, M.M.; Al-Juhaimi, F.Y.; Babikr, E.E.; Azmi, I.U. An evaluation of bioactive compounds, fatty acid composition, and oil quality of chia (*Salvia hispanica* L.) seed roasted at different temperatures. *Food Chem.* **2020**, *333*, 127531. [CrossRef]
44. Luo, Y.W.; Xie, W.H.; Jin, X.X.; Wang, Q.; Zai, X.M. Effects of germination and cooking for enhanced in vitro iron, calcium and zinc bioaccessibility from faba bean, azuki bean, and mung bean sprouts. *CYTA J. Food.* **2013**, *11*, 318–323. [CrossRef]
45. Gonçalves, A.; Goufo, P.; Barros, A.; Domínguez-Perles, R.; Trindade, H.; Rosa, E.A.; Rodrigues, M. Cowpea (*Vigna unguiculata* L. Walp), a renewed multipurpose crop for a more sustainable agri-food system: Nutritional advantages and constraints. *J. Sci. Food Agric.* **2016**, *96*, 2941–2951. [CrossRef] [PubMed]
46. Lestienne, I.; Icard-Vernière, C.; Mouquet, C.; Picq, C.; Trèche, S. Effects of soaking whole cereal and legume seeds on iron, zinc and phytate contents. *Food Chem.* **2005**, *89*, 421–425. [CrossRef]
47. Phillips, R.D.; McWatters, K.H.; Chinnan, M.S.; Hung, Y.C.; Beuchat, L.R.; Sefa-Dedeh, S.; Sakyi-Dawson, E.; Ngoddy, P.; Nnanyelugo, D.; Enwere, J.; et al. Utilization of cowpeas for human food. *Field Crops Res.* **2003**, *8*, 193–213. [CrossRef]
48. Kruger, J.; Minnis-Ndimba, R.; Mtshali, C.; Minnaar, A. Novel in situ evaluation of the role minerals play in the development of the hard-to-cook (HTC) defect of cowpeas and its effect on the in vitro mineral bioaccessibility. *Food Chem.* **2015**, *174*, 365–371. [CrossRef]
49. Dai, F.J.; Chau, C.F. Classification and regulatory perspectives of dietary fiber. *J. Food Drug Anal.* **2017**, *25*, 37–42. [CrossRef]
50. Maghsoudlou, Y.; Asghari, G.M.; Tavasoli, S. Effects of heat treatment on the phenolic compounds and antioxidant capacity of quince fruit and its tisane's sensory properties. *J. Food Sci. Tech.* **2019**, *56*, 2365–2372. [CrossRef]
51. Esparza, I.; Cimminelli, M.J.; Moler, J.A.; Jiménez-Moreno, N.; Ancín-Azpilicueta, C. Stability of phenolic compounds in grape stem extracts. *Antioxidants* **2020**, *9*, 720. [CrossRef]
52. Zapata, J.E.; Sepúlveda, C.T.; Álvarez, A.C. Kinetics of the thermal degradation of phenolic compounds from achiote leaves (*Bixa orellana* L.) and its effect on the antioxidant activity. *Food Sci. Technol.* **2021**, *42*, e30920. [CrossRef]
53. Kumar, G.P.; Battu, G.B.; Kotha, N.S.; Raju, L.R. Isolation and evaluation of tamarind seed polysaccharide being used as a polymer in pharmaceutical dosage forms. *Res. J. Pharm. Biol. Chem. Sci.* **2011**, *2*, 274–290.
54. Turturică, M.; Stănciuc, N.; Bahrim, G.; Râpeanu, G. Effect of thermal treatment on phenolic compounds from plum (Prunus domestica) extracts–A kinetic study. *J. Food Eng.* **2016**, *171*, 200–207. [CrossRef]
55. Lang, G.H.; da Silva Lindemann, I.; Ferreira, C.D.; Hoffmann, J.F.; Vanier, N.L.; de Oliveira, M. Effects of drying temperature and long-term storage conditions on black rice phenolic compounds. *Food Chem.* **2019**, *287*, 197–204. [CrossRef]
56. Sorourian, R.; Khajehrahimi, A.E.; Tadayoni, M.; Azizi, M.H.; Hojjati, M. Ultrasound-assisted extraction of polysaccharides from *Typha domingensis*: Structural characterization and functional properties. *Int. J. Biol. Macromol.* **2020**, *160*, 758–768. [CrossRef]
57. Kozioł, A.; Cybulska, J.; Pieczywek, P.M.; Zdunek, A. Evaluation of structure and assembly of Xyloglucan from Tamarind seed (*Tamarindus indica* L.) with atomic force microscopy. *Food Biophys.* **2015**, *10*, 396e402. [CrossRef] [PubMed]
58. Rezaei, A.; Nasirpour, A.; Tavanai, H. Fractionation and some physicochemical properties of almond gum (*Amygdalus communis* L.) exudates. *Food Hydrocoll.* **2016**, *60*, 461e469. [CrossRef]
59. Starowicz, M.; Zieliński, H. How Maillard Reaction Influences Sensorial Properties (Color, Flavor and Texture) of Food Products? *Food Rev. Int.* **2019**, *35*, 707–725. [CrossRef]
60. Choi, J.; Kim, J.; Srinivasan, P.; Kim, J.; Park, H.; Byun, M.; Lee, J. Comparison of gamma-ray and electron beam irradiation on extraction yield, morphological and antioxidant properties of polysaccharides from tamarind seed. *Radiat. Phys. Chem.* **2009**, *78*, 605–609. [CrossRef]
61. Nishinari, D.; Takemasa, M.; Yamatoya, K.; Shirakawa, M. Xyloglucan. In *Handbook of Hydrocolloids*; Phillips, G.O., Williams, P.A., Eds.; Woodhead Publishing: Boca Raton, FL, USA, 2009; pp. 535–566.

62. Kazemi, M.; Khodaiyan, F.; Labbafi, M.; Saeid Hosseini, S.; Hojjati, M. Pistachio green hull pectin: Optimization of microwave-assisted extraction and evaluation of its physicochemical, structural and functional properties. *Food Chem.* **2019**, *271*, 663–672. [CrossRef]
63. Perrot, T.; Pauly, M.; Ramírez, V. Emerging roles of β-Glucanases in plant development and adaptative responses. *Plants* **2022**, *11*, 1119. [CrossRef]
64. Wilson, D.; Urbanowicz, B. Glycoside Hydrolase Family 9/Plant Endoglucanases. Available online: https://www.cazypedia.org/index.php/Glycoside_Hydrolase_Family_9/Plant_endoglucanases (accessed on 3 September 2024).
65. Pecora, T.M.G.; Ragazzo, B.; Bertin, W.; Ragonese, A.; Mascagni, M.; Maffei, P.; Pignatello, R. Rheological behavior of a new mucoadhesive oral formulation based on sodium chondroitin sulfate, xyloglucan and glycerol. *J. Funct. Biomater.* **2021**, *12*, 28. [CrossRef]
66. Alpizar-Reyes, E.; Román-Guerrero, A.; Gallardo-Rivera, R.; Varela-Guerrero, V.; Cruz-Olivares, J.; Pérez-Alonso, C. Rheological properties of tamarind (*Tamarindus indica* L.) seed mucilage obtained by spray-drying as a novel source of hydrocolloid. *Int. J. Biol. Macromol.* **2018**, *107*, 817–824. [CrossRef] [PubMed]
67. Saha, D.; Bhattacharya, S. Hydrocolloids as thickening and gelling agents in food: A critical review. *J. Food Sci. Tech.* **2010**, *47*, 587–597. [CrossRef] [PubMed]
68. Hwang, J.; Kokini, J. Structure and rheological function of side branches of carbohydrate polymers. *J. Text. Stud.* **1991**, *22*, 123–167. [CrossRef]
69. Dalvi, A.V.; Ravi, P.R.; Uppuluri, C.T.; Mahajan, R.R.; Katke, S.V.; Deshpande, V.S. Thermosensitive nasal in situ gelling systems of rufinamide formulated using modified tamarind seed xyloglucan for direct nose-to-brain delivery: Design, physical characterization, and in vivo evaluation. *Int. J. Pharm. Investig.* **2021**, *51*, 199–211. [CrossRef]
70. Zhang, H.; Cui, H.; Xie, F.; Song, Z.; Ai, L. Tamarind seeds polysaccharide: Structure, properties, health benefits, modification, and food applications. *Food Hydrocol.* **2024**, *155*, 110222. [CrossRef]
71. Santos, N.L.; Braga, R.C.; Bastos, M.S.; Cunha, P.L.; Mendes, F.R.; Galvão, A.M.; Passos, A.A. Preparation and characterization of Xyloglucan films extracted from *Tamarindus indica* seeds for packaging cut-up 'Sunrise Solo' papaya. *Int. J. Biol. Macromol.* **2019**, *132*, 1163–1175. [CrossRef]
72. Esquena-Moret, J. A Review of Xyloglucan: Self-Aggregation, Hydrogel Formation, Mucoadhesion and Uses in Medical Devices. *Macromol* **2022**, *2*, 562–590. [CrossRef]
73. Busato, A.P.; Reicher, F.; Domingues, R.; Meira-Silveira, J.L. Rheological properties of thermally xyloglucan gel from the seeds of *Hymenaea courbaril*. *Mater. Sci. Eng. C* **2009**, *29*, 410–414. [CrossRef]
74. Adhikari, B.; Howes, T.; Bhandari, B.R.; Truong, V. Stickiness in foods: A review of mechanisms and test methods. *Int. J. Food Sci. Technol.* **2001**, *4*, 1–33. [CrossRef]
75. Yan, X.; Jia, Y.; Man, H.; Liu, L.; Sun, S.; Qi, B.; Li, Y. Intermolecular interactions and gel properties of composite agglomerative networks based on oppositely charged polymers: Effects of pH and ionic strength. *Food Hydrocoll.* **2023**, *139*, 108557. [CrossRef]
76. Nishinari, K.; Takemasa, M.; Zhang, H.; Takahashi, R. Storage plant polysaccharides: Xyloglucans, galactomannans, glucomannans. In *Comprehensive Glycoscience from Chemistry to Systems Biology*, 1st. ed.; Hamerling, H., Ed.; Elsevier Science Ltd.: Amsterdam, The Netherlands, 2007; pp. 613–652. [CrossRef]
77. Sezer, N.; Muataz, A.; Atieh, M.; Koç, M. A comprehensive review on synthesis, stability, thermophysical properties, and characterization of nanofluids. *Powder Tech.* **2019**, *344*, 404–431. [CrossRef]
78. Cortés-Camargo, S.; Gallardo-Rivera, R.; Barragán-Huerta, B.; Dublán-García, O.; Román-Guerrero, A.; Pérez-Alonso, C. Exploring the potential of mesquite gum-nopal mucilage mixtures: Physicochemical and functional properties. *J. Food. Sci.* **2017**, *83*, 113–121. [CrossRef] [PubMed]
79. Ersoy, B.; Evcin, A.; Uygunoglu, T.; Akdemir, Z.B.; Brostow, W.; Wahrmund, J. Zeta Potential–Viscosity relationship in Kaolinite slurry in the presence of dispersants. *Arab. J. Sci. Eng.* **2014**, *39*, 5451–5457. [CrossRef]
80. Genovese, D.B.; Lozano, J.E. The effect of hydrocolloids on the stability and viscosity of cloudy apple juices. *Food Hydrocoll.* **2001**, *15*, 1–7. [CrossRef]
81. Zheng, J.; Van der Meeren, P.; Sun, W. New insights into protein-polysaccharide complex coacervation: Dynamics, molecular parameters, and applications. *Aggregate* **2024**, *5*, e449. [CrossRef]
82. Danaei, M.; Dehghankhold, M.; Ataei, S.; Hasanzadeh Davarani, F.; Javanmard, R.; Dokhani, A.; Khorasani, S.; Mozafari, M.R. Impact of particle size and polydispersity index on the clinical applications of lipidic nanocarrier systems. *Pharmaceutics* **2018**, *10*, 57. [CrossRef]
83. Dilbaghi, N.; Ahuja, M.; Bernela, M.; Kumar, S.; Bhardwaj, P.; Kaur, H. Synthesis and characterization of novel amphiphilic Tamarind seed Xylogulcan-Octenyl Succinic Anhydride conjugate. *J. Polym. Res.* **2020**, *27*, 1–8. [CrossRef]
84. Zohuriaan, M.; Shokrolahi, F. Thermal studies on natural and modified gums. *Polym. Test.* **2004**, *23*, 575–579. [CrossRef]
85. Bala, P.; Samantaray, B.K.; Srivastava, S.K. Dehydration transformation in Ca-montmorillonite. *Bull. Mater. Sci.* **2000**, *23*, 61–67. [CrossRef]

86. Indu-Yadav, V.S.; Rathnam, S.; Yogalakshmi, Y.; Chakraborty, S.; Banerjee, I.; Anis, A.; Pal, K. Synthesis and characterization of Polyvinyl Alcohol-Carboxymethyl Tamarind Gum. *Carbohydr. Polym.* **2017**, *165*, 159–168. [CrossRef]
87. Marais, A.; Kochumalayil, J.; Nilsson, A.; Fogeldtrön, L.; Gamstedt, E.K. Toward an Alternative Compatibilizer for PLA/Cellulose Composites: Grafting of Xyloglucan with PLA. *Carbohydr. Polym.* **2012**, *89*, 1038–1043. [CrossRef] [PubMed]
88. del Real, A.; Wallander, D.; Maciel, A.; Cedillo, G.; Loza, H. Graft Copolymerization of Ethyl Acrylate onto Tamarind Kernel Powder, and Evaluation of its Biodegradability. *Carbohydr. Polym.* **2015**, *117*, 11–18. [CrossRef] [PubMed]

polymers

MDPI

Article

Step-by-Step Analysis of a Copper-Mediated Surface-Initiated Atom-Transfer Radical Polymerization Process for Polyacrylamide Brush Synthesis Through Infrared Spectroscopy and Contact Angle Measurements

Leonardo A. Beneditt-Jimenez [1], Isidro Cruz-Cruz [2], Nicolás A. Ulloa-Castillo [3] and Alan O. Sustaita-Narváez [1,*]

1 School of Engineering and Sciences, Tecnologico de Monterrey, Ave. Eugenio Garza Sada Sur 2501, Monterrey 64840, NL, Mexico; leo.beneditt@tec.mx
2 Institute of Advanced Materials for Sustainable Manufacturing, Tecnologico de Monterrey, Ave. Eugenio Garza Sada Sur 2501, Monterrey 64840, NL, Mexico; isidro.cruz@tec.mx
3 Center for Innovation in Digital Technologies, School of Engineering and Sciences, Tecnologico de Monterrey, Ave. Eugenio Garza Sada Sur 2501, Monterrey 64849, NL, Mexico; nicolas.ulloa@tec.mx
* Correspondence: alan.sustaita@tec.mx

Abstract: Polymer brushes (PBs) are transformative surface-modifying nanostructures, yet their synthesis via controlled methods like copper-mediated surface-initiated atom-transfer radical polymerization (Cu^0-SI-ATRP) faces reproducibility challenges due to a lack of understanding of parameter interdependencies. This study systematically evaluates the Cu^0-SI-ATRP process for polyacrylamide brushes (PAM-PBs), aiming to clarify key parameters that influence the synthesis process. This evaluation followed a step-by-step characterization that tracked molecular changes through infrared spectroscopy (IR) and surface development by contact angle (CA) through two different mixing methods: ultrasonic mixing and process simplification (Method A) and following literature-based parameters (Method B). Both methods, consisting of surface activation, 3-aminopropyltriethoxysilane (APTES) deposition, bromoisobutyryl bromide (BiBB) anchoring, and polymerization, were analyzed by varying parameters like concentration, temperature, and time. Results showed ultrasonication during surface activation enhanced siloxane (1139→1115 cm^{-1}) and amine (1531 cm^{-1}) group availability while reducing APTES concentration to 1 Vol% without drying sufficed for BiBB anchoring. BiBB exhibited insensitivity to concentration but benefited from premixing, evidenced by sharp C–Br (~1170 cm^{-1}) and methyl (3000–2800 cm^{-1}) bands. Additionally, it was observed that PAM-PBs improved with Method A, which had reduced variance in polymer fingerprint regions compared to Method B. Adding to the above, CA measurements gave complementary step-by-step information along the modifications of the surface, revealing distinct wettability behaviors between bulk PAM and synthesized PAM-PBs (from 51° to 37°). As such, this work identifies key parameter influence (e.g., mixing, BiBB concentration), simplifies steps (drying omission, lower APTES concentration), and demonstrates a step-by-step, systematic parameter decoupling that reduces variability. In essence, this detailed parameter analysis addresses the PAM-PBs synthesis process with better reproducibility than the previously reported synthesis method and achieves the identification of characteristic behaviors across the step-by-step process without the imperative need for higher-cost characterizations.

Keywords: polyacrylamide; polymer brushes; Cu^0-SI-ATRP synthesis; surface functionalization; parameter interdependency; layer-by-layer analysis; reproducibility in synthesis

1. Introduction

Material science stands as the cornerstone of technological innovation because it enables solutions to global challenges through the development of novel materials, such as alloys, biomimetic materials, hybrid composites, and stimuli-responsive systems, for tailored functionalities [1]. Among these, polymers have emerged as a transformative class due to their versatility, scalability, and adaptability [2], evolving from traditional uses like packaging or electrical/thermal insulation to multifunctional applications in multifaceted materials, as in polymer-based smart devices [3]. This evolution is closely tied to progress and innovation in synthetic methodologies, where the development of more controlled methods has opened new possibilities.

Controlled polymerization techniques now enable precise manipulation of molecular weight, architecture, and composition, unlocking polymers with programmable behaviors [4,5]. The search for innovations has catalyzed the development of polymer brushes (PBs)— nanostructured materials where polymer chains are tethered to substrates with a high surface density [6]—due to their ability to modify surface properties without altering bulk material characteristics [7,8]. Some applications of PBs include medical solutions, microelectronics, environmental sensors, and surface engineering, among many others [9–12]. However, challenges persist in synthesis methods for industrial use due to their lack of reliability when they are applied to PBs [13–15]. As such, the synthesis of polymer brushes typically involves three stages: substrate functionalization with initiators, surface-initiated polymerization, and post-polymerization modifications [16,17], which demand rigorous control over grafting density and chain length dispersity [18,19]. These challenges have spurred the refinement of synthesis techniques like atom-transfer radical polymerization (ATRP) [20,21], which has allowed precise control over PBs architecture and its physicochemical properties by mitigating premature termination reactions, allowing the synthesis of block copolymers, star-shaped polymers, and hyperbranched structures [22–24]. The pursuit to further improve and control architecture and tailor properties led to the development of focused graft-from methods like surface-initiated ATRP (SI-ATRP), which allowed the growth of polymer chains from surface-bound initiators, offering further control over brush density and polymer chain length. Additionally, the use of transition metal catalysts (e.g., Cu^+, Fe^+, Ni^+) offered the ability to mediate radical growth, ensuring low polydispersity and high end-group fidelity [25–27]. Despite these advances, the reproducibility in the synthesis of PBs to form thin films remains elusive due to a lack of complete understanding of every step of the process in current methods [28,29].

Polyacrylamide (PAM), a water-soluble polymer, inspired the development of acrylamide-based polymer brushes due to their potential applications in enhanced oil recovery, hydrogel fabrication, biomedical engineering, surface engineering, microsensors, prevention of biofouling, and wastewater treatment [30–33]. However, the synthesis of PAM polymer brushes faces challenges in polymer conformation, film architecture, and mechanical robustness, leading to the advent of Cu^0-mediated SI-ATRP (Cu^0-SI-ATRP), i.e., an SI-ATRP method that uses copper as a metal mediator [34]. This method significantly lowered catalyst concentrations and mitigated oxidative side reactions that introduced variation in the resulting polymerization [34,35]. In addition, this synthesis process enhances industrial viability [36] because it can be performed under ambient conditions, is compatible with aqueous media without compromising synthesis integrity, and enables the deposition of ultra-thick brushes ($\geq$1 μm) with uniform deposition thickness and controlled polymer chain length [37–39]. However, challenges remain due to persistent drawbacks in the Cu^0-SI-ATRP process, including a lack of clarity on the effect of the surface preparation layers on the synthesis of PBs, an absence of standardized substrate preparation steps, and a lack of reproducible results in surface formations [33,40,41].

With the aim of improving the reproducibility of the Cu^0-SI-ATRP synthesis process, in this study, a systematic step-by-step analysis was conducted to define and clarify the impact of selected parameters on polymer brush formation. This approach aimed at understanding how these parameters hold interdependency from each other and how this affects overall polymer brush synthesis. By decoupling interdependent parameters, this study establishes a framework for reducing batch-to-batch variability. Ex situ step-by-step analysis using Fourier-Transform Infrared Spectroscopy (FTIR) provided molecular-level insights into the progression of the synthesis process, confirming successful initiator immobilization, polymerization, and relevance of some synthesis parameters. In addition, contact angle (CA) measurements allowed for assessing overall surface properties and mapping the uniformity of surface covering through multi-point droplet analysis. Reducing CA from 51° on pristine substrates to 37° after polymer grafting confirmed enhanced wettability. In addition, low standard deviations across measurement sites seem to indicate uniform brush coverage, which is critical for applications requiring interfacial control.

2. Materials and Methods

2.1. Materials

Acrylamide (AM, electrophoresis grade $\geq$ 99%), (3-Aminopropyl) triethoxysilane (APTES, 99%), methanol (MeOH, ACS reagent $\geq$ 99.8%), 2-Bromo-2-methylpropionyl bromide (BIBB, 98%), triethylamine (TEA, $\geq$99.5%), N,N,N′,N′′,N′′-pentamethyldiethylenetriamine (PMDETA, 99%), polyacrylamide (PAM, 40kMn), and anhydrous tetrahydrofuran (THF, $\geq$99.9% inhibitor-free) were obtained from Sigma-Aldrich (Naucalpan de Juarez, México). Ethyl alcohol (EtOH, anhydrous, $\geq$99.5%), hydrochloric acid (36–38%), and distilled water were obtained from JALMEK (San Nicolás de los Garza, México). Silicon wafers were used as substrates with dimensions of 15 × 10 mm^2. As metal mediators, 0.2 mm thick C1100 copper plates from Easycargo (Ciudad de México, México) were used. Spacers and holders were 3D printed out of polyethylene terephthalate glycol (PETG) obtained from Creality (Shenzhen, China). The anhydrous grade for some chemicals was used here as per availability, although most processes involved water washes and ambient conditions.

2.2. Methodology

The experimental process for the deposition of PBs on silicon substrates is schematically depicted in Figure 1. For surface activation, silicon substrates were subjected to three different soaking procedures in a 1:1 (v/v) MeOH:HCl solution for 30 min: without (w/o) mixing, under magnetic stirring (MS), and in an ultrasonic bath (US). Next, the wet-etch process was stopped by flooding the container with distilled water and flushing it for 5 min. This was followed by rinsing and flushing with MeOH. Finally, EtOH was introduced to cover the substrates, and they were rested for 5 min. The container with Si substrates was flushed, and fresh ethanol was added to cover the substrates.

For the deposition of the APTES layer, specific amounts of APTES were dropped into a container using a syringe to obtain one of the three concentrations in ethanol: 1 Vol%, 3 Vol%, and 5 Vol%. This deposition step was conducted at two solution temperatures: room temperature (RT) and 60 °C. Previous studies indicated that higher temperatures helped stabilize and accelerate the process [42]. APTES deposition was carried out for 30 min without (w/o) a mixing process, under magnetic stirring, or in an ultrasonic bath (US). Subsequently, the silicon/APTES samples were rinsed with EtOH and dried under one of the following three conditions: in an oven at 70 °C for 30 min, at room conditions for around 30 min, or simply the substrates were removed from the precursor solution and placed in the following stage (BiBB anchoring) without the drying step.

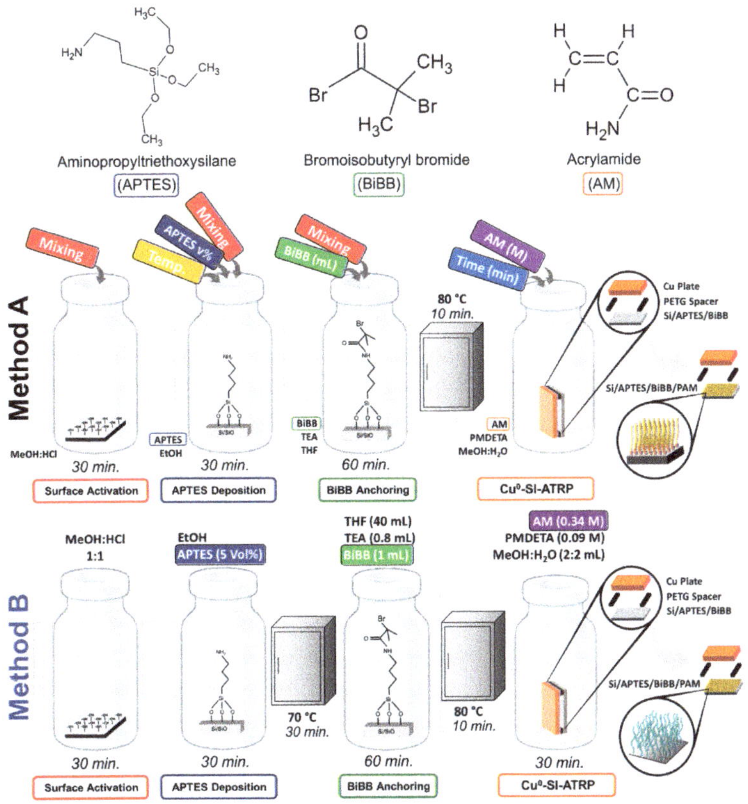

Figure 1. Schematic representation of Cu^0-SI-ATRP process for the synthesis of PAM-PBs for Method A and Method B.

For the BiBB anchoring over the APTES layer, a container was filled with THF, to which TEA was added. Next, BiBB was then dropped down the walls of the container to obtain a concentration of 40:0.8:x (THF:TEA:BiBB) by volume, where x = 1 stands for the base concentration. Other values of x considered in this study were 0.5 and 2. Since there is no consensus in the scientific literature on the method of mixing at this step [43,44], we investigated the effect on BiBB anchoring. Substrates were introduced into the BiBB solutions without mixing, under magnetic stirring, and in an ultrasonic bath for 60 min. Additionally, other substrates were introduced into premixed BiBB solutions (which had been mixed for 5 min) using the same three mixing methods: without (*w*/*o*) mixing, under magnetic stirring, and in an ultrasonic bath. The Si/APTES samples were allowed to rest immersed in the solution for 55 min, resulting in a total immersion time of 60 min. After initiator anchoring, the samples were removed from the BiBB precursor solution and placed in THF to rinse for 1 min to remove any residual reaction products. This rising step was followed by sequential washes with MeOH and distilled water. The Si/APTES/BiBB samples were then dried in an oven at 80 °C for 10 min, allowing for evaporation of any residual reagent.

The Cu pieces were prepared by placing them into a MeOH:HCl solution (1:1, *v*/*v*) for 5 min, followed by sequential washing with distilled water and MeOH, as per previous studies [37]. An aqueous MeOH solution (1:1 v/v) was used to prepare the ATRP solution by AM (at 0.34 M, 0.17 M, or 0.09 M) and PMDETA (0.001 M). Substrates were placed face to face with the prepared Cu pieces, separated by PETG spacers of 0.50 mm. This setup was vertically submerged in the ATRP solution and allowed to react for 15 min, 30 min, or 60 min. Upon elapsed time, the deposited films were rinsed with water and then with MeOH before being left to dry at room temperature (RT). This synthesis process will be labeled as "Method A".

As a reference, a general method previously reported in [36], which adapted Cu^0-SI-ATRP for various monomers, will be considered and denoted here as "Method B". According to this method, silicon substrates were introduced in a MeOH:HCl (1:1 v/v) solution for 30 min, washed, and placed in an alcoholic APTES solution (5 Vol%) for 30 min, from which it was removed and dried in an oven at 70 °C for 30 min. Silicon/APTES samples were introduced into the THF:TEA:BiBB solution (40:0.8:1, by volume) for 60 min and dried in an oven at 80 °C for 10 min. The substrate–spacers–Cu plate arrangement was submerged into the ATRP solution (containing AM at a concentration of 0.34 M) for 30 min before washing and left to dry at room temperature.

2.3. Characterization

Characterization of PAM-PBs thin films on Si wafer substrates, as well as every deposition layer, were performed on an FTIR spectrometer (PerkinElmer, Frontier 8000, México City, Mexico) in attenuated total reflectance (ATR) mode using a universal ATR accessory with a ZnSe crystal. Measurements were performed in the spectral range 4000–750 cm^{-1} by averaging 128 scans at a 2 cm^{-1} resolution. Spectra shown here were normalized to the most intense band in the spectral region of interest after baseline correction performed directly on the equipment's software to highlight relative differences. Regarding the criteria for the spectra in the figures shown here, two alternatives were considered depending on the specific step of interest. Thus, for the previous steps of the polymerization process, that is, from surface activation to BiBB anchoring, the spectra shown here are the average of at least 9 spectra per experimental condition. On the other hand, in the case of PAM and PAM-PBs materials, spectra were chosen considering those with the best signal-to-noise ratio, but the statistics were performed considering 16 measurements for every method.

CA tests were performed using Optical Contact Angle goniometer equipment from Dataphysics. For this, 5 to 6 droplets of 5 µL were automatically dispensed (with the help of the OCA21 software) and spaced throughout every sample to allow individual readings for determining the distribution and homogeneity of the PAM-PBs deposition.

3. Results

Analysis was performed using UV-Vis and Raman spectroscopies, AFM, and SEM. Based on the Alexander–DeGennes model (AdG) [2], the chain-to-chain distance for PAM can be estimated after considering the monomer molecular structure in the range of a few angstroms; therefore, microscopy techniques are incapable of perceiving or detecting polymer brushes accurately. Additionally, the extensive properties of brushes make their characterization difficult when using conventional spectroscopic techniques. Raman spectroscopy could not detect significant differences in the polymer structure. UV-Vis lacked reproducible results as the wavelength was too large compared to deposition thickness (~150 nm, as determined by AFM), and the technique was not susceptible to grafting variability. Therefore, these characterization techniques did not contribute to establishing conclusive differences associated with the specific synthesis process of PBs in this work, similar to what was reported in the literature [45–47]. Thus, these results will not be shown here. In contrast, step-by-step FTIR spectroscopy and CA measurements provided sufficient evidence to obtain new insights into the synthesis of PAM-PBs.

3.1. Surface Activation

The FTIR spectra in Figure 2 show the vibrational modes obtained after the APTES deposition on silicon substrates. The main functional groups associated with the surface activation are the silanol groups (–Si–OH), which facilitate the anchoring of groups related to the APTES molecule deposited from a solution at a fixed concentration of 5 Vol% in ethanol. The analysis focuses on identifying the amine groups that confirm the successful anchoring of APTES [48,49]. The amine vibrational modes were detected in the spectral range of 1800–1600 cm^{-1} and at 1200 cm^{-1}, while the siloxane bands were detected at around 1100 cm^{-1} and at 1000 cm^{-1}. The effect of the mixing process during this stage is also discussed.

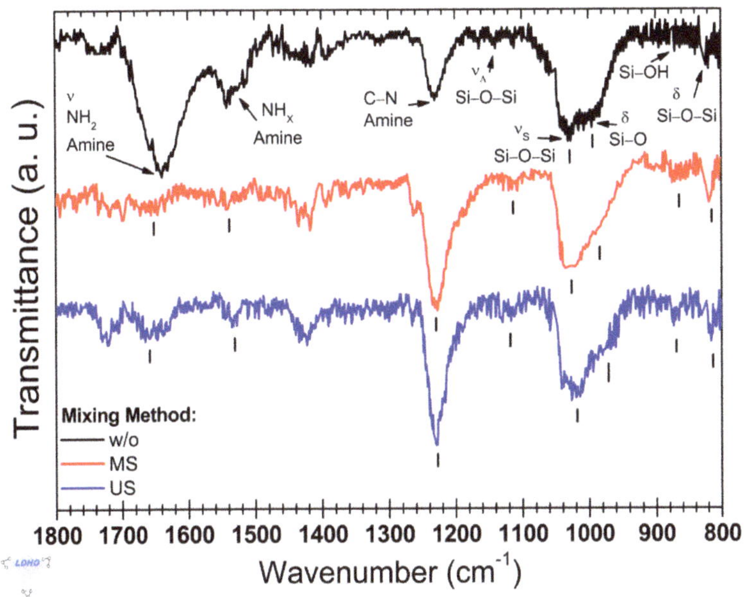

Figure 2. Representative normalized FTIR spectra as obtained after APTES layer deposition on silicon substrates. Surface activation was performed by following one of the three conditions: ultrasonic (US), magnetic stirring (MS), or without any mixing process.

In the sample without (*w/o*) a mixing process, the observed bands associated with siloxane groups are located at 1139 cm^{-1} and 1026 cm^{-1}, while the band corresponding to the silanol group is observed at 994 cm^{-1}. Additionally, primary amines were identified at 1638 cm^{-1}, which corresponds to the most intense band, and secondary amines are present at 1541 cm^{-1}. The band at 1231 cm^{-1} is attributed to the C–N stretching of amine. The primary amino band at 1638 cm^{-1} (being $primary\_A_{1638}$ its relative intensity) was found to be approximately 108% more intense than the secondary band at 1541 cm^{-1} (with $secondary\_A_{1541}$ as its relative intensity), resulting in a $primary\_A_{1638}/secondary\_A_{1541}$ ratio of about 2.1. This intensity is also 28% greater than the siloxane band at 1026 cm^{-1}. Similar values for this ratio have been reported in other studies, where they were attributed to multilayering [50,51]. Although the exact relationship is unclear, this intensity ratio may be influenced by the orientation of APTES attachment. Furthermore, due to the overlap of the amine bands, this ratio can help to confirm the presence of an end-group primary amine termination (NH_2) [52,53]. This topic will be explored in greater detail in a later section, but it is worth noting that the primary to secondary amine intensity ratio is crucial in understanding possible explanations related to multilayering, unfavorable attachment, or lack of attachment altogether [42].

The attachment of APTES under magnetic stirring showed differences in the FTIR spectra compared to the deposition process without mixing. In this stirred condition, the amine band observed at 1654 cm^{-1} appears as a shoulder to a smaller band at 1701 cm^{-1}, which is attributed to a carboxyl group. This carboxyl band is more pronounced when US agitation is applied. The latter can be related to the unattached siloxane groups with ethoxy terminations [54]. On the other hand, the end-termination of the amine group is characterized by the amine functional groups [53,54], where we calculated a primary-to-secondary amine intensity ratio of 1.6. This ratio is favorable for the subsequent anchoring of the BiBB layer. The results shown in Figure 2 reveal that the most prominent IR band appears at 1230 cm^{-1}, with a secondary band at 1031 cm^{-1}, attributed to symmetric Si–O–Si stretching, arising when surface activation is conducted under MS. The presence of this band supports the hypothesis of increased amine availability as the surface end-group. Notably, the silanol shoulder associated with the 1031 cm^{-1} band shifts compared to that of a process without any mixing method. Accompanying this shift is a sharper silanol band observed at 806 cm^{-1}, which is typically linked to the availability of surface silanol groups [42,52]. Also, the shoulder observed at 995 cm^{-1}, for the methodology without mixing (Figure 2, spectrum in black), exhibited a shift to 990 cm^{-1} and a broader width. These findings suggest that there is a reduction in APTES anchorage, as indicated in other studies [55]. Additionally, when the mixing method was utilized, a band with low intensity at 1264 cm^{-1} became present. This band has been attributed to asymmetric stretching of the Si–O–C functional group [42], but this spectral region can also manifest Si-CH_x [52] and CH_2 wagging vibrations [53], all of which are part of APTES backbone propyl chain or ethoxy groups. As the observed behavior of the 1264 cm^{-1} band because of the synthesis process does not contribute to the comprehension of surface coverage or APTES deposition, they will not be considered in this discussion.

On the other hand, when surface activation was performed in an ultrasonic bath (US), its most intense band associated with the amine group of APTES was located at 1233 cm^{-1} (the C–N stretching mode of amine). The asymmetric and symmetric siloxane band positions are centered at 1115 cm^{-1} and 1031 cm^{-1}, respectively. These bands, although muddled by noise, had a relative increase in intensity that could be attributed to a higher functional group ordering, possibly caused by favorable and ordered 3-APTES (the tethering of all siloxane-ethoxy chains to the substrate surface) anchorage on the available surface –Si–OH sites, as reported previously [51,55]. This calls for a particular emphasis on the siloxane band at 1115 cm^{-1}, as siloxane functional groups are responsible for APTES orientation. Moreover, the band associated with the silanol group is shown as a shoulder that shifted towards 982 cm^{-1} from 994 cm^{-1} (as observed in the black spectrum); the band was shifted by 12 cm^{-1}. Additionally, the siloxane bending band at 870 cm^{-1} gained intensity by 90%, along with the silanol band at 809 cm^{-1} gaining intensity by 65%, both compared to the black spectrum in Figure 2. These later bands are generally attributed to the available surface –Si–OH sites, but they could also be related to other vibration modes of –Si–O–Si functional groups. As such, this work notices that incorporating mixing caused siloxane and silanol-attributed bands to have a sharper manifestation and that these may not be exclusive to the surface functional groups but also to APTES molecules with some unattached siloxane groups or multilayering. This interpretation finds ground as the amine bands at 1651 cm^{-1} and at 1531 cm^{-1}, respectively, had a reduced $primary\_A_{1651}/secondary\_A_{1531}$ ratio of 1.15 in comparison to the

$primary\_A_{1638}/secondary\_A_{1541}$ ratio (=2.1) obtained from FTIR spectrum when no mixing method was used. This is further contrasted by the dominant –C–N amine band (~1233 cm^{-1}), which has a relative intensity difference of 87% ($primary\_A_{1651}/secondary\_A_{1233}$ = 7.7). Along with the siloxane band, these bands have been attributed to favorable 3-APTES anchoring by APTES hydrolyzation [42]. The hydrolyzation of APTES causes anchoring where various APTES molecules are connected by siloxane bonds between themselves but leave a siloxane group available for surface tethering. This anchoring process leads to multilayering deposition and is perceived by the manifestation of sharp isolated amine bands [51,52]. In this manner, the FTIR spectrum serves to establish a frame for identifying how APTES is deposited on silicon substrates. Silanol bands with lower ratios to their siloxane bands accentuate a hydrolyzation orientation, with the distinct shoulder at ~980 cm^{-1} being an indicator and siloxane stretching bands at ~1115 cm^{-1} and ~1030 cm^{-1} serving to confirm tethering. Particularly, the asymmetric stretching band at ~1115 cm^{-1} could be the indicator for lateral or hydrolyzed APTES orientation, with ~1030 cm^{-1} siloxane band confirming tethering and the silanol shoulder ~980 cm^{-1} indicating the availability and presence of surface silanol functional groups. Additionally, as observed on the blue spectrum of Figure 2, a $siloxane_{1030}/silanol_{980}$ ratio $\geq$ 1.5 (w/o = 1.2, MS = 1.5, and US = 1.9) could be distinctive in correlating silanol availability and APTES attachment, with a higher ratio indicating either higher attachment or that the hydrolysis process occurred. Hydrolyzation is not unfavorable, as other works have considered that this process favors homogenous film depositions [42,54].

Through the identification of specific APTES bands to analyze and determine how the deposition and anchoring process is performed, it is possible to discern the effect of three mixing methods on the surface activation of silicon substrates. As APTES was used in this section as an indirect assessment of surface activation, it was found that mixing under the influence of an ultrasonic bath seemed to have a positive effect on possibly more effective film formation. This effectiveness is supported by the obtained $primary\_A/secondary\_A$ ratio values (which are affected by the APTES hydrolyzation process), relative siloxane band positions, and $siloxane/silanol$ ratio values as well. This was not the case for mixing under the influence of a magnetic stirrer, where both lineshape changes and a decreased band intensity suggest a less effective surface coverage or unfavorable APTES deposition. Hence, its implementation in the subsequent deposition layers was not continued. Specifically for the APTES layer, while there is a lack of certainty on monolayer or multilayer formation, the discussion will not focus on it but on favorable APTES anchoring, which will be addressed in the next section.

3.2. APTES Deposition

The main FTIR APTES bands, shown in Figure 3, establish benchmarks for analyzing how APTES is attached to the surface under several synthesis conditions (synthesis temperature, APTES concentration, and mixing method during the process) after surface activation in an ultrasonic bath.

The solution temperature has been reported to benefit APTES monolayer deposition, possibly [42]. However, this was not observed here as both synthesis temperatures, room temperature (RT) and 60 °C, led to similar FTIR spectra, as seen in Figure 3a, with a slight distinction for the spectrum in black (RT conditions) that exhibited the previously mentioned band at ~1260 cm^{-1}, which could be attributed to APTES Si–O–C bands and methyl groups present in the molecular backbone. Regardless, as this band has not been found to contribute to discerning the attachment orientation, it can be omitted.

Relative intensities were similar overall, except for the siloxane vibration band at 1113 cm^{-1} for the deposition under RT conditions, which presented a relative intensity increase of 229% over the spectrum for the material deposited from an APTES solution at 60 °C. This band is paired with the siloxane band at 1021 cm^{-1}, attributed to siloxane formation from APTES anchoring, which has a similar relative intensity for materials synthesized under both temperature conditions. There is a lack of certainty regarding the assignment of these bands. Some studies, for instance, assign the band located at 1113 cm^{-1} to the contribution coming from siloxane formation, whereas the band at 1021 cm^{-1} has been attributed to linear (also called transversal) siloxanes, which are directly attached to the silicon surface in contrast to lateral siloxanes [53,55]. Therefore, this could imply that APTES attachment through transversal siloxane groups was favored during APTES deposition at room temperature due to more effective surface functionalization. However, quantification of surface functional groups density is a challenge, and at this point, it is not possible to offer additional evidence to support this better effectiveness in surface functionalization by FTIR measurements only. As mentioned in the previous section, the shoulder at ~990 cm^{-1} can be attributed to the Si–O bending mode in the siloxane functional group, which can be differentiated from the silanol vibration mode observed in the spectral region of ~900–800 cm^{-1} belonging to surface-bound functional groups. The samples synthesized at room temperature presented a $siloxane_{1021}/silanol_{992}$ ratio of 1.4, while those synthesized at 60 °C had a $siloxane_{1019}/silanol_{991}$ ratio of 1.1. It suggests a relatively high density of silanol functional groups, which can be caused by hydrolyzation, unfavorable orientation, or incomplete APTES attachment. However, this spectral region is also susceptible to the presence of surface-bound functional groups, which makes the temperature an inconclusive parameter for the APTES deposition. Despite this, clearer information can be obtained from the observed changes in the FTIR spectra because of the effect of other parameters.

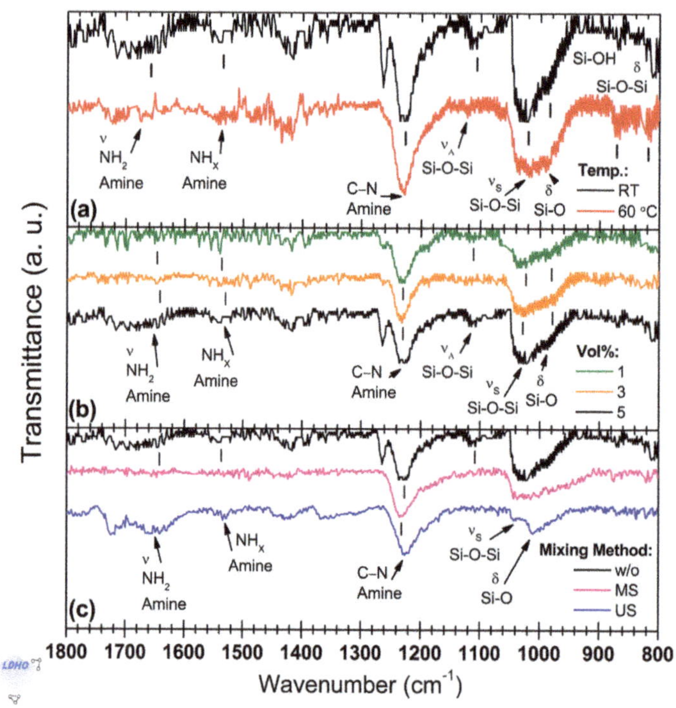

Figure 3. Representative normalized FTIR spectra of APTES deposited on silicon substrates when changing: (**a**) solution temperature, (**b**) APTES concentration, and (**c**) mixing method during deposition.

Concerning the effect of APTES concentration (Figure 3b), using FTIR spectroscopy, it was observed that 1 Vol% was enough to ensure surface attachment of APTES. The other concentrations only promoted differences in the relative intensity of the bands associated with amine and siloxane functional groups, which can be attributed to multilayering. The spectra for 1 and 3 Vol% concentrations showed a $primary\_A_{1652}/secondary\_A_{1535}$ ratio of 0.6, i.e., the secondary amine band around 1535 cm^{-1} possesses a higher relative intensity than that corresponding to the primary amine functional group located at 1652 cm^{-1}. In addition, at an APTES concentration of 5 Vol%, the dominant amine band was shifted towards 1659 cm^{-1}, and the $primary\_A_{1659}/secondary\_A_{1535}$ ratio was modified to 1.6. Despite this observed change in the relative intensities of the bands, the $primary\_A/secondary\_A$ ratio offers partial information; therefore, additional information coming from other FTIR bands is necessary as support. For example, despite similarities in their lineshapes among the spectra of Figure 3b, two bands are easily observable at ~1113 cm^{-1} (siloxane) and ~1260 cm^{-1} (previously discussed) at the highest APTES concentration considered here (5 Vol%, black spectrum in Figure 3b). It could be correlated to favorable attachment or APTES multilayer deposition because of the relatively high APTES concentration. In addition, it could be supported by the $amine_{1230}/siloxane_{1025}$ ratio (=1.3) for the highest APTES concentration, which also changed compared to the other two conditions (2.2 and 1.7 for the APTES concentrations of 1 and 3 Vol%, respectively).

The mixing method is the parameter that promoted a noticeable difference in FTIR spectra, as seen in Figure 3c. As suggested by the FTIR measurements, depending on the specific mixing method, APTES molecule attachment can be hindered. If APTES deposition is performed without any mixing method (like either US or MS), some bands can be observed at 1652 cm^{-1} (primary amine), 1260 cm^{-1}, 1228 cm^{-1} (amine), and at 1025 cm^{-1} (siloxane), as well as the silanol bands at 872 cm^{-1} and 816 cm^{-1}. Otherwise, deposition under US or MS mixing methods led to poor reproducibility, and sometimes, it was not possible to observe the whole set of characteristic FTIR bands of the APTES molecule. Thus, it is difficult to confirm how specifically the APTES molecule is attached to the substrate depending on the mixing method, i.e., it is not possible to distinguish between primary orientation, partial reaction (unfavorable), or crosslinking (due to hydrolysis), for example [51,55]. However, based on some differences in the band positions and some modifications in the spectra's lineshape, it seems that there is sufficient evidence for unfavorable APTES orientation when a mixing method is used. If US is utilized during APTES deposition, for example, the characteristic band for amine is manifested by the convolution of two bands, the first at 1655 cm^{-1} and the second at 1615 cm^{-1}. In addition, the siloxane band, which is usually located at ~1100 cm^{-1}, is shifted to ~1090 cm^{-1}, whereas the band at 1024 cm^{-1} has reduced sharpness and increased FWHM, with a shoulder extending towards 965 cm^{-1}.

In summary, although APTES deposition from the precursor solution at a concentration of 1 Vol% was confirmed, the quality of the thin film is unclear. Additionally, mixing methods were counter-productive for a favorable APTES molecule attachment. Therefore, based on FTIR measurements, APTES deposition from a 5 Vol% precursor solution at room temperature without a mixing process seems to be appropriate for further BiBB anchoring.

3.3. *Effect of the Drying Process on APTES Deposition*

The impact of drying methods on APTES film formation has been previously reported in the scientific literature in the range from 30 to 60 °C but in a different context to PBs synthesis [55]. Figure 4a shows the effect of two drying conditions on the FTIR spectra of APTES thin films. In addition, the effect on BiBB anchoring is also considered as indirect evidence for the APTES drying process (Figure 4b).

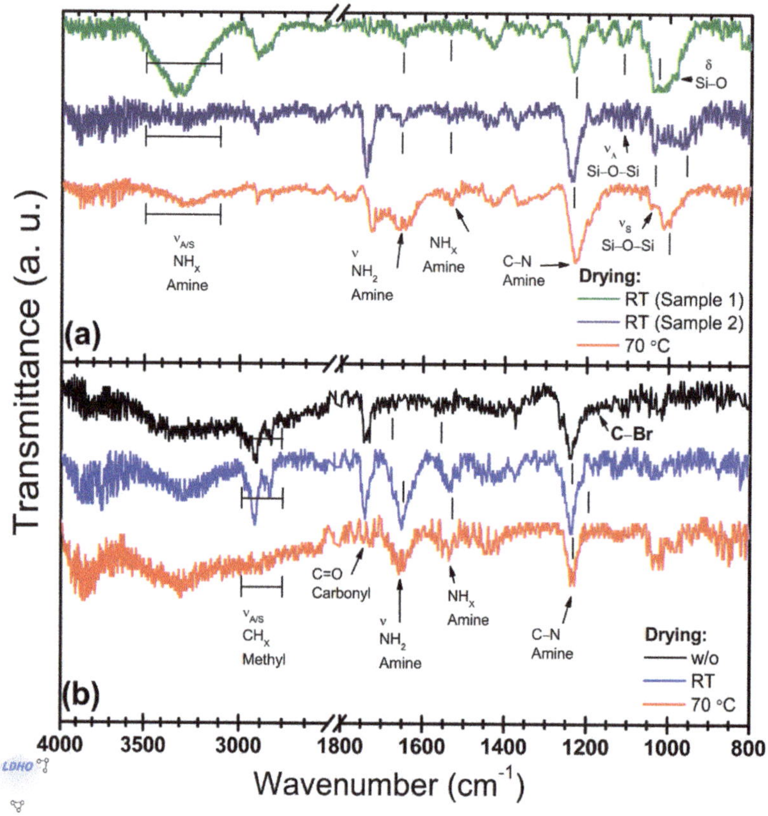

Figure 4. Representative normalized FTIR spectra for APTES thin films dried at two temperatures (**a**) and for BiBB anchored over Si/APTES samples when the APTES layer was dried under two conditions (**b**). In (**a**), spectra for two dried samples at room temperature are shown as evidence of poor reproducibility.

Differences have been observed in the amine bands depending on the drying process. At room temperature, the deposited films presented poor reproducibility, as illustrated in Figure 4a. The amine (NH_2) region at high wavenumbers (3500–3000 cm^{-1}) shows poorly defined bands. RT-dried samples could present, or not, two distinct amine bands located at 3332 cm^{-1} and 3277 cm^{-1}, which are assigned to asymmetric and symmetric stretching vibrations, respectively. The relative intensity of these bands is around 73% higher than that observed on APTES films dried at 70 °C. These amine bands have not been correlated to film quality or surface APTES orientation in the literature, but the observed difference in the FTIR spectra could be attributed to adhered APTES material or residual water in the material due to their relative intensities.

Independently of the drying conditions, APTES thin films exhibit their bands for primary amine and tertiary amine functional groups at the same position (at 1654 cm^{-1} and 1531 cm^{-1}, respectively), but the band position of a carbonyl group was observed for both 70 °C and those dried at RT from possible ethoxy group terminations or trapped ambient CO_2. Additionally, the $primary\_A_{1652}/secondary\_A_{1530}$ ratio was modified from 2.8 (70 °C) to 1.6 for dried samples at RT. Although the drying process has been considered beneficial in the literature, from Figure 4b, it seems that there are no significant differences in the APTES films' properties to hinder further BiBB anchoring. Upon the BiBB anchoring, the amine band of the APTES film (initially located at ~1230 cm^{-1}) was shifted to ~1245 cm^{-1}. Moreover, methyl bands are clearly observed in the 3000–2800 cm^{-1} region due to their increased intensity. It is worth noting that, in comparison to the dried samples, the BiBB samples without any drying process showed a reduction by 73% and 25% in the intensity of their bands attributed to the primary and secondary amine functional groups, respectively. While the presence of secondary amine bands in FTIR spectra is desirable, as they confirm BiBB anchoring, primary amine bands are considered unfavorable as they correlate with exposed end functional groups that were not activated.

With the confirmation of BiBB anchoring, even without any drying process, the requirements for successful BiBB anchorage and adequate surface covering are questioned.

3.4. BiBB Anchoring

Focusing on BiBB anchoring, Figure 5 presents the FTIR bands used to identify the presence of the initiator. The BiBB anchoring is crucial to improve surface preparation and increase the availability of initiators for the PBs synthesis, by which three parameters were explored: initiator concentration (Figure 5a), mixing method (Figure 5b), and mixing time (Figure 5c).

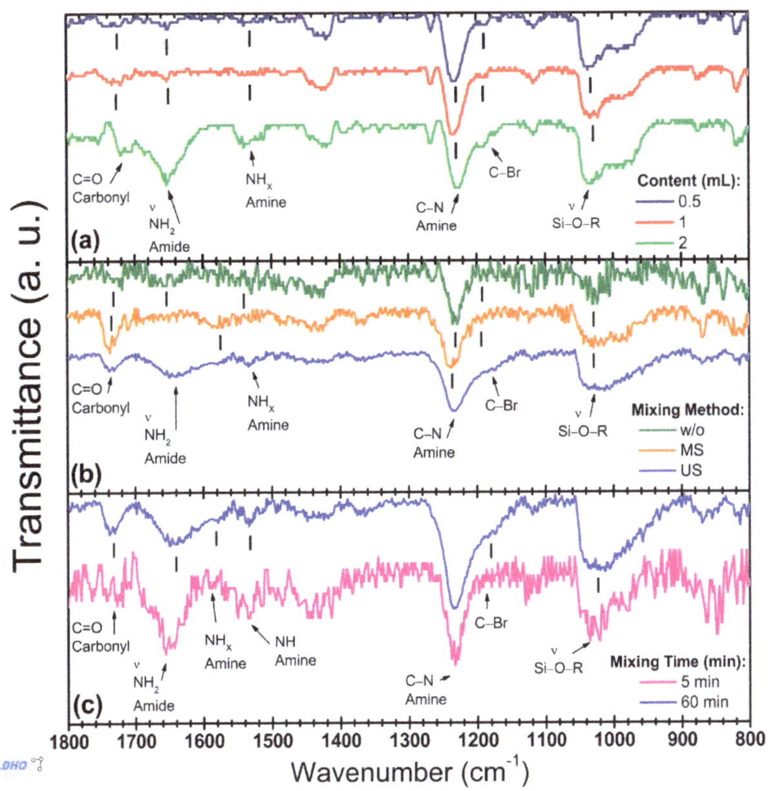

Figure 5. Representative normalized FTIR spectra obtained after BiBB anchoring over Si/APTES samples for different (**a**) BiBB concentrations, (**b**) solution mixing methods, and (**c**) mixing times.

Previous works have identified the products and by-products of the BiBB solution, as well as how the precursor solution enables BiBB anchoring [56–58]. However, it was observed from Section 3.3 that the deposition conditions for the APTES layer had little to no influence on the BiBB anchorage. This could be caused by interactions between the solution by-products and the APTES layer that ensure BiBB anchorage regardless of APTES deposition parameters, but this mechanism is not examined in this work. Regardless, Figure 5 shows bands belonging to Si/APTES and Si/APTES/BiBB samples, with certain bands gaining relevance. Among these, a change in the spectral lineshape is caused by the transition from amine to amide, and a shoulder at ~1170 cm^{-1} is observed among the structural bands. Smith et al. [59] proposed the identification of halogen bands at higher wave numbers by carbon bands such as a C–Br vibrational mode. These vibrational modes were identified in the 1200–1000 cm^{-1} region, overlapping with many other carbon-based bands, making them difficult to identify. However, since the C–Br bond is an end-group termination at the surface level, it could favor their detection, allowing this uncommon band to be identified as confirmation of BiBB anchorage.

By varying the x amount of BiBB from 0.5 to 2 mL in the THF:TEA:BiBB precursor solution (40:0.8:x, by volume), noticeable changes in the relative intensity of amide, bromidyl, and carbonyl bands were observed (Figure 5a). Increasing the amount of BiBB to 2 mL seemed to cause a change in either the BiBB anchoring or the APTES layer as carbonyl and amide bands convoluted at 1709 cm^{-1} (as a shoulder) and 1651 cm^{-1}, respectively, with a relative intensity increase of up to 69%. Additionally, intensity changes were observed in the siloxane band at 1028 cm^{-1} with its silanol shoulder, but neither of these changes has a clear explanation as these bands are mostly attributed to APTES molecule and surface-level interactions [42,52]. However, the manifestation of the C–Br stretch proposed previously [59] was appreciable, becoming more distinct as an independent shoulder at higher concentration. This relative intensity shift may allude to a higher density of initiator availability and supports its correlation to a bromide end-group termination.

The incorporation of mixing methods caused an impact on BiBB anchoring that could be correlated to surface activation, hence molecule end-group fidelity (Figure 5b). Additionally, mixing seems to make the C–Br band at ~1170 cm^{-1} become more noticeable as a shoulder, along with defined sharpness of secondary amide groups tending towards ~1550 cm^{-1} or lower. These bands may correlate with amine site availability and successfully anchored BiB- (activated isopropylbromide by loss of a bromide) molecules. The spectra without mixing differed from mixed sample spectra (Figure 5b), as the former presented a primary amide band at ~1660 cm^{-1}, with a secondary amide at ~1585 cm^{-1}, and a tertiary amide or amine at ~1530 cm^{-1}. Stirred (MS) samples differed as they lacked a distinguishable primary amide band, but spectra for sonicated (US) samples did express amide bands, with some shifts. While these shifts do not change the functional group and molecular vibration interpretation, the primary-to-secondary amide ratio ($primary\_Am_{1637}/secondary\_Am_{1596}$) and the $primary\_Am_{1637}/secondary\_Am_{1530}$ ratios changed to 1.6 and 2.3, respectively, from without mixing's ratios of 2.3 and 1.3. This change poses a distinction and possible correlation of bands as amides (primary and secondary) confirm BiBB anchoring but could also allude to brush density [6,60,61]. The reduction from 2.3 to 1.6 may be attributed to hydrogen bonds or lateral pressure from other BiBB groups or amide sites tethering a larger number of BiBB molecules as the secondary amide (NH_x) stretching vibrational mode is susceptible to these pressures [62,63]. Furthermore, the increase in the relative intensities from 1.3 to 2.3 supports this observation as the tertiary amide, usually non-observed, is identified at lower wavenumbers (~1530 cm^{-1}) and may be attributed to amines from APTES (non-activated sites). US mixing seems to reduce the presence of non-activated amine sites, possibly confirming a higher BiBB anchoring density. However, the relationship between the intensities of the amide and amine functional groups is complex and cannot be used as the sole confirmation. As such, the C–Br band at 1177 cm^{-1} is of particular interest as it can be directly correlated to the BiBB molecule. This band manifested in US method spectra as a distinct shoulder, unlike in the stirred method and the method *w/o* mixing. Additionally, the C–Br band may not be tied only to BiBB anchorage but also to end group fidelity, complementing the analysis of indirect bands and confirming initiator density.

Expanding on mixing impact in BiBB anchoring, "continuous mixing" (i.e., having Si/APTES samples immersed from the start in the precursor solution) differed from "premixed solutions" (where the Si/APTES samples were immersed into the precursor solution after mixing), as denoted by changes in amide–amine and C–Br bands, with all becoming magnified by premixing (Figure 5c). Deposited samples from premixed precursor solutions had amide bands centered at 1596 cm^{-1} and 1558 cm^{-1} (assigned to secondary amide NH_x and tertiary amide, respectively), which could suggest a higher density of activated sites. As the primary amide group at 1641 cm^{-1} can be attributed to an anchored BiBB by the carbonyl overlapping at 1726 cm^{-1}, the secondary amide could confirm the activated sites, and a tertiary amide possibly being a consequence of dense surface activation where an amine may be housing more than one BiBB molecule. Additionally, based on the shoulder widening of the C–Br band and the relative increase in its intensity, BiBB anchoring from premixed precursor solutions proved to be effective and could ensure a higher BiBB surface anchorage density. In this context, both mixing methods showed a band attributed to carbonyl at 1726 cm^{-1} to further confirm the amine-to-amide transition by the BiBB anchoring.

3.5. Cu^0-SI-ATRP Difference

This section identifies the characteristic bands of PAM and focuses on amide functional group bands and methyl bands, as illustrated in Figure 6. Key parameters, such as the concentration of acrylamide (AM) in the ATRP solution (Figure 6a) and the duration of the synthesis process (Figure 6b), can be evaluated to optimize PAM brush formation via the Cu^0-SI-ATRP method. Furthermore, utilizing a commercial PAM as a comparative reference allows for a more thorough analysis and comparison of synthesis methods (Figure 6c). Overall, the PAM bands resulting from the Cu^0-SI-ATRP synthesis were successfully characterized, revealing characteristic amide twin bands at 1656 cm^{-1} and 1610 cm^{-1}, confirmation of alkane methyl groups within the 3000–2800 cm^{-1} range, and asymmetric and symmetric amide bands at 3337 cm^{-1} and 3168 cm^{-1}, respectively.

The concentration of AM in the solution has an influence on managing the polymerization process. When the amount of AM is reduced to a quarter of the original level (0.09 M), it can lead to erratic polymerization. However, using half of the original concentration (0.17 M, down from 0.34 M) is sufficient to facilitate brush formation (as shown in Figure 6a). Both concentrations produced similar spectral bands for the asymmetric and symmetric amide (at 3326 cm^{-1} and 3183 cm^{-1}, respectively), asymmetric and symmetric methyl (2923 cm^{-1} and 2856 cm^{-1}), amide–carbonyl group (1645 cm^{-1}), primary amide (1603 cm^{-1}), and secondary amide (around 1542 cm^{-1}). This indicates that similar PAM formations were achieved in both cases, enabling brush synthesis with less material. However, at a concentration of 0.09 M, there was a loss of bands for amide stretches and noticeable shifts in position, which highlighted issues with the control of PAM synthesis, causing possible voids or aggregates on the surface. Specifically, the primary amide band at 1603 cm^{-1} shifted towards the secondary amide region at 1579 cm^{-1}, and a tertiary amine band appeared at 1530 cm^{-1}. This behavior has been linked in other studies to crosslinking or branching [23,62,64]. While this study did not investigate the chemical balance in the solution that may cause these changes, it is suggested that the lower concentration of AM leads to disordered and non-uniform growth, resulting in the branching of PAM chains.

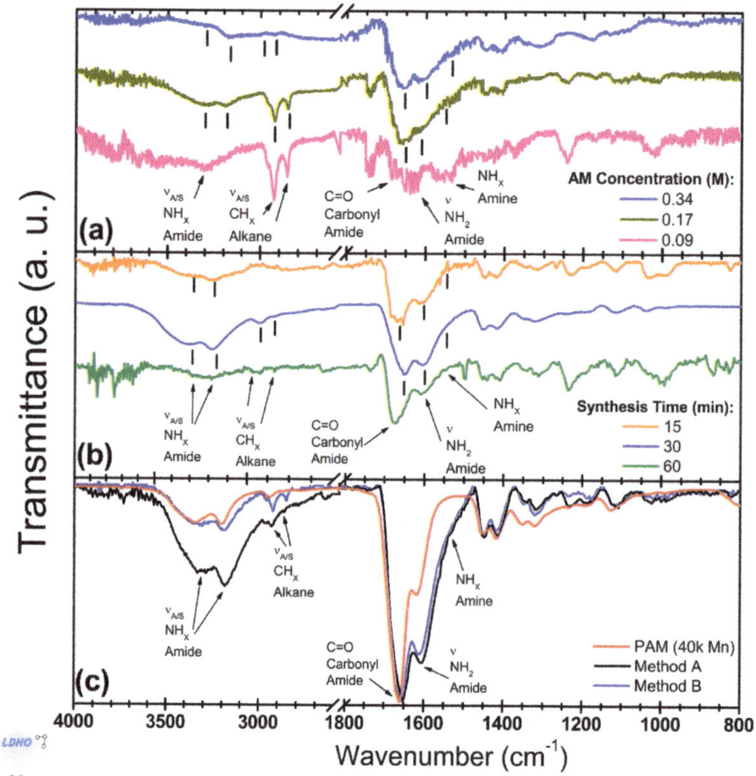

Figure 6. Representative normalized FTIR spectra for Cu^0-SI-ATRP PAM after changes to (**a**) acrylamide (AM) molar content, (**b**) synthesis time, and (**c**) spectra for commercial PAM. This work's method (Method A) and a general Cu^0-SI-ATRP method (Method B) were used to observe similarities and confirm PAM formation.

Synthesis time variations revealed a timeframe for the formation of PAM brushes. The PAM spectra at 15 min and 30 min were similar, but after 30 min, notable deformations in the lineshape indicated excessive and uncontrolled polymerization (see Figure 6b). At 15 min, polymerization was inconsistent; the spectra often lacked evidence of polymerization in the expected amide region (3500–3000 cm^{-1}) and the methyl bands (3000–2800 cm^{-1}). However, when successful, the 15 min synthesis exhibited distinct bands corresponding to saturated alkanes (1668 cm^{-1}), primary amides (1612 cm^{-1}), and a shoulder peak indicative of secondary amides (1570 cm^{-1}), confirming PAM formation like that observed at 30 min. The inconsistencies noted might relate to the polymerization kinetics involved in the initiation process. At 60 min, the twin bands correspond to the amide-carbonyl, and primary amide began to deteriorate. The $carbonyl_{1672}/primary_{1608}$ amide ratio was 1.7, differing from the ratios of 1.2 and 1.4 observed at 30 min (with bands at 1646 cm^{-1} and 1604 cm^{-1}) and 15 min, respectively. Previous studies highlighted this relative intensity relationship concerning PAM conformation [64]. While this study notes a correlation between PAM conformation and chain growth, as observed through these functional group bands, further investigation is necessary to confirm these findings. Overall, there appears to be a timeframe for PAM brush formation, where 15 min is the minimum time needed to initiate brush polymerization. Beyond 60 min, the structure may collapse or undergo multilayer deposition or crosslinking, resembling bulk material. This was further investigated by comparing different PAM films, particularly those produced by the proposed Method A, with those from Method B and bulk commercial PAM.

Commercially bought PAM, as compared to PAM brushes synthesized through Method A and Method B, exhibited similar widths and molecular band positions. However, subtle variations were noted at higher wavenumbers in the asymmetric and symmetric amide bands, as well as in the alkane methyl bands. Notably, at higher wavenumbers, there was an intensity reduction of up to 21% compared to the dominant amide-carbonyl C=O band around 1660 cm^{-1} for all samples. The Cu^0-SI-ATRP methods featured a primary amide band at 1606 cm^{-1}, with an amine shoulder near 1560 cm^{-1}. In contrast, the $carbonyl_{1660}/primary_{1606}$ amide ratio for these methods was 1.2, while for bulk PAM, it was 2.0. While polymerization can be confirmed, these differences between commercial bulk material and the Cu^0-SI-ATRP methods are attributed to the deposition methods themselves. However, the observed disparities in lineshapes suggest conformational differences among the synthesized materials, which have been addressed in previous works by analyzing the amide functional group bands (their presence or absence in the spectra, position, and relative intensity) in such a way that is possible to determine primary (–$CONH_2$), secondary (–CONH–R), or tertiary (–CON–R_2) amide side-groups [64–66].

Considering the spectral region between 3600 and 2600 cm^{-1}, significant differences in the amide asymmetric and symmetric contributions were observed, with peaks at 3329 cm^{-1} and 3337 cm^{-1} for samples synthesized following the Method A and at 3181 cm^{-1} and 3168 cm^{-1} for those samples from Method B. However, no significant differences were noted for the asymmetric and symmetric methyl bands. In the amide spectral range (1800–1500 cm^{-1}), Welch's t-test indicated a difference only for the distinct amide carbonyl bands at 1652 cm^{-1} and 1656 cm^{-1} for Method A and Method B.

Statistical analysis comparing samples from Method A and Method B revealed differences in repeatability through spectral evaluation (Figure 7). For this, the spectra of 16 samples for every method were considered and deconvoluted into molecular groups, following a method like Murugan et al. [67], using an FWHM approach with Lorentz fitting. Data in Figure 7 illustrate the average peak positions for the functional groups in the spectral region of 1800–1500 cm^{-1}, with their corresponding standard deviation. Compared to Method B, Method A exhibits a narrow distribution for the peak positions, which could suggest less variability in the PBs properties when Method A is the chosen synthesis route. In addition, Method A exhibited the expected FTIR bands for the amide functional group in PAM-PBs, whereas Method B exhibited less common additional bands, which can be attributed to some irregularities in the material (like branching or crosslinking) during the polymerization process.

3.6. Effect on the Macro-Scale: Contact Angle

CA measurements evaluated the evolution of the surface characteristics along the whole synthesis process (Figure 8a), and surface homogeneity for each method was analyzed (Figure 8b). It is well-known that PAM is a hydrophilic material; this was corroborated by the low CA value measured ($29.2° \pm 3.0°$) on a thin film of bulk PAM, which was spin-coated over neat silicon substrates. It will be considered an interesting reference for further discussions. Note that this CA value is ~22° below that corresponding to the neat substrate (50.9°), which confirms that the thin film is effectively covering the substrate. As previously reported [63,68], in bulk PAM material, the surface functional groups are in contact with water during CA measurements to become hydrated, creating hydrogen bonds and polar attraction. However, while the hydrophilic nature was kept, grafting

Method A and Method B did not exhibit the same CA value and even had a difference between themselves (by 8.5°, on average). Furthermore, it was observed that Method A presented a different surface coverage as compared to Method B, which could be correlated to potential differences in polymer chain grafting density.

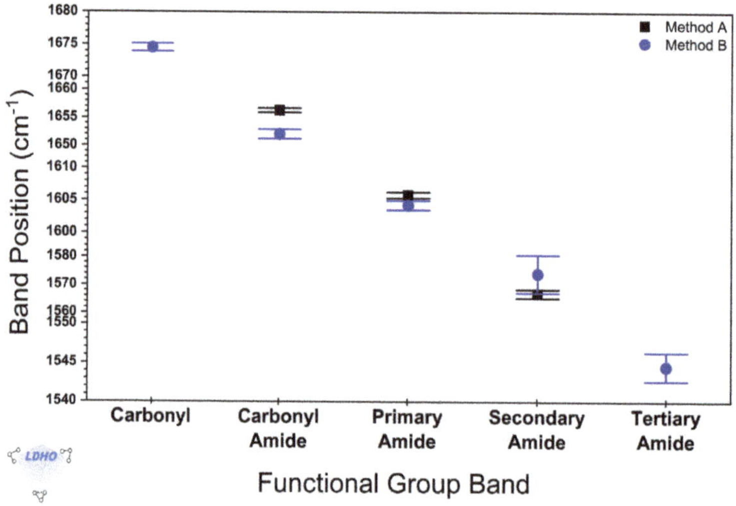

Figure 7. Average peak positions and standard deviation for selected bands in the spectral region of 1800–1500 cm^{-1}.

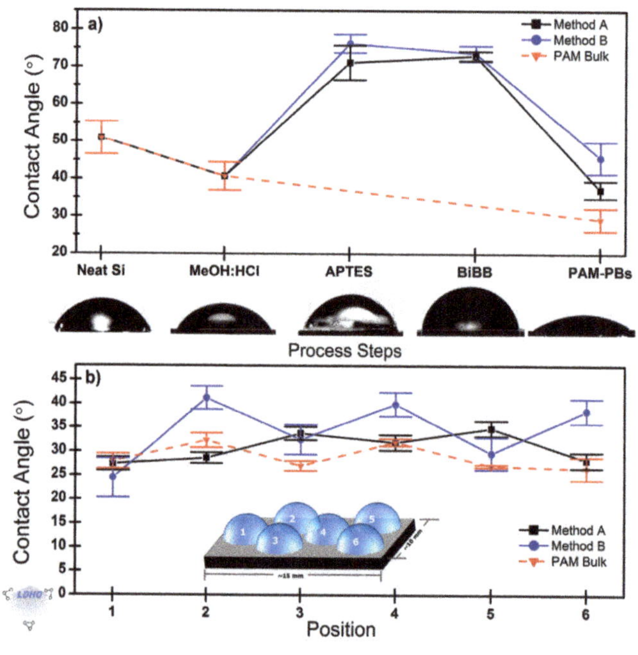

Figure 8. Contact angle measurements for (**a**) Method A, Method B, and commercial bulk PAM film (40k Mn) along the whole synthesis process with representative droplet images and (**b**) CA measurements for each synthesis method considered in this study over 6 regions distributed on the surface, as illustrated.

After Cu^0-SI-ATRP brush polymerization, the CA value for materials synthesized by Method A was 37.3° ± 2.3°, which is lower than that for Method B (45.8° ± 4.3°). These two CA values permit a discernment of how PAM is deposited depending on the synthesis methods. The structural alignment (generally perpendicular to the substrate would limit the number of functional end-groups in the deposited material that interacts with water molecules during the measurement. Depending on the specific PAM (or PAM-PBs)—water interaction, differences in water absorption and further polymer swelling can be observed [69,70]. In this manner, CA values are tied to grafting density. A CA closer to the bulk material may correlate to a higher grafting density by a denser surface coverage over a higher CA value.

To evaluate the surface distribution of the deposition, six equidistant test points for every sample were analyzed by CA measurements, as per the inset in Figure 8b. Through the whole sample, CA values showed a variation of 1.5° ± 0.2° when Method A was used, and for Method B, it was 3.0° ± 0.6°. This supports what was stated above from the FTIR analysis about a more effective surface coverage when Method A is used. CA values follow the order bulk PAM < Method A < Method B. This could suggest that the grafted density and thin film formation by Method A may lead to a densely packed material (high graft density) with surface properties closer to the bulk material. According to the Wenzel-type wetting model, the intermolecular distance between adjacent polymer chains allows water infiltration when the material is synthesized from Method A, but due to the relatively higher brush density, the observed surface response is mainly attributed to PAM-PBs—water interaction. In contrast, if there is a reduced surface density of brushes, as is proposed in the case for Method B, then infiltration may be a dominant behavior, but as brushes are of a nanoscale (~150 nm) dimension, then a droplet of 5 μL would have sufficient volume to saturate the brush film and contact the substrate, allowing for the substrate properties to come into effect. This is observed in Figure 8b, where CA measurements showed variation in their values of up to 7.8°, 16.8°, and 4.9° when Method A, Method B, and bulk PAM deposition were considered, respectively. This could be a consequence of the surface preparation parameters, improving effective area coverage, structure, and orientation of brush formation.

CA proves to be sufficient for a quick and efficient characterization tool both to confirm surface modification and evolution across the PBs synthesis steps. Additionally, it was observed that CA measurements could highlight differences in surface coverage or graft density as Method A and Method B had different CA values from each other in spite of both being Cu^0-SI-ATRP methods. This is further emphasized by the difference in general CA values between a bulk material thin film and Cu^0-SI-ATRP methods, marking an intrinsic difference. Therefore, despite bulk PAM and PAM-PBs having the same functional groups (as discussed in FTIR analysis), their structure presents unique and different properties depending on the synthesis route, as observed through CA measurements.

4. Conclusions

PAM-PBs were synthesized through the Cu^0-SI-ATRP method. Through the step-by-step analysis across the different stages of the synthesis route, the impact, relevance, and dependence (or independence) of some parameters were identified. The analyzed parameters allowed a full understanding, as much as possible, of the Cu^0-SI-ATRP synthesis method, starting from surface preparation.

Surface activation plays an important role in ensuring surface distribution and successful film formation, as it enables favorable molecular groups on the substrate surface that will be used for the subsequent layer-by-layer deposition. As evidenced by FTIR measurements, the use of an ultrasonic bath favored the desired functional group availability on a silicon surface, as further corroborated after APTES deposition. Bands associated with siloxane groups at 1139 cm^{-1} and 1026 cm^{-1} shifted to 1115 cm^{-1} and 1031 cm^{-1} and gained sharpness. Amine bands at 1651 cm^{-1} and at 1531 cm^{-1}, respectively, had a reduced $primary\_A_{1651}/secondary\_A_{1531}$ amine ratio of 1.2 in comparison to the $primary\_A/secondary\_A$ amine ratio (=2.1) obtained when no mixing method was used. This is further contrasted by the dominant amine band –C–N (~1233 cm^{-1}), which has a relative intensity difference of 87% ($primary\_A_{1651}/secundary\_A_{1233}$ = 7.7). Furthermore, silanol bands were found to accentuate a hydrolyzation orientation, as asymmetric stretching band ~1115 cm^{-1} could be the indicator for lateral or hydrolyzed APTES orientation, with ~1030 cm^{-1} siloxane band confirming tethering, and the silanol shoulder ~980 cm^{-1} indicating surface silanol availability and presence. This work proposes that a *siloxane/silanol* ratio $\geq$ 1.5 could be an indicator to determine APTES attachment, with a higher ratio indicating hydrolyzation confirmation, as was observed. This band analysis considers APTES hydrolyzation to be favorable as it promotes homogenous film depositions.

APTES layer formation can be promoted even at the lower precursor solution concentration considered in this study (1 Vol%). Additionally, it was confirmed that the use of mixing methods hindered APTES anchorage. Upon APTES deposition, drying can be omitted, as BiBB anchorage was confirmed and manifested beneficial bands from the C–Br group at 1170 cm^{-1}, methyl bands from 3000–2800 cm^{-1}, and dominant amine bands around 1550 cm^{-1}. It was found that BiBB and its aggressive by-products of the bromide-base serve as self-regulating mechanisms that cleave excessive APTES to ensure both BiB– availability and amine groups (activated amines) for anchoring the initiator molecule. This effect was later confirmed when analyzing BiBB anchorage.

BiBB solutions and anchorage did not show sensitivity to the concentration of the precursor solution. Mixing methods offered the most noticeable changes; ultrasonication seemed to benefit the reactions incurred from solution esterification, causing starch formation. Additionally, it is proposed that the C–Br band correlates to the BiBB molecule and is tied to BiBB anchorage to confirm end-group fidelity. This band was found to not be discernible when no mixing was used, while the US method manifested a shoulder at 1177 cm^{-1}. This is further enhanced by premixing the mixture for 5 min prior to the incorporation of substrates instead of continuously mixing the solution. Particularly, uncommon tertiary amide bands at 1558 cm^{-1} called attention to the amine bands centered at 1596 cm^{-1} as it could suggest a higher density of activated sites, as compared to the "continuous mixing" method. As such, IR spectroscopy could identify changes, with BiBB being mostly defined by the amine-amide group changes, an increase in the prevalence of methyl groups, the manifestation of a tertiary amine group, and the manifestation of the C–Br band ~1170 cm^{-1}.

PAM-PBs synthesized by Cu^0-SI-ATRP were found to be sensible to monomer concentration and reaction time. Reducing AM concentration to half of the original value (0.17 M) preserves brush formation, as evidenced by consistent FTIR bands for amide (1645–1603 cm^{-1}) and methyl (2923–2856 cm^{-1}) groups, mirroring commercial PAM. However, further reduction to 0.09 M disrupts polymerization uniformity, causing shifts in primary amide toward the amine region at 1579 cm^{-1} and tertiary amine that manifested at 1530 cm^{-1}, indicative of possible branching or crosslinking. Similarly, synthesis time reveals a narrow operational window from 15–30 min that yields well-defined PAM brushes, while exceeding 60 min degrades structural integrity, marked by altered $carbonyl\_Am/primary\_Am$ amide band ratios (1.7 vs. 1.2 at 30 min) and could serve as evidence for conformation changes. Balancing Cu^0-SI-ATRP parameters and conditions may lead to further advances in polymerization efficiency and brush fidelity, ensuring reproducible, high-quality PAM brush synthesis for advanced surface applications.

Method A, the Cu^0-SI-ATRP method for PAM-PBs developed here, showed similar spectra to Method B (which was based on the literature). Both methods shared spectra bands that were distinctive for PAM when compared to a commercial-bought bulk PAM of 40k Mn. Moreover, statistical analysis magnified the differences between Method A and Method B in their molecular group contributions. It would seem Method A presents greater reproducibility, whereas in Method B, when peak positions are determined, the higher standard deviations suggest structural irregularities, such as branching or crosslinking. This could explain the differences observed between methods in CA measurements, where surface modification and evolution were successfully observed, as CA also offered a complementary insight.

Contact angle (CA) measurements revealed distinct wettability behaviors between bulk PAM and synthesized PAM-PBs—obtained from Method A and Method B. Bulk PAM exhibited high hydrophilicity (CA: 29.2° ± 3.0°), attributed to its dense hydration via polar amide groups. In contrast, Cu^0-SI-ATRP-grafted brushes showed intermediate contact angles, with Method A (37.3° ± 2.3°) being closer to the bulk material than Method B (45.8° ± 4.3°) in hydrophilicity and surface homogeneity. The reduced CA variability in Method A (1.5° ± 0.2° vs. 3.0° ± 0.6° for Method B) suggests higher grafting density and uniform surface coverage, mimicking bulk-like behavior. This work proposes that structural differences are responsible for these differences, with Method A having densely packed, perpendicularly aligned brushes that limit substrate interaction. We propose water infiltration through the intermolecular spaces of brushes causes a wetting state by which surface tension creates a contact boundary at the brushes, while Method B's irregular grafting permits substrate influence by having larger intermolecular spaces, elevating CA. This difference from bulk PAM calls attention to structure and possible conformation, as PAM-PBs seem to balance hydration and grafting density, whereas bulk PAM exhibits its volumetric hydrophilicity. Additionally, Method A had a hydrophilic behavior closer to the bulk material while presenting a more effective surface coverage, highlighting not only an improvement over Method B but also that CA is a property that can be influenced by synthesis parameters. This can be further explored by future works, as graft density differences can be attributed to the differences between the CA measurement of Method A and Method B. These findings highlight CA as a sensitive indicator of surface quality and synthesis efficacy and prove to be sufficient in differentiating brush-based thin films from bulk thin films.

Collectively, these findings place CA and FTIR as quick and efficient methods by which PAM-PBs can be characterized, identified, and confirmed. Additionally, because Method B, a literature-based method, is flexible and capable of allowing brush growth with other monomers, and due to its similarity with Method A, other monomers can be used to benefit from the increased reproducibility and surface uniformity reported in our work.

Author Contributions: Conceptualization, A.O.S.-N. and L.A.B.-J.; methodology, L.A.B.-J., A.O.S.-N., I.C.-C. and N.A.U.-C.; validation, I.C.-C. and A.O.S.-N.; formal analysis, L.A.B.-J. and I.C.-C.; investigation, L.A.B.-J.; resources, A.O.S.-N.; data curation, L.A.B.-J.; writing—original draft preparation, L.A.B.-J.; writing—review and editing, A.O.S.-N., I.C.-C. and N.A.U.-C.; visualization, I.C.-C.; supervision, A.O.S.-N. and N.A.U.-C.; project administration, L.A.B.-J.; funding acquisition, A.O.S.-N. All authors have read and agreed to the published version of the manuscript.

Funding: The authors gratefully acknowledge the financial support from Tecnológico de Monterrey through Ph.D. scholarship (L.A.B.-J.), resources for materials and fees.

Institutional Review Board Statement: Not applicable.

Data Availability Statement: The datasets presented in this article are available from the first author (or corresponding author) under request.

Acknowledgments: L.A.B.J. thanks CONAHCYT through the maintenance scholarship No. 770933. This work was developed in the Laboratory of Optoelectronic Hybrid Devices (LDHO).

Conflicts of Interest: The authors declare no conflict of interest.

References

1. Roy, D.; Cambre, J.N.; Sumerlin, B.S. Future Perspectives and Recent Advances in Stimuli-Responsive Materials. *Prog. Polym. Sci.* **2010**, *35*, 278–301. [CrossRef]
2. Painter, P.C.; Coleman, M.M. *Fundamentals of Polymer Science*, 2nd ed.; Routledge: New York, NY, USA, 2019; ISBN 9780203755211.
3. Díaz Lantada, A. Systematic Development Strategy for Smart Devices Based on Shape-Memory Polymers. *Polymers* **2017**, *9*, 496. [CrossRef] [PubMed]
4. Yan, J.; Bockstaller, M.R.; Matyjaszewski, K. Brush-Modified Materials: Control of Molecular Architecture, Assembly Behavior, Properties and Applications. *Prog. Polym. Sci.* **2020**, *100*, 101180. [CrossRef]
5. Matyjaszewski, K.; Sumerlin, B.S.; Tsarevsky, N.V.; Chiefari, J. (Eds.) *Controlled Radical Polymerization*; American Chemical Society: Washington, DC, USA, 2015; Volume 1187, ISBN 9780841230484.
6. Milner, S.T. Polymer Brushes. *Science* **1991**, *251*, 905–914. [CrossRef]
7. Ivkov, R.; Butler, P.D.; Satija, S.K.; Fetters, L.J. Effect of Solvent Flow on a Polymer Brush: A Neutron Reflectivity Study of the Brush Height and Chain Density Profile. *Langmuir* **2001**, *17*, 2999–3005. [CrossRef]
8. Idota, N.; Nagase, K.; Tanaka, K.; Okano, T.; Annaka, M. Stereoregulation of Thermoresponsive Polymer Brushes by Surface-Initiated Living Radical Polymerization and the Effect of Tacticity on Surface Wettability. *Langmuir* **2010**, *26*, 17781–17784. [CrossRef]
9. Bonanno, L.M.; Delouise, L.A. Optical Detection of Polyacrylamide Swelling Behavior in a Porous Silicon Sensor. *Mater. Res. Soc. Symp. Proc.* **2008**, *1133*, 31–36. [CrossRef]
10. Kaushik, A.; Kumar, R.; Dev Jayant, R.; Nair, M. Nanostructured Gas Sensors for Health Care: An Overview. *J. Pers. Nanomed.* **2015**, *1*, 10–23.
11. Boisseau, P.; Loubaton, B.N. Nanotechnology in Medicine. *Comptes Rendus Phys.* **2011**, *12*, 620–636. [CrossRef]
12. Wang, S.; Wang, Z.; Li, J.; Li, L.; Hu, W. Surface-Grafting Polymers: From Chemistry to Organic Electronics. *Mater. Chem. Front.* **2020**, *4*, 692–714. [CrossRef]
13. Buhl, K.B.; Agergaard, A.H.; Lillethorup, M.; Nikolajsen, J.P.; Pedersen, S.U.; Daasbjerg, K. Polymer Brush Coating and Adhesion Technology at Scale. *Polymers* **2020**, *12*, 1475. [CrossRef] [PubMed]
14. Conrad, J.C.; Robertson, M.L. Shaping the Structure and Response of Surface-Grafted Polymer Brushes via the Molecular Weight Distribution. *JACS Au* **2023**, *3*, 333–343. [CrossRef] [PubMed]
15. Wang, R.; Wei, Q.; Sheng, W.; Yu, B.; Zhou, F.; Li, B. Driving Polymer Brushes from Synthesis to Functioning. *Angew. Chem. Int. Ed.* **2023**, *62*, e202219312. [CrossRef]
16. Edmondson, S.; Armes, S.P. Synthesis of Surface-Initiated Polymer Brushes Using Macro-Initiators. *Polym. Int.* **2009**, *58*, 307–316. [CrossRef]
17. Teare, D.O.H.; Barwick, D.C.; Schofield, W.C.E.; Garrod, R.P.; Ward, L.J.; Badyal, J.P.S. Substrate-Independent Approach for Polymer Brush Growth by Surface Atom Transfer Radical Polymerization. *Langmuir* **2005**, *21*, 11425–11430. [CrossRef]
18. Wang, C.; Zhao, H. Polymer Brush-Based Nanostructures: From Surface Self-Assembly to Surface Co-Assembly. *Soft Matter* **2022**, *18*, 5138–5152. [CrossRef] [PubMed]
19. Milner, S.T.; Witten, T.A.; Cates, M.E. Theory of the Grafted Polymer Brush. *Macromolecules* **1988**, *21*, 2610–2619. [CrossRef]
20. Ślusarczyk, K.; Flejszar, M.; Chmielarz, P. Less Is More: A Review of ML-Scale of SI-ATRP in Polymer Brushes Synthesis. *Polymer* **2021**, *233*, 124212. [CrossRef]
21. Gazzola, G.; Filipucci, I.; Rossa, A.; Matyjaszewski, K.; Lorandi, F.; Benetti, E.M. Oxygen Tolerance during Surface-Initiated Photo-ATRP: Tips and Tricks for Making Brushes under Environmental Conditions. *ACS Macro Lett.* **2023**, *12*, 1166–1172. [CrossRef]
22. Tom, J.; Brilmayer, R.; Schmidt, J.; Andrieu-Brunsen, A. Optimization of Surface-Initiated Photoiniferter-Mediated Polymerization under Confinement, and the Formation of Block Copolymers in Mesoporous Films. *Polymers* **2017**, *9*, 539. [CrossRef]
23. Klemm, B.; Picchioni, F.; van Mastrigt, F.; Raffa, P. Starlike Branched Polyacrylamides by RAFT Polymerization—Part I: Synthesis and Characterization. *ACS Omega* **2018**, *3*, 18762–18770. [CrossRef] [PubMed]
24. Li, F.; Du, M.; Zheng, Z.; Song, Y.; Zheng, Q. A Facile, Multifunctional, Transparent, and Superhydrophobic Coating Based on a Nanoscale Porous Structure Spontaneously Assembled from Branched Silica Nanoparticles. *Adv. Mater. Interfaces* **2015**, *2*, 1500201. [CrossRef]
25. Nikolajsen, J.P.; Thinnesen, L.V.; Karabiber, O.A.; Daasbjerg, K.; Pedersen, S.U.; Kongsfelt, M.; Lillethorup, M. Surface-Initiated Controlled Radical Polymerization: Going beyond Laboratory Scale. *ACS Appl. Polym. Mater.* **2023**, *5*, 3534–3541. [CrossRef]

26. Sato, T.; Dunderdale, G.J.; Hozumi, A. Threshold of Surface Initiator Concentration for Polymer Brush Growth by Surface-Initiated Atom Transfer Radical Polymerization. *Langmuir* **2023**, *40*, 480–488. [CrossRef]
27. Wu, D.; Li, W.; Zhang, T. Surface-Initiated Zerovalent Metal-Mediated Controlled Radical Polymerization (SI-Mt ^{0} CRP) for Brush Engineering. *Acc. Chem. Res.* **2023**, *56*, 2329–2340. [CrossRef]
28. Besford, Q.A.; Uhlmann, P.; Fery, A. Spatially Resolving Polymer Brush Conformation: Opportunities Ahead. *Macromol. Chem. Phys.* **2023**, *224*, 2200180. [CrossRef]
29. Pester, C.W.; Klok, H.-A.; Benetti, E.M. Opportunities, Challenges, and Pitfalls in Making, Characterizing, and Understanding Polymer Brushes. *Macromolecules* **2023**, *56*, 9915–9938. [CrossRef]
30. Fundeanu, I.; Van Der Mei, H.C.; Schouten, A.J.; Busscher, H.J. Microbial Adhesion to Surface-Grafted Polyacrylamide Brushes after Long-Term Exposure to PBS and Reconstituted Freeze-Dried Saliva. *J. Biomed. Mater. Res. A* **2010**, *94*, 997–1000. [CrossRef]
31. Zhao, B.; Brittain, W.J. Polymer Brushes: Surface-Immobilized Macromolecules. *Prog. Polym. Sci.* **2000**, *25*, 677–710. [CrossRef]
32. Chen, W.-L.; Cordero, R.; Tran, H.; Ober, C.K. Polymer Brushes: Novel Surfaces for Future Materials. *Macromolecules* **2017**, *50*, 4089–4113. [CrossRef]
33. Barbey, R.; Lavanant, L.; Paripovic, D.; Schüwer, N.; Sugnaux, C.; Tugulu, S.; Klok, H.-A. Polymer Brushes via Surface-Initiated Controlled Radical Polymerization: Synthesis, Characterization, Properties, and Applications. *Chem. Rev.* **2009**, *109*, 5437–5527. [CrossRef] [PubMed]
34. Matyjaszewski, K.; Spanswick, J. Copper-Mediated Atom Transfer Radical Polymerization. In *Polymer Science: A Comprehensive Reference*; Elsevier: Amsterdam, The Netherlands, 2012; pp. 377–428.
35. Liarou, E.; Whitfield, R.; Anastasaki, A.; Engelis, N.G.; Jones, G.R.; Velonia, K.; Haddleton, D.M. Copper-Mediated Polymerization without External Deoxygenation or Oxygen Scavengers. *Angew. Chem. Int. Ed.* **2018**, *57*, 8998–9002. [CrossRef] [PubMed]
36. Zhang, T.; Du, Y.; Kalbacova, J.; Schubel, R.; Rodriguez, R.D.; Chen, T.; Zahn, D.R.T.; Jordan, R. Wafer-Scale Synthesis of Defined Polymer Brushes under Ambient Conditions. *Polym. Chem.* **2015**, *6*, 8176–8183. [CrossRef]
37. Zhang, T.; Du, Y.; Müller, F.; Amin, I.; Jordan, R. Surface-Initiated Cu(0) Mediated Controlled Radical Polymerization (SI-CuCRP) Using a Copper Plate. *Polym. Chem.* **2015**, *6*, 2726–2733. [CrossRef]
38. Yan, W.; Fantin, M.; Spencer, N.D.; Matyjaszewski, K.; Benetti, E.M. Translating Surface-Initiated Atom Transfer Radical Polymerization into Technology: The Mechanism of Cu0-Mediated SI-ATRP under Environmental Conditions. *ACS Macro Lett.* **2019**, *8*, 865–870. [CrossRef]
39. Zeng, Y.; Xie, L.; Chi, F.; Liu, D.; Wu, H.; Pan, N.; Sun, G. Controlled Growth of Ultra-Thick Polymer Brushes via Surface-Initiated Atom Transfer Radical Polymerization with Active Polymers as Initiators. *Macromol. Rapid Commun.* **2019**, *40*, e1900078. [CrossRef]
40. Matyjaszewski, K.; Tsarevsky, N.V. Macromolecular Engineering by Atom Transfer Radical Polymerization. *J. Am. Chem. Soc.* **2014**, *136*, 6513–6533. [CrossRef]
41. Matyjaszewski, K. Atom Transfer Radical Polymerization (ATRP): Current Status and Future Perspectives. *Macromolecules* **2012**, *45*, 4015–4039. [CrossRef]
42. Sypabekova, M.; Hagemann, A.; Rho, D.; Kim, S. Review: 3-Aminopropyltriethoxysilane (APTES) Deposition Methods on Oxide Surfaces in Solution and Vapor Phases for Biosensing Applications. *Biosensors* **2022**, *13*, 36. [CrossRef]
43. Uttley, O.F.; Brummitt, L.A.; Worrall, S.D.; Edmondson, S. Accessible and Sustainable Cu(0)-Mediated Radical Polymerisation for the Functionalisation of Surfaces. *Polym. Chem.* **2020**, *11*, 3831–3840. [CrossRef]
44. Che, Y.; Zhang, T.; Du, Y.; Amin, I.; Marschelke, C.; Jordan, R. "On Water" Surface-Initiated Polymerization of Hydrophobic Monomers. *Angew. Chem. Int. Ed.* **2018**, *57*, 16380–16384. [CrossRef] [PubMed]
45. Yasumoro, K.; Fujita, Y.; Arimatsu, H.; Fujima, T. A New Composite Structure of PEDOT/PSS: Macro-Separated Layers by a Polyelectrolyte Brush. *Polymers* **2020**, *12*, 456. [CrossRef] [PubMed]
46. Mai, T.; Wolski, K.; Puciul-Malinowska, A.; Kopyshev, A.; Gräf, R.; Bruns, M.; Zapotoczny, S.; Taubert, A. Anionic Polymer Brushes for Biomimetic Calcium Phosphate Mineralization-A Surface with Application Potential in Biomaterials. *Polymers* **2018**, *10*, 1165. [CrossRef]
47. Yin, X.; Wu, D.; Yang, H.; Wang, J.; Zhang, X.; Li, H.; Zheng, T.; Wang, L.; Zhang, T. Galvanic-Replacement-Assisted Surface-Initiated Atom Transfer Radical Polymerization for Functional Polymer Brush Engineering. *ACS Macro Lett.* **2022**, *11*, 296–302. [CrossRef]
48. Karade, V.C.; Sharma, A.; Dhavale, R.P.; Dhavale, R.P.; Shingte, S.R.; Patil, P.S.; Kim, J.H.; Zahn, D.R.T.; Chougale, A.D.; Salvan, G.; et al. APTES Monolayer Coverage on Self-Assembled Magnetic Nanospheres for Controlled Release of Anticancer Drug Nintedanib. *Sci. Rep.* **2021**, *11*, 5674. [CrossRef]
49. Banu, M.; Radoi, A.; Simion, M.; Kusko, M. Reproducible Functionalization of Silicon Substrates Intended for Biomedical Applications. In Proceedings of the 2016 International Semiconductor Conference (CAS), Sinaia, Romania, 10–12 October 2016; IEEE: New York, NY, USA, 2016; pp. 151–154.

50. Rozyyev, V.; Shevate, R.; Pathak, R.; Murphy, J.G.; Mane, A.U.; Sibener, S.J.; Elam, J.W. Vapor-Phase Grafting of Functional Silanes on Atomic Layer Deposited Al2O3. *J. Vac. Sci. Technol. A* **2023**, *41*, 032408. [CrossRef]
51. Miranda, A.; Martínez, L.; De Beule, P.A.A. Facile Synthesis of an Aminopropylsilane Layer on Si/SiO2 Substrates Using Ethanol as APTES Solvent. *MethodsX* **2020**, *7*, 100931. [CrossRef]
52. Arkles, B. Tailoring Surfaces with Silanes. *Chemtech* **1977**, *7*, 766–778.
53. Vandenberg, E.T.; Bertilsson, L.; Liedberg, B.O.; Uvdal, K.; Erlandsson, R.; Elwing, H. Structure of 3-Aminopropyl Triethoxy Silane on Silicon Oxide. *J. Colloid. Interface Sci.* **1991**, *147*, 103–118. [CrossRef]
54. Asenath Smith, E.; Chen, W. How To Prevent the Loss of Surface Functionality Derived from Aminosilanes. *Langmuir* **2008**, *24*, 12405–12409. [CrossRef]
55. Acres, R.G.; Ellis, A.V.; Alvino, J.; Lenahan, C.E.; Khodakov, D.A.; Metha, G.F.; Andersson, G.G. Molecular Structure of 3-Aminopropyltriethoxysilane Layers Formed on Silanol-Terminated Silicon Surfaces. *J. Phys. Chem. C* **2012**, *116*, 6289–6297. [CrossRef]
56. Belegrinou, S.; Malinova, V.; Masciadri, R.; Meier, W. Efficient Two-Step Synthesis of 11,11′-Dithiobis[1-(2-Bromo-2- Methylpropionyloxy)Undecane], a Conventional Initiator for Grafting Polymer Brushes from Gold Surfaces via ATRP. *Synth. Commun.* **2010**, *40*, 3000–3007. [CrossRef]
57. Bellucci, G.; Berti, G.; Bianchini, R.; Orsini, L. The Reaction of Triethylamine with Bromine. An Evaluation of the Formation Constant of the Transient 1:1 Complex and a Study of Its Conversion into Products. *Tetrahedron Lett.* **1982**, *23*, 3635–3638. [CrossRef]
58. Fried, S.; Kleene, R.D. Tetrahydrofuran Compounds. I. Cleavage by Hydrogen Halides. *J. Am. Chem. Soc.* **1941**, *63*, 2691. [CrossRef]
59. Smith, B.C. Halogenated Organic Compounds. *Spectroscopy* **2023**, *38*, 12–15,42. [CrossRef]
60. Brittain, W.J.; Minko, S. A Structural Definition of Polymer Brushes. *J. Polym. Sci. A Polym. Chem.* **2007**, *45*, 3505–3512. [CrossRef]
61. Boyes, S.G.; Granville, A.M.; Baum, M.; Akgun, B.; Mirous, B.K.; Brittain, W.J. Polymer Brushes—Surface Immobilized Polymers. *Surf. Sci.* **2004**, *570*, 1–12. [CrossRef]
62. Coates, J. Interpretation of Infrared Spectra, A Practical Approach. In *Encyclopedia Analytical Chemistry*; Meyers, R.A., Ed.; John Wiley & Sons Ltd.: Hoboken, NJ, USA, 2000; pp. 10815–10837.
63. Jung, J.; Jang, J.; Ahn, J. Characterization of a Polyacrylamide Solution Used for Remediation of Petroleum Contaminated Soils. *Materials* **2016**, *9*, 16. [CrossRef]
64. Zhang, X.; Liu, W.; Chen, Y.; Gong, A.; Chen, C.; Xi, F. Self-Condensing Vinyl Polymerization of Acrylamide. *Polym. Bull.* **1999**, *43*, 29–34. [CrossRef]
65. Peng, Y.; Rong, Y.; Feng, S.; Jin, D. Quality Control of Polyacrylamide Products with Infrared Spectroscopy. *J. Phys. Conf. Ser.* **2021**, *2009*, 012038. [CrossRef]
66. Gupta, M.K.; Bansil, R. Laser Raman Spectroscopy of Polyacrylamide. *J. Polym. Sci.* **1981**, *19*, 353–360. [CrossRef]
67. Murugan, R.; Mohan, S.; Bigotto, A. FTIR and Polarised Raman Spectra of Acrylamide and Polyacrylamide. *J. Korean Phys. Soc.* **1998**, *32*, 505.
68. Sugiyama, H.; Katoh, K.; Sekine, N.; Sekine, Y.; Watanabe, T.; Ikeda-Fukazawa, T. Effects of Hydrophilic Groups of Polymer on Change in Hydrogen-Bonding Structure of Water in Hydrogels during Dehydration. *Chem. Phys. Lett.* **2024**, *856*, 141655. [CrossRef]
69. Jayaramudu, T.; Ko, H.-U.; Kim, H.C.; Kim, J.W.; Kim, J. Swelling Behavior of Polyacrylamide–Cellulose Nanocrystal Hydrogels: Swelling Kinetics, Temperature, and PH Effects. *Materials* **2019**, *12*, 2080. [CrossRef]
70. Chhetry, G.; Gooh Pattader, P.S. Effect of Patterns on Polyacrylamide Hydrogel Surface towards Enhancement of Water Retention. *Polymer* **2024**, *308*, 127362. [CrossRef]

polymers

MDPI

Article

Multi-Statistical Pragmatic Framework to Study UV Exposure Effects via VIS Reflectance in Automotive Polymer Components

Jose Amilcar Rizzo-Sierra [1,*], Luis Alvaro Montoya-Santiyanes [2], Cesar Isaza [1], Karina Anaya [1], Cristian Felipe Ramirez-Gutierrez [1] and Jonny Paul Zavala de Paz [1,*]

1 Cuerpo Académico de Tecnologías de la Información y Comunicación Aplicada, Universidad Politécnica de Querétaro (UPQ), Carretera Estatal 420 SN, El Marqués, Santiago de Querétaro 76240, Mexico; cesar.isaza@upq.edu.mx (C.I.); karina.anaya@upq.mx (K.A.); cristian.ramirez@upq.edu.mx (C.F.R.-G.)

2 Unidad de Alta Tecnología, Facultad de Ingeniería, Universidad Nacional Autónoma de México (UNAM), Fray Antonio de Monroy e Hijar 260, Villas del Mesón, Juriquilla, Santiago de Querétaro 76230, Mexico; montoyas@unam.edu

* Correspondence: jose.rizzo@upq.edu.mx (J.A.R.-S.); jonny.zavala@upq.edu.mx (J.P.Z.d.P.)

Abstract

This study evaluates the cosmetic degradation of polyethylene (PE) and polypropylene (PP) automotive components under four exposure scenarios—no exposure, outdoor exposure with and without glass shielding, and accelerated UV chamber weathering (ASTM G154)—through the evolution of visible (VIS) reflectance. Thirty-two samples (16 PE, 16 PP) were monitored over five time points; surface reflectance was recorded at 21 wavelengths and summarized into seven VIS bands, and hardness (Shore D) was measured pre/post-exposure. Repeated-measures univariate and multivariate analyses consistently revealed significant effects of Condition, Time, and their interaction on reflectance, with initial-reflectance adjustment improving inference stability across bands. PE exhibited more gradual and coherent reflectance decay, whereas PP showed greater band-to-band variability—most notably under UV chamber exposure. Additionally, hardness decreased in most exposed groups, aligning with optical changes. To place spectral trajectories in a kinetic context, a family of exponential models with small-sample information criterion selection was fitted, yielding $\eta(t)$—a dimensionless degradation efficiency summarizing spectral change. The contribution of this work is a multi-statistical framework—combining VIS-band-aware summaries with covariate-adjusted univariate/multivariate testing—that supports comparisons across materials and exposure conditions, underscoring the practical value of UV chamber protocols as surrogates for outdoor weathering. In sum, the study demonstrates the effectiveness of multivariate and covariate-adjusted models in quantifying optical degradation of polyolefins, offering pragmatic guidance for assessing mid- to long-term performance in automotive applications.

Keywords: automotive polymers; UV degradation; reflectance spectroscopy; statistical analyses

1. Introduction

Polyolefins, such as polyethylene (PE) and polypropylene (PP), i.e., thermoplastic polymers, are extensively utilized in automotive applications due to their favorable mechanical properties, cost-effectiveness, and ease of processing [1,2]. However, their susceptibility to environmental degradation, particularly under ultraviolet (UV) radiation, poses significant challenges for long-term durability in terms of mechanical performance and aesthetic/cosmetic qualities [3,4]. Degradation processes such as thermal aging and mechanical wear can significantly alter their physical and chemical properties over time [5–7]. In particular, exposure to UV light initiates photo-oxidative reactions in these polymers, leading to chain scission, cross-linking, and the formation of carbonyl groups, which collectively result in discoloration/cosmetic degradation, embrittlement, and loss of mechanical integrity [8–12]. These last subjects being critical for the automotive industry.

The automotive industry often employs accelerated aging tests, such as UV exposure chambers, in order to predict the service life of polymeric components [13–17]. These tests simulate prolonged environmental exposure within a condensed timeframe, allowing for the assessment of material performance under controlled conditions [16,18–20]. Standards such as ASTM D5208-14 and ISO 4892-3:2024 provide guidelines for conducting such evaluations [21,22]. Additionally, real-world outdoor exposures, both with and without protective barriers like glass, provide insights into the environmental durability of these materials [23–26].

On the other hand, reflectance spectroscopy in the visible range, employed in the present study, serves as a non-destructive method to monitor the surface degradation of polymers. For example, Goddin et al. reviewed the effectiveness of VIS–SWIR (visible to short-wave infrared) spectroscopy in distinguishing plastic litter on and below water surfaces. It supports the method's ability to non-invasively monitor polymer presence and condition, reinforcing its utility in environmental diagnostics through spectral reflectance analysis [27]. Goustoridis et al. [28] showcase white light reflectance spectroscopy (WLRS) as a rapid, accurate, and contactless tool for determining the thickness and transformation of films and layers, including polymer coatings. It highlights WLRS's versatility in tracking real-time surface changes like dissolution or degradation, without damaging the sample. Makki et al. [29] employed reflectance spectroscopy to quantify weathering-induced changes in polymer films. The technique proved effective in identifying surface color shifts and chemical transformations due to UV exposure, demonstrating its reliability for non-destructive monitoring. Chaudhary et al. [30] reported on using UV-VIS reflectance analysis to examine polymer degradation. The method was essential for identifying shifts in optical properties related to molecular structural changes, underscoring its role in non-destructive diagnostics. Matuana et al. [31] used reflectance spectroscopy to assess UV-induced surface degradation in wood–plastic composites. Changes in reflectance spectra, especially within the visible range, were directly linked to color fading and chemical deterioration. Choi et al. [32] investigated whitening in UV-exposed elastomers, and linked surface degradation to spectral variations in the visible region. Reflectance data allowed for clear, non-destructive identification of photo-induced chemical changes on the material surface. Exposito et al. [33] detail how UV-VIS spectroscopy detects photodegradation by monitoring new absorption bands and reflectance losses. It is especially useful for mapping oxidative changes and morphological deterioration over time. Wijdekop et al. [34] applied reflectance spectroscopy to PVC-coated steel undergoing weathering. The method successfully captured gradual discoloration due to UV and environmental stress, showing strong potential for long-term, non-invasive surface quality evaluation. Gulmine et al. [35] tracked the chemical breakdown and surface oxidation of polyethylene during UV aging. Variations in spectral intensity served as indicators of both physical and chemical deterioration. Komiya et al. assessed cosmetic degradation and microstructural alterations via reflectance data in polymers, with clear visibility of changes in the VIS spectrum due to photo-oxidation and cross-linking reactions [36].

Recent work has also emphasized how polymer formulation—such as resin density, stabilizer dosage, pigment loading, and incorporation of nanoparticles—directly affects the rate and type of UV-induced degradation. For example, Du et al. [21] quantitatively analyzed LDPE sheets with varying thicknesses and pro-oxidant metal stearate contents, revealing significant effects on carbonyl index, molecular weight reduction, and VIS reflectance shifts under ASTM UV weathering. Zhang et al. [37] analyzed polyethylene samples with differing densities and nanoparticle fillers, establishing correlation between formulation parameters and surface degradation kinetics assessed via reflectance-based techniques. Andrady et al. [38] investigated LDPE laminates exposed to UV radiation in both air and seawater. Their quantitative FTIR depth-profiling and tensile property assessment revealed diffusion-controlled photo-oxidation localized in a thin surface layer and slower degradation kinetics in seawater due to limited oxygen solubility. Weizman et al. [39] quantitatively demonstrated that incorporating nanoscale UV stabilizers into LDPE films significantly affects photodegradation kinetics. Using UV-VIS reflectance and FTIR analysis, they showed that nanoparticle-enhanced formulations reduce stabilizer loss and delay surface oxidation, emphasizing the importance of additive design in modulating optical aging under UV exposure. Brostow et al. [40] systematically studied polypropylene films containing HALS and nano-ZnO under accelerated outdoor UV exposure, demonstrating that combinations of stabilizers confer a measurable increase in UV resistance and tensile strength retention. Cavalcanti & Rabello [41] evaluated polypropylene films with different levels of red pigment and UV-stabilizers (UV absorbers and HALS), showing that pigment concentration quantitatively alters color shift (ΔE^*) and mechanical degradation trajectories. Grigoriadou et al. [42] found that HDPE nanocomposites containing SiO_2 or montmorillonite nanoparticles catalyze UV degradation, resulting in increased carbonyl and vinyl groups, surface cracking, and altered optical properties detectable via spectroscopy. Regarding other recent polymer-based industrial applications, such as CNT/epoxy resins with PCL as fillers, Xian et al. also quantitatively link formulation/preparation parameters to design-driven performance metrics [43], validating the perspective applied here to the rate and mode of UV-driven degradation in formulated polyolefins intended for outdoor use.

This range of quantitative findings directly informs the experimental design contemplated in this study, excepting the pigment/filler/stabilizer composition factor, outside of its scope. By systematically varying polymer type (PE vs. PP) and exposure modes (outdoor vs. artificial UV), we aim to investigate their combined influence on VIS reflectance degradation trajectories. This integration of formulation technology and controlled experimental design helps to ensure a rigorous quantitative comparison of cosmetic stability in automotive polymer components. In addition, particularly regarding reflectance–transmittance data gathering and analysis, these previous works indicate that changes in polymer spectra can signal alterations in surface chemistry and morphology due to weathering/cosmetic degradation [44–46]. Specifically, the emergence of new absorption bands or shifts in existing ones can be correlated with the formation of degradation products or sub-products such as carbonyl compounds [47–50]. In summary, characterizing and understanding the degradation behavior of polymers such as PE and PP is crucial for predicting their service life and ensuring reliability and cosmetic standards compliance for automotive industry components.

In this study, we investigate the degradation behavior of eight automotive plastic parts—four made of PE and four of PP—subjected to four distinct environmental exposure conditions: (C1) no exposure (control), (C2) outdoor exposure with glass protection, (C3) outdoor exposure without protection, and (C4) accelerated aging via UV chamber simulation. In particular, ASTM G154 protocol ([51]) was implemented. Reflectance measurements were conducted at five time points (T0 to T4) across 21 wavelengths grouped into seven VIS-bands (Violet to Red), with each band comprising three closely spaced wavelengths in order to serve as statistical replicates. The experimental design incorporates both between-subjects factors (Material and Exposure Condition) and a within-subjects factor (Time). Given the repeated measures and the hierarchical structure of the data, statistical analyses were performed using repeated-measures ANOVA and ANCOVA for univariate responses and MANOVA and MANCOVA for multivariate responses. Initial reflectance values at T0 were included as covariates to adjust for baseline differences between samples. This comprehensive approach aims to consider the effects of material type, exposure environmental condition, and time on the optical properties via VIS reflectance of automotive polymers (PP and PE) over time, providing a framework with valuable insights for the development of more durable and cosmetically stable materials in automotive applications.

2. Materials and Methods

2.1. Samples and Experimental Design

The study was conducted using 8 distinct automotive plastic part samples, labeled as subjects, from Subject 1 to 8. The first four (Subject 1 to Subject 4) were made of polyethylene (PE), while the remaining four (Subject 5 to Subject 8) were made of polypropylene (PP). Each of these subjects was exposed to one of four distinct environmental exposure conditions:

C1: No exposure (control).
C2: Outdoor environmental exposure with glass protection.
C3: Outdoor environmental exposure without glass protection.
C4: Accelerated aging via UV chamber simulation (ASTM G154 protocol [51]).

Reflectance was measured for each subject at five time points:

- **T0:** Initial time (before exposure).
- **T1–T4:** After 1/4, 2/4, 3/4, and 4/4 of the experiment duration, respectively. In this particular respect, the experimentation run extended for a period of 4 months. A uniform temporal spacing was selected to ensure consistent tracking of mid- to long-term cumulative environmental exposure under outdoor and artificial weathering/aging scenarios for the type of materials considered, as supported by studies such as ref. [52] regarding mechanical degradation, and ref. [53] on UV aging. While front-loaded sampling designs may be preferable when early degradation dynamics are dominant, the mentioned literature precedents indicate that reflectance variation is gradual, justifying neither denser early sampling nor reflectance rate-of-change analyses in this study. On the other hand, fixed-interval data were required to develop repeated-measures statistical modeling frameworks. Tests were conducted on the campus of *Universidad Politécnica de Querétaro* (20°32′48″ N, 100°16′28″ W), *El Marqués, Querétaro, México*. The time lapse went from July to October 2024, during the local rainy season. Climatological data from Querétaro Intercontinental Airport indicate that, over these four months, daily mean temperatures typically ranged from 14 °C (overnight lows) to 25 °C (daytime highs), with relative humidity between 64% and 69%. Monthly accumulated precipitation was approximately 96 mm in July, 86 mm in August, 87 mm in September, and 30 mm in October [54].

Reflectance measurements were taken in the visible light spectrum using 21 specific wavelengths grouped into 7 VIS-bands, each composed of 3 closely spaced wavelengths:

- **Violet:** 400, 410, 420 [nm].
- **Blue:** 450, 460, 470 [nm].
- **Cyan:** 480, 490, 500 [nm].
- **Green:** 520, 530, 540 [nm].
- **Yellow:** 560, 570, 580 [nm].
- **Orange:** 600, 610, 620 [nm].
- **Red:** 640, 650, 660 [nm].

That meant that each wavelength within a band was treated as a replicate for statistical purposes. The Subject ID and material are not independent, as the type of material is systematically associated with specific subject IDs.

Shore D Hardness Measurements: Surface hardness was evaluated at two stages (T0 and T4) using the Shore D scale, a standardized metric for assessing the rigidity of polymers. For each material and exposure condition, measurements were obtained from four independent samples (subjects), each tested once per stage. This allowed quantification of mechanical degradation due to UV exposure. The main features of the Shore D digital durometer used are present in Table 1.

Table 1. Features of the Gain Express Holdings Ltd. 560-10D Shore D digital durometer used.

Feature	Description
Sharp cone point	SR 0.1 [mm], 30° including angle.
Measure sample thickness	>6 [mm] (no more than 3 layers).
Measured area	Not less than the pressure edge of 15 [mm].
LCD display	4 digits, 40 × 18 [mm] (1.57×) 0.71 [inch].
Measuring range/Graduation	0–100 HD.
Indentor running	0–2.2 [mm].
Indentor press on telos	A, C type 0.55–8.06 [N], D type 0–44.5 [N].
Resolution	0.5 HD.
Power supply	1 × SR44 button cell battery.
Item size	87 [mm] × 56 [mm] × 25 [mm] (L × W × D) (3.43″ × 2.2″ × 0.98″)
Item weight	36 [g] (4.79 [oz]).

Reflectance Measurements: Reflectances were realized with a Konica Minolta CM-700d spectrophotometer, whose main features are described in Table 2.

Table 2. Features of the Konica Minolta CM-700d spectrophotometer used.

Feature	Description
Illumination/viewing system	Diffused illumination, 8° viewing angle. Specular component included/excluded. Conforms to CIE No. 15 ISO 7724/1, DIN5033 Teil7, ASTM E 1164, and JIS Z 8722.
Size of integrating sphere/diameter	40 [mm].
Spectral separation device	Diffraction grating.
Wavelength range	400 to 700 [nm].
Wavelength pitch	10 [nm].
Half bandwidth	≃10 [nm].
Reflectance range	0 to 175%, display resolution: 0.01%
Light source	Pulsed Xenon Lamp (with UV cut filter).
Measurement time	≃1 [s].
Minimum measurement interval	≃2 [s] (in SCI or SCE mode).

2.2. Statistical Structure of the Experiment

A structured experimental design with both between-subject and within-subject factors was employed. The statistical analysis framework rests upon clear identification and distinction between the following:

- **Independent variables (factors):** Variables manipulated or categorized in the experiment to assess their influence on the outcomes.
- **Dependent variables (responses):** Quantities that are measured as outcomes and are hypothesized to be affected by the independent variables.

- **Between-subjects factors:** Factors for which each subject (i.e., each individual subject) is observed under only one level. These define groups of subjects.
- **Within-subjects factors (repeated measures):** Factors for which each subject is observed at multiple levels. These correspond to repeated measurements over time or under different internal conditions.

The experimental structure employed can be described as follows:

2.2.1. Dependent (Response) Variables

- **Univariate case:** Mean reflectance per VIS-band (7 VIS-bands total), computed by averaging the reflectance over the 3 closely spaced wavelengths within each band.
- **Multivariate case:** The 7-dimensional reflectance vector per observation, with each component corresponding to the mean reflectance in one VIS-band (Violet to Red).

2.2.2. Independent Variables (Factors)

- **Between-subject factors:**
  - **Material (2 levels):** Polyethylene (PE), polypropylene (PP). Note: Material is not independent from Subject ID, as each subject is uniquely assigned a material type.
  - **Condition of Exposure (4 levels):** C1 (control), C2 (glass-protected), C3 (unprotected), C4 (UV chamber).
- **Within-subject factors:**
  - **Time (5 levels):** T0 (initial), T1, T2, T3, T4. All subjects were measured repeatedly at the same 5 time points.

2.2.3. Covariates (for ANCOVA/MANCOVA)

- **Initial Reflectance (T0):** For each VIS-band (or the 7-dimensional vector), the reflectance at time T0 is used as a covariate. This adjustment allows the models to control for baseline variability among subjects that may arise due to inherent color differences, pigmentation, or surface features.

2.3. *Experimental Units and Repetition*

- Each of the 8 automotive part samples is treated as a subject.
- Each subject is assigned a unique combination of material and exposure condition.
- Reflectance is recorded for each subject at 5 time points and for 21 wavelengths grouped into 7 VIS-bands, with 3 replicate wavelengths per band.
- For statistical analysis, the mean reflectance across the 3 replicates within a band is treated as the response variable (univariate) or part of a vector of responses (multivariate).

This design allows the analysis of the following:

- Temporal evolution of reflectance (within-subject effects),
- Material and exposure-related differences (between-subject effects),
- Adjustments for initial reflectance variability (via covariates).

The repeated-measures design ensures that each subject serves as its own control over time, increasing the statistical power and sensitivity to changes in reflectance behavior.

2.4. *Statistical Analyses*

Regarding this present experiment, the following analyses were conducted using MATLAB R2024b and appropriate statistical packages. Custom scripts were developed specifically, and the structure of the analyses was guided by the nature of the reflectance data (longitudinal, repeated-measures, multiband), the presence of both between- and within-subject factors, and the interest in controlling for initial reflectance variability. The primary inferential route of the present study is a repeated-measures MANCOVA on the 7-band reflectance vector with T0 as covariate, testing Material, Condition, Time, and their interactions. When multivariate significance arises, band-wise ANCOVAs (with T0 as covariate) are used to localize effects and aid interpretation within the visible spectrum. In addition, RM-MANOVA and unadjusted ANOVA are retained and reported as well; as complementary, coarse-sensitivity benchmarks related to statistical analyses/models without baseline adjustment. Shore D hardness was analyzed via one-way/two-way ANOVA with appropriate post-hoc tests as a material performance endpoint. Finally, bounded exponential summaries $\eta(t)$ (first-order, stretched/lag, constrained bi-exponential; AICc selection) were developed, enabling descriptive kinetics (η_{inf}, $t1/2$) that complement the inferential framework.

1. **Shore D Hardness Analysis**
   Mechanical degradation was assessed using Shore D hardness values obtained before (T0) and after (T4) exposure. Two complementary statistical approaches were employed:
   - *Paired Tests:* For each material and exposure condition, post-exposure hardness was compared to reference values using paired t-tests or Wilcoxon signed-rank tests, depending on the normality of the paired differences (tested via Lilliefors test).
   - *One-Way ANOVA with Tukey Post Hoc:* For each material separately, a one-way ANOVA was conducted to compare hardness values across the four exposure conditions. Tukey's test was used for multiple comparisons.

   These tests were used to detect significant mechanical degradation due to environmental exposure and to compare the resilience of each polymer under the different aging scenarios.
2. **Material Non-Specific ANOVA and ANCOVA**
   Mean reflectance per VIS-band was used as the dependent variable. The effects of material, exposure condition, and time were evaluated using repeated-measures ANOVA and ANCOVA. The initial reflectance (T0) was included as a covariate in ANCOVA. For each of the 7 VIS-bands (Violet to Red), the mean reflectance (averaged over its 3 wavelengths) was analyzed independently using univariate repeated-measures ANOVA. The between-subjects factors were *Material* (PE vs. PP) and *Condition of Exposure* (C1–C4), and the within-subjects factor was *Time* (T0–T4). In ANCOVA models, the initial reflectance value at T0 for each band was included as a covariate to adjust for baseline differences.
3. **Material-Specific ANOVA and ANCOVA**
   For polyethylene and polypropylene materials, respectively, mean reflectance per VIS-band was analyzed with exposure condition and time as factors. Again, ANCOVA models included T0 reflectance as a covariate. The analyses in the previous item were repeated separately for the PE group and the PP group to explore material-specific effects of exposure condition and time. This allowed for better isolation of degradation behavior and interaction effects within each material class.
4. **Material-Non-specific MANOVA and MANCOVA**
   Identical structure to the previous item, to confirm reproducibility and consistency under alternative implementations or assumptions. The reflectance data were aggregated into 7-dimensional vectors (one per VIS-band) per subject per time point. These vectors were used as multivariate responses in a repeated-measures MANOVA framework, with *Material* and *Condition* as between-subject factors and *Time* as the within-subject factor. In MANCOVA, the corresponding 7-dimensional vector at T0 was used as a covariate to remove initial heterogeneity.

5. **Material-Specific MANOVA and MANCOVA**
For each material type, multivariate reflectance responses (7 bands) were analyzed as a function of exposure condition and time, with or without the corresponding initial reflectance vector as covariate. The multivariate analysis approach was applied separately to PE and PP groups, allowing the multivariate trajectory of degradation across the 7 VIS-bands to be explored as a function of exposure condition and time. MANCOVA models again included initial (T0) reflectance vectors as covariates.

The statistical workflow was designed to evaluate how visible reflectance evolves across environmental conditions, sampling times, and material types while accounting for initial differences between samples. All analyses were conducted in MATLAB R2024b using both univariate and multivariate approaches. On the other hand, prior to hypothesis testing, the global normality of the reflectance and hardness data was confirmed by Shapiro–Wilk tests on each group defined by Material, Condition, and Time ($p > 0.05$ in all cases).

For the Shore D hardness data, paired comparisons against the reference condition were evaluated by first testing the differences for normality using the Lilliefors test (MATLAB's `lillietest`). When normality was not rejected, paired t-tests were applied; otherwise, Wilcoxon signed–rank tests were used. Homogeneity of variances across conditions was assessed by Levene's test (`vartestn` with 'LeveneAbsolute', $\alpha = 0.05$). One-way ANOVA models were then fitted separately for polyethylene and polypropylene, followed by Tukey–Kramer multiple comparisons (`multcompare`, 'CType' = "tukey-kramer").

Now, in order to explore global reflectance variation over time, a repeated-measures MANOVA (RM-MANOVA) using the full reflectance spectra (31 wavelengths from 400 to 700 nm at 10 nm intervals) was conducted first. This analysis included the within-subject factor **Time** (T_0 to T_4) and the between-subject factors **Material** (polyethylene or polypropylene) and **Condition** (C1 to C4). The response matrix was formed by concatenating all wavelength values for each sample across the five time points.

A complementary multivariate ANCOVA (MANCOVA) was performed to assess the statistical effects of **Material, Condition, Time** (T1–T4), and their interactions on the average reflectance within each of the seven visible VIS-bands (violet to red). The initial reflectance at T_0 was included as a covariate to control for baseline differences across samples. Covariates were adjusted via linear regression, and residuals were analyzed through MANOVA. The effects tested included all second-order interactions and the third-order interaction **Material** $\times$ **Condition** $\times$ **Time**. Multivariate significance was evaluated using Wilks's λ, Pillai's trace, Hotelling–Lawley trace, and Roy's largest root statistics.

Additionally, separate ANCOVA models were fitted for polyethylene and polypropylene to detect potential material-specific effects. Each model included **Condition, Time,** the initial reflectance covariate, and their second- and third-order interactions. Univariate p-values were used to determine the significance of each term in each VIS-band.

All significance thresholds were set at $\alpha = 0.05$. Results are reported in terms of statistical significance per factor and interaction, and tables summarize the outcomes of all multivariate and adjusted univariate models.

Summarizing this section's rationale, in order to comprehensively assess the effects of material, exposure condition, and time on spectral reflectance, both univariate and multivariate statistical approaches were employed in this study. Univariate models (ANOVA, ANCOVA) were used to evaluate reflectance changes at each individual VIS-band (e.g., Violet, Blue), allowing for band-specific significance testing. However, given that reflectance values across adjacent VIS-bands are spectrally correlated, multivariate methods (MANOVA, MANCOVA) were also applied to evaluate global degradation patterns. These methods leverage the covariance structure among multiple dependent variables to enhance statistical power, reduce the risk of Type I error associated with multiple testing, and detect interaction effects (e.g., Material × Condition × Time) that may not be evident in isolated univariate analyses. This dual approach enables more robust and interpretable modeling of spectral degradation dynamics across the visible range, as supported by [55], when discussing how multivariate methods can detect global effects that univariate ones may overlook, and [56], by explaining how differences in covariance matrices influence validity and descriptive power features of MANOVAs over ANOVAs when variable dependency could be considered.

2.5. Data Processing

Prior to statistical modeling, the raw reflectance measurements were pre-processed to account for replicated wavelengths and categorical structure. Each measurement of reflectance was originally recorded at 21 discrete wavelengths (3 per spectral color) for each sample at each time point.

2.5.1. Reflectance Data

Reflectance was originally recorded at 21 discrete wavelengths for each sample at each time point. These were grouped into seven VIS-bands (Violet to Red), each composed of three closely spaced wavelengths considered as technical replicates. The following steps were taken:

- **Averaging of Technical Replicates.** To minimize redundancy and stabilize the variance, the three reflectance measurements corresponding to each VIS-band (e.g., 400, 410, and 420 [nm] for Violet) were averaged. This yielded one representative reflectance value per color per time point per sample, resulting in a data structure of 7 colors × 5 time points × 32 samples (8 subjects × 4 conditions). These wavelengths within each VIS-band were selected based on spectral proximity (10 nm spacing), forming narrow spectral windows where degradation dynamics are expected to be smooth and correlated. Studies on polymer photo-oxidation and reflectance spectroscopy have consistently shown that UV-induced degradation manifests as gradual, continuous shifts in spectral properties over short wavelength ranges, rather than abrupt changes at narrow intervals [21,41]. Thus, averaging across each band preserves the underlying spectral behavior while improving signal stability. Now, from a statistical perspective, treating each triplet of closely spaced wavelengths as technical replicates reduces random noise and intra-band variability. This not only stabilizes the error variance across repeated-measures and ANCOVA models, but also reinforces key assumptions such as homoscedasticity and normality. Moreover, averaging helps avoid overparameterization in multivariate models, enabling more interpretable and robust estimation of effects across VIS-bands. Consequently, this approach enhances model reliability without compromising spectral resolution in the context of polymer degradation tracking.
- **Reshaping to Longitudinal Format:** The averaged reflectance values were organized into a subject-wise longitudinal data structure across five time points (T0–T4).
- **Covariate Definition:** Reflectance values at T0 were stored separately and used as covariates in ANCOVA and MANCOVA models to adjust for baseline differences in initial color or pigmentation.

2.5.2. Hardness Data

Shore D hardness was measured once at the beginning (T0) and once at the end of the exposure period (T4) for each sample. The data were pre-processed as follows:

- **Delta Hardness Calculation:** For each subject, the change in hardness (ΔHardness = T4 − T0) was computed to quantify mechanical degradation over time.

- **Structuring by Experimental Condition:** The ΔHardness values were grouped by material type and exposure condition to facilitate one-way and two-way ANOVA analyses.

2.5.3. Categorical Encoding

Categorical variables were coded as follows:

- `Material`: Binary variable (0 = polyethylene, 1 = polypropylene), inferred directly from sample ID.
- `Condition`: Four-level categorical factor corresponding to exposure condition (C1–C4).
- `Time`: Five-level within-subject factor (T0–T4), treated as repeated measure.
- `Color`: Seven-level categorical factor representing the visible spectrum bands (Violet to Red).
- `Subject_ID`: Unique identifier for each of the 8 distinct subjects (Subject 1 to 8), reused across all conditions.

2.5.4. Dependency Note

It is critical to note that the variables `Material` and `Subject_ID` are not statistically independent, as each subject is uniquely associated with either polyethylene or polypropylene. Therefore, in multivariate models involving both factors, collinearity was addressed by excluding one when necessary or by nesting appropriately in hierarchical models.

2.5.5. Software

All preprocessing and statistical modeling were conducted in MATLAB R2024b. Categorical variables were encoded using the `categorical()` function and all analyses used either built-in functions such as `fitrm`, `manova`, `fitlm`, or custom scripts to handle repeated-measures structures.

3. Results and Discussion

3.1. Overview of the Dataset

Reflectance data were collected from 32 automotive plastic samples (8 unique subjects × 4 exposure conditions), with each sample measured at 5 time points (T0–T4) and across 31 wavelengths in the visible spectrum (400–700 [nm] at 10 [nm] intervals). Each subject corresponds uniquely to either polyethylene (PE) or polypropylene (PP) and to one specific exposure condition. Reflectance values were pre-processed by averaging the 3 technical replicates per VIS-band when required. The resulting dataset comprises one reflectance value per wavelength, per time point, per sample.

3.2. Preliminary Visualization of Reflectance Trends

To explore global reflectance behavior over time and conditions, average reflectance curves were computed for each exposure condition, grouping samples by material (polyethylene or polypropylene). The average reflectance at each wavelength was calculated across all corresponding samples at each time point. Accordingly, Figures 1–4 show the average reflectance spectra for both polyethylene (solid lines) and polypropylene (dashed lines) measured at five different exposure times (T0 to T4) under four distinct environmental conditions: no exposure, glass-covered exposure, uncovered exposure, and accelerated exposure in a UV chamber (ASTM G154 protocol). Consequently, each figure presents the evolution of reflectance as a function of wavelength (400–700 [nm]), grouped by material and time.

Under control conditions (Figure 1), both PE and PP maintain stable reflectance profiles across the spectrum throughout the entire exposure period. No significant shifts or trends are observed, confirming this condition as a reliable baseline. Under the glass-protected exposure condition (Figure 2), PE exhibits decrement reflectance variation over time, more distinguishable from 470 [nm] on to longer wavelengths. In the same condition, PP shows distinguishable variations but without a well-defined trend over time. These patterns suggest the presence of photo-oxidative effects, possibly mitigated by the glass barrier, that are more prevalent for PE than PP. On the other hand, in the unprotected condition (Figure 3), PE showed reflectance variation with no definite tendency over time from the entire VIS-band. In contrast, PP data allows for the observation of an incremental variation in reflectance from T0 to T4, prevalent at shorter wavelengths (specially from 400 to 580 [nm] approximately). Finally, regarding the UV chamber condition (Figure 4), PE reflectance variation followed a decremental trend for shorter wavelengths (400 to 520 [nm] approximately), followed by an incremental one at longer wavelengths (660 to 700 [nm]). Meanwhile, PP shows reflectance variations with no discernible trend along the entire VIS-band.

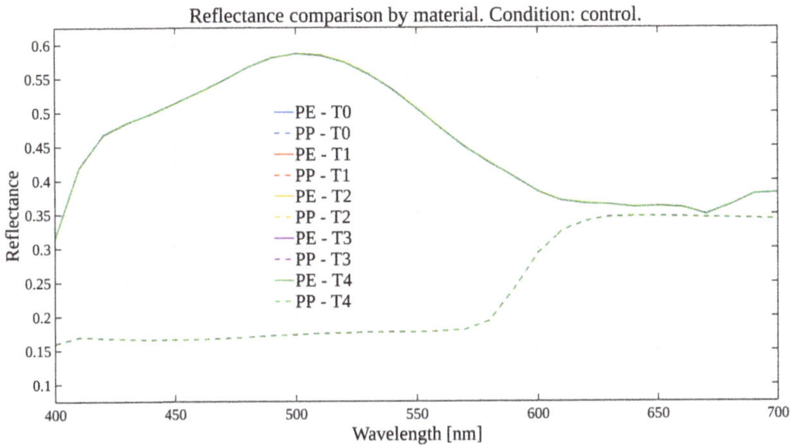

Figure 1. Mean reflectance curves for both polyethylene (solid lines) and polypropylene (dashed lines) under the control condition. One curve per time point (T0–T4).

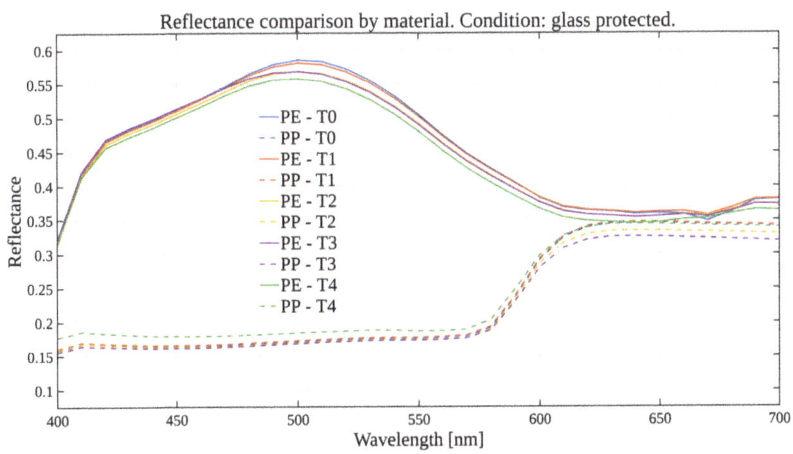

Figure 2. Mean reflectance curves for both polyethylene (solid lines) and polypropylene (dashed lines) under the glass-protected condition. One curve per time point (T0–T4).

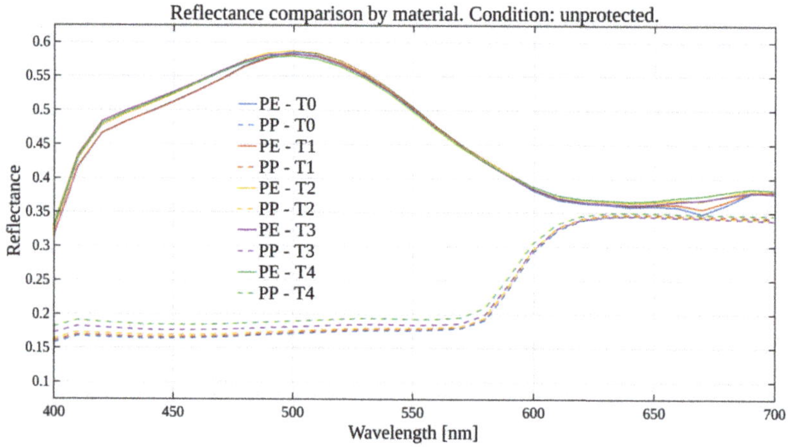

Figure 3. Mean reflectance curves for both polyethylene (solid lines) and polypropylene (dashed lines) under the unprotected condition (no glass coverage). One curve per time point (T0–T4).

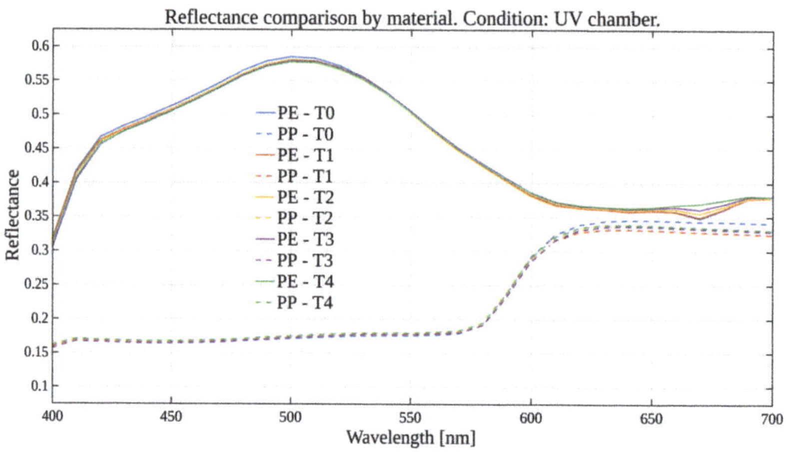

Figure 4. Mean reflectance curves for both polyethylene (solid lines) and polypropylene (dashed lines) under the UV chamber condition (ASTM G154 protocol). One curve per time point (T0–T4).

Overall, polyethylene exhibited higher reflectance values across all wavelengths and conditions compared to polypropylene. In both materials, the most noticeable changes occurred under the glass-protected and unprotected exposure conditions. These findings indicate that material type and exposure condition influence the optical behavior of the polymers over time.

On the other hand, these visualizations serve as an initial qualitative inspection of reflectance variation/degradation over time, revealing both wavelength-dependent and material-dependent differences, and motivating the subsequent statistical analysis aimed at formally quantifying these effects through univariate and multivariate statistical models present in the following sections.

3.3. Temporal Evolution by VIS-Band and Exposure Condition

In order to examine how surface optical properties evolve under different exposure conditions, reflectance trends were analyzed across seven visible VIS-bands. The analysis compares polyethylene and polypropylene behaviors over five time points (T0–T4) under four exposure conditions.

- Violet (400–420 [nm]), Blue (450–470 [nm]), and Cyan (480–500 [nm]) bands (Figures 5–7).

In the **UV chamber** condition, both PE and PP exhibited minimal variation in reflectance across time. Under **unprotected** exposure, PE showed a noticeable increase, particularly between T3 and T4, whereas PP remained stable. For the **control** condition, reflectance remained steady for both materials throughout the test period, whereas for the **glass-protected** condition, slight decreases and increases for PE and PP are observed, respectively.

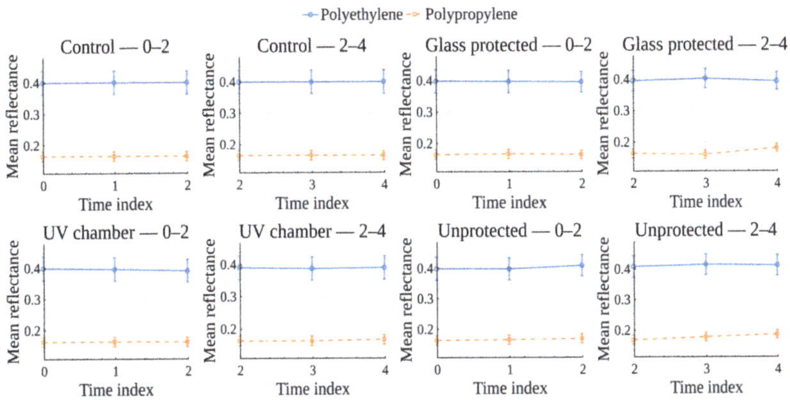

Figure 5. Temporal evolution of mean reflectance in the Violet band (400–420 nm) for polyethylene (solid lines) and polypropylene (dashed lines) under four conditions (Control, Glass-protected, Unprotected, and UV chamber). Panels split the time course into early (0–2) and late (2–4) intervals; markers denote integer times (T0–T4). Error bars show visually scaled (×0.30) SEM—standard error of the mean—applied uniformly across panels.

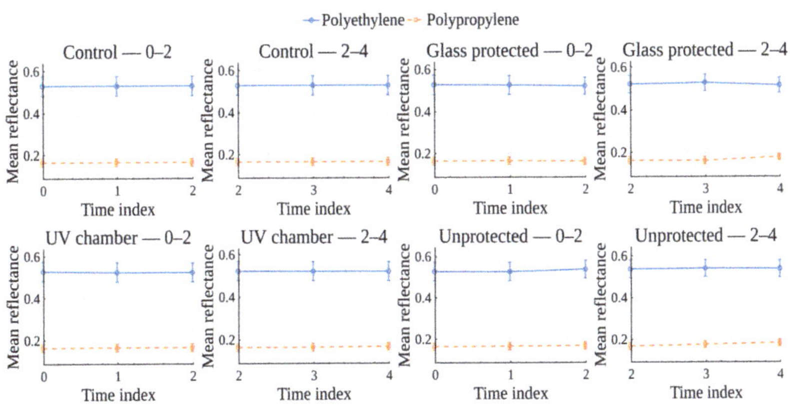

Figure 6. Temporal evolution of mean reflectance in the Blue band (450–470 nm) for polyethylene (solid) and polypropylene (dashed) across the four conditions (Control, Glass-protected, Unprotected, and UV chamber). Panels split the time course into early (0–2) and late (2–4) intervals; markers denote integer times (T0–T4). Error bars show visually scaled (×0.30) SEM—standard error of the mean—applied uniformly across panels.

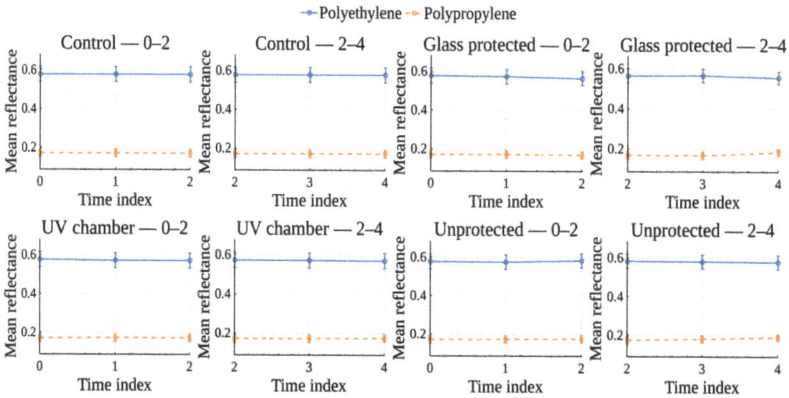

Figure 7. Temporal evolution of mean reflectance in the Cyan band (480–500 nm) comparing polyethylene (solid) versus polypropylene (dashed) under Control, Glass-protected, Unprotected, and UV chamber conditions. Panels split the time course into early (0–2) and late (2–4) intervals; markers denote integer times (T0–T4). Error bars show visually scaled (×0.30) SEM—standard error of the mean—applied uniformly across panels.

- Green (520–540 [nm]), and Yellow (560–580 [nm]) bands (Figures 8 and 9).

Under the **UV chamber** condition, PE shows minimal (almost indistinguishable in the image) decrease, while that was not the case for PP, with no apparent variation. Regarding **unprotected** exposure, PP does not show noticeable change, but PE showed an increase between T3 and T4. For the **control** condition, reflectance did not noticeably change for either material throughout the test period. Simultaneously, for the **glass-protected** condition, slight variations for PE and PP are observed, but with no overall discernible reflectance increase or decrease trend.

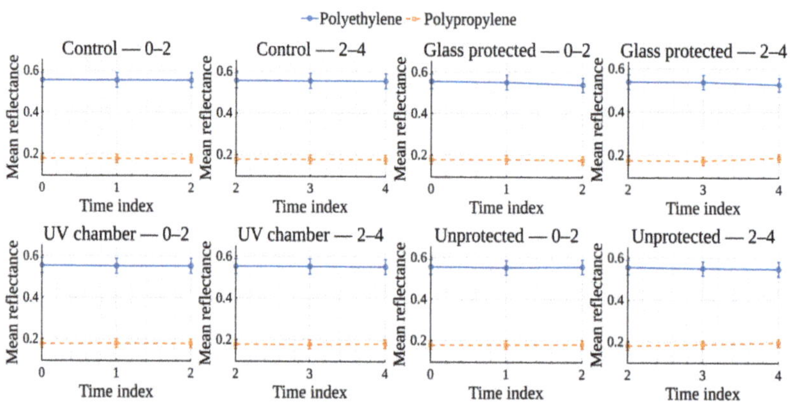

Figure 8. Temporal evolution of mean reflectance in the Green band (520–540 nm) for polyethylene (solid) and polypropylene (dashed) under the four illumination/exposure conditions. Panels split the time course into early (0–2) and late (2–4) intervals; markers denote integer times (T0–T4). Error bars show visually scaled (×0.30) SEM—standard error of the mean—applied uniformly across panels.

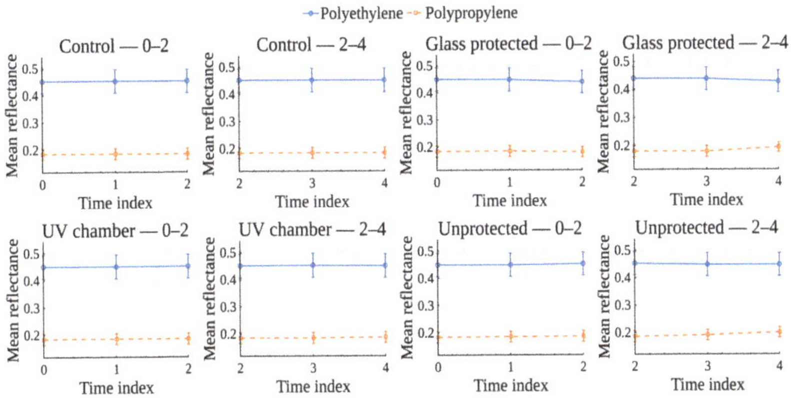

Figure 9. Temporal evolution of mean reflectance in the Yellow band (560–580 nm) for polyethylene (solid) and polypropylene (dashed), by condition (Control, Glass-protected, Unprotected, and UV chamber). Panels split the time course into early (0–2) and late (2–4) intervals; markers denote integer times (T0–T4). Error bars show visually scaled (×0.30) SEM—standard error of the mean—applied uniformly across panels.

- Orange (600–620 [nm]) and Red (640–660 [nm]) bands (Figures 10 and 11).

UV chamber condition shows no variation for either material. **Unprotected** exposure depicts variation for PE, but no definite trend towards increase or decrease, and no change for PP. Now, for the **control** condition, reflectance did not show noticeable change for either material throughout the test period. With regards to the **glass-protected** condition, no variation is discernible for PP. In contrast, very small variations are detectable for PE, but with no definite trend for the Orange band and a very small incremental trend for the Red band.

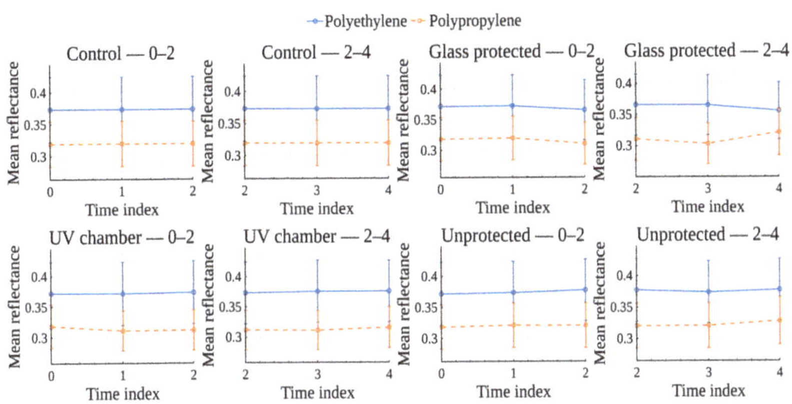

Figure 10. Temporal evolution of mean reflectance in the Orange band (600–620 nm) for polyethylene (solid) versus polypropylene (dashed), shown per condition. Time segmentation (0–2 vs. 2–4) and scaled (×0.30) SEM—standard error of the mean bars.

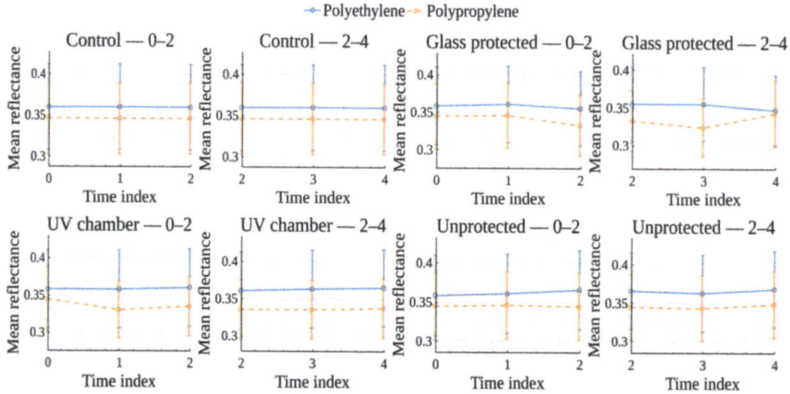

Figure 11. Temporal evolution of mean reflectance in the Red band (640–660 nm) contrasting polyethylene (solid) and polypropylene (dashed) under Control, Glass-protected, Unprotected, and UV chamber conditions. Panels: 0–2 and 2–4; intervals: markers at integer times. Error bars: scaled SEM (×0.30) applied uniformly across bands and conditions.

In summary, due to the relative scale used in these previously discussed visualizations (Figures 5–11), which is necessary to allow for material comparison/joint-visualization, variations are arduous to observe. However, polyethylene consistently demonstrated more sensitivity to environmental exposure—particularly in glass-protected and unprotected conditions—through gradual reflectance changes, especially in shorter wavelengths. In contrast, polypropylene remained more stable across all bands and conditions, highlighting its higher resistance to photodegradation under the tested scenarios.

3.4. Time Evolution of Subject-Normalized Degradation Efficiency

Figures 12 and 13 show the temporal evolution of the subject-normalized degradation efficiency, $\eta(t)$, computed on seven reflectance bands (three wavelengths per band, as described in Section 2.1) for PE and PP. For each material–condition–time cell, the reflectance of each individual specimen was first normalized to its own baseline spectrum at $t = 0$ (per-specimen $||R_0||$ normalization across the relevant bands), computed the per-specimen efficiency $\eta_i(t)$, and then aggregated across specimens to obtain the mean and uncertainty at each time point. In detail, $\eta(t)$ is defined in terms of the reflectance vectors measured at multiple wavelengths. For each specimen i (piece), material m, and exposure condition c, $R_i(t_k, \lambda_j)$ is the reflectance at time index t_k and wavelength λ_j ($j = 1, \ldots, J$). The specimen's own first measurement was taken as its baseline,

$$t_{0,i} = \min\{t_k \text{ observed for specimen } i\}, \qquad \mathbf{R}_{0,i} = \left[R_i(t_{0,i}, \lambda_1), \ldots, R_i(t_{0,i}, \lambda_J)\right].$$

Defining the ℓ_2 norm $\|\mathbf{x}\|_2 = (\sum_j x_j^2)^{1/2}$, the subject-normalized efficiency for specimen i is

$$\eta_i(t_k) = \frac{\left\|\mathbf{R}_i(t_k) - \mathbf{R}_{0,i}\right\|_2}{\left\|\mathbf{R}_{0,i}\right\|_2}, \qquad \mathbf{R}_i(t_k) = \left[R_i(t_k, \lambda_1), \ldots, R_i(t_k, \lambda_J)\right]. \tag{1}$$

This definition uses all J wavelengths ($J = 31$) measured with the spectrophotometer. Now, in order to obtain a color-banded summary, wavelengths were grouped into seven triplets B_b (violet, blue, cyan, green, yellow, orange, red) and averaged within each band:

$$\bar{R}_i(t_k, b) = \frac{1}{|B_b|} \sum_{\lambda \in B_b} R_i(t_k, \lambda), \qquad \bar{\mathbf{R}}_i(t_k) = \left[\bar{R}_i(t_k, 1), \ldots, \bar{R}_i(t_k, 7)\right],$$

$\eta_i(t_k)$ was defined as in (1), but replacing $\mathbf{R}_i, \mathbf{R}_{0,i}$ by their band-averaged versions $\bar{\mathbf{R}}_i, \bar{\mathbf{R}}_{0,i}$. These last are the ones presented in Figures 12 and 13. For each material–condition group (m, c), specimens were aggregated at each time t_k:

$$\bar{\eta}_{m,c}(t_k) = \frac{1}{n_{m,c}(t_k)} \sum_{i\in(m,c)} \eta_i(t_k), \qquad s_{m,c}(t_k) = \sqrt{\frac{1}{n_{m,c}(t_k)-1} \sum_{i\in(m,c)} \left(\eta_i(t_k) - \bar{\eta}_{m,c}(t_k)\right)^2}. \quad (2)$$

Values $\bar{\eta}_{m,c}(t_k)$ were plotted with error bars chosen as either SD ($\pm s_{m,c}$), SEM ($\pm s_{m,c}/\sqrt{n}$), or a 95% confidence interval ($\pm 1.96\, s_{m,c}/\sqrt{n}$). The error bars presented in Figures 12 and 13 are the first choice.

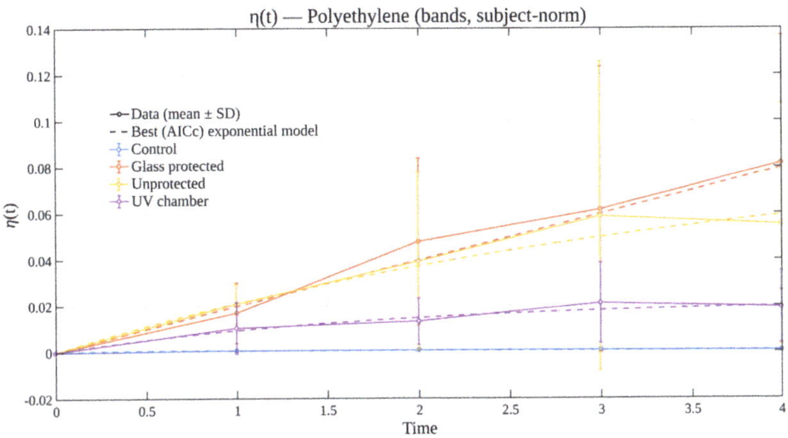

Figure 12. Subject-normalized degradation efficiency $\eta(t)$ for **polyethylene** computed from spectral reflectance as $\eta_i(t) = \|R_i(t) - R_i(t_0)\|_2 / \|R_i(t_0)\|_2$, and then averaged across specimens within each condition (markers show group mean; error bars denote ±SD). Dashed curves are the best exponential fits selected by AICc among the candidate family (first-order, stretched, lag/offset variants, and bi-exponential). Time is shown as the discrete sampling index (t_0–t_4).

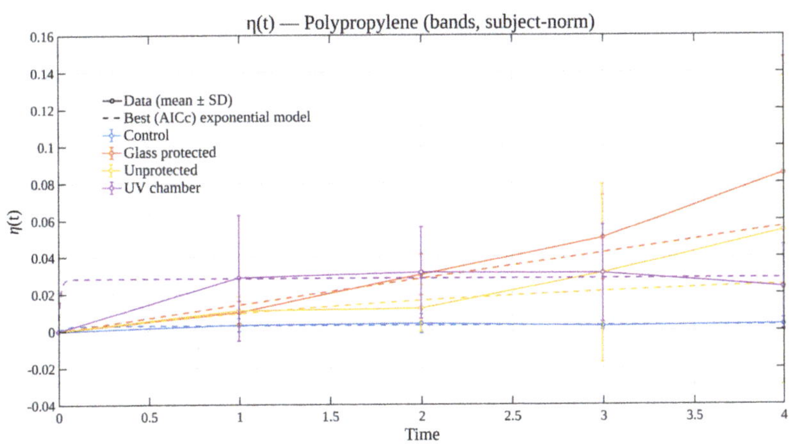

Figure 13. Subject-normalized degradation efficiency $\eta(t)$ for **polypropylene** under the four exposure conditions (mean ± SD across specimens). Dashed lines indicate the AICc-selected best exponential model in each condition. Time is the discrete sampling index (t_0–t_4).

- *Kinetic models and parameter interpretation.*

$\bar{\eta}_{m,c}(t)$ values were fit with physically interpretable monotonic exponentials (nonlinear least squares) using SD-based weights $w_k = 1/s_{m,c}(t_k)^2$ when available. Candidate models considered were

$$\text{First-order:}\quad \eta(t) = \eta_\infty\Big(1 - e^{-t/\tau}\Big), \qquad \eta_\infty \geq 0,\ \tau > 0, \tag{3}$$

$$\text{Stretched:}\quad \eta(t) = \eta_\infty\Big(1 - e^{-(t/\tau)^\beta}\Big), \qquad \eta_\infty \geq 0,\ \tau > 0,\ 0.2 \leq \beta \leq 2.5, \tag{4}$$

$$\text{First-order with lag:}\quad \eta(t) = \eta_\infty\Big(1 - e^{-\max(t-\delta,0)/\tau}\Big), \qquad \delta \geq 0, \tag{5}$$

$$\text{Stretched with lag:}\quad \eta(t) = \eta_\infty\Big(1 - e^{-\left(\max(t-\delta,0)/\tau\right)^\beta}\Big), \tag{6}$$

$$\text{First-order with offset:}\quad \eta(t) = \eta_0 + \eta_\infty\Big(1 - e^{-t/\tau}\Big), \qquad \eta_0 \in [-0.05, 0.05]. \tag{7}$$

Half-times can be considered directly. For Equations (3)–(7),

$t_{1/2} = \tau \ln 2$ (no lag), $t_{1/2} = \delta + \tau \ln 2$ (with lag), *and* $t_{1/2} = \tau(\ln 2)^{1/\beta}$ for stretched forms.

- *Model selection (AICc).*

For each candidate with *k* free parameters, the weighted residual sum of squares is

$$\text{SSE} = \sum_{k=1}^{n} w_k \Big[\bar{\eta}_{m,c}(t_k) - \hat{\eta}(t_k)\Big]^2,$$

and the following computation was performed:

$$\text{AIC} = n \ln\!\left(\frac{\text{SSE}}{n}\right) + 2k, \qquad \text{AIC}_\text{c} = \text{AIC} + \frac{2k(k+1)}{n-k-1}.$$

The model with the smallest AIC_c was reported as "best" and drawn as the main dashed curve in the overlays shown by Figures 12 and 13.

- *Interpretation.*

Since $\eta(t)$ measures the *relative* spectral distance from each piece's own baseline (Equation (1)), values near zero indicate minimal change; larger values quantify stronger spectral evolution (e.g., due to oxidation, yellowing, or microcracking). Subject normalization removes differences (such as color and finish, which were not an issue from the get-go) at $t = 0$ and makes groups comparable; error bars reflect between-piece dispersion within each material–condition–time cell and directly inform weighting in the fits. Within Figures 12 and 13, dashed curves correspond to the best exponential model selected by AICc among a small family with physical interpretability (first-order, stretched exponential, and lag/offset variants), fitted to the band-wise mean trajectories with optional inverse-variance weighting; solid symbol–lines connect the observed means for readability. For both materials, the *Control* condition remains essentially flat at $\eta(t) \approx 0$ over the 4-month window, as expected. The *UV chamber* condition exhibits a modest early increase followed by a shallow plateau, consistent with a rapid initial response and limited cumulative dose. In contrast, the outdoor *Unprotected* and *Glass-protected* conditions show monotonic growth of $\eta(t)$, with *Glass-protected* typically presenting the largest values at each time point. This ordering is consistent with a scenario where the glass filter modifies the incident spectrum and thermal load, yielding measurable—but still bounded—cumulative change over the study period. Across both materials, the AICc favored the simple first-order model in all band-wise cases presented, indicating that over this limited exposure interval the data are well described by a single characteristic timescale without requiring additional stretching or lag terms. Taken together, these trajectories indicate negligible drift in controls, modest chamber-induced change, and progressive outdoor degradation with spectral filtering having a measurable effect.

- *Related context to prior works.*

Table 3 situates the present study within representative literature on polyolefin degradation. Prior studies typically quantify ageing using chemical (e.g., FTIR carbonyl index—CI), optical (UV-VIS color/yellowness), or mechanical proxies and then discuss trend shapes qualitatively or with single-model fits tied to specific test setups. In contrast, this study systematically links broadband optical response to time-resolved damage across multiple illuminations and for two different materials, and does so with a unified, bounded response variable and a transparent model-selection protocol. Concretely, a time series of broadband reflectance (both full-spectrum and seven VIS-bands) is analyzed and fitted into physically interpretable kinetics (first-order, stretched, lagged, and bi-exponential), selecting the description by AICc with uncertainty-weighted regression. This yields directly comparable summary parameters (e.g., $t_{1/2}$, asymptotic response) across conditions, enabling the possibility of cross-material and cross-illumination ranking rather than case-by-case narratives. In particular, regarding the subject of degradation efficiency, while previous reports often emphasize a single proxy and a single kinetic hypothesis, the present framework (i) treats the optical signal as an *integrated* damage reporter over the spectrum (not merely a narrow spectral index), (ii) compares several mechanistic exponentials under the same informational/theoretical framework, and (iii) provides compact visual summaries (such as overlays and heatmaps (these latter not included currently in the manuscript) that indicates when simple first-order behavior suffices and when multi-timescale (bi-exponential) dynamics could be more appropriate. For completeness, the bounded response was denoted as $\eta(t) \in [0, 1]$ to summarize progression and was conceived primarily as a convenient reporting scale rather than as a claim of main contribution.

Table 3. Representative literature on polyolefin degradation quantification contrasted with the present VIS–reflectance framework. Prior work mostly quantifies damage via chemical or color indices (e.g., carbonyl index, yellowness), gas evolution, or mechanical retention; here, we use band-aggregated visible reflectance tracked over time and visualize a bounded, model-based trajectory $\eta(t)$ as a compact descriptor of spectral change. Kinetic constants ($t_{1/2}$, η_∞) are discussed conceptually but are not tabulated in this version. Metrics and models are as reported by each source.

Study	Material(s)	Illumination/Exposure	Metric (Proxy)	Reported Kinetics/ Key Outcome
Gulmine et al. (2003) [35]	PE (LDPE)	Artificial accelerated weathering	FTIR carbonyl index; UV-VIS; morphology	CI increases with dose; qualitative kinetics; links spectral growth to surface oxidation.
Fernando et al. (2007) [49]	PE, PP	Photodegradation (UV)	CO_2 evolution; CI	Oxidation/gas-evolution correlates with CI; effective first-order-like regimes discussed.
Shyichuk et al. (2005) [57]	PE, PP, E–P copolymer	UV exposure; depth profiles	Depth-resolved CI	Near-surface oxidation dominates; gradients consistent with short-λ absorption, implicit multi-timescale behavior.

Table 3. *Cont.*

Study	Material(s)	Illumination/Exposure	Metric (Proxy)	Reported Kinetics/Key Outcome
Tocháček & Vrátničková (2014) [17]	PP and copolymers	UV accelerated; temperature effects	Mechanical retention; color/CI	Temperature accelerates damage; Arrhenius-type behavior; practical lifetime prediction.
Rodríguez et al. (2020) [8]	LDPE	UV-aging	Tensile/fracture metrics	Monotonic loss of toughness; time-to-threshold comparisons; no explicit spectral efficiency.
Grause et al. (2020) [58]	LDPE, HDPE, PP	Natural/outdoor and accelerated UV weathering (survey across setups)	FTIR, UV-VIS, DSC, mechanical testing	Carbonyl Index (CI), yellowness index, crystallinity increase, tensile retention.
Bourgogne et al. (2022) [44]	Polymers (incl. polyolefins context)	LED weathering, wavelength-resolved	Chemical/CI changes	Shorter wavelengths drive faster photooxidation; wavelength selectivity highlighted.
Du et al. (2024) [21]	PE sheets (types)	UV; varied parameters	Multi-technique spectral/material indices	Factor sensitivity (grade, thickness, stabilizers); qualitative kinetics under parameter sweeps.
Arese et al. (2025) [3]	Recycled PP	UV ageing	Morphological, thermal, mechanical	Stabilization/processing history modulate ageing; no unified bounded spectral metric.
This work	PE, PP	VIS: unprotected vs. glass-protected; UV chamber	**Band-aggregated VIS reflectance** (7 bands; full spectrum inspected); $\eta(t) \in [0,1]$ visualized	**Compact, model-based spectral trajectories** selected by small-sample information criteria; graphical comparison across materials/conditions/bands; kinetic constants discussion.

In summary, the approach presented in this work with respect to prior polyolefin studies—predominantly relying on chemical/color indices (e.g., carbonyl index, yellowness), CO_2 evolution, or mechanical retention—contrasts with the tracking of time-resolved visible reflectance aggregated into seven VIS-bands and summarizing of its evolution through a bounded, model-based trajectory $\eta(t)$. This representation provides a compact, visually comparable signature across materials and exposures without requiring destructive assays. We restrict $\eta(t)$ to graphical summaries—consistent with the scope of the study—and defer tabulated kinetic constants (e.g., $t_{1/2}, \eta_\infty$) to future, higher-powered datasets. This keeps the emphasis on comparability (PE vs. PP; outdoor vs. glass-protected vs. UV chamber) while maintaining methodological consistency.

3.5. Effect of Environmental Exposure on Shore D Hardness

The statistical analysis of Shore D hardness summarized in Table 4 portrays significant degradation in both materials under specific exposure conditions. For polyethylene, paired hypothesis tests showed significant decreases in hardness after exposure to outdoor conditions without glass protection (OEW/OG) and to the UV chamber (UVC), while no significant change was observed under outdoor exposure with glass (OEWG). In polypropylene, all post-exposure conditions led to statistically significant hardness losses relative to the reference state (RH), indicating higher susceptibility to both direct and filtered UV exposure.

On the other hand, the one-way ANOVA confirmed overall differences among conditions for polyethylene ($p < 0.05$), with Tukey's test identifying significant pairwise differences between the reference and the OEW/OG and UVC groups. In contrast, for polypropylene, the ANOVA result was not statistically significant ($p > 0.05$), likely due to larger within-group variability; however, the post hoc test still detected consistent pairwise differences between the reference and all exposure conditions. In detail, as shown in Table 4, while all pairwise comparisons for polypropylene samples showed significant differences ($p < 0.05$), the global one-way ANOVA test did not reach statistical significance ($p > 0.05$). This arises due to the inherent differences between these two statistical approaches. Pairwise tests compare each post-exposure condition to the reference individually, using fewer degrees of freedom while being sensitive to small effect sizes. In contrast, the ANOVA global test evaluates the variance among all groups simultaneously, where high within-group variability and relatively small sample sizes can reduce statistical power. Additionally, since ANOVA relies on the overall variance structure, it may fail to detect significance when the differences are not uniformly distributed across all groups. For clarity, standard deviation error bars have been added to Figures 14 and 15 to visually illustrate the within-group variability in Shore D hardness for polypropylene. This visualization supports the statistical findings and helps explain why the ANOVA test did not reach significance despite consistent pairwise results.

These results indicate that polypropylene is more vulnerable to hardness loss under environmental exposure, including when protected by glass, whereas polyethylene exhibits partial resilience when shielded from direct UV radiation.

Table 4. Statistical analysis summary of Shore D hardness measurements for polyethylene and polypropylene samples before and after environmental exposure. Paired tests (parametric or non-parametric depending on normality) compared each post-exposure condition to the reference (unexposed, *reference hardness*, i.e., *RH*) state. Complementarily, a one-way ANOVA with Tukey's post hoc comparisons was conducted separately for each material to assess overall differences across exposure conditions. Each result indicates whether the difference was statistically significant at the $\alpha = 0.05$ level.

Material	Test Type	Results
Polyethylene	Paired comparisons (vs. *RH*)	Outdoor exposure without glass (*OEW/OG*): $p < 0.05$ Outdoor exposure with glass (*OEWG*): $p > 0.05$ Simulated UV chamber (*UVC*): $p < 0.05$
	One-way ANOVA + Tukey	Global test: $p < 0.05$ Significant differences: *RH-OEW/OG*, *RH-UVC*
Polypropylene	Paired comparisons (vs. *RH*)	Outdoor exposure without glass (*OEW/OG*): $p < 0.05$ Outdoor exposure with glass (*OEWG*): $p < 0.05$ Simulated UV chamber (*UVC*): $p < 0.05$
	One-way ANOVA + Tukey	Global test: $p > 0.05$ Significant differences: *RH-OEW/OG*, *RH-OEWG*, *RH-UVC*

3.6. *Material Non-Specific ANOVA and ANCOVA*

3.6.1. Effect of Material, Condition, and Time on Reflectance: Three-Way ANOVA

The ANOVA results show that the factor **Material** had a statistically significant effect ($p < 0.05$) on average reflectance across the spectrum, particularly for the shorter wavelength bands (Violet to Yellow). This indicates that polyethylene and polypropylene differ systematically in their optical response under environmental exposure.

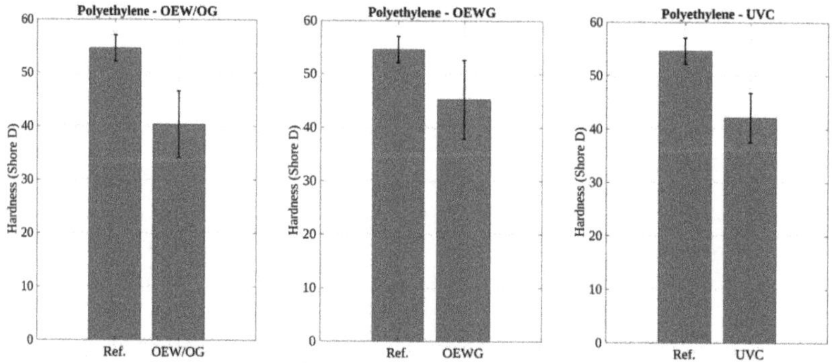

Figure 14. Hardness comparison for polyethylene components before and after exposure. Bar plots show mean Shore D hardness with standard deviation. Three exposure conditions are compared against the reference: OEW/OG (open environment without glass), OEWG (open environment with glass), and UVC (simulated ultraviolet exposure). Significant differences are highlighted from pairwise tests.

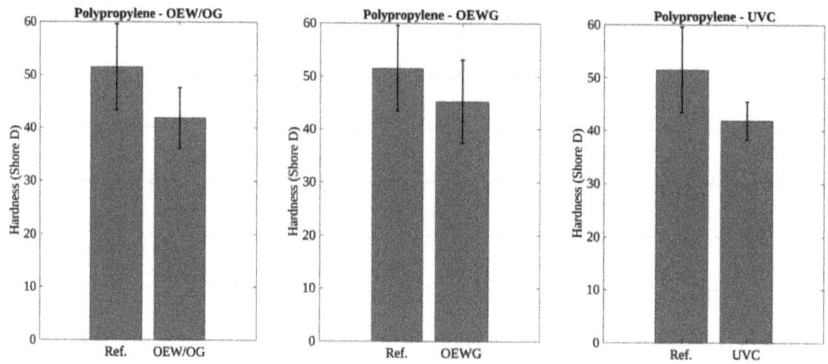

Figure 15. Hardness comparison for polyethylene components before and after exposure. Bar plots show mean Shore D hardness with standard deviation. Three exposure conditions are compared against the reference: OEW/OG (open environment without glass), OEWG (open environment with glass), and UVC (simulated ultraviolet exposure). Significant differences are highlighted from pairwise tests.

No statistically significant effects ($p > 0.05$) were found for the factors **Condition** or **Time**, nor for any second- or third-order interaction terms. These results suggest that, when pooling all materials together, neither the specific exposure condition nor the measurement time produced detectable differences in reflectance within each VIS-band.

It is noteworthy that only the material type consistently influenced reflectance for the first five VIS-bands (Violet to Yellow), while no significant effects were detected in the longer wavelength regions (Orange and Red). These findings are consistent with the physical expectation that UV-induced degradation may manifest more strongly in the short-wavelength reflectance behavior.

3.6.2. Controlling for Baseline Differences: Full-Factorial ANCOVA Model

Table 5 summarizes the ANCOVA significance outcomes for each visible VIS-band, considering the effects of **Material**, **Condition**, **Time**, and the covariate *initial reflectance*, as well as their interactions up to the four-way level.

The covariate was statistically significant ($p < 0.05$) in all seven VIS-bands, confirming that baseline differences in reflectance strongly influence the response during exposure. Condition exhibited a significant effect in six out of seven bands (all except Orange), indicating that environmental exposure type plays a central role in the reflectance degradation process. Time was also significant in Violet, Blue, Cyan, and Yellow, suggesting that degradation dynamics evolve significantly over time, especially in the shorter wavelengths of the visible spectrum.

The effect of Material was only significant in the Blue and Cyan bands, implying that polyethylene and polypropylene exhibit distinct degradation behavior primarily in the central spectral range. Among the interaction terms, Condition × Covariate was significant in all bands, pointing to a dependency of reflectance response on both baseline reflectance and environmental scenario. Similarly, Time × Covariate interactions were significant in Blue, Cyan, Green, and Yellow, highlighting that the rate or direction of change over time is modulated by initial reflectance values.

The Material × Condition interaction reached significance in Blue, Orange, and Red, suggesting that the effect of exposure condition differs between materials, particularly at the spectral extremes. Moreover, third-order interactions involving the covariate, specifically Material × Condition × Covariate and Material × Time × Covariate, were significant in the Orange and Red bands, pointing towards potential complex dependencies at longer wavelengths. However, no four-way interaction (Material × Condition × Time × Covariate) was significant in any VIS-band, indicating that the combined effect of all experimental factors and the covariate does not significantly explain additional variance beyond the lower-order terms.

Overall, Blue and Cyan emerged as the most responsive VIS-bands to both main effects and interactions, while Orange and Red exhibited more nuanced, higher-order interaction patterns. These findings underscore the importance of considering spectral specificity and covariate adjustment when modeling reflectance degradation in polymeric materials exposed to environmental stressors.

3.7. Material Specific ANOVA and ANCOVA

3.7.1. Separate Effects of Exposure and Time Within Each Material: Two-Way ANOVA

The two-way ANOVA results summarized in Table 6 indicate that none of the evaluated factors—Condition, Time, or their interaction—had a statistically significant effect on the average reflectance within any visible VIS-band ($p > 0.05$ in all cases). This finding held consistently across both material types, polyethylene (PE) and polypropylene (PP). These results suggest that, for the considered statistical framework, under the current experimental conditions and timescale, the reflectance of the tested automotive plastic specimens remained stable regardless of the type of environmental exposure or the duration of the exposure period.

In order to explore that perspective, three steps were taken for each band and material in Table 6 with the interaction model *Condition, Time, Condition × Time*. First, eta-squared (*partial effect size*) and Cohen's f^2, were calculated, respectively:

$$\eta_p^2 = \frac{\mathrm{SS}_{\mathrm{effect}}}{\mathrm{SS}_{\mathrm{effect}} + \mathrm{SS}_{\mathrm{error}}} \tag{8}$$

$$f^2 = \frac{\eta_p^2}{1 - \eta_p^2} \tag{9}$$

Table 5. ANCOVA significance summary for mean reflectance in each visible VIS-band. The model includes **Material** (polyethylene or polypropylene), **Condition** (four environmental exposure scenarios), **Time** (T1–T4), and the covariate *initial reflectance* measured at T0. All main effects and interaction terms up to the four-way level were tested. Cells indicate whether each effect was statistically significant ($p < 0.05$) or not ($p > 0.05$). This full-factorial model controls for initial reflectance differences and evaluates the extent to which reflectance changes depend on material type, exposure conditions, and time.

VIS-Band	Material	Condition	Time	Covariate	M × C	M × T	M × Cov	C × T	C × Cov	T × Cov	M × C × T	M × C × Cov	M × T × Cov	C × T × Cov	M × C × T × Cov
Violet	$p > 0.05$	$p < 0.05$	$p < 0.05$	$p < 0.05$	$p > 0.05$	$p > 0.05$	$p > 0.05$	$p > 0.05$	$p < 0.05$	$p > 0.05$	$p > 0.05$	$p > 0.05$	$p > 0.05$	$p > 0.05$	$p > 0.05$
Blue	$p < 0.05$	$p < 0.05$	$p < 0.05$	$p < 0.05$	$p < 0.05$	$p > 0.05$	$p > 0.05$	$p > 0.05$	$p < 0.05$	$p < 0.05$	$p > 0.05$	$p > 0.05$	$p > 0.05$	$p > 0.05$	$p > 0.05$
Cyan	$p < 0.05$	$p < 0.05$	$p < 0.05$	$p < 0.05$	$p > 0.05$	$p > 0.05$	$p > 0.05$	$p > 0.05$	$p < 0.05$	$p < 0.05$	$p > 0.05$	$p > 0.05$	$p > 0.05$	$p > 0.05$	$p > 0.05$
Green	$p > 0.05$	$p < 0.05$	$p > 0.05$	$p < 0.05$	$p > 0.05$	$p > 0.05$	$p > 0.05$	$p > 0.05$	$p < 0.05$	$p < 0.05$	$p > 0.05$	$p > 0.05$	$p > 0.05$	$p > 0.05$	$p > 0.05$
Yellow	$p > 0.05$	$p < 0.05$	$p < 0.05$	$p < 0.05$	$p > 0.05$	$p > 0.05$	$p > 0.05$	$p > 0.05$	$p < 0.05$	$p < 0.05$	$p > 0.05$	$p > 0.05$	$p > 0.05$	$p > 0.05$	$p > 0.05$
Orange	$p > 0.05$	$p > 0.05$	$p > 0.05$	$p < 0.05$	$p < 0.05$	$p > 0.05$	$p > 0.05$	$p > 0.05$	$p < 0.05$	$p > 0.05$	$p > 0.05$	$p < 0.05$	$p < 0.05$	$p > 0.05$	$p > 0.05$
Red	$p > 0.05$	$p < 0.05$	$p > 0.05$	$p < 0.05$	$p < 0.05$	$p > 0.05$	$p > 0.05$	$p > 0.05$	$p < 0.05$	$p > 0.05$	$p > 0.05$	$p < 0.05$	$p < 0.05$	$p > 0.05$	$p > 0.05$

Table 6. Two-way ANOVA (p values) evaluating the effects of exposure **Condition** (C1: no exposure, C2: exposure under glass, C3: exposure without glass, C4: artificial UV exposure; ASTM G154 protocol), **Time** (T0 to T4), and their interaction on the average visible reflectance of automotive plastic specimens. The response variable is the mean reflectance within each visible VIS-band, calculated as the average reflectance over three adjacent wavelengths for each VIS-band (Violet to Red). The analysis was conducted separately for each material: polyethylene (PE) and polypropylene (PP). Each cell indicates whether the effect was statistically significant at the $\alpha = 0.05$ level.

Color	PE Cond	PE Time	PE Cond × Time	PP Cond	PP Time	PP Cond × Time
Violet	$p > 0.05$	$p > 0.05$	$p > 0.05$	$p > 0.05$	$p > 0.05$	$p > 0.05$
Blue	$p > 0.05$	$p > 0.05$	$p > 0.05$	$p > 0.05$	$p > 0.05$	$p > 0.05$
Cyan	$p > 0.05$	$p > 0.05$	$p > 0.05$	$p > 0.05$	$p > 0.05$	$p > 0.05$
Green	$p > 0.05$	$p > 0.05$	$p > 0.05$	$p > 0.05$	$p > 0.05$	$p > 0.05$
Yellow	$p > 0.05$	$p > 0.05$	$p > 0.05$	$p > 0.05$	$p > 0.05$	$p > 0.05$
Orange	$p > 0.05$	$p > 0.05$	$p > 0.05$	$p > 0.05$	$p > 0.05$	$p > 0.05$
Red	$p > 0.05$	$p > 0.05$	$p > 0.05$	$p > 0.05$	$p > 0.05$	$p > 0.05$

Second, to relate non-significant results to sensitivity, a post hoc power analysis for a small "smallest effect size of interest" (SESOI) $f^2 = 0.02$ (Cohen's convention), with $\alpha = 0.05$ and target power $1 - \beta = 0.80$ was also performed. Power for each test was calculated under the noncentral F distribution with noncentrality

$$\lambda = f^2 \cdot df_{\text{error}}, \tag{10}$$

so that

$$\text{Power} = 1 - F_{\text{nc}F}(F_{1-\alpha}; df_1, df_2, \lambda), \tag{11}$$

where $F_{\text{nc}F}(\cdot; df_1, df_2, \lambda)$ is the (noncentral) cumulative distribution function (CDF) of F, $F_{1-\alpha}$ is the $(1-\alpha)$ critical value of the central $F(df_1, df_2)$ with numerator and denominator degrees of freedom df_1 and df_2 respectively, and λ is the noncentrality parameter. Here, df_1 corresponds to the effect's degrees of freedom (e.g., levels of *Condition*, *Time*, or their interaction minus one), and df_2 corresponds to the model's error degrees of freedom. The noncentrality parameter λ quantifies the true signal strength relative to noise. As previously mentioned, under the general linear model (GLM) formulation used, $\lambda = f^2 \cdot df_{\text{error}}$.

Third, the minimum detectable effect (MDE) f^2 at 80% power was also reported by numerically solving the above expression for f^2:

$$\text{MDE}_{0.80} = \min\{f^2 : \text{Power}(f^2) \geq 0.80\}. \tag{12}$$

Time was treated as categorical (equally spaced levels but not assumed linear), matching Table 6. In this context, η_p^2 gauges the share of variance uniquely attributable to a term after partialling out the error term; f^2 rescales η_p^2 to a signal-to-noise metric useful for power.

Results indicated that across bands, eta-squared (*partial effect size*) estimates for *Condition*, *Time*, and their interaction were near zero (typical medians ≤ 0.001). For the SESOI $f^2 = 0.02$, post-hoc power was below 0.80 for all terms. Consequently, the median *MDE* at 80% power was ≈0.19 for *Condition*, ≈0.22 for *Time*, and ≈0.34 for the interaction (values similar in PE and PP). In summary, these results indicate that the analysis behind Table 6 (two-way ANOVA with Time categorical and omnibus tests) is less sensitive to small effects, since the design would have reliably detected medium–large effects (MDEs ≈ 0.19–0.34), but not very small ones. As the observed η_p^2 are uniformly tiny, the null results are consistent with the absence of medium/large effects rather than with broad under-sampling. On the other hand, this interpretation also aligns with other sections of the present manuscript, since alternative analyses did identify effects. Thus, it is within reason considering that the dataset is not generally under-powered, but that the specific Table 6 framework is inherently insensitive to very small effects.

3.7.2. Material-Specific ANCOVA: Separate Models for Polyethylene and Polypropylene

Table 7 presents the ANCOVA results for each VIS-band analyzed separately by material. For **polyethylene**, the covariate (initial reflectance) was consistently significant ($p < 0.05$) across all VIS-bands, indicating strong baseline dependence in the observed reflectance patterns. Condition was also significant in all bands, whereas Time showed significance only in the blue band. Notably, interactions involving the covariate were frequent: Condition × Covariate and Time × Covariate were both significant in seven, six out of seven bands respectively, and the three-way interaction (Condition × Time × Covariate) appeared only in yellow and orange.

In contrast, for **polypropylene**, only the covariate was significant in the first five VIS-bands (Violet to Yellow), with no significant effects of Condition or Time, except for Orange and Red in the former, nor any interactions in those bands. Significant effects of Condition and Condition × Covariate emerged only in the Orange and Red bands, accompanied by a significant Time × Covariate interaction in Orange. This suggests that reflectance variation in polypropylene is less sensitive to experimental factors than in polyethylene and largely governed by initial reflectance levels except at longer wavelengths.

Overall, the results indicate material-dependent reflectance behavior: Polyethylene shows broader sensitivity to environmental exposure and time effects, while polypropylene's variation is more restricted and primarily driven by baseline properties.

Table 7. ANCOVA significance summary for average reflectance in each visible VIS-band, analyzed separately for **polyethylene** and **polypropylene**. Each model included the fixed factors **Condition** (four environmental exposure scenarios) and **Time** (four sampling points: T_1 to T_4), as well as the covariate *Initial Reflectance* measured at baseline (T_0). All second-order interactions (Condition × Time, Condition × Covariate, Time × Covariate) and the third-order interaction (Condition × Time × Covariate) were also tested. The table reports whether each effect was statistically significant ($p < 0.05$) or not ($p > 0.05$), based on individual ANCOVA models fitted for each visible VIS-band.

Material	VIS-Band	Condition	Time	Covariate	C × T	C × Cov	T × Cov	C × T × Cov
Polyethylene	Violet	$p < 0.05$	$p > 0.05$	$p < 0.05$	$p > 0.05$	$p < 0.05$	$p < 0.05$	$p > 0.05$
Polyethylene	Blue	$p < 0.05$	$p < 0.05$	$p < 0.05$	$p > 0.05$	$p < 0.05$	$p < 0.05$	$p > 0.05$
Polyethylene	Cyan	$p < 0.05$	$p > 0.05$	$p < 0.05$	$p > 0.05$	$p < 0.05$	$p < 0.05$	$p > 0.05$
Polyethylene	Green	$p < 0.05$	$p > 0.05$	$p < 0.05$	$p > 0.05$	$p < 0.05$	$p > 0.05$	$p > 0.05$
Polyethylene	Yellow	$p < 0.05$	$p > 0.05$	$p < 0.05$	$p > 0.05$	$p < 0.05$	$p < 0.05$	$p < 0.05$
Polyethylene	Orange	$p < 0.05$	$p > 0.05$	$p < 0.05$	$p > 0.05$	$p < 0.05$	$p < 0.05$	$p < 0.05$
Polyethylene	Red	$p < 0.05$	$p > 0.05$	$p < 0.05$	$p > 0.05$	$p < 0.05$	$p < 0.05$	$p > 0.05$
Polypropylene	Violet	$p > 0.05$	$p > 0.05$	$p < 0.05$	$p > 0.05$	$p > 0.05$	$p > 0.05$	$p > 0.05$
Polypropylene	Blue	$p > 0.05$	$p > 0.05$	$p < 0.05$	$p > 0.05$	$p > 0.05$	$p > 0.05$	$p > 0.05$
Polypropylene	Cyan	$p > 0.05$	$p > 0.05$	$p < 0.05$	$p > 0.05$	$p > 0.05$	$p > 0.05$	$p > 0.05$
Polypropylene	Green	$p > 0.05$	$p > 0.05$	$p < 0.05$	$p > 0.05$	$p > 0.05$	$p > 0.05$	$p > 0.05$
Polypropylene	Yellow	$p > 0.05$	$p > 0.05$	$p < 0.05$	$p > 0.05$	$p > 0.05$	$p > 0.05$	$p > 0.05$
Polypropylene	Orange	$p < 0.05$	$p > 0.05$	$p < 0.05$	$p > 0.05$	$p < 0.05$	$p < 0.05$	$p > 0.05$
Polypropylene	Red	$p < 0.05$	$p > 0.05$	$p < 0.05$	$p > 0.05$	$p < 0.05$	$p > 0.05$	$p > 0.05$

3.8. Material Non-Specific MANOVA and MANCOVA

3.8.1. Multivariate Assessment of Experimental Factors on Reflectance Response: Full MANOVA Model

As shown in Table 8, a multivariate analysis of variance (MANOVA) was conducted in order to evaluate the effects of *Material, Condition, Time,* and their interactions on the multivariate response of average reflectance across seven VIS-bands in the visible spectrum (Violet, Blue, Cyan, Green, Yellow, Orange, Red). The analysis showed that only the main effect of *Material* was statistically significant ($p < 0.05$) across all multivariate test statistics (Wilks's λ, Pillai's trace, Hotelling–Lawley trace, and Roy's greatest root). In contrast, neither *Condition*, nor *Time*, nor any of their two-way or three-way interactions with *Material*, showed significant effects ($p > 0.05$). These results indicate that the material type (polyethylene or polypropylene) significantly influences the overall reflectance profile, independently of the environmental exposure conditions or duration of exposure.

Table 8. Multivariate analysis of variance (MANOVA) assessing the effects of **Material** (polyethylene, polypropylene), **Condition** (C1: no exposure; C2: environmental exposure under glass; C3: environmental exposure without protection; C4: UV chamber, ASTM G154 protocol), **Time** (T0 to T4), and their interactions on the multivariate response of average spectral reflectance. The table summarizes the statistical significance of each factor and interaction across four multivariate test statistics: Wilks's λ, Pillai's trace, Hotelling–Lawley trace, and Roy's greatest root.

Effect	Test	Significance
Material	Wilks's λ	$p < 0.05$
	Pillai's trace	$p < 0.05$
	Hotelling–Lawley trace	$p < 0.05$
	Roy's greatest root	$p < 0.05$
Condition	Wilks's λ	$p > 0.05$
	Pillai's trace	$p > 0.05$
	Hotelling–Lawley trace	$p > 0.05$
	Roy's greatest root	$p > 0.05$
Time	Wilks's λ	$p > 0.05$
	Pillai's trace	$p > 0.05$
	Hotelling–Lawley trace	$p > 0.05$
	Roy's greatest root	$p > 0.05$
Material × Condition	Wilks's λ	$p > 0.05$
	Pillai's trace	$p > 0.05$
	Hotelling–Lawley trace	$p > 0.05$
	Roy's greatest root	$p > 0.05$
Material × Time	Wilks's λ	$p > 0.05$
	Pillai's trace	$p > 0.05$
	Hotelling–Lawley trace	$p > 0.05$
	Roy's greatest root	$p > 0.05$
Condition × Time	Wilks's λ	$p > 0.05$
	Pillai's trace	$p > 0.05$
	Hotelling–Lawley trace	$p > 0.05$
	Roy's greatest root	$p > 0.05$
Material × Condition × Time	Wilks's λ	$p > 0.05$
	Pillai's trace	$p > 0.05$
	Hotelling–Lawley trace	$p > 0.05$
	Roy's greatest root	$p > 0.05$

3.8.2. MANCOVA Analysis of Visible Reflectance Adjusted by Initial Conditions

The MANCOVA results adjusted by initial reflectance (T0) in Table 9 reveal that none of the main effects—**Material**, **Condition**, or **Time**—were statistically significant according to any of the four multivariate test statistics. Among the second-order interactions, only **Material × Condition** showed significance under the Hotelling–Lawley trace, suggesting a possible dependence of reflectance variation on the specific combination of material and environmental exposure. The third-order interaction **Material × Condition × Time** was also statistically significant under the Hotelling–Lawley criterion, indicating that temporal changes in reflectance may be differentially affected depending on both material type and exposure condition. These significances must be considered within the context of other multivariate statistics such as the ones also discussed in this manuscript, needing thoughtful interpretation.

Table 9. Multivariate significance summary from a MANCOVA model adjusted by initial reflectance (T0) for each average visible VIS-band. The response matrix includes the average reflectance values for seven spectral bands (Violet, Blue, Cyan, Green, Yellow, Orange, and Red) across 224 observations. The model tests the main effects of **Material** (polyethylene or polypropylene), **Condition** (four levels of environmental exposure), and **Time** (four sampling stages from T1 to T4), as well as second- and third-order interactions between these factors. Each cell indicates whether the corresponding effect was statistically or not according to four multivariate test statistics: Wilks's λ, Pillai's trace, Hotelling–Lawley trace, and Roy's largest root.

Effect	Wilks's λ	Pillai's Trace	Hotelling–Lawley	Roy's Largest Root
Material	$p > 0.05$	$p > 0.05$	$p > 0.05$	$p > 0.05$
Condition	$p > 0.05$	$p > 0.05$	$p > 0.05$	$p > 0.05$
Time	$p > 0.05$	$p > 0.05$	$p > 0.05$	$p > 0.05$
Material × Condition	$p > 0.05$	$p > 0.05$	$p < 0.05$	$p > 0.05$
Material × Time	$p > 0.05$	$p > 0.05$	$p > 0.05$	$p > 0.05$
Condition × Time	$p > 0.05$	$p > 0.05$	$p > 0.05$	$p > 0.05$
Material × Condition × Time	$p > 0.05$	$p > 0.05$	$p < 0.05$	$p > 0.05$

3.9. *Material-Specific MANOVA and MANCOVA*

3.9.1. Repeated-Measures MANOVA: Within-Material Analysis of Exposure and Time

The multivariate results in Table 10 show that, for polyethylene, the factor *Condition* had a statistically significant effect on the mean reflectance across the visible spectrum, as evidenced by all four multivariate tests ($p < 0.05$). In contrast, for polypropylene, no consistent multivariate significance was found for *Condition*, although Roy's test did indicate marginal significance. For both materials, the factor *Time* and the interaction *Condition × Time* were not significant according to Wilks's, Pillai's, and Hotelling–Lawley criteria; however, Roy's test did detect significance for *Time* in both materials and for the interaction in polyethylene. This pattern suggests that differences between exposure conditions had a more pronounced and consistent effect on polyethylene reflectance, whereas time-dependent changes were weaker and dependent on the test statistic used.

Table 10. Multivariate analysis of variance (MANOVA) results for the repeated-measures effects of *Condition*, *Time*, and their interaction on the mean reflectance across seven visible VIS-bands (Violet to Red). The analysis was conducted separately for polyethylene and polypropylene samples. Each entry indicates whether the corresponding multivariate test (Wilks's Lambda, Pillai's Trace, Hotelling–Lawley Trace, Roy's Greatest Root) yielded statistical significance ($p < 0.05$) or not ($p > 0.05$). Both *Condition* and *Time* were treated as within-subjects factors in the repeated-measures design.

Effect	Polyethylene				Polypropylene			
	Wilks	Pillai	Hotelling	Roy	Wilks	Pillai	Hotelling	Roy
Condition	$p < 0.05$	$p < 0.05$	$p < 0.05$	$p < 0.05$	$p > 0.05$	$p > 0.05$	$p > 0.05$	$p < 0.05$
Time	$p > 0.05$	$p > 0.05$	$p > 0.05$	$p < 0.05$	$p > 0.05$	$p > 0.05$	$p > 0.05$	$p < 0.05$
Condition × Time	$p > 0.05$	$p > 0.05$	$p > 0.05$	$p < 0.05$	$p > 0.05$	$p > 0.05$	$p > 0.05$	$p > 0.05$

3.9.2. Separate MANCOVA Models by Material Adjusted for Initial Reflectance

The MANCOVA results in Table 11 adjusted by initial reflectance (T0) and stratified by material reveal distinct multivariate response patterns between polyethylene and polypropylene samples. For **polyethylene**, the effects of *Condition* and the *Condition × Time* interaction reached statistical significance only under the Hotelling–Lawley trace and Roy's largest root, while Wilks's λ and Pillai's trace remained nonsignificant. This suggests that variability in reflectance among exposure conditions was mainly driven by strong effects in a reduced number of response directions. In contrast, no multivariate effects reached significance for **polypropylene**, except for the *Condition* × *Time* interaction detected via Hotelling–Lawley trace alone. The limited significance observed across multiple statistics indicates that reflectance changes in polypropylene were less structured or exhibited higher intra-group variability. Overall, these results highlight material-dependent differences in how environmental exposure modulates spectral reflectance across time.

Table 11. Multivariate significance summary from two separate MANCOVA models conducted for polyethylene and polypropylene, respectively. The response matrix consists of the average reflectance values computed for each of the seven visible VIS-bands measured at four post-exposure stages (T1 to T4). Each model includes two fixed between-subject factors: **Condition** (four levels of environmental exposure) and **Time** (T1 to T4), as well as their second-order interaction (**Condition × Time**). The initial reflectance (T0) was included as a covariate to control for baseline variability. Statistical significance ($p < 0.05$ or $p > 0.05$) was evaluated using four classical multivariate test statistics: Wilks's λ, Pillai's trace, Hotelling–Lawley trace, and Roy's largest root.

Material	Effect	Wilks's λ	Pillai's Trace	Hotelling–Lawley Trace	Roy's Largest Root
Polyethylene	Condition	$p > 0.05$	$p > 0.05$	$p < 0.05$	$p < 0.05$
	Time	$p > 0.05$	$p > 0.05$	$p > 0.05$	$p > 0.05$
	Condition × Time	$p > 0.05$	$p > 0.05$	$p < 0.05$	$p < 0.05$
Polypropylene	Condition	$p > 0.05$	$p > 0.05$	$p > 0.05$	$p > 0.05$
	Time	$p > 0.05$	$p > 0.05$	$p > 0.05$	$p > 0.05$
	Condition × Time	$p > 0.05$	$p > 0.05$	$p < 0.05$	$p > 0.05$

3.10. General Discussion

The statistical analyses performed in this study provide strong evidence that the evolution of reflectance in automotive plastic components is significantly influenced by the type of polymer, the environmental exposure conditions, and the duration of exposure.

First, three- and two-way ANOVA analyses (Tables 6 and 12, respectively) show only a significant effect across almost all VIS-bands for Material (excepting Orange and Red) for the material non-specific ANOVA in Table 12. For the material-specific ANOVA summarized in Table 6, no significant effects were detected. These results seem reasonable considering the statistical applicability to the experimental setup under discussion of ANOVA models. Given the multivariate nature of the reflectance data, in which measurements across several spectral bands are inherently correlated, the use of multivariate analysis of variance (MANOVA) is statistically more appropriate than performing multiple separate ANOVAs. Conducting independent ANOVAs for each wavelength or VIS-band increases the risk of inflated type I error rates and fails to capture the underlying covariance structure between dependent variables, as discussed in [59]. MANOVAs address these limitations by testing for group differences across the full vector of dependent variables simultaneously, thereby providing a more comprehensive approach to detecting multivariate effects [60]. In our case, the use of MANOVAs allows for the detection of global degradation patterns in reflectance, which may not be evident when examining each wavelength independently, which seems to be the case with Table 6 results. This strategy is especially valuable when effects are subtle yet consistent across bands, or when interactions affect the multivariate profile rather than individual components. Overall, the application of MANOVA in the present study procures control of statistical error and maximizes the interpretive power of the reflectance dataset by considering both the magnitude and the interdependence of changes across the visible spectrum.

Table 12. ANOVA results (p values) for average reflectance per VIS-band, including the interaction between Material (**M**), Condition (**C**), and Time (**T**). Each cell indicates whether the effect was statistically significant at the $\alpha = 0.05$ level. Summary of significance thresholds (p values) from a three-way ANOVA performed on the average reflectance per VIS-band. The response variable is the mean reflectance computed for each band of the visible spectrum (Violet to Red), based on the average of three adjacent wavelengths. The between-subject factors considered are **Material** (polyethylene or polypropylene), **Condition** (four levels of environmental exposure), and **Time** (five measurement times). Each cell indicates whether the corresponding main effect or interaction was statistically significant at $\alpha = 0.05$.

Color	Material	Condition	Time	M × C	M × T	C × T	M × C × T
Violet	$p < 0.05$	$p > 0.05$	$p > 0.05$	$p > 0.05$	$p > 0.05$	$p > 0.05$	$p > 0.05$
Blue	$p < 0.05$	$p > 0.05$	$p > 0.05$	$p > 0.05$	$p > 0.05$	$p > 0.05$	$p > 0.05$
Cian	$p < 0.05$	$p > 0.05$	$p > 0.05$	$p > 0.05$	$p > 0.05$	$p > 0.05$	$p > 0.05$
Green	$p < 0.05$	$p > 0.05$	$p > 0.05$	$p > 0.05$	$p > 0.05$	$p > 0.05$	$p > 0.05$
Yellow	$p < 0.05$	$p > 0.05$	$p > 0.05$	$p > 0.05$	$p > 0.05$	$p > 0.05$	$p > 0.05$
Orange	$p > 0.05$	$p > 0.05$	$p > 0.05$	$p > 0.05$	$p > 0.05$	$p > 0.05$	$p > 0.05$
Red	$p > 0.05$	$p > 0.05$	$p > 0.05$	$p > 0.05$	$p > 0.05$	$p > 0.05$	$p > 0.05$

Apropos, the MANOVA models performed in this study (Tables 8 and 10), the material non-specific MANOVA in Table 8, being a broader scoped analysis than the other, indicated only Material effects, while the one reported in Table 10, broad as well but material-specific to our study case, shows there are indeed discernible effects for Condition, Time, and for Condition × Time in regards to PE, and for Condition and Time for PP. These results sanction us being in the right track to start the general study terms of interest proposed in this paper, with literature antecedents such as [1,3,61,62].

The ANCOVA models (Tables 5 and 7), which incorporated the initial reflectance as a covariate, demonstrated that baseline differences among samples played a significant role in explaining the observed variability. In the material non-specific analysis shown in Table 5, the covariate was statistically significant in all VIS-bands, suggesting that initial optical characteristics strongly modulate the degradation trajectory. In the same analysis, Condition and Time factors were also significant in most VIS-bands, in complete agreement with what one should expect for environmental exposure. The material specific analysis in Table 7 confirms the previous findings, additionally showing increased prevalence of significant results for PE, which is reasonable considering what has been reported in the literature regarding PE having greater UV exposition susceptibility than PP [57,63–65].

In regards to the MANCOVA models (Tables 9 and 11), material non-specific results in Table 9 portrait significant reflectance effects only through Material × Condition and Material × Condition × Time factors. Material specific results in Table 11 allow us to appreciate significant effects in Condition, Condition × Time for PE, and Condition × Time for PP. These results further reinforce the pattern detected by the previously discussed MANOVA analyses.

In addition to the multivariate test statistics performed, η^2 (*partial effect size*) for the *Time* factor using repeated-measures ANCOVA adjusted for the baseline (T0) reflectance of each VIS-band was conducted as well, with *Material* and *Condition* as between-subject factors. The resulting effect-size profile shown in Figure 16 was consistently positive across all bands, indicating that *Time* indeed explained a proportion of variance in reflectance changes regardless of the multivariate statistic contemplated. Effect magnitudes were highest in the Violet and Blue bands ($\eta^2 \approx 0.24$–0.28), gradually decreasing towards the green and yellow regions, and smallest in the orange and red bands ($\eta^2 \approx 0.02$–0.03). This gradual spectral pattern supports the interpretation that temporal degradation effects are systematic and wavelength-dependent, even if multivariate p-values differ between

Pillai's Trace, Wilks's Lambda, Hotelling's Trace, and Roy's Largest Root. Therefore, the observed variability in statistical significance reflects differences in test sensitivity rather than adversarial/contradictory temporal trends.

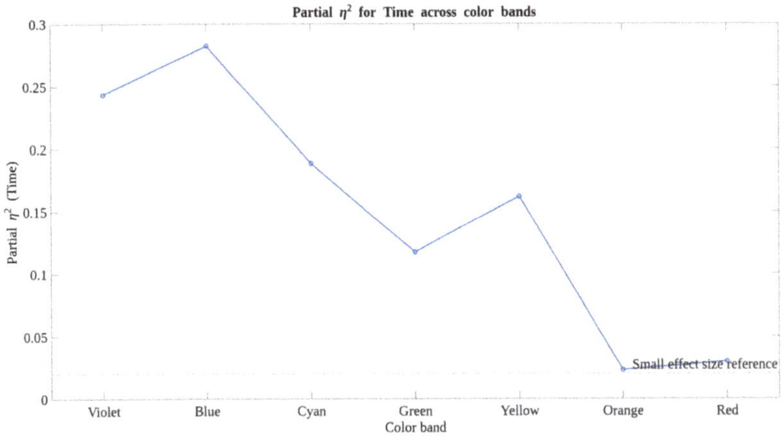

Figure 16. η^2 (*partial effect size*) for the Time factor across VIS-bands. Values were obtained from repeated-measures ANCOVA adjusted for the baseline (T0) reflectance of each band, with *Material* and *Condition* as between-subject factors. The dashed horizontal line marks the threshold for a small effect size ($\eta^2 = 0.02$). Error bars are omitted for clarity.

On the other hand, Figure 17 shows the variability in slope values (ΔReflectance per time unit) compared between polyethylene (PE) and polypropylene (PP) across the three exposure conditions (C2–C4). In 14 out of the 21 possible cases (7 spectral bands × 3 conditions), the standard deviation for PE exceeded that for PP. This pattern supports the notion that PE generally exhibited greater variability in reflectance change rates under these exposure conditions. The inclusion of both positive and negative slope values in the analysis reflects the actual direction of change in reflectance, rather than only its magnitude. Therefore, this observation supports the interpretation that PE tends to be more sensitive to the tested environmental exposure conditions, whereas PP generally demonstrates greater stability in its optical response over time. This is consistent with established knowledge of polyolefin optical degradation (reflectance changes), since it has been documented that polyethylene is more susceptible to UV exposure optical effects at longer wavelength bands than polypropylene due to the presence of tertiary carbon atoms along its backbone, which serve as preferential sites for radical formation and subsequent chain scission, among other photo-oxidation mechanisms [58]. Again, Figure 17 allows us to appreciate that indeed PE has greater variability in 5 out of 9 occasions (∼60%) than PE for the longer wavelength bands (i.e., Yellow, Orange, and Red). Overall, the present study reinforces the importance of accounting for material-specific degradation pathways and test conditions when designing proper accelerated UV-radiation weathering protocols.

Finally, exposure settings used in this work were designed to modulate the spectral content—chiefly the ultraviolet (UV) component—rather than the intensity of the visible (VIS) band. Polyethylene photodegradation is initiated predominantly by UV photons that generate radicals and carbonyls; by contrast, VIS photons ($E < 3.1$ eV) are typically insufficient to cleave C–C/C–H bonds in the absence of sensitizers. In our data this is reflected by the subject-normalized $\eta(t)$, which remains ≈ 0 for the Control (ambient VIS with minimal UV) while increasing under UV-richer conditions. Since VIS intensity was not varied independently, and irradiance was not logged, the present study cannot resolve a dose–response to VIS intensity per se. Future work should decouple spectrum and intensity (e.g., neutral-density attenuation at fixed spectrum, separate UV/VIS channels, and on-site dosimetry) to explicitly test VIS-intensity effects on $\eta(t)$, as well as accompanying statistical treatment.

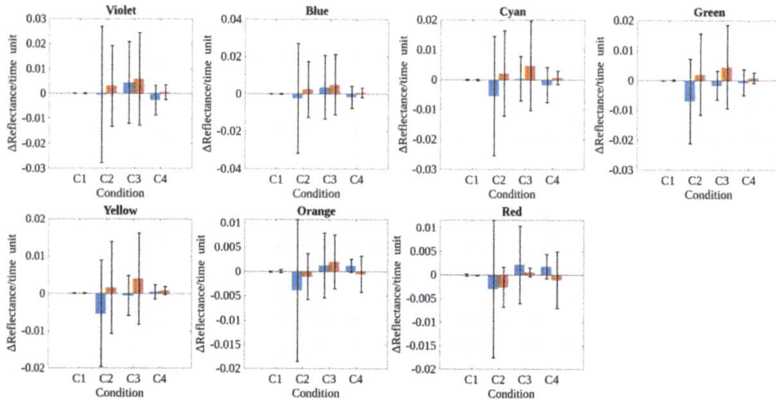

Figure 17. Mean reflectance slopes with respect to time (i.e., ΔReflectance/time unit) for each VIS-band, comparing polyethylene (PE) and polypropylene (PP) under the four exposure conditions (C1–C4). Bars represent the group mean, and error bars indicate the standard deviation across replicates.

3.11. Integrated Synthesis of the Results and Discussion

Overall aim. A multi-statistical framework was established to quantify optical degradation of polyolefins by combining multivariate descriptors and covariate-adjusted models, and by cross-validating spectral outcomes against mechanical and morphological readouts.

Data structure and covariates. The analysis accounts for material (PE, PP), exposure condition (control, glass-protected, unprotected, UV chamber), wavelength/VIS-band, and repeated measurements across indistinguishable specimens (used as a random/covariate factor when appropriate). This enables inference on condition effects while controlling between-specimen variability.

Multivariate contrasts. Multivariate comparisons of spectral summaries and colorimetric indices (MANOVA/MANCOVA when covariates apply) consistently separated exposure conditions, with the largest effects under UV chamber and unprotected VIS. Covariate adjustment (e.g., baseline spectral level and specimen) improved power and reduced residual variance relative to unadjusted ANOVA.

Time-resolved modeling. For wavelength-resolved and VIS-band-aggregated responses, physically motivated kinetics (first-order, stretched, lag, offset, and bi-exponential) were compared using information criteria (AICc). First-order sufficed at the VIS-band level, while bi-exponential became competitive where early/late processes coexisted. Half-times $t_{1/2}$ and asymptotes could also be extracted to enable condition ranking and cross-material comparisons.

Spectral change as an efficiency signal. A bounded, broadband reflectance-based trajectory $\eta(t) \in [0,1]$ was used as a compact summary of spectral change. $\eta(t)$ supports the multivariate findings by providing interpretable growth-to-plateau behavior and uncertainty-weighted fits for visualization and meta-comparison.

Uncertainty and robustness. Uncertainty via weighted fitting (using within-time dispersion) was reported, considering effect sizes and confidence bounds. Sensitivity analyses (band vs. full-spectrum aggregation, and inclusion/exclusion of lags/offsets) showed stable condition ordering and similar kinetics classes.

Mechanical consistency (hardness). Trends in mechanical hardness changes under UV chamber and unprotected VIS paralleled the spectral degradation ranks, reinforcing that the optical pathway captured by the multivariate/covariate-adjusted models is consistent with functional property loss.

Material-specific notes. PP typically showed faster early-stage evolution (smaller $t_{1/2}$) than PE under comparable irradiance histories, while PE exhibited clearer late-time plateaus. Glass protection attenuated both the rate and the asymptotic level across materials, consistent with short-λ screening.

Limitations and scope. The present study used a nested design in which subjects (Specimen IDs) are uniquely associated with one material (four specimens per material, eight in total). Thus, subjects are *nested* within material rather than crossed across materials. This implies that material effects are estimated from between-subject contrasts within each material, and cannot be disentangled from unobserved subject-level idiosyncrasies beyond those controlled by our normalization. Our subject-normalized spectral efficiency, $\eta(t)$, reduces between-subject offsets and scaling by (i) using the subject's own baseline ($t = 0$) within each condition and (ii) scaling by the subject's baseline spectral norm, thereby emphasizing within-subject temporal change. While this improves comparability and precision for time trends and condition contrasts, it does not change the nesting structure: inference on material differences should be interpreted as conditional on the sampled subjects (four per material) and considered preliminary/exploratory. We therefore report effect sizes and model-selected kinetics (AICc) with appropriate caution, and we refrain from population-wide claims about materials beyond the scope of this nested cohort.

The contribution of this work is then an integrated, covariate-aware, multivariate framework that (i) separates condition effects while controlling specimen heterogeneity, (ii) maps spectral change onto interpretable kinetic summaries for cross-study comparison, and (iii) corroborates optical signatures with mechanical evidence. This synthesis provides a portable analysis blueprint across polyolefin grades, exposure protocols, and instrument settings.

3.12. Linking Spectral Trends to Polymer Degradation Mechanisms

The spectral patterns reported here are consistent with established pathways of polyolefin photo-oxidation and weathering. First, the band-wise reflectance decay—more pronounced at shorter VIS wavelengths under UV chamber exposure—agrees with wavelength selectivity of photo-oxidation and the higher quantum efficiency of shorter-λ irradiation [4,10,44]. This selectivity is compatible with early-stage chromophore formation and surface carbonyl build-up that drive increased absorption/scattering and thus reduced reflectance [35,47].

Second, the observed multi-timescale kinetics (cases where a constrained bi-exponential competed with first-order fits) align with concurrent fast surface processes (radical initiation/oxygen uptake) and slower propagation/oxidative embrittlement, as reported for PE/PP under accelerated and natural weathering [49,58]. Depth-sensitive studies have shown near-surface oxidation gradients [57]; such gradients can manifest kinetically as an apparent lag or stretched response in bulk optical signals, consistent with our occasional $\delta > 0$ (lag) or $\beta < 1$ (stretched) selections.

Third, condition-wise contrasts match practical expectations: The UV chamber (UVA/UVB fluorescent lamps) accelerates damage relative to outdoor exposure, while glass shielding attenuates high-energy radiation and slows spectral change [17,21]. Material-wise, PP's larger hardness losses alongside a more variable reflectance trajectory point to morphology/additive dependencies frequently seen in automotive grades and recycled streams [3].

Overall, the statistical evidence (RM-MANCOVA/ANCOVA and kinetic summaries) provides a mechanistic narrative: *(i)* early, wavelength-sensitive oxidation dominates the visible response; *(ii)* multi-pathway kinetics are plausible in some configurations; and *(iii)* exposure protocol and shielding modulate rates and asymptotes in line with known chemistry of PE/PP photoaging. These connections strengthen the materials-science interpretation of the findings without altering the inferential framework.

4. Conclusions

The results of this study demonstrate that visible—VIS—reflectance behavior of automotive polyolefins under UV-radiation stress is strongly influenced by environmental exposure, material type, and time. The reflectance dataset, measured across seven visible VIS-bands and analyzed using ANOVA, MANOVA, ANCOVA, and MANCOVA statistical models, revealed that both PE and PP respond distinctly to UV-radiation weathering/aging conditions. In addition, statistical models conducted revealed that both materials underwent changes in hardness as well, an indicator of accompanying mechanical degradation. In particular, hardness decreased in most exposed groups, with UV chamber and unprotected outdoor exposure showing the greatest reductions.

Generally speaking, PE showed more gradual and consistent reflectance reduction across VIS-bands, whereas PP presented more variable responsiveness, particularly under UV chamber (ASTM G154 protocol) exposure. The repeated-measures MANOVA models identified present main effects for Time and Condition, while the ANCOVA models confirmed the importance of adjusting for initial reflectance to account for its spectral evolution over time, besides highlighting a range of both main or second order material-differentiated effects. MANCOVA results manifested that the interaction between Condition and Time was the most robust predictor of spectral evolution in the dataset. These trends were mutually reinforced by the set of conducted analyses, which validated the statistical robustness of the findings across VIS-bands. These results reinforce the necessity of covariate adjustment in optical degradation/aging studies and spotlight the practical value of using UV chamber tests to emulate short/medium/long-term outdoor effects. The results also align with prior literature documenting the molecular mechanisms of polyolefin degradation, including chain scission and oxidation in both PE and PP automotive components.

Therefore, findings support the use of multivariate and covariate-adjusted statistical frameworks for evaluating optical degradation/aging in polymers, combining approaches to capture both individual and collective spectral behavior, and suggest that PP may be more sensitive to exposure condition variability, while PE exhibits steadier degradation patterns under similar stressors. Additionally, they are relevant in the fields of material selection and accelerated testing protocols design and standardization regarding the automotive industry and governmental regulatory agencies, supporting the use of VIS reflectance analysis as a non-destructive tool for evaluating photo-induced degradation/aging/weathering in polymer components.

4.1. Preliminary Implications of the Results

Results presented provide concrete guidance for both materials selection and degradation modeling in automotive applications:

(i) **Balance between appearance and mechanical retention.** Across exposure scenarios, PP tended to show *smaller net changes in VIS reflectance* (cosmetic change) to PE while exhibiting a *larger decrease in Shore D hardness*. In practice, PP can be preferable for parts where color/appearance in the visible range dominates acceptance criteria (e.g., trims within the field of view), whereas PE is comparatively safer where mechanical retention under ageing is the primary constraint. Designers should trade off these trends according to part function.

(ii) **Exposure management and shielding.** The "glass-protected" condition consistently attenuated VIS reflectance change relative to "unprotected" outdoor exposure, and the UV chamber produced faster and larger spectral shifts. For under-glazing components or where coatings/films are feasible, adopting transparent shields or UV-absorbing layers reduces cosmetic drift. When specifying laboratory testing, the chamber protocol is a *conservative* surrogate for outdoor weathering to rank alternatives.

(iii) **Band-aware monitoring.** Summarising reflectance into seven VIS-bands offers a compact, production-friendly QC signal. Tracking the bands most relevant to brand colour targets helps detect early drift without full spectral processing. The figures show the time course per band (mean $\pm$ SEM, scaled for clarity), enabling side-by-side comparison of materials and conditions.

(iv) **Kinetic reading of the curves.** The fitted exponential family supplies a kinetic lens over spectral trajectories. Although $\eta(t)$ is reported here *graphically* (with optional half-time $t_{1/2}$ and asymptote η_∞ discussed but not tabulated), the relative ordering of time scales and plateaus can be inferred from the curves: faster initial changes under chamber exposure along with more coherent (smoother) trajectories for PE and more band-specific responses for PP.

(v) **Test and inspection planning.** Given the segmented evolution (earlier drift within the first interval and slower change thereafter), front-loaded inspections (e.g., during early service) are advisable for appearance-critical PP parts, while mechanical-critical PE parts merit periodic hardness checks throughout service life. Chamber-to-field factors can be established by matching the early $\eta(t)$ rise in the chamber to acceptable outdoor drift.

(vi) **Modelling pragmatics.** Information-criterion selection consistently favored simple exponentials per VIS-band, with stretched or bi-exponential models only where justified by data. For routine programs we recommend the following: (1) per-band first-order fits as default; (2) AICc-based escalation only if residual structure persists; (3) reporting mean $\pm$ SEM and the selected model per condition for transparency.

4.2. Limitations and Scope

As stated earlier in this manuscript, these implications are preliminary given the sample size and the material–subject linkage in this dataset. Estimates should be read as patterns and rankings rather than population parameters. Nevertheless, the patterns are internally consistent across bands; conditions and hardness and provide actionable guidance for part selection and qualification testing.

Author Contributions: Conceptualization, J.A.R.-S.; data curation, J.A.R.-S.; formal analysis J.A.R.-S., L.A.M.-S. and C.F.R.-G.; funding acquisition, J.P.Z.d.P. and K.A.; investigation, C.I. and K.A.; methodology, J.A.R.-S., L.A.M.-S. and C.F.R.-G.; resources, C.I., K.A. and J.P.Z.d.P.; software, J.A.R.-S., C.I. and K.A.; supervision, C.F.R.-G. and L.A.M.-S.; validation, J.A.R.-S. and C.F.R.-G.; visualization, J.A.R.-S., C.I. and C.F.R.-G.; writing—original draft preparation, J.A.R.-S., C.I., K.A. and C.F.R.-G.; writing—review and editing, J.A.R.-S., C.I., K.A., C.F.R.-G. and J.P.Z.d.P. All authors have read and agreed to the published version of the manuscript.

Funding: This research received no external funding.

Institutional Review Board Statement: Not applicable.

Informed Consent Statement: Not applicable.

Data Availability Statement: The raw data supporting the conclusions of this article will be made available by the authors on request.

Acknowledgments: The authors gratefully acknowledge the support received through scholarships from the *Sistema Nacional de Investigadoras e Investigadores (SNI)* program, granted by the *Secretaría de Ciencia, Humanidades, Tecnología e Innovación (SECIHTI-México).*

Conflicts of Interest: The authors declare no conflicts of interest.

Abbreviations

The following abbreviations are used in this manuscript:

ANOVA	Analysis of Variance
ANCOVA	Analysis of Covariance
MANCOVA	Multivariate Analysis of Covariance
MANOVA	Multivariate Analysis of Variance
PE	Polyethylene
PP	Polypropylene
UV	Ultraviolet
VIS	Visible

References

1. McKeen, L.W. *The Effect of UV Light and Weather on Plastics and Elastomers*; William Andrew: Amsterdam, The Netherlands, 2019
2. Gilbert, A. *Photochemistry*; Royal Society of Chemistry: London, UK, 2002; Volume 33.
3. Arese, M.; Cavallo, B.; Ciaccio, G.; Brunella, V. Characterization of Morphological, Thermal, and Mechanical Performances and UV Ageing Degradation of Post-Consumer Recycled Polypropylene for Automotive Industries. *Materials* **2025**, *18*, 1090 [CrossRef]
4. Feldman, D. Polymer Weathering: Photo-Oxidation. *J. Polym. Environ.* **2002**, *10*, 163–173. [CrossRef]
5. Redjala, S.; Ferhoum, R.; Aït Hocine, N.; Azem, S. Degradation of polycarbonate properties under thermal aging. *J. Fail. Anal Prev.* **2019**, *19*, 536–542. [CrossRef]
6. Pospíšil, J.; Horák, Z.; Kruliš, Z.; Nešpůrek, S.; Kuroda, S.i. Degradation and aging of polymer blends I. Thermomechanical and thermal degradation. *Polym. Degrad. Stab.* **1999**, *65*, 405–414. [CrossRef]
7. Andrady, A.L. Assessment of environmental biodegradation of synthetic polymers. *J. Macromol. Sci. Part C* **1994**, *34*, 25–76.
8. Rodriguez, A.K.; Mansoor, B.; Ayoub, G.; Colin, X.; Benzerga, A.A. Effect of UV-aging on the mechanical and fracture behavior of low density polyethylene. *Polym. Degrad. Stab.* **2020**, *180*, 109185. [CrossRef]
9. Yousif, E.; Haddad, R. Photodegradation and photostabilization of polymers, especially polystyrene: A review. *SpringerPlus* **2013**, *2*, 1–32. [CrossRef]
10. Larché, J.F.; Bussière, P.O.; Therias, S.; Gardette, J.L. Photooxidation of polymers: Relating material properties to chemical changes. *Polym. Degrad. Stab.* **2012**, *97*, 25–34. [CrossRef]
11. Gugumus, F. Mechanisms of photooxidation of polyolefins. *Angew. Makromol. Chem. Appl. Macromol. Chem. Phys.* **1990**, *176*, 27–42.
12. Carlsson, D.; Wiles, D. The photodegradation of polypropylene films. III. Photolysis of polypropylene hydroperoxides. *Macromolecules* **1969**, *2*, 597–606. [CrossRef]
13. Lukachevskaya, I.; Vasilyeva, E.; Ivanov, A.; Kychkin, A. Design of a UV Chamber for Studying the Destruction of Polymer Composite Materials. *Theor. Found. Chem. Eng.* **2024**, *58*, 670–676. [CrossRef]
14. Frigione, M. Assessment of the Ageing and Durability of Polymers. *Polymers* **2022**, *14*, 1934. [CrossRef]
15. Frigione, M.; Rodríguez-Prieto, A. Can accelerated aging procedures predict the long term behavior of polymers exposed to different environments? *Polymers* **2021**, *13*, 2688. [CrossRef]
16. Qin, J.; Jiang, J.; Tao, Y.; Zhao, S.; Zeng, W.; Shi, Y.; Lu, T.; Guo, L.; Wang, S.; Zhang, X.; et al. Sunlight tracking and concentrating accelerated weathering test applied in weatherability evaluation and service life prediction of polymeric materials: A review. *Polym. Test.* **2021**, *93*, 106940. [CrossRef]
17. Tocháček, J.; Vrátníčková, Z. Polymer life-time prediction: The role of temperature in UV accelerated ageing of polypropylene and its copolymers. *Polym. Test.* **2014**, *36*, 82–87. [CrossRef]
18. Pickett, J.E. Weathering of plastics. In *Handbook of Environmental Degradation of Materials*; Elsevier: Amsterdam, The Netherlands, 2018; pp. 163–184.
19. Chin, J.; Byrd, E.; Embree, N.; Garver, J.; Dickens, B.; Finn, T.; Martin, J. Accelerated UV weathering device based on integrating sphere technology. *Rev. Sci. Instruments* **2004**, *75*, 4951–4959. [CrossRef]

0. Mallon, P.; Li, Y.; Zhang, R.; Chen, H.; Wu, Y.; Sandreczki, T.; Jean, Y.; Suzuki, R.; Ohdaira, T. Durability of polymeric coatings: Effects of natural and artificial weathering. *Appl. Surf. Sci.* **2002**, *194*, 176–181. [CrossRef]
1. Du, B.; Lee, C.; Ji, Y. Study of Factors Affecting UV-Induced Photo-Degradation in Different Types of Polyethylene Sheets. *Polymers* **2024**, *16*, 2709. [CrossRef]
2. *ISO 4892-3:2024*; Plastics–Methods of Exposure to Laboratory Light Sources–Part 3: Fluorescent UV Lamps. International Organization for Standardization: Geneva, Switzerland, 2024.
3. Zhang, Y.; Jiang, W. Effective strategies to enhance ultraviolet barrier ability in biodegradable polymer-based films/coatings for fruit and vegetable packaging. *Trends Food Sci. Technol.* **2023**, *139*, 104139.
4. Riahinezhad, M.; Hallman, M.; Masson, J.F. Critical Review of Polymeric Building Envelope Materials: Degradation, Durability and Service Life Prediction. *Buildings* **2021**, *11*, 299. [CrossRef]
5. Syamsir, A.; Ishak, Z.A.M.; Yusof, Z.M.; Salwi, N.; Nadhirah, A. Durability control of UV radiation in glass fiber reinforced polymer (GFRP)-A review. *AIP Conf. Proc.* **2018**, *2031*, 020033.
6. Jacques, L. Accelerated and outdoor/natural exposure testing of coatings. *Prog. Polym. Sci.* **2000**, *25*, 1337–1362. [CrossRef]
7. Goddijn-Murphy, L.; Martínez-Vicente, V.; Dierssen, H.M.; Raimondi, V.; Gandini, E.; Foster, R.; Chirayath, V. Emerging technologies for remote sensing of floating and submerged plastic litter. *Remote Sens.* **2024**, *16*, 1770. [CrossRef]
8. Goustouridis, D.; Raptis, I.; Mpatzaka, T.; Fournari, S.; Zisis, G.; Petrou, P.; Beltsios, K.G. Non-destructive characterization of selected types of films and other layers via white light reflectance spectroscopy (WLRS). *Micro* **2022**, *2*, 495–507. [CrossRef]
9. Makki, H.; Adema, K.N.; Peters, E.A.; Laven, J.; van der Ven, L.G.; van Benthem, R.A.; de With, G. Quantitative spectroscopic analysis of weathering of polyester-urethane coatings. *Polym. Degrad. Stab.* **2015**, *121*, 280–291. [CrossRef]
30. Chaudhary, R.G.; Juneja, H.D.; Gharpure, M.P. Chelate polymer compounds with bis (bidentate) ligand: Synthesis, spectral, morphological and thermal degradation studies. *J. Chin. Adv. Mater. Soc.* **2013**, *1*, 121–133. [CrossRef]
31. Matuana, L.M.; Jin, S.; Stark, N.M. Ultraviolet weathering of HDPE/wood-flour composites coextruded with a clear HDPE cap layer. *Polym. Degrad. Stab.* **2011**, *96*, 97–106. [CrossRef]
32. Choi, S.S.; Chung, H.S.; Joo, Y.T.; Min, B.K.; Lee, S.H. Analysis of the whitening phenomenon of a thermoplastic elastomer article by UV weathering. *Polym. Test.* **2011**, *30*, 415–419. [CrossRef]
33. Exposito, J.; Becker, C.; Ruch, D.; Aubriet, F. Study of polymer material aging by laser mass spectrometry, UV-visible spectroscopy, and environmental scanning electron microscopy. *Adv. Phys. Chem.* **2007**, *2007*, 095753. [CrossRef]
34. Wijdekop, M.; Arnold, J.; Evans, M.; John, V.; Lloyd, A. Monitoring with reflectance spectroscopy the colour change of PVC plastisol coated strip steel due to weathering. *Mater. Sci. Technol.* **2005**, *21*, 791–797. [CrossRef]
35. Gulmine, J.; Janissek, P.; Heise, H.; Akcelrud, L. Degradation profile of polyethylene after artificial accelerated weathering. *Polym. Degrad. Stab.* **2003**, *79*, 385–397. [CrossRef]
36. Komiya, T.; Nakayama, H.; Shimaoka, T.; Inoue, K. Non-destructive test for geomembranes by visible near-infrared spectroscopy. In Proceedings of the 8th International Conference on Geosynthetics (8ICG), Yokohama, Japan, 18–22 September 2006; pp. 373–377.
37. Zhang, S.; Lin, T.; Zhang, C.; Yang, W.; Zhou, X.; Cable, R.; Choi, J.; Michaelson, E.; Thakre, P.; Serrat, C.; et al. Analysis of Accelerated Weathering Effect on Polyethylene With Varied Parameters Using a Combination of Analytical Techniques. *ChemistrySelect* **2024**, *9*, e202404334. [CrossRef]
38. Andrady, A.L.; Law, K.L.; Donohue, J.; Koongolla, B. Accelerated degradation of low-density polyethylene in air and in sea water. *Sci. Total Environ.* **2022**, *811*, 151368. [CrossRef]
39. Weizman, O.; Mead, J.; Dodiuk, H.; Ophir, A.; Kenig, S. The effect of nanoparticles on the loss of UV stabilizers in polyethylene films. *Polym. Degrad. Stab.* **2022**, *195*, 109811. [CrossRef]
40. Brostow, W.; Lu, X.; Gencel, O.; Osmanson, A.T. Effects of UV stabilizers on polypropylene outdoors. *Materials* **2020**, *13*, 1626. [CrossRef]
41. Cavalcanti, R.S.d.F.B.; Rabello, M.S. The effect of red pigment and photo stabilizers on the photo degradation of polypropylene films. *Mater. Res.* **2019**, *22*, e20180708. [CrossRef]
42. Grigoriadou, I.; Paraskevopoulos, K.; Chrissafis, K.; Pavlidou, E.; Stamkopoulos, T.G.; Bikiaris, D. Effect of different nanoparticles on HDPE UV stability. *Polym. Degrad. Stab.* **2011**, *96*, 151–163. [CrossRef]
43. Xian, G.; Qi, X.; Shi, J.; Tian, J.; Xiao, H. Toughened and self-healing carbon nanotube/epoxy resin composites modified with polycaprolactone filler for coatings, adhesives and FRP. *J. Build. Eng.* **2025**, *111*, 113207. [CrossRef]
44. Bourgogne, D.; Therias, S.; Gardette, J.L. Wavelength effect on polymer photooxidation under LED weathering conditions. *Polym. Degrad. Stab.* **2022**, *202*, 110021. [CrossRef]
45. van der Pal, K.J.; Maric, M.; Van Bronswijk, W.; Lewis, S.W. Ultraviolet-visible spectroscopic characterisation of automotive window tints for forensic purposes. *Anal. Methods* **2015**, *7*, 5391–5395. [CrossRef]
46. Adema, K.N.; Makki, H.; Peters, E.A.; Laven, J.; van der Ven, L.G.; van Benthem, R.A.; de With, G. Depth-resolved infrared microscopy and UV-VIS spectroscopy analysis of an artificially degraded polyester-urethane clearcoat. *Polym. Degrad. Stab.* **2014**, *110*, 422–434. [CrossRef]

47. Celina, M.; Gillen, K.T. Carbonyl identification and quantification uncertainties for oxidative polymer degradation. *Polym. Degrad. Stab.* **2021**, *183*, 109–117. [CrossRef]
48. Raia, V.; Mukherjeeb, C.; Jainc, B. UV-Vis and FTIR spectroscopy of gamma irradiated polymethyl methacrylate. *Indian J. Pure Appl. Phys.* **2017**, *55*, 775–785.
49. Fernando, S.S.; Christensen, P.A.; Egerton, T.A.; White, J.R. Carbon dioxide evolution and carbonyl group development during photodegradation of polyethylene and polypropylene. *Polym. Degrad. Stab.* **2007**, *92*, 2163–2172. [CrossRef]
50. Kaczmarek, H.; OŁdak, D.; Podg, A. Photochemical Properties of Polyethylene Modified by Low-Molecular Organic Compounds. *Polym. J.* **2003**, *35*, 634–639. [CrossRef]
51. *ASTM G154-16*; Standard Practice for Operating Fluorescent Ultraviolet (UV) Lamp Apparatus for Exposure of Nonmetallic Materials. ASTM International: West Conshohocken, PA, USA, 2016.
52. Kim, S.; Lee, Y.; Kim, C.; Choi, S. Analysis of mechanical property degradation of outdoor weather-exposed polymers. *Polymers* **2022**, *14*, 357. [CrossRef]
53. Ainali, N.M.; Bikiaris, D.N.; Lambropoulou, D.A. Aging effects on low-and high-density polyethylene, polypropylene and polystyrene under UV irradiation: An insight into decomposition mechanism by Py-GC/MS for microplastic analysis. *J. Anal. Appl. Pyrolysis* **2021**, *158*, 105207. [CrossRef]
54. WeatherSpark. Historical Weather in July 2024 at Querétaro Intercontinental Airport. Available online: https://weatherspark.com/h/m/145777/2024/7/Historical-Weather-in-July-2024-at-Quer%C3%A9taro-Intercontinental-Airport-Mexico (accessed on 3 July 2025).
55. Pérez-Cova, M.; Platikanov, S.; Stoll, D.R.; Tauler, R.; Jaumot, J. Comparison of multivariate ANOVA-based approaches for the determination of relevant variables in experimentally designed metabolomic studies. *Molecules* **2022**, *27*, 3304. [CrossRef] [PubMed]
56. Friendly, M.; Sigal, M. Visualizing tests for equality of covariance matrices. *Am. Stat.* **2020**, *74*, 144–155. [CrossRef]
57. Shyichuk, A.; White, J.; Craig, I.; Syrotynska, I. Comparison of UV-degradation depth-profiles in polyethylene, polypropylene and an ethylene–propylene copolymer. *Polym. Degrad. Stab.* **2005**, *88*, 415–419. [CrossRef]
58. Grause, G.; Chien, M.F.; Inoue, C. Changes during the weathering of polyolefins. *Polym. Degrad. Stab.* **2020**, *181*, 109364. [CrossRef]
59. Din, I.u.; Hayat, Y. ANOVA or MANOVA for correlated traits in agricultural experiments. *Sarhad J. Agric.* **2021**, *37*, 1250–1259. [CrossRef]
60. Akbay, L.; Akbay, T.; Erol, O.; Kılınç, M. Inadvertent use of ANOVA in educational research: ANOVA is not a surrogate for MANOVA. *J. Meas. Eval. Educ. Psychol.* **2019**, *10*, 302–314. [CrossRef]
61. Maag, K.; Lenhard, W.; Löffles, H. New UV curing systems for automotive applications. *Prog. Org. Coatings* **2000**, *40*, 93–97. [CrossRef]
62. Kuka, E.; Cirule, D.; Andersone, I.; Vasiljevs, L.O.; Merna, J.; Sarakovskis, A.; Kurnosova, N.; Sansonetti, E.; Vevere, L.; Andersons, B. A step to microplastic formation: Microcracking and associated surface transformations of recycled LDPE, LLDPE, HDPE, and PP plastics exposed to UV radiation. *Polym. Degrad. Stab.* **2024**, *229*, 110967. [CrossRef]
63. Raab, M.; Kotulak, L.; Kolařík, J.; Pospíšil, J. The effect of ultraviolet light on the mechanical properties of polyethylene and polypropylene films. *J. Appl. Polym. Sci.* **1982**, *27*, 2457–2466. [CrossRef]
64. Holländer, A.; Klemberg-Sapieha, J.; Wertheimer, M.R. Vacuum-ultraviolet induced oxidation of the polymers polyethylene and polypropylene. *J. Polym. Sci. Part A Polym. Chem.* **1995**, *33*, 2013–2025. [CrossRef]
65. Lestari, P.; Trihadiningrum, Y.; Warmadewanthi, I. Simulated degradation of low-density polyethylene and polypropylene due to ultraviolet radiation and water velocity in the aquatic environment. *J. Environ. Chem. Eng.* **2022**, *10*, 107553. [CrossRef]

MDPI AG
Grosspeteranlage 5
4052 Basel
Switzerland
Tel.: +41 61 683 77 34

Polymers Editorial Office
E-mail: polymers@mdpi.com
www.mdpi.com/journal/polymers

www.ingramcontent.com/pod-product-compliance
Lightning Source LLC
Chambersburg PA
CBHW061817210926
51850CB00006B/729
* 9 7 8 3 7 2 5 8 7 3 0 6 7 *